国际材料前沿丛书
International Materials Frontier Series

ELSEVIER

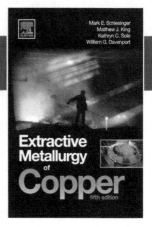

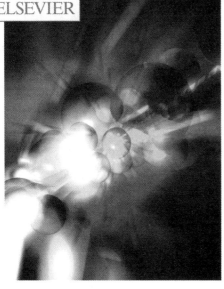

Mark E. Schlesinger,
Matthew J. King,
Kathryn C. Sole,
William G. Davenport 著

铜提取冶金（第5版）

Extractive Metallurgy of Copper (Fifth Edition)

影印版

·长沙·

国字:18-2017-160

Extractive Metallurgy of Copper, Fifth Edition
Mark E. Schlesinger, Matthew J. King, Kathryn C. Sole, William G. Davenport
ISBN: 978008096789 (ISBN of original edition)
Copyright © 2011 by Elsevier Ltd. All rights reserved.
Authorized English language reprint edition published by the Proprietor.
Copyright © 2017 by Elsevier (Singapore) Pte Ltd. All rights reserved.

Elsevier (Singapore) Pte Ltd.
3 Killiney Road
#08-01 Winsland House I
Singapore 239519
Tel: (65) 6349-0200
Fax: (65) 6733-1817
First Published <2017>
<2017>年初版

Printed in China by Central South University Press under special arrangement with Elsevier (Singapore) Pte Ltd. This edition is authorized for sale in China only, excluding Hong Kong SAR, Macao SAR and Taiwan. Unauthorized export of this edition is a violation of the Copyright Act. Violation of this Law is subject to Civil and Criminal Penalties.

本书英文影印版由Elsevier(Singapore) Pte Ltd.授权中南大学出版社在中国境内独家发行。本版仅限在中国境内(不包括香港、澳门以及台湾)出版及标价销售。未经许可之出口,视为违反著作权法,将受民事及刑事法律之制裁。

本书封底贴有Elsevier防伪标签,无标签者不得销售。

内容简介

本书主要介绍了从矿石和废料中提取铜的基本原理和先进技术。借鉴大量的国际先进企业经验，全面阐述了从原矿和二次资源提取铜再到铜材产品的系统生产工艺。

本书实用性强，特别介绍了目前广泛应用的粗铜精炼技术，增加的内容有：硫化铜矿澳斯麦特炉和艾萨炉先进熔炼技术、硫化精矿湿法冶金技术、副产品（贵金属）回收和环境保护、电子废料回收、风能和太阳能传输用铜电缆生产。

本书适用于从事冶金、化工和环境等专业研究人员以及企业工程师使用。

作者简介

Mark E. Schlesinger 博士，美国密苏里科技大学环境科学与技术中心、材料研究中心教授。1980 年获密苏里大学罗拉分校冶金工程学士学位，1985 年获亚利桑那大学材料科学与工程硕士学位，1988 年获亚利桑那大学材料科学与工程博士学位。曾在美国犹他大学、密苏里大学罗拉分校任教。曾荣获多项研究及教育奖励。2014 年起担任 TMS - AIME 指导委员会成员。

目 录

1 概述 ·· 1
  1.1 前言 ·· 1
  1.2 黄铜矿提取铜 ······························ 2
    1.2.1 铜矿浮选(3,4) ······················· 4
    1.2.2 造锍熔炼(5,6,9) ··················· 4
    1.2.3 铜锍吹炼(8,9) ······················ 5
    1.2.4 直接炼铜(10) ························ 7
    1.2.5 粗铜火法精炼和电解精炼(13,14) ·············· 7
  1.3 湿法炼铜 ···································· 8
    1.3.1 溶剂萃取(16) ························ 8
    1.3.2 电积(17) ····························· 9
  1.4 阴极铜熔化和浇铸 ························ 10
    1.4.1 铜产品种类 ···························· 10
  1.5 铜和铜合金废料回收(18,19) ············ 11
  1.6 小结 ·· 12
  参考文献 ·· 12
  建议阅读 ·· 12

2 产量与用途 ····································· 13
  2.1 铜矿石和边界品位 ······················· 14
  2.2 铜冶炼厂分布 ····························· 17
  2.3 铜价格 ····································· 29
  2.4 小结 ·· 29
  参考文献 ·· 29

3 铜精矿生产——简介与粉碎 ··············· 31
  3.1 富集工艺流程 ····························· 31
  3.2 粉碎工序 ·································· 31
  3.3 爆破 ·· 32
    3.3.1 矿石粒度的确定 ····················· 34
    3.3.2 矿石硬度的自动化测量 ··········· 34
  3.4 粉碎 ·· 35
  3.5 磨矿 ·· 35
    3.5.1 磨矿粒度和铜矿物解离 ··········· 35
    3.5.2 磨矿设备 ···························· 36
    3.5.3 浮选给料粒度控制 ················· 36
    3.5.4 仪表和控制 ·························· 43
  3.6 粉碎的最新发展 ·························· 46
    3.6.1 高压辊式破碎 ······················· 46
    3.6.2 矿物学自动分析 ···················· 47
  3.7 小结 ·· 47
  参考文献 ·· 48
  建议阅读 ·· 48

4 铜精矿的生产 ·································· 51
  4.1 泡沫浮选 ·································· 51
  4.2 浮选化学 ·································· 52
    4.2.1 捕收剂 ······························· 52
    4.2.2 浮选选择性 ·························· 53
    4.2.3 调节剂 ······························· 54
    4.2.4 起泡剂 ······························· 55
  4.3 铜矿石的特殊浮选工艺 ················· 55
  4.4 浮选槽 ····································· 56
    4.4.1 柱形槽 ······························· 56

4.5 传感器、操作和控制 …… 64
    4.5.1 工艺物质流连续化学分析 …… 65
    4.5.2 设备的可视系统 …… 67
4.6 浮选产品 …… 67
    4.6.1 浓缩和脱水 …… 67
    4.6.2 尾矿 …… 68
4.7 其他矿物浮选分离 …… 68
    4.7.1 黄金浮选 …… 68
4.8 小结 …… 69
参考文献 …… 69
建议阅读 …… 70

5 造锍熔炼基础 …… 73

5.1 为什么熔炼？ …… 73
5.2 铜锍和炉渣 …… 74
    5.2.1 炉渣 …… 74
    5.2.2 铁酸钙和橄榄石渣 …… 79
    5.2.3 铜锍 …… 81
5.3 造锍熔炼的化学反应 …… 82
5.4 熔炼过程概述 …… 83
5.5 熔炼产物：铜锍、炉渣和尾气 …… 84
    5.5.1 铜锍 …… 84
    5.5.2 炉渣 …… 84
    5.5.3 烟气 …… 84
5.6 小结 …… 86
参考文献 …… 86
建议阅读 …… 88

6 闪速熔炼 …… 89

6.1 奥图泰闪速熔炼炉 …… 89
    6.1.1 结构 …… 90
    6.1.2 冷却水套 …… 94
    6.1.3 精矿喷嘴 …… 95
    6.1.4 辅助燃料喷嘴 …… 95
    6.1.5 铜锍口和渣口 …… 96
6.2 附属设备 …… 96
    6.2.1 精矿混料系统 …… 96
    6.2.2 固体给料干燥器 …… 97
    6.2.3 给料系统 …… 97
    6.2.4 制氧站 …… 98
    6.2.5 热风炉(可选) …… 98
    6.2.6 余热锅炉 …… 98
    6.2.7 烟尘回收利用 …… 98
6.3 闪速熔炼炉操作 …… 99
    6.3.1 开炉和停炉 …… 99
    6.3.2 正常运行 …… 99
6.4 控制 …… 100
    6.4.1 精矿处理量和铜锍品位控制 …… 100
    6.4.2 炉渣成分控制 …… 101
    6.4.3 温度控制 …… 101
    6.4.4 反应塔和炉缸控制 …… 101
6.5 杂质行为 …… 102
    6.5.1 灰尘中不可回收杂质 …… 102
    6.5.2 其他控制杂质的工业方法 …… 103
6.6 奥图泰闪速熔炼最新发展及未来趋势 …… 103
6.7 因科闪速熔炼 …… 103
    6.7.1 炉体细节 …… 104
    6.7.2 精矿喷嘴 …… 104
    6.7.3 水冷系统 …… 104
    6.7.4 铜锍口和渣口 …… 105
    6.7.5 直升烟道 …… 105
    6.7.6 辅助设备 …… 105
    6.7.7 固体物料干燥器 …… 106

6.7.8 精矿喷嘴进料系统 …… 106
6.7.9 尾气冷却和收尘系统 …… 106
6.8 因科闪速熔炼炉总结 …… 106
6.9 因科和奥图泰闪速炉熔炼对比 …… 107
6.10 小结 …… 107
参考文献 …… 107
建议阅读 …… 110

7 浸没式风嘴熔炼：诺兰达炉、特尼恩特炉和瓦纽科夫炉 …… 111
7.1 诺兰达工艺 …… 111
7.2 反应原理 …… 114
 7.2.1 铜锍渣分离 …… 114
 7.2.2 铜锍品位选择 …… 115
 7.2.3 杂质行为 …… 115
 7.2.4 废弃物及残渣的熔炼 …… 115
7.3 操作和控制 …… 116
 7.3.1 控制 …… 116
7.4 提高产能 …… 117
7.5 特尼恩特熔炼 …… 117
 7.5.1 熔化铜锍 …… 117
7.6 工艺过程 …… 118
7.7 操作 …… 118
7.8 控制 …… 120
 7.8.1 温度控制 …… 120
 7.8.2 炉渣和铜锍成分控制 …… 120
 7.8.3 铜锍和炉渣深度控制 …… 120
7.9 杂质分布 …… 120
7.10 讨论 …… 121
 7.10.1 铜锍品位与 SO_2 捕集率关系 …… 121
 7.10.2 炉子寿命和热风嘴维修 …… 121
 7.10.3 炉体冷却 …… 121
 7.10.4 尾气余热回收 …… 122
7.11 瓦纽科夫浸没式风嘴熔炼 …… 122
7.12 小结 …… 123
参考文献 …… 124
建议阅读 …… 125

8 铜锍吹炼 …… 127
8.1 化学原理 …… 127
 8.1.1 造铜反应 …… 129
 8.1.2 吹炼除杂 …… 134
8.2 工业化 P-S 转炉吹炼操作 …… 134
 8.2.1 风嘴与尾气收集 …… 136
 8.2.2 温度控制 …… 137
 8.2.3 温度选择 …… 138
 8.2.4 温度测量 …… 138
 8.2.5 炉渣和流量控制 …… 139
 8.2.6 渣形成速率 …… 139
 8.2.7 熔炼终点检测 …… 139
8.3 富氧空气强化 P-S 转炉 …… 140
8.4 转炉生产率最优化 …… 140
 8.4.1 固体料熔化最优化 …… 141
 8.4.2 铜精矿转炉熔炼 …… 142
 8.4.3 炉体寿命最优化 …… 142
8.5 P-S 转炉新进展 …… 142

8.5.1 集束式喷射进料技术 …… 142
8.5.2 废料喷射进料技术 …… 143
8.5.3 转炉外壳设计 …… 143
8.6 其他铜锍吹炼方法 …… 143
　　8.6.1 霍博肯转炉 …… 144
　　8.6.2 闪速吹炼 …… 144
　　8.6.3 浸没风嘴式诺兰达连续吹炼 …… 147
　　8.6.4 吹炼技术新进展 …… 150
8.7 小结 …… 150
参考文献 …… 151
建议阅读 …… 153

9 熔池熔炼技术：澳斯麦特法/艾萨法和三菱法 …… 155

9.1 基本操作 …… 155
9.2 进料 …… 156
9.3 TSL 炉和喷枪 …… 156
9.4 熔炼原理 …… 163
　　9.4.1 除杂 …… 163
9.5 开炉和停炉 …… 163
9.6 当前装备现状 …… 164
9.7 TSL 铜锍吹炼技术 …… 164
9.8 三菱法工艺 …… 165
　　9.8.1 简介 …… 165
　　9.8.2 三菱法工艺 …… 165
　　9.8.3 熔炼炉细节 …… 166
　　9.8.4 电贫化炉细节 …… 166
　　9.8.5 吹炼炉细节 …… 168
　　9.8.6 最佳铜锍品位 …… 169
　　9.8.7 三菱法熔炼/吹炼过程控制 …… 169
9.9 21世纪的三菱法工艺 …… 174

9.10 小结 …… 175
参考文献 …… 176
建议阅读 …… 177

10 闪速直接炼铜 …… 179

10.1 优缺点 …… 179
10.2 理想的直接炼铜工艺 …… 179
10.3 工业化单炉直接炼铜 …… 182
10.4 化学原理 …… 184
10.5 炉渣成分对渣含铜的影响 …… 185
10.6 产业化细节 …… 185
10.7 过程控制 …… 186
　　10.7.1 目标：无铜锍层避免泡沫渣产出 …… 186
　　10.7.2 无铜锍层熔炼导致渣含铜升高 …… 186
10.8 炉渣电炉贫化回收铜 …… 186
　　10.8.1 格沃古夫 …… 187
　　10.8.2 奥林匹克坝 …… 187
10.9 直接炼铜技术受限于渣含铜 …… 187
10.10 直接炼法杂质行为 …… 187
10.11 小结 …… 188
参考文献 …… 189
建议阅读 …… 189

11 炉渣中铜的损失 …… 191

11.1 炉渣含铜 …… 191
11.2 降低渣含铜Ⅰ：减少炉渣的产生 …… 193
11.3 降低渣含铜Ⅱ：减少炉渣中铜含量 …… 193
11.4 降低渣含铜Ⅲ：炉渣高温澄清或还原 …… 194

11.5 降低渣含铜Ⅳ：炉渣选矿处理 …… 197
11.6 总结 …………………… 201
参考文献 ……………………… 201
建议阅读 ……………………… 203

12 硫的捕集与固化 ………… 205
12.1 熔炼和吹炼过程的尾气 … 206
  12.1.1 硫捕集率 ………… 207
12.2 硫酸生产 ………………… 208
12.3 冶炼厂尾气治理 ……… 208
  12.3.1 烟气冷却和余热回收 …… 210
  12.3.2 电收尘 …………… 211
  12.3.3 水力洗涤除尘和冷却 …… 211
  12.3.4 脱汞 ……………… 211
  12.3.5 污酸处理 ………… 212
12.4 气体干燥 ………………… 212
  12.4.1 干燥塔 …………… 212
  12.4.2 制酸厂主风机 …… 213
12.5 制酸过程化学反应 …… 214
  12.5.1 SO_2氧化为SO_3 … 214
12.6 工业硫酸生产 ………… 218
  12.6.1 催化转化器 ……… 224
  12.6.2 $SO_2 \rightarrow SO_3$转换反应途径 ………… 224
  12.6.3 反应途径特征 …… 225
  12.6.4 吸收塔 …………… 226
  12.6.5 气气热交换器和酸冷却器 …………… 227
  12.6.6 硫酸产品等级 …… 227
12.7 其他硫回收方法 ……… 227
  12.7.1 WSA法 …………… 227
  12.7.2 硫酸铵法 ………… 228
12.8 硫酸生产现状及未来发展 …………………… 229
  12.8.1 SO_2进气浓度最优化 …………… 229
  12.8.2 余热回收最优化 … 230
12.9 硫的替代产品 ………… 231
12.10 硫捕集的改进方向 …… 231
12.11 总结 …………………… 231
参考文献 ……………………… 232
建议阅读 ……………………… 234

13 火法精炼和阳极铸造 ……… 237
13.1 火法精炼工业方法 …… 237
  13.1.1 转炉精炼 ………… 238
  13.1.2 反射炉精炼 ……… 240
13.2 火法精炼的化学原理 … 240
  13.2.1 脱硫：Cu－O－S体系 ………………… 240
  13.2.2 脱氧：Cu－C－H－O体系 ………………… 241
13.3 脱氧碳氢化合物的选择 …………………… 241
13.4 阳极铸造 ………………… 241
  13.4.1 阳极模具 ………… 243
  13.4.2 阳极一致性 ……… 243
  13.4.3 阳极制备 ………… 243
13.5 阳极连续浇铸 ………… 244
13.6 废阳极和残极再熔铸 … 245
13.7 火法精炼其他杂质脱除 …………………… 245
13.8 小结 …………………… 247
参考文献 ……………………… 247
建议阅读 ……………………… 248

14 电解精炼 ……… 251

- 14.1 电解精炼过程 ……… 251
- 14.2 电解精炼化学反应和阳极杂质行为 ……… 252
  - 14.2.1 金和铂族金属 ……… 253
  - 14.2.2 硒和碲 ……… 253
  - 14.2.3 铅和锡 ……… 254
  - 14.2.4 砷、铋、钴、铁、镍、硫和锑 ……… 254
  - 14.2.5 银 ……… 255
  - 14.2.6 氧 ……… 255
  - 14.2.7 杂质行为小结 ……… 256
- 14.3 设备 ……… 257
  - 14.3.1 阳极 ……… 258
  - 14.3.2 阴极 ……… 258
  - 14.3.3 电解槽 ……… 259
  - 14.3.4 电器部件 ……… 260
- 14.4 典型电解精炼循环 ……… 260
- 14.5 电解液 ……… 261
  - 14.5.1 添加剂 ……… 262
  - 14.5.2 电解液温度 ……… 266
  - 14.5.3 电解液过滤 ……… 266
  - 14.5.4 电解液净化除杂 ……… 266
- 14.6 阴极铜纯度影响因素 ……… 267
  - 14.6.1 物理因素 ……… 267
  - 14.6.2 化学因素 ……… 267
  - 14.6.3 电化学因素 ……… 268
- 14.7 能源消耗的最小化 ……… 269
- 14.8 工业化电解精炼 ……… 269
- 14.9 铜电解精炼的近期发展及趋势 ……… 274
- 14.10 小结 ……… 275
- 参考文献 ……… 275
- 建议阅读 ……… 279

15 湿法炼铜：概述和浸出 ……… 281

- 15.1 湿法工艺回收铜 ……… 281
- 15.2 铜矿浸出化学原理 ……… 282
  - 15.2.1 氧化铜矿的浸出 ……… 282
  - 15.2.2 硫化铜矿的浸出 ……… 283
- 15.3 浸出方法 ……… 285
- 15.4 堆浸和就地浸出 ……… 287
  - 15.4.1 堆浸和就地浸出化学原理 ……… 288
  - 15.4.2 工业化堆浸 ……… 290
  - 15.4.3 工业化就地浸出 ……… 301
- 15.5 槽浸 ……… 301
- 15.6 搅拌浸出 ……… 303
  - 15.6.1 氧化矿 ……… 303
  - 15.6.2 硫化矿 ……… 304
- 15.7 氧压浸出 ……… 304
  - 15.7.1 黄铜矿湿法冶金的经济和技术驱动 ……… 304
  - 15.7.2 高温和高压浸出 ……… 308
- 15.8 未来发展方向 ……… 315
- 15.9 小结 ……… 316
- 参考文献 ……… 317
- 建议阅读 ……… 322

16 溶剂萃取 ……… 323

- 16.1 溶剂萃取过程 ……… 323
- 16.2 铜溶剂萃取化学原理 ……… 324
- 16.3 有机相组成 ……… 325
  - 16.3.1 萃取剂 ……… 325
  - 16.3.2 稀释剂 ……… 327
- 16.4 减小杂质迁移和提高电解液纯度 ……… 328
- 16.5 设备 ……… 329
  - 16.5.1 混合室设计 ……… 329

16.5.2 澄清室设计 ……… 330
16.6 循环配置 ……………… 331
　16.6.1 串联回路 ………… 331
　16.6.2 并联和串并联回路
　　　　　…………………… 333
　16.6.3 洗涤除杂段 ……… 333
16.7 串联回路的级数设计 … 333
　16.7.1 萃取剂浓度测定 … 333
　16.7.2 萃取与反萃等温线
　　　　 测定 ……………… 334
　16.7.3 萃取率测定 ……… 334
　16.7.4 反萃有机相铜平衡
　　　　 浓度测定 ………… 334
　16.7.5 有机相萃取过程铜
　　　　 的迁移 …………… 335
　16.7.6 铜反萃电解液流量
　　　　 测定 ……………… 335
　16.7.7 其他方法 ………… 336
16.8 串联和串并联循环的定量
　　　比较 …………………… 336
16.9 操作注意事项 ………… 336
　16.9.1 操作的稳定性 …… 336
　16.9.2 三相形成 ………… 337
　16.9.3 相连续性 ………… 339
　16.9.4 有机相损失和回收
　　　　 …………………… 339
16.10 工业化溶剂萃取厂 … 339
16.11 小结 ………………… 344
参考文献 …………………… 344
建议阅读 …………………… 346

17 电积 ……………………… 349
17.1 电积工艺 ……………… 349
17.2 铜电积化学原理 ……… 349
17.3 电能消耗 ……………… 350

17.4 设备和操作实践 ……… 351
　17.4.1 阴极 ……………… 351
　17.4.2 阳极 ……………… 351
　17.4.3 电解槽设计 ……… 353
　17.4.4 电流密度 ………… 355
　17.4.5 酸雾抑制 ………… 356
　17.4.6 电解液 …………… 356
　17.4.7 电解液添加剂 …… 360
17.5 阴极铜纯度优化 ……… 360
17.6 电流效率优化 ………… 361
17.7 现代工业电积车间 …… 362
17.8 浸出液直接电积 ……… 362
17.9 电解沉积现状及未来发展
　　　趋势 …………………… 368
17.10 小结 ………………… 369
参考文献 …………………… 369
建议阅读 …………………… 371

18 再生铜的回收和加工 …… 373
18.1 原料循环 ……………… 373
　18.1.1 厂内废料 ………… 373
　18.1.2 新废料 …………… 374
　18.1.3 旧废料 …………… 375
18.2 再生铜分级和定义 …… 379
18.3 废料处理和选矿 ……… 380
　18.3.1 电线电缆处理 …… 380
　18.3.2 汽车用铜回收 …… 382
　18.3.3 电子废料处理 …… 384
18.4 小结 …………………… 385
参考文献 …………………… 385
建议阅读 …………………… 387

19 铜循环利用的化学冶金 … 389
19.1 再生铜的特点 ………… 389
19.2 原生铜熔炼过程中的废料

    处理 ·············· 389
  19.2.1 熔炼炉用废料
    ·············· 390
  19.2.2 转炉和阳极炉用
    废料 ············ 391
 19.3 再生铜冶炼厂 ········ 391
  19.3.1 高品位再生铜熔炼
    ·············· 391
  19.3.2 熔炼黑铜 ········ 391
  19.3.3 吹炼黑铜 ········ 393
  19.3.4 火法精炼和电解精炼
    ·············· 394
 19.4 小结 ·············· 394
 参考文献 ················ 395
 建议阅读 ················ 396

20 熔化和浇铸 ············ 397
 20.1 产品等级和质量 ······ 397
 20.2 熔化技术 ············ 399
  20.2.1 炉型 ············ 399
  20.2.2 氢和氧的测定和控制
    ·············· 403
 20.3 浇铸设备 ············ 403
  20.3.1 浇铸坯材 ········ 404
  20.3.2 浇铸线材和棒材 ··· 404
  20.3.3 浇铸无氧铜 ······ 409
  20.3.4 浇铸带材 ········ 409
 20.4 小结 ·············· 410
 参考文献 ················ 411
 建议阅读 ················ 412

21 副产品和废物走向 ······ 415
 21.1 钼的回收和处置 ······ 415
 21.2 浮选药剂 ············ 415
 21.3 操作 ·············· 415

 21.4 最优化 ············· 417
 21.5 阳极泥 ············· 418
  21.5.1 阳极泥组成 ······ 418
  21.5.2 阳极泥处理流程 ··· 421
 21.6 烟尘处理 ············ 422
 21.7 炉渣利用或无害化处理
   ·················· 423
 21.8 小结 ·············· 425
 参考文献 ················ 425
 建议阅读 ················ 426

22 铜生产成本 ············ 427
 22.1 总体投资成本：从矿山到
   精炼厂 ············· 427
  22.1.1 投资成本的变化 ··· 429
  22.1.2 冶炼厂的合理规模
    ·············· 429
 22.2 总直接运营成本：从矿山
   到精炼厂 ··········· 429
  22.2.1 直接运营成本变化影响
    ·············· 430
 22.3 总生产成本、销售价格和
   盈利能力 ··········· 430
  22.3.1 副产品效益 ······ 431
 22.4 选矿成本 ············ 431
 22.5 冶炼成本 ············ 433
 22.6 电解精炼成本 ········ 435
 22.7 废料回收铜生产成本 ··· 435
 22.8 铜湿法冶金成本 ······ 436
 22.9 盈利能力 ············ 438
 22.10 小结 ············· 438
 参考文献 ················ 438
 建议阅读 ················ 439

索引 ···················· 441

Contents

Preface — xv
Preface to the Fourth Edition — xvii
Preface to the Third Edition — xix
Preface to the Second Edition — xxi
Preface to the First Edition — xxiii

1. **Overview** — 1
   1.1. Introduction — 1
   1.2. Extracting Copper from Copper–Iron–Sulfide Ores — 2
      1.2.1. Concentration by Froth Flotation — 4
      1.2.2. Matte Smelting — 4
      1.2.3. Converting — 5
      1.2.4. Direct-to-Copper Smelting — 7
      1.2.5. Fire Refining and Electrorefining of Blister Copper — 7
   1.3. Hydrometallurgical Extraction of Copper — 8
      1.3.1. Solvent Extraction — 8
      1.3.2. Electrowinning — 9
   1.4. Melting and Casting Cathode Copper — 10
      1.4.1. Types of Copper Product — 10
   1.5. Recycle of Copper and Copper-Alloy Scrap — 11
   1.6. Summary — 12
   Reference — 12
   Suggested Reading — 12

2. **Production and Use** — 13
   2.1. Copper Minerals and Cut-off Grades — 14
   2.2. Location of Extraction Plants — 17
   2.3. Price of Copper — 29
   2.4. Summary — 29
   References — 29

3. **Production of High Copper Concentrates — Introduction and Comminution** — 31
   3.1. Concentration Flowsheet — 31
   3.2. The Comminution Process — 31
   3.3. Blasting — 32
      3.3.1. Ore-size Determination — 34
      3.3.2. Automated Ore-toughness Measurements — 34
   3.4. Crushing — 35
   3.5. Grinding — 35
      3.5.1. Grind Size and Liberation of Copper Minerals — 35

v

	3.5.2.	Grinding Equipment	36
	3.5.3.	Particle-Size Control of Flotation Feed	36
	3.5.4.	Instrumentation and Control	43
3.6.	Recent Developments in Comminution		46
	3.6.1.	High Pressure Roll Crushing	46
	3.6.2.	Automated Mineralogical Analysis	47
3.7.	Summary		47
References			48
Suggested reading			48

4. Production of Cu Concentrate from Finely Ground Cu Ore — 51

4.1.	Froth Flotation		51
4.2.	Flotation Chemicals		52
	4.2.1.	Collectors	52
	4.2.2.	Selectivity in Flotation	53
	4.2.3.	Differential Flotation – Modifiers	54
	4.2.4.	Frothers	55
4.3.	Specific Flotation Procedures for Cu Ores		55
4.4.	Flotation Cells		56
	4.4.1.	Column Cells	56
4.5.	Sensors, Operation, and Control		64
	4.5.1.	Continuous Chemical Analysis of Process Streams	65
	4.5.2.	Machine Vision Systems	67
4.6.	The Flotation Products		67
	4.6.1.	Thickening and Dewatering	67
	4.6.2.	Tailings	68
4.7.	Other Flotation Separations		68
	4.7.1.	Gold Flotation	68
4.8.	Summary		69
References			69
Suggested Reading			70

5. Matte Smelting Fundamentals — 73

5.1.	Why Smelting?		73
5.2.	Matte and Slag		74
	5.2.1.	Slag	74
	5.2.2.	Calcium Ferrite and Olivine Slags	79
	5.2.3.	Matte	81
5.3.	Reactions During Matte Smelting		82
5.4.	The Smelting Process: General Considerations		83
5.5.	Smelting Products: Matte, Slag and Offgas		84
	5.5.1.	Matte	84
	5.5.2.	Slag	84
	5.5.3.	Offgas	86
5.6.	Summary		86
References			86
Suggested Reading			88

6. Flash Smelting — 89

6.1.	Outotec Flash Furnace		89
	6.1.1.	Construction Details	90

Contents

	6.1.2.	Cooling Jackets	94
	6.1.3.	Concentrate Burner	95
	6.1.4.	Supplementary Hydrocarbon Fuel Burners	95
	6.1.5.	Matte and Slag Tapholes	96
6.2.	Peripheral Equipment		96
	6.2.1.	Concentrate Blending System	96
	6.2.2.	Solids Feed Dryer	97
	6.2.3.	Bin and Feed System	97
	6.2.4.	Oxygen Plant	98
	6.2.5.	Blast Heater (optional)	98
	6.2.6.	Heat Recovery Boiler	98
	6.2.7.	Dust Recovery and Recycle System	98
6.3.	Flash Furnace Operation		99
	6.3.1.	Startup and Shutdown	99
	6.3.2.	Steady-state Operation	99
6.4.	Control		100
	6.4.1.	Concentrate Throughput Rate and Matte Grade Controls	100
	6.4.2.	Slag Composition Control	101
	6.4.3.	Temperature Control	101
	6.4.4.	Reaction Shaft and Hearth Control	101
6.5.	Impurity Behavior		102
	6.5.1.	Non-recycle of Impurities in Dust	102
	6.5.2.	Other Industrial Methods of Controlling Impurities	103
6.6.	Outotec Flash Smelting Recent Developments and Future Trends		103
6.7.	Inco Flash Smelting		103
	6.7.1.	Furnace Details	104
	6.7.2.	Concentrate Burner	104
	6.7.3.	Water Cooling	104
	6.7.4.	Matte and Slag Tapholes	105
	6.7.5.	Gas Uptake	105
	6.7.6.	Auxiliary Equipment	105
	6.7.7.	Solids Feed Dryer	106
	6.7.8.	Concentrate Burner Feed System	106
	6.7.9.	Offgas Cooling and Dust Recovery Systems	106
6.8.	Inco Flash Furnace Summary		106
6.9.	Inco vs. Outotec Flash Smelting		107
6.10.	Summary		107
References			107
Suggested Reading			110

7. Submerged Tuyere Smelting: Noranda, Teniente, and Vanyukov — 111

7.1.	Noranda Process		111
7.2.	Reaction Mechanisms		114
	7.2.1.	Separation of Matte and Slag	114
	7.2.2.	Choice of Matte Grade	115
	7.2.3.	Impurity Behavior	115
	7.2.4.	Scrap and Residue Smelting	115
7.3.	Operation and Control		116
	7.3.1.	Control	116
7.4.	Production Rate Enhancement		117
7.5.	Teniente Smelting		117
	7.5.1.	Seed Matte	117

	7.6.	Process Description	118
	7.7.	Operation	118
	7.8.	Control	120
		7.8.1. Temperature Control	120
		7.8.2. Slag and Matte Composition Control	120
		7.8.3. Matte and Slag Depth Control	120
	7.9.	Impurity Distribution	120
	7.10.	Discussion	121
		7.10.1. Super-high Matte Grade and SO_2 Capture Efficiency	121
		7.10.2. Campaign Life and Hot Tuyere Repairing	121
		7.10.3. Furnace Cooling	121
		7.10.4. Offgas Heat Recovery	122
	7.11.	Vanyukov Submerged Tuyere Smelting	122
	7.12.	Summary	123
	References		124
	Suggested Reading		125
8.	**Converting of Copper Matte**		127
	8.1.	Chemistry	127
		8.1.1. Coppermaking Reactions	129
		8.1.2. Elimination of Impurities During Converting	134
	8.2.	Industrial Peirce—Smith Converting Operations	134
		8.2.1. Tuyeres and Offgas Collection	136
		8.2.2. Temperature Control	137
		8.2.3. Choice of Temperature	138
		8.2.4. Temperature Measurement	138
		8.2.5. Slag and Flux Control	139
		8.2.6. Slag Formation Rate	139
		8.2.7. End Point Determinations	139
	8.3.	Oxygen Enrichment of Peirce—Smith Converter Blast	140
	8.4.	Maximizing Converter Productivity	140
		8.4.1. Maximizing Solids Melting	141
		8.4.2. Smelting Concentrates in the Converter	142
		8.4.3. Maximizing Campaign Life	142
	8.5.	Recent Improvements in Peirce—Smith Converting	142
		8.5.1. Shrouded Blast Injection	142
		8.5.2. Scrap Injection	143
		8.5.3. Converter Shell Design	143
	8.6.	Alternatives to Peirce—Smith Converting	143
		8.6.1. Hoboken Converter	144
		8.6.2. Flash Converting	144
		8.6.3. Submerged-Tuyere Noranda Continuous Converting	147
		8.6.4. Recent Developments in Peirce—Smith Converting Alternatives	150
	8.7.	Summary	150
	References		151
	Suggested Reading		153
9.	**Bath Matte Smelting: Ausmelt/Isasmelt and Mitsubishi**		155
	9.1.	Basic Operations	155
	9.2.	Feed Materials	156
	9.3.	The TSL Furnace and Lances	156
	9.4.	Smelting Mechanisms	163
		9.4.1. Impurity Elimination	163

	9.5.	Startup and Shutdown	163
	9.6.	Current Installations	164
	9.7.	Copper Converting Using TSL Technology	164
	9.8.	The Mitsubishi Process	165
		9.8.1. Introduction	165
		9.8.2. The Mitsubishi Process	165
		9.8.3. Smelting Furnace Details	166
		9.8.4. Electric Slag-Cleaning Furnace Details	167
		9.8.5. Converting Furnace Details	168
		9.8.6. Optimum Matte Grade	169
		9.8.7. Process Control in Mitsubishi Smelting/Converting	169
	9.9.	The Mitsubishi Process in the 2000s	174
	9.10.	Summary	175
	References		176
	Suggested Reading		177
10.	**Direct-To-Copper Flash Smelting**		179
	10.1.	Advantages and Disadvantages	179
	10.2.	The Ideal Direct-to-Copper Process	179
	10.3.	Industrial Single Furnace Direct-to-Copper Smelting	182
	10.4.	Chemistry	184
	10.5.	Effect of Slag Composition on % Cu-in-Slag	185
	10.6.	Industrial Details	185
	10.7.	Control	186
		10.7.1. Target: No Matte Layer to Avoid Foaming	186
		10.7.2. High % Cu-in-Slag from No-Matte-Layer Strategy	186
	10.8.	Electric Furnace Cu-from-Slag Recovery	186
		10.8.1. Głogów	187
		10.8.2. Olympic Dam	187
	10.9.	Cu-in-Slag Limitation of Direct-to-Copper Smelting	187
	10.10.	Direct-to-Copper Impurities	187
	10.11.	Summary	188
	References		189
	Suggested Reading		189
11.	**Copper Loss in Slag**		191
	11.1.	Copper in Slags	191
	11.2.	Decreasing Copper in Slag I: Minimizing Slag Generation	193
	11.3.	Decreasing Copper in Slag II: Minimizing Copper Concentration in Slag	193
	11.4.	Decreasing Copper in Slag III: Pyrometallurgical Slag Settling/Reduction	194
	11.5.	Decreasing Copper in Slag IV: Slag Minerals Processing	197
	11.6.	Summary	201
	References		201
	Suggested Reading		203
12.	**Capture and Fixation of Sulfur**		205
	12.1.	Offgases From Smelting and Converting Processes	206
		12.1.1. Sulfur Capture Efficiencies	207
	12.2.	Sulfuric Acid Manufacture	208

	12.3.	Smelter Offgas Treatment	208
		12.3.1. Gas Cooling and Heat Recovery	210
		12.3.2. Electrostatic Precipitation of Dust	211
		12.3.3. Water Quenching, Scrubbing, and Cooling	211
		12.3.4. Mercury Removal	211
		12.3.5. The Quenching Liquid, Acid Plant Blowdown	212
	12.4.	Gas Drying	212
		12.4.1. Drying Tower	212
		12.4.2. Main Acid Plant Blowers	213
	12.5.	Acid Plant Chemical Reactions	214
		12.5.1. Oxidation of SO_2 to SO_3	214
	12.6.	Industrial Sulfuric Acid Manufacture	218
		12.6.1. Catalytic Converter	224
		12.6.2. $SO_2 \rightarrow SO_3$ Conversion Reaction Paths	224
		12.6.3. Reaction Path Characteristics	225
		12.6.4. Absorption Towers	226
		12.6.5. Gas to Gas Heat Exchangers and Acid Coolers	227
		12.6.6. Grades of Product Acid	227
	12.7.	Alternative Sulfuric Acid Manufacturing Methods	227
		12.7.1. Haldor Topsøe WSA	227
		12.7.2. Sulfacid®	228
	12.8.	Recent and Future Developments in Sulfuric Acid Manufacture	229
		12.8.1. Maximizing Feed Gas SO_2 Concentrations	229
		12.8.2. Maximizing Heat Recovery	230
	12.9.	Alternative Sulfur Products	231
	12.10.	Future Improvements in Sulfur Capture	231
	12.11.	Summary	231
	References		232
	Suggested Reading		234
13.	**Fire Refining (S and O Removal) and Anode Casting**		237
	13.1.	Industrial Methods of Fire Refining	237
		13.1.1. Rotary Furnace Refining	238
		13.1.2. Hearth Furnace Refining	240
	13.2.	Chemistry of Fire Refining	240
		13.2.1. Sulfur Removal: the Cu–O–S System	240
		13.2.2. Oxygen Removal: the Cu–C–H–O System	241
	13.3.	Choice of Hydrocarbon for Deoxidation	241
	13.4.	Casting Anodes	241
		13.4.1. Anode Molds	243
		13.4.2. Anode Uniformity	243
		13.4.3. Anode Preparation	243
	13.5.	Continuous Anode Casting	244
	13.6.	New Anodes from Rejects and Anode Scrap	245
	13.7.	Removal of Impurities During Fire Refining	245
	13.8.	Summary	247
	References		247
	Suggested Reading		248
14.	**Electrolytic Refining**		251
	14.1.	The Electrorefining Process	251

	14.2.	Chemistry of Electrorefining and Behavior of Anode Impurities	252
		14.2.1. Au and Platinum-group Metals	253
		14.2.2. Se and Te	253
		14.2.3. Pb and Sn	254
		14.2.4. As, Bi, Co, Fe, Ni, S, and Sb	254
		14.2.5. Ag	255
		14.2.6. O	255
		14.2.7. Summary of Impurity Behavior	256
	14.3.	Equipment	257
		14.3.1. Anodes	258
		14.3.2. Cathodes	258
		14.3.3. Cells	259
		14.3.4. Electrical Components	260
	14.4.	Typical Refining Cycle	260
	14.5.	Electrolyte	261
		14.5.1. Addition Agents	262
		14.5.2. Electrolyte Temperature	266
		14.5.3. Electrolyte Filtration	266
		14.5.4. Removal of Impurities from the Electrolyte	266
	14.6.	Maximizing Copper Cathode Purity	267
		14.6.1. Physical Factors Affecting Cathode Purity	267
		14.6.2. Chemical Factors Affecting Cathode Purity	267
		14.6.3. Electrical Factors Affecting Cathode Purity	268
	14.7.	Minimizing Energy Consumption	269
	14.8.	Industrial Electrorefining	269
	14.9.	Recent Developments and Emerging Trends in Copper Electrorefining	274
	14.10.	Summary	275
	References		275
	Suggested Reading		279
15.	Hydrometallurgical Copper Extraction: Introduction and Leaching		281
	15.1.	Copper Recovery by Hydrometallurgical Flowsheets	281
	15.2.	Chemistry of the Leaching of Copper Minerals	282
		15.2.1. Leaching of Copper Oxide Minerals	282
		15.2.2. Leaching of Copper Sulfide Minerals	283
	15.3.	Leaching Methods	285
	15.4.	Heap and Dump Leaching	287
		15.4.1. Chemistry of Heap and Dump Leaching	288
		15.4.2. Industrial Heap Leaching	290
		15.4.3. Industrial Dump Leaching	301
	15.5.	Vat Leaching	301
	15.6.	Agitation Leaching	303
		15.6.1. Oxide Minerals	303
		15.6.2. Sulfide Minerals	304
	15.7.	Pressure Oxidation Leaching	304
		15.7.1. Economic and Process Drivers for a Hydrometallurgical Process for Chalcopyrite	304
		15.7.2. Elevated Temperature and Pressure Leaching	308
	15.8.	Future Developments	315

		15.9.	Summary	316
		References		317
		Suggested Reading		322

16. Solvent Extraction — 323
- 16.1. The Solvent-Extraction Process — 323
- 16.2. Chemistry of Copper Solvent Extraction — 324
- 16.3. Composition of the Organic Phase — 325
  - 16.3.1. Extractants — 325
  - 16.3.2. Diluents — 327
- 16.4. Minimizing Impurity Transfer and Maximizing Electrolyte Purity — 328
- 16.5. Equipment — 329
  - 16.5.1. Mixer Designs — 329
  - 16.5.2. Settler Designs — 330
- 16.6. Circuit Configurations — 331
  - 16.6.1. Series Circuit — 331
  - 16.6.2. Parallel and Series-parallel Circuits — 333
  - 16.6.3. Inclusion of a Wash Stage — 333
- 16.7. Quantitative Design of a Series Circuit — 333
  - 16.7.1. Determination of Extractant Concentration Required — 333
  - 16.7.2. Determination of Extraction and Stripping Isotherms — 334
  - 16.7.3. Determination of Extraction Efficiency — 334
  - 16.7.4. Determination of Equilibrium Stripped Organic Cu Concentration — 334
  - 16.7.5. Transfer of Cu Extraction into Organic Phase — 335
  - 16.7.6. Determination of Electrolyte Flowrate Required to Strip Cu Transferred — 335
  - 16.7.7. Alternative Approach — 336
- 16.8. Quantitative Comparison of Series and Series-parallel Circuits — 336
- 16.9. Operational Considerations — 336
  - 16.9.1. Stability of Operation — 336
  - 16.9.2. Crud — 337
  - 16.9.3. Phase Continuity — 339
  - 16.9.4. Organic Losses and Recovery — 339
- 16.10. Industrial Solvent-Extraction Plants — 339
- 16.11. Summary — 344
- References — 344
- Suggested Reading — 346

17. Electrowinning — 349
- 17.1. The Electrowinning Process — 349
- 17.2. Chemistry of Copper Electrowinning — 349
- 17.3. Electrical Requirements — 350
- 17.4. Equipment and Operational Practice — 351
  - 17.4.1. Cathodes — 351
  - 17.4.2. Anodes — 351
  - 17.4.3. Cell Design — 353
  - 17.4.4. Current Density — 355
  - 17.4.5. Acid Mist Suppression — 356
  - 17.4.6. Electrolyte — 356
  - 17.4.7. Electrolyte Additives — 360

	17.5.	Maximizing Copper Purity	360
	17.6.	Maximizing Energy Efficiency	361
	17.7.	Modern Industrial Electrowinning Plants	362
	17.8.	Electrowinning from Agitated Leach Solutions	362
	17.9.	Current and Future Developments	368
	17.10.	Summary	369
	References		369
	Suggested Reading		371

18. Collection and Processing of Recycled Copper — 373

	18.1.	The Materials Cycle	373
		18.1.1. Home Scrap	373
		18.1.2. New Scrap	374
		18.1.3. Old Scrap	375
	18.2.	Secondary Copper Grades and Definitions	379
	18.3.	Scrap Processing and Beneficiation	380
		18.3.1. Wire and Cable Processing	380
		18.3.2. Automotive Copper Recovery	382
		18.3.3. Electronic Scrap Treatment	384
	18.4.	Summary	385
	References		385
	Suggested Reading		387

19. Chemical Metallurgy of Copper Recycling — 389

	19.1.	Characteristics of Secondary Copper	389
	19.2.	Scrap Processing in Primary Copper Smelters	389
		19.2.1. Scrap Use in Smelting Furnaces	390
		19.2.2. Scrap Additions to Converters and Anode Furnaces	391
	19.3.	The Secondary Copper Smelter	391
		19.3.1. High-grade Secondary Smelting	391
		19.3.2. Smelting to Black Copper	391
		19.3.3. Converting Black Copper	393
		19.3.4. Fire Refining and Electrorefining	394
	19.4.	Summary	394
	References		395
	Suggested Reading		396

20. Melting and Casting — 397

	20.1.	Product Grades and Quality	397
	20.2.	Melting Technology	399
		20.2.1. Furnace Types	399
		20.2.2. Hydrogen and Oxygen Measurement/Control	403
	20.3.	Casting Machines	403
		20.3.1. Billet Casting	404
		20.3.2. Bar and Rod Casting	404
		20.3.3. Oxygen-free Copper Casting	409
		20.3.4. Strip Casting	409
	20.4.	Summary	410
	References		411
	Suggested Reading		412

21. Byproduct and Waste Streams — 415
- 21.1. Molybdenite Recovery and Processing — 415
- 21.2. Flotation Reagents — 415
- 21.3. Operation — 415
- 21.4. Optimization — 417
- 21.5. Anode Slimes — 418
  - 21.5.1. Anode Slime Composition — 418
  - 21.5.2. The Slime Treatment Flowsheet — 421
- 21.6. Dust Treatment — 422
- 21.7. Use or Disposal of Slag — 423
- 21.8. Summary — 425
- References — 425
- Suggested Reading — 426

22. Costs of Copper Production — 427
- 22.1. Overall Investment Costs: Mine through Refinery — 427
  - 22.1.1. Variation in Investment Costs — 429
  - 22.1.2. Economic Sizes of Plants — 429
- 22.2. Overall Direct Operating Costs: Mine through Refinery — 429
  - 22.2.1. Variations in Direct Operating Costs — 430
- 22.3. Total Production Costs, Selling Prices, Profitability — 430
  - 22.3.1. Byproduct Credits — 431
- 22.4. Concentrating Costs — 431
- 22.5. Smelting Costs — 433
- 22.6. Electrorefining Costs — 435
- 22.7. Production of Copper from Scrap — 435
- 22.8. Leach/Solvent Extraction/Electrowinning Costs — 436
- 22.9. Profitability — 438
- 22.10. Summary — 438
- References — 438
- Suggested Reading — 439

Index — 441

Preface

The preceding nine years have seen significant changes in the way that copper metal is produced. The most important change is the continuing high price of copper. It has been some time since the industry as a whole has enjoyed such success. This has led to an expansion of the industry, and the list of mines and production facilities is considerably changed from the previous edition.

Technology changes have occurred as well. The most notable of these include:

- The continuing adoption of high-intensity bath smelting for both primary concentrates and secondary materials;
- Significant increase in the use of hydrometallurgical processing technology for sulfide concentrates;
- New technology to mitigate the environmental impact of copper ore mining and processing.

These changes are reflected in this edition, along with expanded coverage of byproduct recovery (important when gold prices are US $1400/oz!) and secondary copper recovery and processing. The continuing search for a replacement for the Peirce–Smith converter is also featured. The day is getting closer when this venerable device will belong to the past!

There has also been a change in the group of conspirators behind this book. We welcome to the group Dr. Kathryn C. Sole from South Africa. Kathy specializes in hydrometallurgical process development, and the addition of her expertise is important to three old-fashioned pyrometallurgists. She improves the diversity of the group in more ways than one!

As with previous efforts, this edition of *Extractive Metallurgy of Copper* is largely a product of the copper industry as a whole, since so many engineers and scientists volunteered their time and expertise (along with photographs and drawings) to make sure we got it right. Our mentors and assistants included:

- Peter Amelunxen and Roger Amelunxen (Amelunxen Mineral Engineering)
- Mike Bernard (Terra Nova Technologies)
- Elizabeth Bowes, Shingefei Gan, Ursula Mostert, Sandip Naik, Ian Ralston, Bernardo Soto, and Gabriel Zárate (Anglo American)
- Alistair Burrows (ISASMELT)
- Leszek Byszyński (KGHM)
- Connie Callahan, James Davis III, David Jones, Steve Koski, Michael Lam, John Quinn, and Cory Stevens (Freeport-McMoRan)
- John Joven S. Chiong (PASAR)
- Felix Conrad (Aurubis AG)
- Frank Crundwell (Crundwell Management Solutions)

- Angus Feather (BASF, formerly Cognis)
- Phil Donaldson (Xstrata Technologies)
- Jim Finch (McGill University)
- Mark Firestein (Bateman-Litwin)
- David George-Kennedy, Art Johnston, David Kripner, and Mark Taylor (Rio Tinto)
- Tom Gonzalez (Hatch)
- J. Brent Hiskey (University of Arizona)
- Nicholas Hogan (Incitec Pivot)
- Kun Huang (Chinese Academy of Sciences)
- John Hugens (Fives North American Combustion)
- Cyrus Kets (Moonshine Advertising)
- Hannu Laitala, Robert Matusewicz, Lauri Palmu, and Markus Reuter (Outotec Oyj)
- Pascal Larouche (Xstrata Copper)
- Theo Lehner (Boliden Mineral AB)
- Pierre Louis (PEL Consulting)
- Guangsheng Luo (Tsinghua University)
- Phil Mackey (P.J. Mackey Technology)
- Dennis Marschall (Kazakhmys PLC)
- Michael Nicol (Murdoch University)
- Jan W. Matousek (Chlumsky, Armbrust and Meyer)
- Graeme Miller (Miller Metallurgical Services)
- Michael Moats (University of Utah)
- Tony Moore (MMG Sepon)
- Enock Mponda and Charles Shonongo (Konkola Copper Mines)
- Keith Mathole and Tlengelani Muhlare (Palabora Mining)
- V. Ramachandran (RAM Consultants)
- Tim Robinson (Republic Alternative Technologies)
- Helge Rosenberg (Haldor Topsøe)
- Stefan Salzmann (Bamag)
- Heguri Shinichi (Sumitomo)
- Edwin Slonim (Turbulent Technologies)
- Matthew Soderstrom and Owen Tinkler (Cytec Industries)
- Gary Spence (Encore Wire)
- Nathan Stubina (Barrick)
- Ilya Terentiev (Consultant)
- Mark Vancas (Bateman Engineering)
- Craig van der Merwe (First Quantum Minerals)
- Rob West (BHP Billiton)
- Yutaka Yasuda (Hibi Kyodo Smelting)
- Roe-Hoan Yoon (Virginia Polytechnic Institute)

The fifth edition of *Extractive Metallurgy of Copper* is dedicated to the memory of Prof. Akira Yazawa (1926–2010). Professor Yazawa performed pioneering research on the fundamentals of copper smelting and converting in the 1950s, and continued to make contributions well into this century. His original work is still referenced in this volume, and will likely continue to be in editions to come.

Preface to the Fourth Edition

This edition contains more-than-ever industrial information, all of it provided generously by our industrial friends and colleagues. We thank them profusely for their help and generosity over the years.

The publication we consulted most for this edition was *Copper 99/Cobre99* (TMS, Warrendale, PA [six volumes]). For a near-future update, we direct the reader to *Copper 03/Cobre 03* being held in Santiago, Chile, November 30, 2003 (www.cu2003.cl).

As with previous editions, Margaret Davenport read every word of our manuscript. After 27 years of proofreading, she may well know more than the authors.

Preface to the Third Edition

This edition chronicles the changes which have taken place in copper extraction over the last 20 years. The major changes have been the shrinkage of reverberatory smelting, the continued growth of flash smelting and the remarkable (and continuing) growth of solvent extraction/electrowinning. The use of stainless steel cathodes (instead of copper starting sheets) in electrorefining and electrowinning has also been a significant development.

These industrial growth areas receive considerable attention in this edition as do SO_2 collection and sulphuric acid manufacture. SO_2 capture has continued to grow in importance — only a few smelters now emit their SO_2 to the atmosphere.

Several important volumes on copper extraction have appeared recently, namely: *Copper 91/Cobre 91* (Pergamon Press, New York [four volumes]) and *Extractive Metallurgy of Copper, Nickel and Cobolt* (TMS, Warrendale, Pennsylvania [two volumes]). A volume on Converting, *Fire-refining and Casting* is scheduled to appear in 1994 (TMS) and the proceedings of *Cobre95/Copper 95* will appear in 1995. The reader is directed to these publications for updated information.

We wish to thank our colleagues in the copper industry for their many contributions to this edition. They have responded to our questions, encouraged us to visit their plants and engaged us in rigorous debate regarding extraction optimization. We would particularly like to thank Brian Felske (Felske and Associates), David Jones (Magma Copper Company) and Eric Partelpoeg (Phelps Dodge Mining Company). Without them this edition would not have been possible.

The manuscript was prepared and proofed by Patricia Davenport and Margaret Davenport. Their perseverance, skill and enthusiasm are happily acknowledged.

Preface to the Second Edition

For this edition we have concentrated mainly on bringing the operating data and process descriptions of the first edition up to date. Typographical errors have been corrected and several passages have been rewritten to avoid misinterpretation. Since most of the new data have come directly from operating plants, very few new references have been added. For collections of recent published information, the reader is directed to the excellent symposium publications: Extractive Metallurgy of Copper, Volumes I and II, Yannopoulos, J. C. and Agarwal, J. C. editors, A.I.M.E., New York, 1976, Copper and Nickel Converters, Johnson, R.E., A.I.M.E., New York, 1979, and to the reviews of copper technology and extractive metallurgy published annually in the *Journal of Metals* (A.I.M.E., New York). Most of the credit for this edition should go to the many industrial engineers and scientists who almost without exception responded to our requests for new information on their processes. We would like in particular to single out Jan Matousek of INCO, Keith Murden of Outokumpu Oy and John Schloen of Canadian Copper Refiners (now a metallurgical consultant) for their help.

September 1979

A. K. Biswas
W. G. Davenport

Preface to the First Edition

This book describes the extraction of copper from its ores. The starting point is with copper ores and minerals and the finishing point is the casting and quality control of electrical grade copper. Techniques for recovering copper from recycled scrap are also discussed.

The main objectives of the book are to describe the extractive metallurgy of copper as it is today and to discuss (qualitatively and quantitatively) the reasons for using each particular process. Arising from these descriptions and discussions are indications as to how copper-extraction methods will develop in the future. Control of air and water pollution is of tremendous importance when considering future developments and these are discussed in detail for each process. Likewise, the energy demands of each process are dealt with in detail. Costs are mentioned throughout the text and they are considered in depth in the final chapter.

The book begins with an introductory synopsis (for the generalist reader) of the major copper-extraction processes. It then follows copper extraction in a stepwise fashion beginning with mineral benefication and advancing through roasting, smelting, converting, refining, casting and quality control. Hydrometallurgy and its associated processes are introduced just before electrorefining so that electrowinning and electrorefining can be discussed side by side and the final products of each method compared. The last two chapters are not in sequence — they are devoted to the sulphur pollution problem and to economics.

As far as possible, the length of each chapter is commensurate with the relative importance of the process it describes. Blast-furnace copper smelting is, for example, given a rather brief treatment because it is a dying process while newer techniques such a continuous copper-making and solvent extraction are given extensive coverage because they may assume considerable importance in the near future.

A word about units: the book is metric throughout, the only major exception to the Standard International Unit System being that energy is reported in terms of kilocalories and kilowatt-hours. The principal units of the book are metric tons (always written tonnes in the text), kilograms and metres. A conversion table is provided in Appendix I. A knowledge of thermodynamics is assumed in parts of the book, particularly with respect to equilibrium constants. For concise information on the thermodynamic method as applied to metallurgy, the reader is directed to *Metallurgical Thermochemistry* by O. Kubaschewski, E. L. Evans and C. B. Alcock, an earlier volume in this series.

The text of the book is followed by four appendixes which contain units and conversion factors: stoichiometric data; enthalpy and free energy data; and a summary of the properties of electrolytic tough pitch copper.

Copper is one of man's most beautiful and useful materials. It has given us great satisfaction to describe and discuss the methods by which it is obtained. Both of our universities have had a long association with the copper industries of our countries, and it is hoped that, through this book, this association will continue.

A. K. Biswas
University of Queensland

W. G. Davenport
McGill University

Chapter 1

Overview

1.1. INTRODUCTION

Copper is most commonly present in the earth's crust as copper−iron−sulfide and copper sulfide minerals, such as chalcopyrite ($CuFeS_2$) and chalcocite (Cu_2S). The concentration of these minerals in an ore body is low. Typical copper ores contain from 0.5% Cu (open pit mines, Fig. 1.1) to 1 or 2% Cu (underground mines). Pure copper metal is mostly produced from these ores by concentration, smelting, and refining (Fig. 1.2).

Copper also occurs to a lesser extent in oxidized minerals (carbonates, oxides, hydroxy-silicates, sulfates). Copper metal is usually produced from these minerals by leaching, solvent extraction, and electrowinning (Fig. 1.3). These processes are also used to treat chalcocite (Cu_2S).

A third major source of copper is scrap copper and copper alloys. Production of copper from recycled used objects is 10 or 15% of mine production. In addition, there is considerable re-melting/re-refining of scrap generated during fabrication and manufacture. Total copper production in 2010 (mined and from end-of-use scrap) was ~20 million tonnes.

FIGURE 1.1 Open pit Cu mine. Note the new blast holes, top right, and blasted ore to the left of them. The shovel is placing blasted ore in the truck from where it will go to processing. The water truck is suppressing dust. The front end loader is cleaning up around the shovel. The shovel is electric. Its power wire mostly lies on the surface except over the wire bridge under which all vehicles travel to and from the shovel. (*Photo courtesy of Freeport-McMoRan Copper & Gold Inc.*).

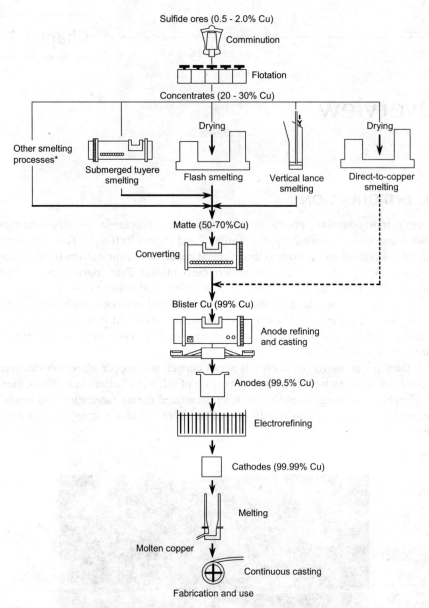

FIGURE 1.2 Main processes for extracting copper from sulfide ores. Parallel lines indicate alternative processes. *Principally Mitsubishi and Vanyukov smelting.

This chapter introduces the principal processes by which copper is extracted from ore and scrap. It also indicates the relative industrial importance of each.

1.2. EXTRACTING COPPER FROM COPPER−IRON−SULFIDE ORES

About 80% of the world's copper-from-ore originates in Cu−Fe−S ores. Cu−Fe−S minerals are not easily dissolved by aqueous solutions, so the vast majority of copper extraction from these minerals is pyrometallurgical. The extraction entails:

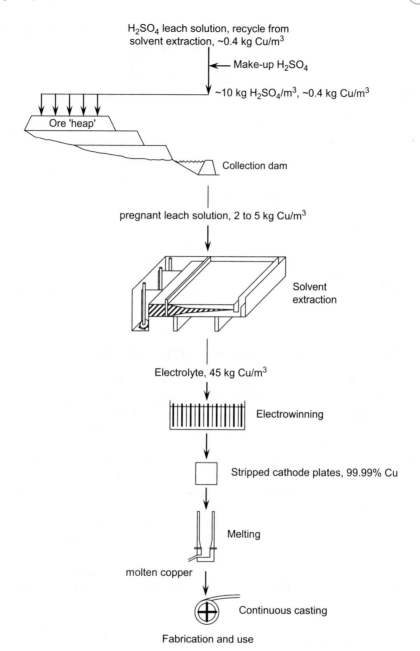

FIGURE 1.3 Flowsheet for leaching oxide and Cu_2S ores. The dissolved Cu is recovered by solvent extraction purification/strengthening then electrowinning. Leaching accounts for ~20% of primary (from ore) copper production.

(a) Isolating the Cu−Fe−S and Cu−S mineral particles in an ore to a concentrate by froth flotation
(b) Smelting this concentrate to molten high-Cu sulfide matte
(c) Converting (oxidizing) this molten matte to impure molten copper
(d) Fire- and electrorefining this impure copper to ultra-pure copper.

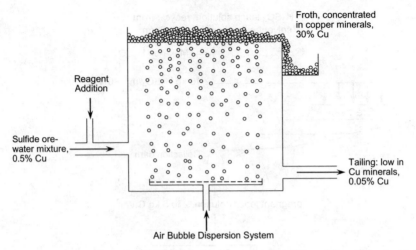

FIGURE 1.4 Schematic view of flotation cell. Reagents cause Cu—Fe sulfide and Cu sulfide minerals in the ore to attach to rising air bubbles, which are then collected in a short-lived froth. This froth is de-watered to become concentrate. The un-floated waste passes through several cells before being discarded as a final tailing. Many types and sizes (up to 300 m^3) of cell are used.

1.2.1. Concentration by Froth Flotation (Chapters 3 and 4)

The copper ores being mined in 2010 are too lean in copper (0.5—2% Cu) to be smelted directly. Heating and melting their huge quantity of waste rock would require prohibitive amounts of hydrocarbon fuel. Fortunately, the Cu—Fe—S and Cu—S minerals in an ore can be isolated by physical means into high-Cu *concentrate*, which can then be smelted economically.

The most effective method of isolating the Cu minerals is froth flotation. This process causes the Cu minerals to become selectively attached to air bubbles rising through a slurry of finely ground ore in water (Fig. 1.4). Selectivity of flotation is created by using reagents, which make Cu minerals water repellent while leaving waste minerals wetted. In turn, this water repellency causes Cu minerals to float on rising bubbles while the other minerals remain un-floated. The floated Cu-mineral particles overflow the flotation cell in a froth to become concentrate containing ~30% Cu.

Flotation is preceded by crushing and grinding the mined Cu ore into small (~50 μm) particles. Its use has led to adoption of smelting processes which efficiently smelt finely ground material.

1.2.2. Matte Smelting (Chapters 5, 6, and 9)

Matte smelting oxidizes and melts flotation concentrate in a large, hot (1250 °C) furnace (Figs. 1.2 and 1.5). The objective of the smelting is to oxidize S and Fe from the Cu—Fe—S concentrate to produce a Cu-enriched molten sulfide phase (matte). The oxidant is almost always oxygen-enriched air.

Example reactions are:

$$2CuFeS_2(s) + 3.25O_2(g) \xrightarrow[\text{in oxygen enriched air}]{30\ °C} \underset{\text{molten matte}}{\overset{1220\ °C}{Cu_2S - 0.5FeS(l)}} + 1.5FeO(s) + 2.5SO_2(g) \quad (1.1)$$

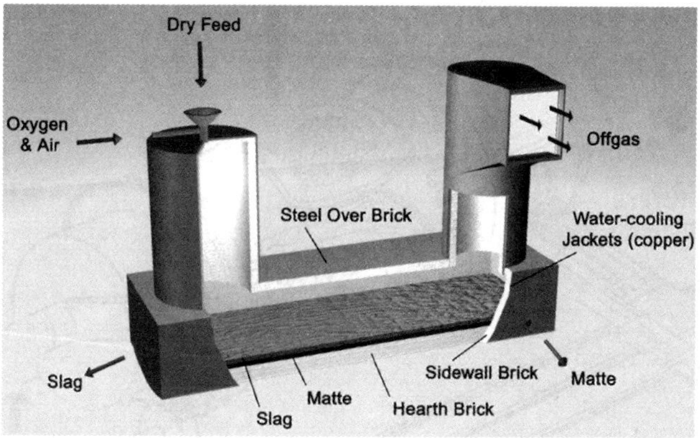

FIGURE 1.5 Outokumpu oxygen-enriched air flash furnace. Flash furnaces are typically 20 m long and 7 m wide. They smelt 1000–3000 tonnes of concentrate per day.

$$\underset{1220\,°C}{2FeO(s)} + \underset{\substack{30\,°C \\ \text{in quartz flux}}}{SiO_2(s)} \rightarrow \underset{\substack{1250\,°C \\ \text{molten slag}}}{Fe_2SiO_4(l)} \tag{1.2}$$

Simultaneously, the gangue minerals in the concentrate dissolve in the molten slag. Reactions (1.1) and (1.2) are both exothermic – they supply considerable heat for the smelting process.

The products of smelting are (a) molten sulfide matte (45–75% Cu) containing most of the copper in the concentrate, and (b) molten oxide slag with as little Cu as possible. The molten matte is subsequently converted (oxidized) in a converting furnace to form impure molten copper. The slag is treated for Cu recovery, then discarded or sold (Chapter 11).

SO_2-bearing offgas (10–60% SO_2) is also generated. SO_2 is harmful to the flora and fauna so it must be removed before the offgas is released to the atmosphere. This is almost always done by capturing the SO_2 as sulfuric acid (Chapter 14).

An important objective of matte smelting is to produce a slag which contains as little Cu as possible. This is done by (a) including SiO_2 flux in the furnace charge to promote matte–slag immiscibility, and (b) keeping the furnace hot so that the slag is molten and fluid.

Matte smelting is most often done in flash furnaces (Fig. 1.5). It is also done in top lance and submerged tuyere furnaces (Chapters 7 and 9). Three smelters also smelt concentrate directly to molten copper (Chapter 10).

1.2.3. Converting (Chapters 8 and 9)

Copper converting is an oxidation of the molten matte from smelting with air or oxygen-enriched air. It removes Fe and S from the matte to produce crude (99% Cu) molten copper. This copper is then sent to fire- and electrorefining. Converting is mostly carried out in cylindrical Peirce–Smith converters (Fig. 1.6).

Liquid matte (1220 °C) is transferred from the smelting furnace in large ladles and poured into the converter through a large central mouth (Fig. 1.6b). The oxidizing blast is then started and the converter is rotated, forcing air into the matte through a line of tuyeres along the length of the vessel. The heat generated in the converter by Fe and S oxidation is sufficient to make the process autothermal.

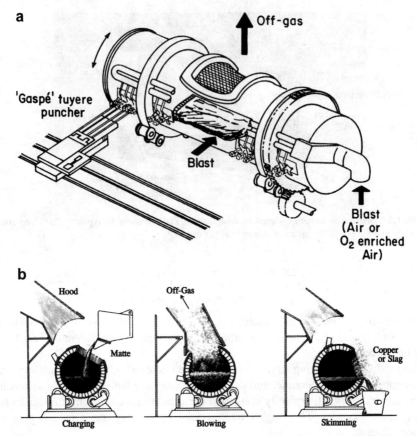

FIGURE 1.6 a. Peirce–Smith converter for producing molten 'blister' copper from molten Cu–Fe–S matte, typical production rate 200–600 tonnes of copper per day. Oxygen-enriched air or air 'blast' is blown into the matte through submerged tuyeres. Silica flux is added through the converter mouth or by air gun through an endwall. Offgas is collected by means of a hood above the converter mouth. (After Boldt & Queneau, 1967 courtesy Vale). b. Positions of Peirce–Smith converter for charging, blowing, and skimming (Boldt & Queneau, 1967 courtesy Vale). SO_2 offgas escapes the system unless the hooding is tight. A converter is typically 4 or 4.5 m diameter.

The converting takes place in two sequential stages:

(a) The FeS elimination or slag forming stage:

$$\underset{\text{in molten matte}}{\underset{1220\,°C}{2FeS(l)}} + \underset{\text{in blast}}{\underset{30\,°C}{3O_2(g)}} + \underset{\text{in quartz flux}}{\underset{30\,°C}{SiO_2(s)}} \rightarrow \underset{\text{molten slag}}{\underset{1200\,°C}{Fe_2SiO_4(l)}} + \underset{\text{in offgas}}{2SO_2(g)} + \text{heat} \qquad (1.3)$$

(b) The copper forming stage:

$$\underset{\text{liquid matte}}{\underset{1200\,°C}{Cu_2S(l)}} + \underset{\text{in blast}}{\underset{30\,°C}{O_2(g)}} \rightarrow \underset{\text{in impure molten copper}}{\underset{1200\,°C}{2Cu(l)}} + \underset{\text{in offgas}}{\underset{1200\,°C}{SO_2(g)}} + \text{heat} \qquad (1.4)$$

Coppermaking (b) occurs only after the matte contains less than about 1% Fe, so that most of the Fe can be removed from the converter (as slag) before copper production begins. Likewise, significant oxidation of copper does not occur until the sulfur content of the copper falls below ~0.02%. Blowing is terminated near this sulfur end point. The resulting molten *blister* copper (1200 °C) is sent to refining.

Because conditions in the converter are strongly oxidizing and agitated, converter slag inevitably contains 4–8% Cu. This Cu is recovered by settling or froth flotation. The slag is then discarded or sold (Chapters 11 and 21).

SO_2, 8–12 vol.-% in the converter offgas, is a byproduct of both converting reactions. It is combined with smelting furnace gas and captured as sulfuric acid. There is, however, some leakage of SO_2 into the atmosphere during charging and pouring (Fig. 1.6b). This problem is encouraging development of continuous converting processes (Chapters 8 and 9).

1.2.4. Direct-to-Copper Smelting (Chapter 10)

Smelting and converting are separate steps in oxidizing Cu–Fe–S concentrates to metallic copper. It would seem natural that these two steps should be combined to produce copper directly in one furnace. It would also seem natural that this should be done continuously rather than by batch-wise Peirce–Smith converting.

In 2010, copper is made in a single furnace at only three places — Olympic Dam, Australia; Glogow, Poland; and Chingola, Zambia — all using a flash furnace. The strongly oxidizing conditions in a direct-to-copper furnace produce a slag with 14–24% oxidized Cu. The expense of reducing this Cu back to metallic copper has so far restricted the process to low-Fe concentrates, which produce little slag.

Continuous smelting/converting, even in more than one furnace, has energy, SO_2 collection, and cost advantages. Mitsubishi lance, Outokumpu flash, and Noranda submerged tuyere smelting/converting all use this approach (Chapters 7–9).

1.2.5. Fire Refining and Electrorefining of Blister Copper (Chapters 13 and 14)

The copper from the above processing is electrochemically refined to high-purity cathode copper. This final copper contains less than 20 ppm undesirable impurities. It is suitable for electrical and almost all other uses.

Electrorefining requires strong, flat thin anodes to interleave with cathodes in a refining cell (Fig. 1.6). These anodes are produced by removing S and O from molten blister copper, and casting the resulting *fire-refined* copper in open, anode shape molds (occasionally in a continuous strip caster).

Copper electrorefining entails (a) electrochemically dissolving copper from impure anodes into $CuSO_4$–H_2SO_4–H_2O electrolyte, and (b) electrochemically plating pure copper (without the anode impurities) from the electrolyte onto stainless steel (occasionally copper) cathodes.

Copper is deposited on the cathodes for 7–14 days. The cathodes are then removed from the cell. Their copper is stripped, washed, and (a) sold or (b) melted and cast into useful products (Chapter 20).

The electrolyte is an aqueous solution of H_2SO_4 (150–200 kg/m^3) and $CuSO_4$ (40–50 kg Cu/m^3). It also contains impurities and trace amounts of chlorine and organic *addition agents*.

Many anode impurities are insoluble in this electrolyte (Au, Pb, Pt metals, Sn). They do not interfere with the electrorefining. They are collected as solid *slimes* and treated for Cu and byproduct recovery (Chapter 21).

Other impurities such as As, Bi, Fe, Ni and Sb are partially or fully soluble. Fortunately, they do not plate with the copper at the low voltage of the electrorefining cell (~0.3 V). They must, however, be kept from accumulating in the electrolyte to avoid physical contamination of the cathode copper. This is done by continuously bleeding part of the electrolyte through a purification circuit.

1.3. HYDROMETALLURGICAL EXTRACTION OF COPPER

About 80% of copper-from-ore is obtained by flotation, smelting, and refining. The other 20% is obtained hydrometallurgically. Hydrometallurgical extraction entails:

(a) Sulfuric acid leaching of Cu from broken or crushed ore to produce impure Cu-bearing aqueous solution
(b) Transfer of Cu from this impure solution to pure, high-Cu electrolyte via solvent extraction
(c) Electroplating pure cathode copper from this pure electrolyte.

The ores most commonly treated this way are (a) oxide copper minerals, including carbonates, hydroxy-silicates, sulfates, and (b) chalcocite (Cu_2S).

The leaching is mostly done by dripping dilute sulfuric acid on top of heaps of broken or crushed ore (~0.5% Cu) and allowing the acid to trickle through to collection ponds (Fig. 1.3). Several months of leaching are required for efficient Cu extraction.

Oxidized minerals are rapidly dissolved by sulfuric acid by reactions like:

$$\underset{\text{in ore}}{CuO(s)} + \underset{\text{in sulfuric acid}}{H_2SO_4(l)} \xrightarrow{30°C} \underset{\text{pregnant leach solution}}{Cu^{2+}(aq) + SO_4^{2-}(aq) + H_2O(l)} \quad (1.5)$$

Sulfide minerals, on the other hand, require oxidation, schematically:

$$\underset{\text{in ore}}{Cu_2S(s)} + \underset{\text{in air}}{2.5O_2(g)} + \underset{\text{in sulfuric acid}}{H_2SO_4(l)} \xrightarrow[30°C]{\text{bacterial enzyme catalyst}}$$

$$\underset{\text{pregnant leach solution}}{2Cu^{2+}(aq) + 2SO_4^{2-}(aq) + H_2O(l)} \quad (1.6)$$

As shown, sulfide leaching is greatly speeded up by bacterial action (Chapter 15).

Leaching is occasionally applied to Cu-bearing flotation tailings, mine wastes, old mines, and fractured ore bodies. Leaching of ore heaps is the most important process.

1.3.1. Solvent Extraction (Chapter 16)

The solutions from heap leaching contain 1–6 kg Cu/m^3 and 0.5–5 kg H_2SO_4/m^3 plus dissolved impurities such as Fe and Mn. These solutions are too dilute in Cu and too impure for direct electroplating of pure copper metal. Their Cu must be transferred to pure, high-Cu electrolyte.

Chapter | 1 Overview

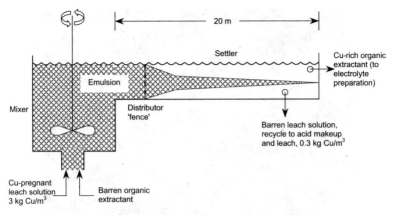

FIGURE 1.7 Schematic view of solvent extraction mixer/settler for extracting Cu from pregnant leach solution into organic extractant. The Cu-loaded organic phase goes forward to another mixer/setter (stripper) where Cu is stripped from the organic into pure, strongly acidic, high-Cu electrolyte for electrowinning. The process is continuous.

The transfer is done by:

(a) Extracting Cu from an impure leach solution into a Cu-specific liquid organic extractant
(b) Separating (by gravity) the Cu-loaded extractant from the Cu-depleted leach solution
(c) Stripping Cu from the loaded extractant into 185 kg H_2SO_4/m^3 electrolyte.

Extraction and stripping are carried out in large mixer–settlers (Fig. 1.7).

The solvent extraction process is represented by the reaction:

$$\underset{\substack{\text{in pregnant}\\\text{leach solution}}}{Cu^{2+}(aq)} + SO_4^{2-} + \underset{\substack{\text{in organic}\\\text{solvent}}}{2RH} \xrightarrow{30\,°C} \underset{\substack{\text{in organic}\\\text{solvent}}}{R_2Cu} + 2H^+ + \underset{\substack{\text{in aqueous solution,}\\\text{recycle to leach}}}{SO_4^{2-}} \qquad (1.7)$$

It shows that a low-acid (i.e. low H^+) aqueous phase causes the organic extractant to *load* with Cu (as R_2Cu). It also shows that a high acid solution causes the organic to unload (*strip*). Thus, when organic extractant is contacted with weak acid pregnant leach solution [step (a) above], Cu is loaded into the organic phase. Then when the organic phase is subsequently put into contact with high acid electrolyte [step (c) above], the Cu is stripped from the organic into the electrolyte at high Cu^{2+} concentration, suitable for electrowinning.

The extractants absorb considerable Cu but almost no impurities. They give electrolytes, which are strong in Cu but dilute in impurities.

1.3.2. Electrowinning (Chapter 17)

The Cu in the above electrolytes is universally recovered by electroplating pure metallic cathode copper (Fig. 1.8). This electrowinning is similar to electrorefining except that the anode is inert (usually lead, but increasingly iridium oxide-coated titanium).

The cathode reaction is:

$$\underset{\text{in sulfate electrolyte}}{Cu^{2+}(aq)} + \underset{\text{electrons from external power supply}}{2e^-} \xrightarrow{60\,°C} \underset{\text{pure metal deposit on cathode}}{Cu(s)} \qquad (1.8)$$

FIGURE 1.8 Plates of electrowon copper on stainless steel cathodes after removal from an electrowinning cell. These cathodes are ~1 m wide and 1.3 m deep. They are carefully washed to remove electrolyte. Their copper plates are then stripped off in automatic machines and sent to market. They typically contain <20 ppm impurities. Note the polymer side strips — they prevent copper from plating around the edges. (*Photo courtesy of Freeport-McMoRan Copper & Gold Inc.*).

The anode reaction is:

$$\underset{\text{in electrolyte}}{H_2O(l)} \xrightarrow{60\,°C} \underset{\substack{\text{evolved on} \\ \text{inert anode}}}{0.5O_2(g)} + \underset{\substack{\text{in electrolyte,} \\ \text{recycle to} \\ \text{organic stripping}}}{2H^+(aq)} + \underset{\substack{\text{electrons to external} \\ \text{power supply}}}{2e^-} \quad (1.9)$$

About 2.0 V are required. Pure metallic copper (less than 20 ppm undesirable impurities) is produced at the cathode and gaseous O_2 at the anode.

1.4. MELTING AND CASTING CATHODE COPPER

The first steps in making products from electrorefined and electrowon copper are melting and casting. The melting is mostly done in vertical shaft furnaces, in which descending cathode sheets are melted by ascending hot combustion gases. Low-sulfur fuels prevent sulfur pickup. Reducing flames prevent excessive oxygen pickup.

The molten copper is mostly cast in continuous casting machines from where it goes to rolling, extrusion, and manufacturing. An especially significant combination is continuous bar casting/rod rolling (Chapter 20). The product of this process is 1 cm diameter rod, ready for drawing to wire.

1.4.1. Types of Copper Product

The copper described above is *electrolytic tough pitch* copper. It contains ~0.025% oxygen and less than 20 ppm unwanted impurities. It is far and away the most common type of copper. A second type is oxygen-free copper (<5 ppm O). It is used for highly demanding applications, such as high-end audio equipment. About 25% of copper is used in alloy form as brasses, bronzes, etc. Much of the copper for these alloys comes from recycle scrap.

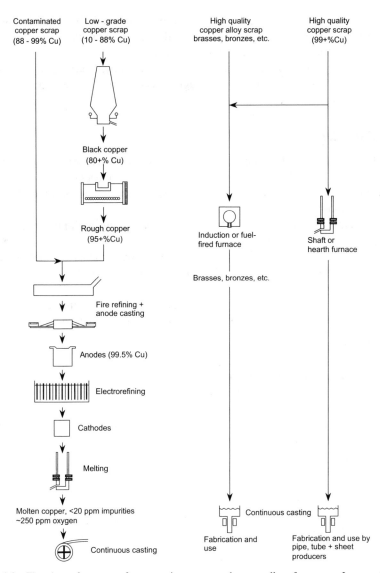

FIGURE 1.9 Flowsheet of processes for recovering copper and copper alloys from scrap. Low-grade scrap is usually smelted in shaft furnaces but other furnaces (e.g. electric) are also used.

1.5. RECYCLE OF COPPER AND COPPER-ALLOY SCRAP (CHAPTERS 18 AND 19)

Recycle of copper and copper-alloy scrap used objects (old scrap) accounts for 10–15% of pre-manufacture copper production. Considerable manufacturing waste is also recirculated.

Production of copper from scrap (a) requires considerably less energy than mining and processing copper ore, (b) avoids mine, concentrator, leach, and smelter wastes, and (c) helps to ensure the availability of copper for future generations.

The treatment given to copper scrap depends on its purity (Fig. 1.9). The lowest grade scrap is smelted and refined like concentrate in a primary or secondary (scrap) smelter/

refinery. Higher-grade scrap is fire-refined, then electrorefined. The highest-grade scrap (mainly manufacturing waste) is often melted and cast without refining. Its copper is used for non-electrical products, including tube, sheet, and alloys.

Alloy scrap (brass, bronze) is melted and cast as alloy. There is no advantage to smelting/refining it to pure copper. Some slagging is done during melting to remove dirt and other contaminants.

1.6. SUMMARY

About 80% of the world's copper-from-ore is produced by concentration/smelting/refining of sulfide ores. The other 20% is produced by heap leaching/solvent extraction/electrowinning of oxide, and chalcocite ores. Copper production from recycled end-of-use objects is about 15% of copper-from-ore copper production.

Electrochemical processing is always used in producing high-purity copper: electrorefining in the case of pyrometallurgical extraction and electrowinning in the case of hydrometallurgical extraction. The principal final copper product is electrolytic tough pitch copper (~250 ppm oxygen and 20 ppm unwanted impurities). It is suitable for nearly all uses.

The tendency in copper extraction is toward the processes that do not harm the environment and which consume little energy and water. This has led to:

(a) Energy- and pollution-efficient oxygen-enriched smelting
(b) Solvent extraction/electrowinning
(c) Increased re-circulation of water
(d) Increased recycle of end-of-use scrap.

REFERENCE

Boldt, J. R., & Queneau, P. (1967). *The winning of nickel*. Toronto: Longmans Canada Ltd.

SUGGESTED READING

Copper, (2010). *Proceedings of the seventh international conference, Vol. I—VII*. Hamburg: GDMB.
Davenport, W. G., Jones, D. M., King, M. J., & Partelpoeg, E. H. (2010). *Flash smelting, analysis, control and optimization* (2nd ed.), Wiley.

Chapter 2

Production and Use

Metallic copper occurs occasionally in nature. For this reason, it was known to man about 10,000 B.C (CDA, 2010a). Its early uses were in jewelry, utensils, tools, and weapons. Its use increased gradually over the years, and then dramatically in the 20th century with mass adoption of electricity. This dramatic growth continues in the 21st century with the rapid industrialization of China (Fig. 2.1).

Copper is an excellent conductor of electricity and heat. It resists corrosion. It is easily fabricated into wire, pipe, and other forms, and easily joined. Electrical conductivity, thermal conductivity, and corrosion resistance are its most exploited properties (Table 2.1).

This chapter discusses production and use of copper around the world. It gives production, use, and price statistics, and identifies and locates the world's largest copper-producing plants. It shows that the Andes mountain region of South America (Chile and Peru) is the world's largest source of copper (Table 2.2). The remaining production is scattered around the world. The world's 20 largest mines are listed in Table 2.3 and plotted in Fig. 2.2.

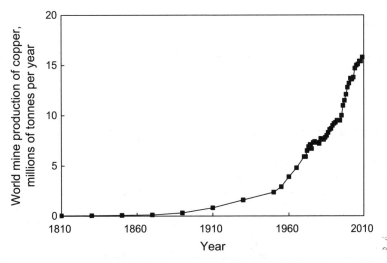

FIGURE 2.1 World mine production of copper in the 19th through 21st centuries (Butts, 1954; Edelstein & Porter, 2010). The steep rise after 1995 is notable.

TABLE 2.1 United States Usage of Copper by Exploited Property and by Application (CDA, 2010b, 2010c)

	% of total use
Exploited property	
Electrical conductivity	63
Corrosion resistance	18
Thermal conductivity	11
Mechanical and structural properties	6
Aesthetics	2
Application	
Building wire	23
Plumbing and heating	13
Automotive	10
Air conditioning, refrigeration & natural gas	10
Power utilities	9
Telecommunications	6
In-plant equipment	5
Ordnance	3
Business electronics	2
Lighting and wiring devices	2
Other	17

2.1. COPPER MINERALS AND CUT-OFF GRADES

Table 2.4 lists the main copper minerals. These minerals occur at low concentrations in ores. The remainder consists mostly of oxide rock such as andesite or granite, with a small amount of iron minerals such as pyrite. It is now rare to find a large copper deposit averaging more than 1 or 2% Cu. Copper ores containing down to 0.4% Cu (average) are being mined from open pits, while ores down to 1% (average) are being taken from underground mines.

The average grade of ore being extracted from any given mine is determined by the *cut-off grade* (% Cu), which separates *ore* from *waste*. Material with less than the cut-off grade (when combined with all the ore going to concentration or leaching) cannot be profitably treated for copper recovery. This material is waste. It is removed to large waste dumps.

TABLE 2.2 World Production of Copper in 2008, Kilotonnes of Contained Copper (Edelstein & Porter, 2010)

Country	Mine production (kt/a Cu)	Smelter production (kt/a Cu)	Refinery production (kt/a Cu)	Electrowon production (kt/a Cu)
Argentina	157		16	
Armenia	19	6		
Australia	886	449	503	53
Austria		67	107	
Belgium		125	394	
Botswana	29	23		
Brazil	206	220	223	
Bulgaria	105	278	127	
Burma	7			
Canada	607	486	442	
Chile	5330	1369	1087	1974
China	960	3370	3900	20
Colombia	3			
Congo	2335			3
Cyprus	11			11
Egypt			4	
Finland	13	174	122	
Georgia	8			
Germany		588	690	
Hungary			12	
India	28	662	670	
Indonesia	633	261	181	
Iran	249	248	200	7
Italy			24	
Japan		1625	1540	
Kazakhstan	420	430	398	
Korea, North	12	15	15	
Korea, South		544	531	

(*Continued*)

TABLE 2.2 World Production of Copper in 2008, Kilotonnes of Contained Copper (Edelstein & Porter, 2010)—cont'd

Country	Mine production (kt/a Cu)	Smelter production (kt/a Cu)	Refinery production (kt/a Cu)	Electrowon production (kt/a Cu)
Laos	89			64
Mexico	247	205	221	75
Mongolia	130			3
Morocco	6			
Namibia	8	19		
Norway		27	32	
Oman		25	24	
Pakistan		18		
Papua New Guinea	201			
Peru	1268	360	304	160
Philippines	21	247	174	
Poland	429	461	527	
Portugal	89			
Romania	2	19	15	
Russia	750	865	860	
Saudi Arabia	1			
Serbia	29	48	34	
Slovakia		28	20	
South Africa	109	95	93	
Spain	7	260	308	
Sweden	57	225	228	
Turkey	83	35	100	
United States	1310	574	767	507
Uzbekistan	95	92	90	
Zambia	546	232	240	175
Zimbabwe	2	10	3	2
Total	17,500	14,700	16,600	3100

TABLE 2.3 The World's 20 Highest Production Capacity Cu Mines (ICSG, 2010)

	Mine	Country	Capacity, kt/a contained Cu	Products
1	Escondida	Chile	1330	Concentrates & leach feed
2	Codelco Norte	Chile	950	Concentrates & leach feed
3	Grasberg	Indonesia	780	Concentrates
4	Collahuasi	Chile	518	Concentrates & leach feed
5	El Teniente	Chile	457	Concentrates
6	Morenci	United States	440	Concentrates & leach feed
7	Taimyr Peninsula	Russia	430	Concentrates
8	Antamina	Peru	400	Concentrates
9	Los Pelambres	Chile	380	Concentrates
10	Bingham Canyon	United States	280	Concentrates
11	Batu Hijau	Indonesia	280	Concentrates
12	Andina	Chile	280	Concentrates
13	Kansanshi	Zambia	270	Concentrates & leach feed
14	Zhezkazgan	Kazakhstan	230	Concentrates
15	Los Bronces	Chile	228	Concentrates & leach feed
16	Olympic Dam	Australia	225	Concentrates & leach feed
17	Rudna	Poland	220	Concentrates
18	Cananea	Mexico	210	Concentrates & leach feed
19	Sarcheshmeh	Iran	204	Concentrates & leach feed
20	Spence	Chile	200	Leach feed

Cut-off grade depends on the copper selling price and mining and extraction costs. If, for example, the price of copper rises and costs are constant, it may become profitable to treat lower grade material. This means that cut-off grade (and average ore grade) will decrease. Lower copper prices and increased costs have the opposite effect.

2.2. LOCATION OF EXTRACTION PLANTS

The usual first stage of copper extraction from sulfide ores is production of high-grade (30% Cu) concentrate from low-grade (~1% Cu) ore (Chapter 3). This is always done at the mine site to avoid transporting worthless rock.

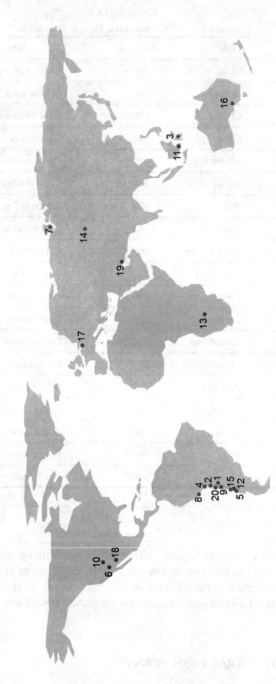

FIGURE 2.2 The world's 20 highest production capacity Cu mines (ICSG, 2010). See Table 2.3 for names and tonnages mined.

TABLE 2.4 Principal Commercial Copper Minerals

Type	Common minerals	Chemical formulas	Theoretical % Cu
Primary sulfide minerals			
Hypogene sulfides	Chalcopyrite	$CuFeS_2$	34.6
	Bornite	Cu_5FeS_4	63.3
Secondary minerals			
Supergene sulfides	Chalcocite	Cu_2S	79.9
	Covellite	CuS	66.5
	Digenite	$Cu_{1.8}S$	78.1
Native copper	Metal	Cu	100.0
Carbonates	Malachite	$CuCO_3 \cdot Cu(OH)_2$	57.5
	Azurite	$(CuCO_3)_2 \cdot Cu(OH)_2$	55.3
Hydroxy-silicates	Chrysocolla	$CuO \cdot SiO_2 \cdot 2H_2O$	36.2
Oxides	Cuprite	Cu_2O	88.8
	Tenorite	CuO	79.9
Sulfates	Antlerite	$CuSO_4 \cdot 2Cu(OH)_2$	53.7
	Brochantite	$CuSO_4 \cdot 3Cu(OH)_2$	56.2

The resulting concentrate is smelted near the mine or in coastal smelters around the world. The trend in recent years has been toward the latter. Coastal smelters can conveniently receive concentrates from around the world, rather than being tied to a single, local, depleting concentrate source (mine). The world's largest smelters are listed in Table 2.5, and their locations shown in Fig. 2.3.

Copper electrorefineries are usually built near the smelter that supplies them with anodes. They may also be built in a convenient central location among several smelters. The world's largest electrorefineries are listed in Table 2.6, and shown in Fig. 2.4.

TABLE 2.5 The World's Largest Cu Smelters in Descending Order of Smelting Capacity (AME, 2010; Antaike, 2009; ICSG, 2010)

	Name	Country	Furnace	Production (kt/a)
1	Guixi	China	F	800
2	Onsan	Korea	F, M	565
3	Birla	India	AS, F, M	500
4	Saganoseki	Japan	F	470

(Continued)

TABLE 2.5 The World's Largest Cu Smelters in Descending Order of Smelting Capacity (AME, 2010; Antaike, 2009; ICSG, 2010)—cont'd

	Name	Country	Furnace	Production (kt/a)
5	Hamburg	Germany	F	450
6	Glogow	Poland	Fcu, B	427
7	Chuquicamata	Chile	F, T	424
8	Norilsk	Russia	F, V	400
9	Caletones	Chile	T	365
10	Jinlong	China	F	350
11	Tuticorin	India	IS	334
12	La Caridad	Mexico	F, T	320
13	Ilo	Peru	IS	317
14	Chingola	Zambia	Fcu	311
15	Huelva	Spain	F	290
16	Tamano	Japan	F	280
17	Yunnan Copper	China	IS	278
18	Kennecott	USA	Fs, Fc	272
19	Altonorte	Chile	N	268
20	Toyo	Japan	F	260
21	Zhezkazgan	Kazakhstan	E	254
22	Balkash	Kazakhstan	V	250
23	Ronnskar	Sweden	E, F	250
24	Caraiba/Camacari	Brazil	F	240
25	Gresik	Indonesia	M	240
26	Olympic Dam	Australia	Fcu	235
27	Mufulira	Zambia	IS	230
28	Naoshima	Japan	M	220
29	Onahama	Japan	R, M	220
30	Mount Isa	Australia	IS	214
31	Pirdop	Bulgaria	F	210
32	Harjavalta	Finland	F	210
33	Jinchuan	China	R, K, AS	200

TABLE 2.5 The World's Largest Cu Smelters in Descending Order of Smelting Capacity (AME, 2010; Antaike, 2009; ICSG, 2010)—cont'd

	Name	Country	Furnace	Production (kt/a)
34	Yanggu Xiangguang	China	Fs, Fc	200
35	Leyte	Philippines	F	200
36	Hayden	USA	IF	200
37	Miami	USA	IS	200
38	Potrerillos	Chile	T	190
39	Daye	China	N	186
40	Horne	Canada	Ns, Nc	164
41	Chuxiong	China	IS	150
42	Jinchang	China	R, AS	150
43	Nchanga	Zambia	IS	150
44	Sarchesmah	Iran	R, F	145
45	Chagres	Chile	F	140
46	Sudbury Copper Cliff	Canada	IF	135
47	Las Ventanas	Chile	T	130
48	Chifeng Jinjian	China	AS	120
49	Huludao	China	AS	120
50	Almalyk	Uzbekistan	IF	120
51	Brixlegg	Austria	B	110
52	Palabora	South Africa	R	110
53	Baiyin	China	F	100
54	Dongfang	China	AS	100
55	Feishang	China	B	100
56	Kangxi	China	B	100
57	Yantai Penghui	China	B	100
58	Yunan Tin	China	AS	100
59	Legnica	Poland	B	100
60	Nadezhda	Russia	F	100

Key: AS: Ausmelt Furnace; R: Reverberatory Furnace; B: Blast Furnace; S: Shaft Furnace; E: Electric Furnace; T: Teniente Furnace; F: Outotec Flash Furnace; V: Vanyukov Furnace; IS: Isasmelt Furnace; s: smelting; IF: Inco Flash Furnace; c: converting; K: Kaldo (TBRC); cu: direct to copper furnace; N: Noranda Furnace.

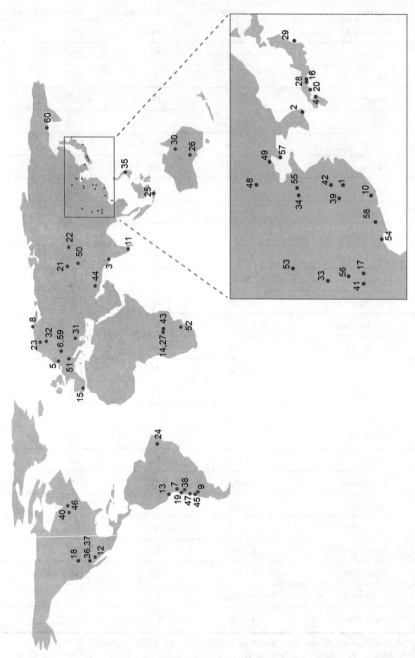

FIGURE 2.3 Location of the world's largest Cu smelters (AME, 2010; Antaike, 2009; ICSG, 2010). See Table 2.5 for details.

TABLE 2.6 The World's Largest Cu Refineries in Descending Order of Copper Production Capacity (AME, 2010; Antaike, 2009; ICSG, 2010; Robinson et al., 2007)

	Name	Country	Production (kt/a)
1	Guixi	China	702
2	Zhongtiaoshan	China	655
3	Onsan	Korea	519
4	Birla	India	500
5	Chucquicamata	Chile	490
6	Yanggu Xiangguang	China	400
7	Glogow	Poland	398
8	Pyshma	Russia	390
9	Las Ventanas	Chile	385
10	Yunnan	China	385
11	Silvassa	India	380
12	Hamburg	Germany	374
13	Jinlong	China	350
14	Olen	Belgium	322
15	Norilsk	Russia	306
16	Garfield	USA	300
17	Nkana	Zambia	300
18	El Paso	USA	295
19	Gresik	Indonesia	289
20	Ilo	Peru	285
21	Jinchuan	China	284
22	Montreal East CCR	Canada	280
23	Townsville	Australia	277
24	Daye	China	265
25	Huelva	Spain	263
26	Tamano	Japan	260
27	Zhezkazgan	Kazakhstan	254
28	Balkash	Kazakhstan	250
29	Saganoseki	Japan	233
30	Naoshima	Japan	230
31	Onahama	Japan	230

(*Continued*)

TABLE 2.6 The World's Largest Cu Refineries in Descending Order of Copper Production Capacity (AME, 2010; Antaike, 2009; ICSG, 2010; Robinson et al., 2007)—cont'd

	Name	Country	Production (kt/a)
32	Ronnskar	Sweden	223
33	Kyshtym Copper	Russia	220
34	Zhangjiagang	China	200
35	Camacari	Brazil	199
36	Pirdop	Bulgaria	197
37	Lunen	Germany	197
38	Amarillo	USA	189
39	Olympic Dam	Australia	182
40	Mufulira	Zambia	180
41	Leyte	Philippines	175
42	Dongying Fangyuan	China	173
43	Hitachi	Japan	170
44	Sarcheshmeh	Iran	168
45	Jinchang	China	150
46	Cobre de Mexico	Mexico	150
47	La Caridad	Mexico	150
48	Toyo/Niihama	Japan	146
49	Tuticorin	India	144
50	Harjavalta	Finland	122
51	Chifeng Jinjian	China	120
52	Fuchunjiang	China	120
53	Huludao	China	120
54	Almalyk	Uzbekistan	120
55	Chifeng Jinfeng	China	110
56	Brixlegg	Austria	107
57	Legnica	Poland	105
58	Feishang	China	100
59	Tianjing	China	100
60	Yantai Penghui	China	100
61	Monchegorsk	Russia	100

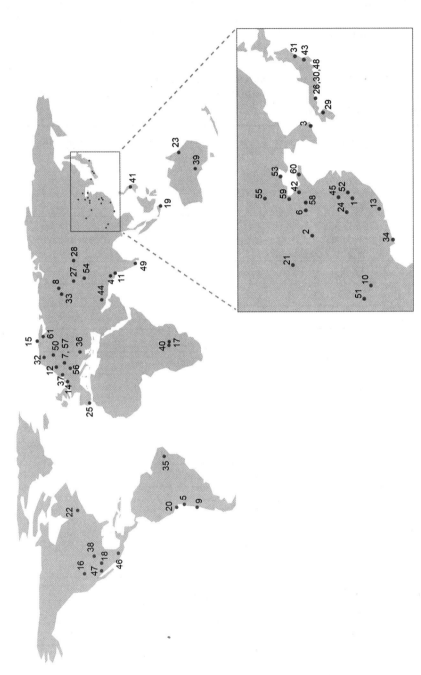

FIGURE 2.4 Location of the world's largest Cu refineries (AME, 2010; Antaike, 2009; ICSG, 2010; Robinson, Davenport, Moats, Karcas, & Demetrio, 2007). See Table 2.6 for details.

Leach/solvent extraction/electrowinning operations are always located next to their mines. This is because ores that are suitable for leaching are low in copper grade, hence uneconomic to transport. The world's largest copper leach/solvent extraction/electrowinning plants are listed in Table 2.7, and their locations shown in Fig. 2.5. This technology dominates in Chile.

TABLE 2.7 The World's Largest Cu Solvent Extraction and Electrowinning (SXEW) Plants in Descending Order of Cathode Production for 2008 (Huang, 2010; ICSG, 2010; Robinson et al., 2008)

	Name	Country	Cathode (kt/a)
1	Jiangxi	China	443
2	Yunnan	China	380
3	Anhui	China	321
4	Escondida	Chile	310
5	Radomiro Tomic	Chile	300
6	Daye	China	275
7	Morenci	USA	267
8	Jinchuan	China	205
9	Sarcheshmeh	Iran	203
10	Spence	Chile	180
11	El Abra	Chile	170
12	Gaby	Chile	150
13	Chuquicamata	Chile	143
14	Tenke Fungurume	D R Congo	135
15	Kansanshi	Zambia	120
16	Zaldivar	Chile	110
17	Cerro Colorado	Chile	100
18	Ningbo JinTian	China	98
19	Zhongtiaoshan	China	93
20	Cerro Verde	Peru	90
21	Quebrada Blanca	Chile	80
22	El Tesoro	Chile	80
23	Safford	USA	70
24	Baiyin	China	67
25	Shanghai Dachang	China	65

TABLE 2.7 The World's Largest Cu Solvent Extraction and Electrowinning (SXEW) Plants in Descending Order of Cathode Production for 2008 (Huang, 2010; ICSG, 2010; Robinson et al., 2008)—cont'd

	Name	Country	Cathode (kt/a)
26	Mantoverde	Chile	62
27	Lomas Bayas	Chile	60
28	Sepon	Laos	60
29	Tyrone	USA	50
30	Colllahuasi	Chile	50
31	Mantos Blancos	Chile	42
32	Michilla	Chile	40
33	Cobre Las Cruces	Spain	40
34	Kinsevere	D R Congo	40
35	Toquepala	Peru	38
36	Los Bronces	Chile	37
37	Ruashi	D R Congo	36
38	Ray	USA	35
39	Chino	USA	35
40	Chemaf	D R Congo	35
41	Nkana	Zambia	32
42	Milpillas	Mexico	30
43	Konkola	Zambia	28
44	Tintaya	Peru	25
45	Mufulira	Zambia	25
46	La Caridad	Mexico	22
47	Carlota	USA	20
48	Silver Bell	USA	20
49	Sierrita	USA	20
50	El Salvador	Chile	20
51	Catemu	Chile	20
52	Mutanda	D R Congo	20
53	Carmen de Andacollo	Chile	18
54	Olympic Dam	Australia	18

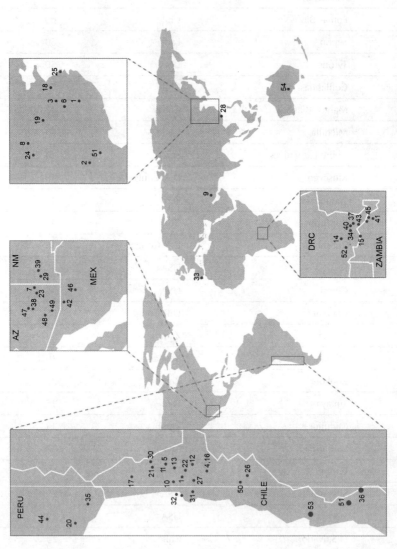

FIGURE 2.5 Location of the world's largest Cu solvent extraction and electrowinning facilities (Huang, 2010; ICSG, 2010; Robinson et al., 2008). See Table 2.7 for details.

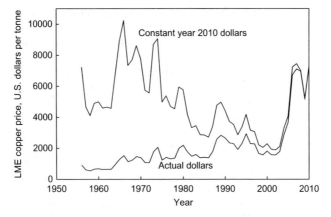

FIGURE 2.6 London Metal Exchange cash price for Grade A copper since 1956. The rapid increases are due to the ongoing rapid industrialization of China. Specifications for Grade A copper are given at www.lme.com and in Chapter 20.

2.3. PRICE OF COPPER

Fig. 2.6 presents selling prices of copper from 1956 to 2010. It shows that the constant dollar price fell until 2002 then rose rapidly. The actual price also rose rapidly after 2002, to record highs.

These high prices are due to the huge demand for copper in China, for its rapid industrialization and infrastructure development.

2.4. SUMMARY

Copper is produced around the world. About 40% is mined in the western mountain region of South America.

Concentrators and leach/solvent extraction/electrowinning plants are located near their mines. Smelters and refineries, on the other hand, are increasingly being located on sea coasts so that they can receive concentrates from mines around the world.

Copper's most exploited property is its high electrical conductivity, in conjunction with its excellent corrosion resistance, formability, and joinability. Its high thermal conductivity and corrosion resistance are also exploited in many heat transfer applications.

Worldwide, about 22 million tonnes of copper come into use per year. About 18 million tonnes of this comes from new mine production and about 4 million tonnes from recycled end-of-use objects and manufacturing wastes.

REFERENCES

AME, (2010). *Copper smelters and refineries.* http://www.ame.com.au.
Antaike, (2009). *Chinese copper industry in 2010–2015. Industry Report.* Beijing, China: Beijing Antaike Information Development Co. Ltd.
Butts, A. (1954). *Copper, the science and technology of the metal, its alloys and compounds.* New York, USA: Reinhold Publishing Corp.
Copper Development Association, (2010a). *Copper in the USA: Bright future, glorious past.* http://www.copper.org.

Copper Development Association, (2010b). *Copper and copper alloy consumption in the United States by functional use — 2009.* http://www.copper.org.

Copper Development Association, (2010c). *U.S. top copper markets — 2009.* http://www.copper.org.

Edelstein, D. L., & Porter, K. D. (2010). *Copper statistics 1900—2008.* Washington, DC, USA: United States Geological Survey. http://minerals.usgs.gov/ds/2005/140/copper.pdf.

Huang, K. (2010). *Main leach-solvent extraction-electrowinning plants in China and their production.* Beijing, China: Institute of Process Engineering, Chinese Academy of Sciences.

ICSG. (2010). *Directory of copper mines and plants 2008 to 2013.* Lisbon, Portugal: International Copper Study Group. http://www.icsg.org.

Robinson, T., Davenport, W., Moats, M., Karcas, G., & Demetrio, S. (2007). Electrolytic copper electrowinning — 2007 world tankhouse operating data. In G. E. Houlachi, J. D. Edwards & T. G. Robinson (Eds.), *Copper 2007, Vol. V: Copper electrorefining and electrowinning* (pp. 375—423). Montreal: CIM.

Robinson, T., Moats, M., Davenport, W., Karcas, G., Demetrio, S., & Domic, E. (2008). Copper solvent extraction — 2007 world operating data. In B. A. Moyer (Ed.), *Solvent extraction: Fundamentals to industrial applications. ISEC 2008, Vol. 1* (pp. 435—440). Montreal: CIM.

Chapter 3

Production of High Copper Concentrates – Introduction and Comminution

The grade of copper ores is typically too low (0.5–2% Cu) for economic direct smelting. Heating and melting the huge quantity of largely worthless rock would require too much energy and too much furnace capacity. For this reason, all ores destined for pyrometallurgical processing are physically concentrated before smelting. The product of this step is *concentrate*, which contains ~30% Cu (virtually all as sulfide minerals).

Ores destined for hydrometallurgical processing are rarely concentrated. Cu is usually extracted from these ores by direct leaching of crushed or milled ore (Chapter 15).

This chapter and Chapter 4 describe the production of high-grade concentrate from low-grade ore. Processing of sulfide minerals is emphasized because these minerals account for virtually all Cu concentration.

3.1. CONCENTRATION FLOWSHEET

Concentration of Cu ores consists of isolating the copper-containing minerals from the rest of the ore. It entails:

(a) Blasting, crushing, and grinding the ore to a size where the Cu mineral grains are liberated from the non-Cu mineral grains, known as *comminution*, and
(b) Physically separating the liberated Cu minerals from non-Cu minerals by *froth flotation* to generate a Cu-rich concentrate and Cu-barren *tailings* (Chapter 4).

Figure 1.2 shows these processes in relation to the overall copper-making flowsheet. Figures 3.1 and 4.1 describe them in detail.

Copper concentrators typically treat 10,000–150,000 tonnes of ore per day, depending on the production rate of their mines (ICSG, 2010).

3.2. THE COMMINUTION PROCESS

To liberate the copper-containing minerals from the gangue materials in the ore (Chapter 2) and enable them to be collected by flotation into a concentrate (Chapter 4), the ore should be finely ground.

Comminution is performed in three stages:

(a) *Breaking* the ore by explosions in the mine (blasting);
(b) *Crushing* large ore pieces by compression in gyratory or roll crushers;

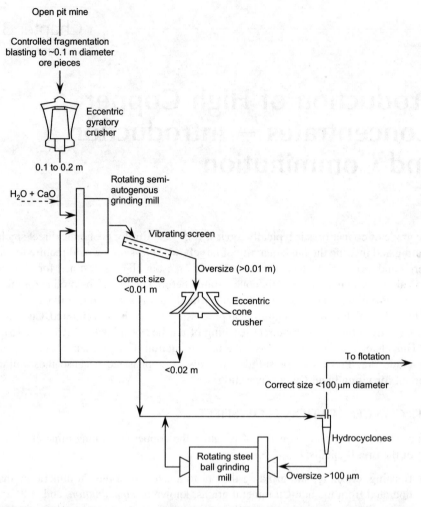

FIGURE 3.1 Flowsheet for preparing small flotation feed particles from as-mined ore fragments. One stage of crushing and two stages of grinding are shown. The crushing is open-circuit, i.e. there is no recycle loop. The two grinding circuits are closed circuit; oversize material is recycled for re-crushing or re-grinding to specified particle sizes.

(c) Wet *grinding* of the crushed ore in rotating tumbling mills, where abrasion, impact, and compression all contribute to breaking the ore.

The final fineness of grind is mainly determined by the number of times an ore particle passes through the grinding mills. Separate crushing and grinding steps are necessary because it is not possible to break massive run-of-mine ore pieces while at the same time controlling the fineness of grind that is necessary for the flotation step.

3.3. BLASTING

Blasting entails drilling holes in a mine bench with automatic equipment, filling the holes with explosive granules, and then electronically igniting the explosive. This explosively

FIGURE 3.2 Blast hole drilling machine and newly drilled holes. The holes will be filled with explosive slurries, then exploded to produce ore fragments. Already exploded ore is shown in the forefront of the photograph. *Photo courtesy of Freeport-McMoRan Copper & Gold Inc.*

fragments the ore near the mine wall (Figs. 1.1 and 3.2). The explosions send shock waves through the ore, cracking the rock, and releasing multiple fragments.

Many Cu open-pit mines use closer drill holes and larger explosive charges to produce smaller ore fragments (Fig. 3.3) and/or uniform size fragments from ores of different toughness (Brandt, Martinez, French, Slattery, & Baker, 2010; Titichoca, Magne, Pereira, Andrades, & Molinet, 2007). By optimizing blasting and fragmentation conditions in the mine, the subsequent crushing requirements can be reduced and throughputs with existing crushing equipment can be increased. This practice also reduces the electrical energy requirement per tonne of ore for communition.

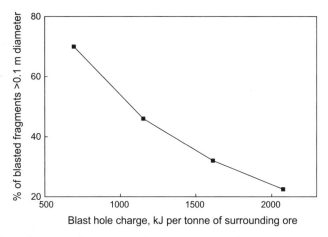

FIGURE 3.3 Size of post-blast fragments as a function of amount of explosive charged to a mine's blast holes. Increasing the explosive charge per tonne of ore decreases the amount of oversize fragments. This graph is for a specific ore. Tougher ore would require more explosive energy to achieve the same result. Weaker ore would require less.

Improved fragmentation has been aided by two recent technologies: automated real-time ore-size determinations and automated real-time ore-toughness determinations.

3.3.1. Ore-size Determination

Automated ore-size determination technology consists of (a) digitally photographing ore particle assemblages at various points in the comminution flowsheet and (b) calculating particle-size distributions from these digital photographs (Maerz, 2001; Split, 2011). Examples of targets for the digital camera are blasted ore as it is being dumped into haul trucks (Fig. 1.1) or ore as it is being dropped into the gyratory crusher (Fig. 3.4). The real-time particle-size measurements are then used to determine whether the amount of explosives needed for subsequent blast holes (the *explosive load*) should be increased or decreased to ensure optimal fragmentation (Fig. 3.3).

3.3.2. Automated Ore-toughness Measurements

Modern blast hole drilling machines are equipped to measure the amount of energy (from torque and drilling time) required to drill a blast hole (Aardvark Drilling, 2010; Kahraman, Bilgin, & Feridunoglu, 2003). The higher the energy requirement, the tougher the ore at the location of the hole. This knowledge enables the blasting engineer to adjust his explosive load in each blast hole to obtain the desired level of fragmentation.

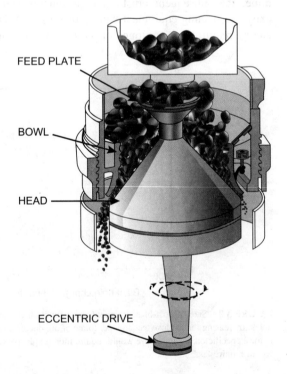

FIGURE 3.4 Gyratory crusher for crushing as-mined ore to ~0.2 m diameter pieces. The crushing is done by compression of ore pieces between the eccentrically rotating spindle and the fixed crusher wall. The crushing surface on the spindle can be up to 3 m high, 1.5 m diameter. Crushing rates are 10,000–150,000 tonnes of ore per day. *Drawing from Boldt and Queneau (1967) courtesy of Vale.*

3.4. CRUSHING

Crushing is mostly done in the mine, mostly using *gyratory crushers* (Fig. 3.4). This permits ore to be transported out of an open-pit mine by conveyor (MacPhail & Richards, 1995). It also permits easy hoisting of ore out of an underground mine. The crushed ore is stored in a coarse-ore stockpile, from which it is sent by conveyor to a grinding mill. The ore from crushing is then sent for grinding.

3.5. GRINDING

3.5.1. Grind Size and Liberation of Copper Minerals

To isolate the copper-containing minerals into a concentrate, the ore should be ground finely enough to liberate the Cu mineral grains from the non-Cu mineral grains. The extent of grinding required to do this is determined by the size of mineral grains in the ore. Laboratory-scale flotation tests on materials of different particle sizes are usually required to ascertain the grind size that is required to *liberate* the copper minerals.

Figure 3.5(a) shows the effect of grind size on recovery of Cu into the concentrate, while Fig. 3.5(b) shows the corresponding Cu concentration in the tailings. There is an optimum grind size for maximum recovery of Cu to the concentrate in the subsequent flotation step (Chapter 4): too large a grind size causes some Cu mineral grains to remain combined with or *occluded* by non-Cu mineral grains, preventing their flotation; too fine

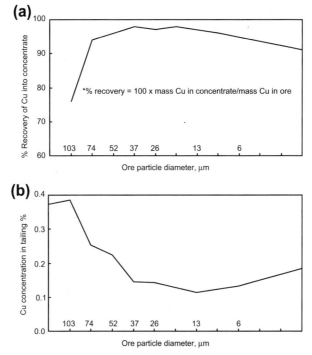

FIGURE 3.5 Effect of grind particle size on (a) Cu recovery to concentrate and (b) % Cu in tailings. The presence of an optimum is shown (Taggart, 1954). Recent (2010) visits to industrial concentrators confirm these patterns.

a grind size causes the rising bubbles in the flotation process to push the very tiny Cu-containing particles aside, preventing their contact with the bubble, and thereby reducing their recovery to the concentrate. In addition, very fine rock and Fe-sulfide mineral particles may agglomerate with very fine Cu-sulfide particles, preventing them from floating.

Liberation of mineral grains from each other generally requires grinding to ~100 μm diameter particle size. Flotation begins to be adversely affected when particles less than ~10 μm are produced.

Grinding requires considerable electrical energy (Table 3.1 (a,b,c)). This is another reason to avoid overgrinding.

3.5.2. Grinding Equipment

Grinding produces ore particles of sufficient fineness for Cu mineral recovery by flotation. The most common grinding mills are *semi-autogenous (SAG)* and *autogenous (AG)* mills (Fig. 3.6) and *ball* mills (Jones & Pena, 1999). A grinding circuit usually consists of one SAG or AG mill and one or two ball mills. Grinding is continuous and fully integrated with the subsequent flotation operation. Grinding is always done wet, with mixtures of ~70 mass% solids in water.

3.5.2.1. Autogenous and Semi-autogenous Mills

The crusher product is ground in a SAG or AG mill. Autogenous mills crush the ore without the need for iron or steel grinding media. They are used when the ore is hard enough for the tumbling ore to grind itself. In SAG milling, ~0.15 m^3 of 13 cm diameter iron or steel balls are added into the mill per 0.85 m^3 of ore (i.e. 15 vol.-% 'steel') to assist grinding. SAG mills are much more common.

The mill product is usually passed over a large vibrating screen to separate oversize *pebbles* from ore particles of the correct size. The correct-size material is sent forward to a ball mill for final grinding. The oversize pebbles are recycled through a small *eccentric* (cone) crusher, then back to the SAG or AG mill (Jones & Pena, 1999; Markkola, Soto, Yañez, & Jimenez, 2007). This procedure maximizes ore throughput and minimizes electrical energy consumption.

3.5.2.2. Ball Mills

The ball mill accepts the SAG or AG mill product. Ball mills give a controlled final grind and produce flotation feed of a uniform size. Ball mills tumble iron or steel balls with the ore. The balls are initially 5–10 cm diameter but gradually wear away as grinding of the ore proceeds. The feed to ball mills (dry basis) is typically 75 vol.-% ore and 25% steel.

The ball mill is operated in closed circuit with a particle-size measurement device and size-control cyclones. The cyclones send correct-size material on to flotation and direct oversize material back to the ball mill for further grinding.

3.5.3. Particle-Size Control of Flotation Feed

A critical step in grinding is ensuring that its product particles are fine enough for efficient flotation. Coarser particles must be isolated and returned for further grinding.

TABLE 3.1a Industrial Crushing and Grinding Data for Three Copper Concentrators, 2010. Flotation Details are Given in Table 4.1

Concentrator	Candaleria, Chile	El Soldado, Chile	Los Bronces Chile
Ore treated per year, tonnes	25,000,000 (2001)	7,700,000 (2010)	21,000,000 (2010)
Ore grade, %Cu	0.9–1.0	0.63	1.055
Crushing		Chalcopyrite	Chalcopyrite
Primary gyratory crusher	One	One	One
Diameter × height, m	1.52 × 2.26	1.1 × 1.65	1.4 × 1.9
Power rating, kW	522	300	430
Rotation speed, RPM			6.5
Product size, m	0.1–0.13	0.2	0.2
Energy consumption, kWh per tonne of ore	0.3 (estimate)	0.2	0.25 (0.18 m diameter product
Secondary crushers	No	Yes, 4	No
First stage grinding			
Mill type	Semi-autogenous	Semi-autogenous	Semi-autogenous
Number of mills	2	1	2
Diameter × length, m	11 × 4.6	10.4 × 5.2	8.5 × 4.3
Power rating each mill, kWh	12,000	11,300	5590
Rotation speed, RPM	9.4–9.8	10.2	11.3
Vol.-% 'steel' in mill	12–15	15	13–15%
Ball size, initial	12.5 cm	12.7	12.7
Ball consumption, kg per tonne of ore	0.3	0.32	0.35
Feed	70% ore, 30% H_2O	80% ore 20% water	70% ore 30% water
Product size	80% < 140 µm	200 µm	3 cm
Oversize treatment	22% ore recycle through two 525 kW crushers		Oversize to pebble crushers

(Continued)

TABLE 3.1a Industrial Crushing and Grinding Data for Three Copper Concentrators, 2010. Flotation Details are Given in Table 4.1—cont'd

Concentrator	Candaleria, Chile	El Soldado, Chile	Los Bronces Chile
Energy consumption, kWh per tonne of ore	7.82		6.5
Second stage grinding			
Mill type	Ball mills	Rod + ball mills	Ball mills
Number of mills	4	4 rod, 7 ball	3
Diameter × length, m	6 × 9	3.6 × 2.5	7 × 11
Power rating each mill, kW	5600	Rod 350 Ball 550	10,800
Rotation speed, RPM		28	12
Vol.-% 'steel' in mill		47	35
Feed		70% ore, 30% w	65% ore, 35% water
Product size			80% < 212 µm
Energy consumption, kWh per tonne of ore	7 (estimate)	~8	7.2
Hydrocyclones	14 Krebs (0.5 m diameter)	14 (~0.6 m diameter)	3 (Krebs + Cavex) (0.66 m diameter)
Particle size monitor		Yes PS1200	Yes PS1200

TABLE 3.1b Industrial Crushing and Grinding Data for Three Copper Concentrators, 2010. Flotation Details are Given in Table 4.1

Concentrator	Africa, open pit	Africa, u-ground	Mantos Blancos, Chile
Ore treated per year, tonnes	4,000,000	900,000	4,500,000
Ore grade, %Cu	0.15	0.45	1.13
Crushing	Chalcocite, chalcopyrite	Chalcopyrite	Chalcocite

Chapter | 3 Production of High Copper Concentrates

TABLE 3.1b Industrial Crushing and Grinding Data for Three Copper Concentrators, 2010. Flotation Details are Given in Table 4.1—cont'd

Concentrator	Africa, open pit	Africa, u-ground	Mantos Blancos, Chile
Primary gyratory crusher	1 jaw crusher	1 jaw crusher	One
Diameter × height, m	4.2 × 6.5	8.5 × 11	1.1 × 1.65
Power rating, kW	250	132	300
Rotation speed, RPM			
Product size, m	80% < 0.25	80% < 0.25	0.15
Energy consumption, kWh per tonne of ore	5.4	4.5	0.4
Secondary crushers	None	Yes, jaw crusher	Yes, 4
First stage grinding			
Mill type	semi-autogenous	semi-autogenous	ball mills
Number of mills	1	1	2
Diameter × length, m	8.53 × 4.35	6.1 × 2.75	3.8 × 4.6
Power rating each mill, kWh	6000	1300	1120
Rotation speed, RPM	14	12	21.6
Vol.-% 'steel' in mill	22%	25%	38–40
Ball size, initial, cm	12.5	12.5	
Ball consumption, kg per tonne of ore	0.41	0.9	
Feed	85% ore, 15% water	85% ore, 15% water	84% ore, 16% water
Product size	80% < 800 μm	80% < 600 μm	75% < 212 μm
Oversize treatment	Pebble crushers, SAG	Recycle to SAG	
Energy consumption, kWh per tonne of ore	8.5	6.5	13.5
Second stage grinding			
Mill type	Ball mill	Ball mill	Ball mill
Number of mills	1	1	1
Diameter × length, m	6.1 × 9.05	4.27 × 7.01	3.5 × 5.2

(Continued)

TABLE 3.1b Industrial Crushing and Grinding Data for Three Copper Concentrators, 2010. Flotation Details are Given in Table 4.1—cont'd

Concentrator	Africa, open pit	Africa, u-ground	Mantos Blancos, Chile
Power rating each mill, kW	6000	1860	940
Rotation speed, RPM	18	16	23
Vol.-% 'steel' in mill	35%	32	38–40
Feed	85% ore, 15% water	85% ore, 15% water	85% ore, 15% water
Product size	80% < 125 μm	80% < 100 μm	75% < 212 μm
Energy consumption, kWh per tonne of ore	16.8	15	20
Hydrocyclones	8 (1.1 m dia)	8 (0.25 m dia)	5 (0.5 m diameter)
Particle size monitor	Outotec PSI	Outotec PSI	

TABLE 3.1c Industrial Crushing and Grinding Data for Three Copper Concentrators, 2010. Flotation Details are Given in Table 4.1

Concentrator	Cerro Verde Peru	Ray, U.S.A.	Sierrita, U.S.A.
Ore treated per year, tonnes	3,900,000	10,000,000 0.453 in	34,000,000 0.23% Cu in cpy
Ore grade, %Cu	0.6 (chalcopyrite)	Chalcopyrite & chalcocite	0.03 Mo in MoS_2
Crushing			
Primary gyratory crusher	One	One	Two
Diameter × height, m	1.5 × 2.87	1.5 × 2.3	1.5 × 2.3
Power rating, kW	750		600
Rotation speed, RPM			
Product size, m	0.125	0.152	80% < 0.1
Energy consumption, kWh per tonne of ore	5		0.08
Secondary crushers	Yes	No	4

TABLE 3.1c Industrial Crushing and Grinding Data for Three Copper Concentrators, 2010. Flotation Details are Given in Table 4.1—cont'd

Concentrator	Cerro Verde Peru	Ray, U.S.A.	Sierrita, U.S.A.
First stage grinding			
Mill type	Ball mills	Semi-autogenous	Ball mills
Number of mills	4	1	16
Diameter × length, m	7.3 × 11	10.4 × 5.2	5 × 5.8
Power rating each mill, kWh	13,000	12,500	3000
Rotation speed, RPM	12.2	8.3–10.8	13.8
Vol.-% 'steel' in mill	38	13	39
Ball size, initial, cm	6.35	14	10, 7.5
Ball consumption, kg per tonne of ore	0.67	0.15	0.39
Feed	75% ore/25% water	83% ore/17% H_2O	98% ore, 2% H_2O
Product size	80% < 150 μm	6.35 cm	80% < 350
Oversize treatment		Cone crushers	
Energy consumption, kWh per tonne of ore	10	12–18	8.4
Second stage grinding			None
Mill type		Ball mills	
Number of mills		2	
Diameter × length, m		5.5 × 9.34	
Power rating each mill, kW		4500	
Rotation speed, RPM		13	
Vol.-% 'steel' in mill		35–40	
Feed		85% ore/15% H_2O	
Product size		80% < 147 μm	
Energy consumption, kWh per tonne of ore		8	

(Continued)

TABLE 3.1c Industrial Crushing and Grinding Data for Three Copper Concentrators, 2010. Flotation Details are Given in Table 4.1—cont'd

Concentrator	Cerro Verde Peru	Ray, U.S.A.	Sierrita, U.S.A.
Hydrocyclones	9 per ball mill (0.84 m diameter)	16 (1.5 m diameter)	2 per ball mill (0.84 m dia.)
Particle size monitor	Yes	No	No

Size control is universally done by *hydrocyclones* (Fig. 3.7; Olson & Turner, 2003). A hydrocyclone uses the principle that, under the influence of a force field, large ore particles in a water-ore mixture (slurry) tend to move faster than small ore particles.

This principle is put into practice by pumping the grinding mill discharge into a hydrocyclone at high speed, 5−10 m/s. The slurry enters tangentially (Fig. 3.7), giving it a rotational motion inside the cyclone. This creates a centrifugal force, which

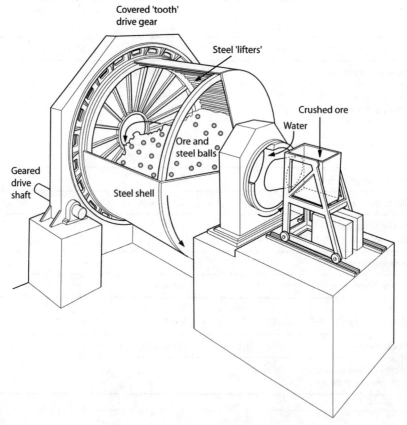

FIGURE 3.6 Semi-autogenous grinding mill.

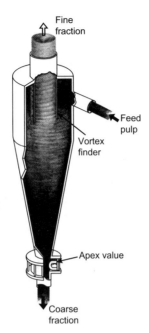

FIGURE 3.7 Cutaway view of hydrocyclone showing (a) tangential input of water–ore particle feed and (b) separation of the feed into fine particle and coarse particle fractions. *Drawing from Boldt and Queneau (1967) courtesy Vale.*

accelerates ore particles toward the cyclone wall. The water content of the slurry, typically ~60% by mass, is adjusted so that (a) the oversize particles are able to reach the wall, where they are dragged out by water flow along the wall and through the apex of the cyclone, and (b) the correct (small) size particles do not have time to reach the wall before they are flushed out with the main slurry flow through the vortex finder.

The principal control parameter for the hydrocyclone is the water content of the incoming slurry. An increase in the water content of the slurry gives less hindered movement of particles, and allows a greater fraction of the input particles to reach the wall and pass through the apex. This increases the fraction of particles being recycled for re-grinding, and ultimately leads to a more finely ground final product. A decrease in water content has the opposite effect.

3.5.4. Instrumentation and Control

Grinding circuits are extensively instrumented and closely controlled (Fig. 3.8, Table 3.2). The objectives of the control are to:

(a) Produce particles of appropriate size for efficient flotation recovery of Cu minerals;
(b) Produce these particles at a rapid rate;
(c) Produce these particles with a minimum consumption of energy.

The most common control strategy is to ensure that the sizes of particles in the final grinding product are within predetermined limits, as sensed by an on-stream particle-size analyzer (Outotec, 2010), and then optimize production rate and energy consumption while maintaining this correct size. Figure 3.8 and the following describe one such control system.

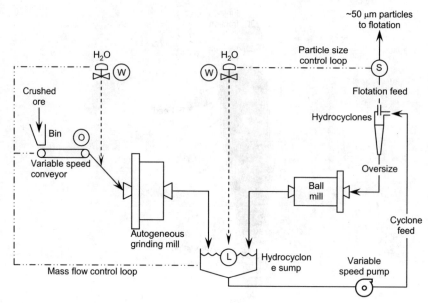

FIGURE 3.8 Control system for grinding mill circuit (———— ore flow, ------ water flow, electronic control signals). The objective is to quickly and cheaply produce correct-size ore particles for efficient Cu mineral flotation. The circled symbols refer to the sensing devices in Table 3.2.

3.5.4.1. Particle-size Control

The particle-size control loop in Fig. 3.8 controls the particle size of the grinding product by automatically adjusting the rate of water addition to the hydrocyclone feed sump. If, for example, the flotation feed contains too many large particles, an electronic signal from the particle-size analyzer (S) automatically activates water valves to increase the water content of the hydrocyclone feed. This increases the fraction of the ore being recycled to the ball mills, and gives a finer grind.

Conversely, too fine a flotation feed automatically cuts back on the rate of water addition to the hydrocyclone feed sump. This decreases ore recycle to the grinding mills, thereby increasing the particle size of the flotation feed. It also permits a more rapid initial feed to the ball mills, and minimizes grinding energy consumption.

3.5.4.2. Ore-throughput Control

The second control loop in Fig. 3.8 gives maximum ore throughput rate without overloading the ball mill. Overloading might become a problem if, for example, the ball mill receives tough and large particles, which require extensive grinding to achieve the small particle size needed by flotation.

The simplest mass-flow control scheme is to use the hydrocyclone sump slurry level to adjust the ore feed rate to the grinding plant. If, for example, slurry level sensor (L) detects that the slurry level is rising (due to tougher ore and more hydrocyclone recycle), it automatically slows the input ore feed conveyor. This decreases flow rates throughout the plant, and stabilizes ball mill loading and sump level.

TABLE 3.2 Sensing and Control Devices for Fig. 3.8 Grinding Control Circuit

Sensing instruments	Symbol in Fig. 3.8	Purpose	Type of device	Use in automatic control system
Ore input rate weightometer	O	Senses feed rate of ore into grinding circuit	Load cells, conveyor speed	Controls ore feed rate
Water flow gages	W	Sense water addition rates	Rotameters	Control water-to-ore ratio in grinding mill feed
On-stream particle-size analyzer	S	Senses a critical particle-size parameter (e.g. percent minus 70 mm) on the basis of calibration curves for the specific ore	Measures diffraction of laser beam by particles in an automatically taken slurry sample (Outotec, 2010)	Controls water addition rate to hydrocyclone feed (which controls recycle & the size of the final grinding circuit product)
Hydrocyclone feed sump level indicator	L	Senses changes of slurry level in sump	Bubble pressure tubes; electric contact probes; ultrasonic echoes; nuclear beam	Controls rate of ore input into grinding circuit (prevents overloading of ball mills)

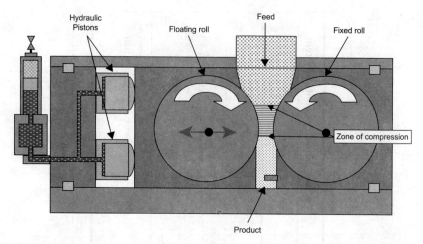

FIGURE 3.9 Sketch of high-pressure roll crusher (Rana et al., 2010). Ore is force fed between the two rotating, hard-surfaced rolls. Constant crushing pressure is provided by hydraulic pistons pressing on the movable left roll housing. Representative details are: 2.4 m roll diameter × 1.65 m length; 21 rotations per minute; 2500–3000 tonnes/hour ore feed rate; <0.05 m feed diameter; <0.0055 m product diameter; 2 × 25 kW power requirement; 130–160 bar pressure on moveable roll housing.

Detection of a falling sump level, on the other hand, automatically increases ore feed rate to the grinding plant, to a prescribed rate or to the maximum capacity of another part of the concentrator, such as the flotation circuit.

There is, of course, a time delay (5–10 min) before the change in ore feed rate is felt in the hydrocyclone feed sump. The size of the sump must be large enough to accommodate further build-up (or draw-down) of slurry during this delay.

3.6. RECENT DEVELOPMENTS IN COMMINUTION

Major recent developments in comminution include:

(a) Improved fragmentation during blasting (Section 3.3);
(b) Incorporation of high-pressure roll crushing in the comminution flowsheet;
(c) Use of scanning electron microscopy to optimize the grind size.

3.6.1. High Pressure Roll Crushing

Figure 3.9 sketches a high-pressure roll crusher. This equipment is used extensively in the cement, diamond, and iron ore industries. First deployed in the copper industry in 2006 at the Cerro Verde concentrator in Peru, it has since been adopted by several copper producers as a replacement for the cone crushers used to grind oversize product from the SAG mill (Fig. 3.1; Burchardt, Patzelt, Knecht, & Klymowsky, 2010). It is increasingly seen as a potential replacement for the SAG mill itself (Rana, Chandrasekaran, & Wood, 2010).

Its perceived advantages are 15–25% lower electrical energy consumption per tonne of ore, and its ability to fracture ore at the mineral grain boundaries (although this is probably more advantageous for leaching than for froth flotation). Its major disadvantage appears to be its difficulty in handling wet, sticky, and high clay ores, and a 10–25% higher initial capital cost (Rana et al., 2010).

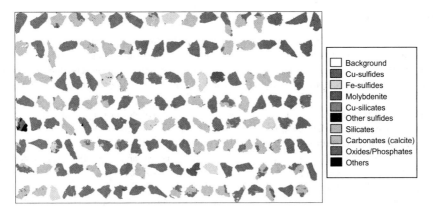

FIGURE 3.10 Scanning electron microscope image of Cu sulfide concentrate particle assemblage. The key is shown at right. The bottom right particle is ~70 μm across. As expected, most of the particles are Cu-sulfide minerals. However, the concentrate contains many silicate particles, which indicates that concentrate cleaning (Section 4.3) could be improved. Fe sulfides have also floated so that depression could also be improved. Further grinding might liberate a few more Cu-sulfide particles, but for the most part grind size is about optimum. *Courtesy of Anglo American: Research.*

3.6.2. Automated Mineralogical Analysis

An important modern tool for optimizing concentrate production is quantitative scanning electron microscopy (SEM) which, for any assemblage of ore particles (a) indicates the mineral grain makeup of every particle in the assemblage, showing *liberated*, *partially liberated*, and *unliberated* copper mineral grains, and (b) suggests whether more or less grinding will give improved Cu recovery to the concentrate (Hamilton, Martin, & Zhou, 2007; Zahn, Baum, Magnuson, Jensen, & Thompson, 2007). Figure 3.10 shows an example concentrate assemblage. This can be done routinely or specifically when Cu recovery to concentrate unexpectedly decreases.

Automated mineralogical analysis can also be useful in the initial design of a new concentrator by indicating the minerals and mineral grain associations of ore samples from around the entire mineral deposit.

3.7. SUMMARY

Copper ores typically contain 0.5−2% Cu. They must be concentrated to ~30% Cu before smelting. The universal concentration technique is froth flotation of finely ground ore particles (Chapter 4).

The feed to froth flotation is produced by comminution, which includes:

(a) Blasting ore fragments from the mine walls (~0.1 m diameter);
(b) Crushing these fragments in eccentric and roll crushers (~0.01 m diameter);
(c) Grinding the crushed ore in rotating tumbling mills (<100 μm diameter).

The resulting finely ground particles are then sent continuously to froth flotation.

A recent development is the formation of finer more uniform fragments during blasting by means of larger explosive loadings in blast holes. This increases crushing rate and lowers crushing electricity requirements. It has been aided by (a) automatic ore-toughness measurements during blast hole drilling and (b) continuous measurement of fragment sizes by electronic imaging of blasted ore during delivery to crushing.

REFERENCES

Aardvark Drilling. (2010). *Aardvark Drilling, Inc.* www.aardvarkdrillinginc.com/pdfs/aardvark.pdf.

Boldt, J. R., & Queneau, P. (1967). *The winning of nickel.* Toronto: Longmans Canada, Ltd.

Brandt, D., Martinez, I., French, T., Slattery, L., & Baker, G. (2010). Better fragmentation through team work at Dos Pobres Mine Safford, Arizona. In *Conference on explosives and blasting technique (36th), Vol. 1* (pp. 13–22). Solon, OH: ISEE.

Burchardt, E., Patzelt, N., Knecht, J., & Klymowsky, R. (2010). HPGRs in copper ore comminution – a technology broke barriers. In *Copper 2010, Vol. 7: Plenary, mineral processing, recycling, posters* (pp. 2621–2635). Clausthal-Zellerfeld, Germany: GDMB.

Hamilton, C., Martin, C., & Zhou, J. (2007). Ore characterization and the value of QEMSCAN. In R. del Villar, J. E. Nesset, C. O. Gomez & A. W. Stradling (Eds.), *Copper 2007, Vol. II; Mineral processing* (pp. 137–147). Montreal: CIM.

ICSG. (2010). *Directory of copper mines and plants 2008 to 2013.* Lisbon: International Copper Study Group.

Jones, S. M., & Pena, R. F. (1999). Milling for the millennium. In B. A. Hancock & M. R. L. Pon (Eds.), *Copper 99–Cobre 99, Vol. II: Mineral processing/environment, health and safety* (pp. 191–204). Warrendale, PA: TMS.

Kahraman, S., Bilgin, N., & Feridunoglu, C. (2003). Dominant rock properties affecting the penetration rate of percussive drills. *International Journal of Rock Mechanics and Mining Sciences, 40,* 711–723.

MacPhail, A. D., & Richards, D. M. (1995). In-pit crushing and conveying at Highland Valley Copper. *Transactions of Society for Mining, Metallurgy, and Exploration, 296,* 1186–1190.

Maerz, N. H. (2001). Automated online optical sizing analysis. In A. R. Mular & D. J. Barratt (Eds.), *International autogenous and semiautogenous grinding technology 2001, Vol. 2* (pp. 250–269). Vancouver: University of British Columbia. web.mst.edu/~norbert/pdf/r__E0110047.pdf.

Markkola, K. G., Soto, J., Yañez, G., & Jimenez, H. (2007). SABC circuit energy consumption optimization. In R. del Villar, J. E. Nesset, C. O. Gomez & A. W. Stradling (Eds.), *Copper 2007, Vol. II: Mineral processing* (pp. 301–312). Montreal: CIM.

Olson, T. J., & Turner, P. A. (2003). *Hydrocyclone selection for plant design.* www.rockservices.net/75_hydrocyclone_selection_for_plant_design-mppd_oct_2002.pdf.

Outotec. (2010). *Grinding control solutions.* www.outotec.com/36358.epibrw.

Rana, I., Chandrasekaran, K., & Wood, K. (2010). HGPR versus SAG milling technology in hard-rock mining – review and analysis. In *Copper 2010, Vol. 7: Plenary, mineral processing, recycling, posters* (pp. 2621–2636). Clausthal-Zellerfeld, Germany: GDMB.

Split. (2011). *Split engineering digital imaging systems.* www.spliteng.com.

Taggart, A. F. (1954). *Handbook of mineral dressing.* New York: John Wiley and Sons.

Titichoca, G., Magne, L., Pereira, G., Andrades, G., & Molinet, P. (2007). Effect of blasting modifications in the semi-autogenous grinding plant of Codelco Chile Andina. In R. del Villar, J. E. Nesset, C. O. Gomez & A. W. Stradling (Eds.), *Copper 2007, Vol. II: Mineral processing* (pp. 265–276). Montreal: CIM.

Zahn, R. P., Baum, W., Magnuson, R., Jensen, D., & Thompson, P. (2007). QEMSCAN characterization of selected ore types for plant optimization. In R. del Villar, J. E. Nesset, C. O. Gomez & A. W. Stradling (Eds.), *Copper 2007, Vol. II: Mineral processing* (pp. 147–156). Montreal: CIM.

SUGGESTED READING

Burger, B., McCaffery, K., McGaffin, I., Jankovic, A., Valery, W., & LaRosa, D. (2006). Batu Hijau model for throughput forecast, mining and milling optimization and expansion studies. In S. K. Kawatra (Ed.), *Advances in comminution* (pp. 461–479). Littleton, CO: SME.

Hustrulid, W. (1999). Blasting principles for open pit mining. In *General design concepts and theoretical foundations. Vols. 1–2,* Boca Raton, Florida: CRC Press.

Jones, S. M., & Pena, R. F. (1999). Milling for the millennium. In B. A. Hancock, & M. R. L. Pon (Eds.), *Copper 99–Cobre 99, Vol. II: Mineral processing/environment, health and safety* (pp. 191–204). Warrendale, PA: TMS.

Kendrick, M., Baum, W., Thompson, P., Wilkie, G., & Gottlieb, P. (2003). The use of the QemSCAN automated mineral analyzer at the Candelaria concentrator. In C. O. Gomez, & C. A. Barahona (Eds.), *Copper/Cobre 2003, Vol. III: Mineral processing* (pp. 415–430). Montreal: CIM.

McCaffery, K., Mahon, J., Arif, J., & Burger, B. (2006). Batu Hijau — controlled mine blasting and blending to optimise process production at Batu Hijau. In M. J. Allan (Ed.), *International autogenous and semi-autogenous grinding technology 2006, Vol. 1*. Vancouver: University of British Columbia.

Wills, B. A., & Napier-Munn, T. J. (2006). *Mineral processing technology* (7th ed.). Oxford: Elsevier.

Chapter 4

Production of Cu Concentrate from Finely Ground Cu Ore

Chapter 3 describes blasting, crushing, and grinding of Cu-sulfide ore (0.5–1% Cu). This chapter describes production of Cu-rich sulfide concentrate (~30% Cu) from this finely ground ore.

4.1. FROTH FLOTATION

The indispensable tool for Cu concentrate production is *froth flotation* (Fuerstenau, Jameson, & Yoon, 2007). This technique is used to upgrade an *ore* to a *concentrate* by selectively *floating* copper-containing minerals away from non-copper minerals. The principles of froth flotation are as follows:

(a) Sulfide minerals are normally wetted by water (*hydrophilic*) but they can be conditioned with reagents (known as *collectors*) that cause them to become water-repellent (*hydrophobic*).
(b) Cu-sulfide minerals can selectively be made hydrophobic by their interaction with the collector, leaving other minerals wetted.
(c) Collisions between ~1 mm diameter rising air bubbles and the now water-repellent Cu minerals result in attachment of the Cu mineral particles to the bubbles (Fig. 4.1). The Cu minerals are *floated* to the surface of the slurry by the air bubbles.
(d) The still-wetted non-copper mineral particles do not attach to the rising bubbles and remain in the slurry.

Industrially, the process entails (a) conditioning a water–ore mixture (*slurry*) to make the Cu minerals water-repellent while leaving the non-Cu minerals hydrophilic and (b) passing a dispersed stream of small bubbles up through the slurry. These procedures cause the Cu mineral particles to attach to the rising bubbles, which carry them to the top of the flotation cell (Fig. 4.2). The other minerals are left behind. They depart the cell through an underflow system. These are mostly non-sulfide *gangue* with a small amount of pyrite.

The last step in the flotation process is creating a strong but short-lived froth by adding *frother* when the bubbles reach the surface of the slurry. This froth prevents bursting of the bubbles and release of the Cu mineral particles back into the slurry. The froth overflows the flotation cell, often with the assistance of paddles (Fig. 4.2), and into a trough. There it collapses and flows into a collection tank.

A sequence of flotation cells is designed to optimize Cu recovery and the Cu grade in the concentrate (Fig. 4.3). The froth from the last set of flotation cells is, after water removal, the *Cu concentrate*.

FIGURE 4.1 Photograph of water-repellent mineral particles attached to rising bubbles. The input bubbles in industrial flotation are ~1 mm diameter, which is about the size of the largest bubble in this photograph. Photograph by H. Rush Spedden.

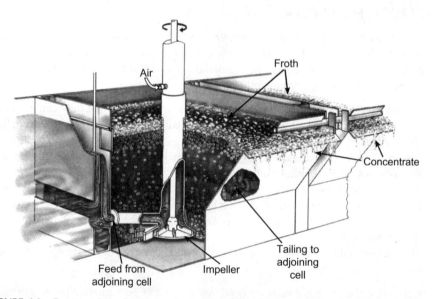

FIGURE 4.2 Cutaway view of cubic mechanical flotation cell. The methods of producing bubbles and gathering froth are shown (Boldt & Queneau, 1967 courtesy Vale). Mechanical flotation cells in recent-design Cu concentrator are up to 300 m^3 cylindrical tanks. They operate continuously.

4.2. FLOTATION CHEMICALS (NAGARAJ & RAVISHANKAR, 2007; WOODCOCK, SPARROW, BRUCKARD, JOHNSON, & DUNNE, 2007)

4.2.1. Collectors

The reagents (collectors), which create the water-repellent surfaces on sulfide minerals are heteropolar molecules. They have a polar (charged) end and a non-polar (hydrocarbon) end. They attach their polar (charged) end to the mineral surface (which is itself polar), leaving the non-polar hydrocarbon end extended outwards (Fig. 4.4). It is the

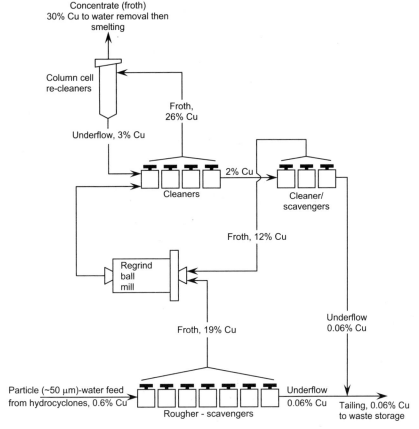

FIGURE 4.3 Flowsheet for floating Cu sulfide concentrate from gangue minerals and Fe sulfides. Residence times in each sector (e.g., rougher–scavenger cells) are 10–20 minutes. Representative solid mass flows in tonnes/day are: Feed from hydrocyclones 40,000; Concentrate (re-cleaner froth) 720; Tailings 39,280; Rougher–scavenger froth 1140; Cleaner–scavenger feed 500; Re-cleaner feed 850.

orientation that imparts the water-repellent character to the conditioned mineral surfaces.

A commonly used collector is xanthate, e.g.:

$$\underset{K^+}{\overset{S}{\underset{S^-}{>}}}C-O-CH_2-CH_2-CH_2-CH_2-CH_3$$

Other sulfur-containing molecules, particularly dithiophosphates, mercaptans, and thionocarbamates, are also used. Commercial collectors are often blends of several of these reagents (Klimpel, 2000). About 10–50 g of collector are required per tonne of ore feed.

4.2.2. Selectivity in Flotation

The simplest froth flotation separation are that of sulfide minerals from waste oxide minerals, such as andesite, granodiorite, and granite. It uses collectors which, when

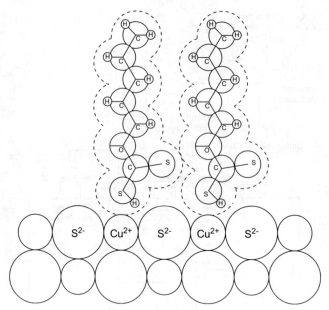

FIGURE 4.4 Sketch of attachment of amyl xanthate ions to covellite. There is a hydrogen atom hidden behind each carbon of the hydrocarbon chain (after Hagihara, 1952).

dissolved in a water–ore slurry, preferentially attach themselves to sulfides. These collectors mostly have a sulfur group at the polar end, which attaches to sulfide minerals but ignores oxides.

4.2.3. Differential Flotation – Modifiers

Separating sulfide minerals, such as chalcopyrite from pyrite, is somewhat more complex. It relies on modifying the surfaces of non-Cu sulfides so that the collector does not attach to them while still attaching to Cu sulfides.

The most common *modifier* is the OH^- (hydroxyl) ion. Its concentration is varied by adjusting the basicity of the slurry with burnt lime (CaO), occasionally with sodium carbonate (Na_2CO_3). The effect is demonstrated in Fig. 4.5, which shows how chalcopyrite, galena, and pyrite can be floated from each other. Each line on the graph marks the boundary between float and non-float conditions for the specific mineral — the mineral floats to the left of its curve, but not to the right.

This graph shows that:

(a) Up to pH 5 (acid slurry), $CuFeS_2$, PbS, and FeS_2 all float;
(b) Between pH 5 and pH 7.5 (neutral slurry), $CuFeS_2$ and PbS float while FeS_2 is depressed;
(c) Between pH 7.5 and pH 10.5 (basic slurry), only $CuFeS_2$ floats.

A bulk Pb–Cu sulfide concentrate could therefore be produced by flotation at pH 6.5. The Pb and Cu sulfides could then be separated at pH 9, which occurs after addition of more CaO.

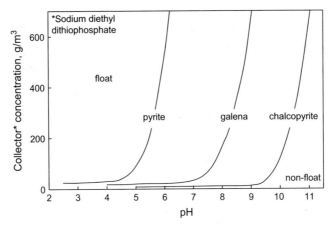

FIGURE 4.5 Effects of collector concentration and pH on the floatability of pyrite, galena and chalcopyrite. Each line marks the boundary between float and non-float conditions for the specific mineral (Wark & Cox, 1934). Precise float/non-float boundary positions depend on collector, mineral, and water compositions.

The modifying effect of OH^- is due to its competition with collector anions, such as xanthates, for a place on the mineral surface. OH^- ions are, for example, selectively adsorbed on pyrite. This prevents appreciable xanthate adsorption on the pyrite, selectively *depressing* it. However, too many OH^- ions will also depress chalcopyrite, so excess CaO must be avoided.

4.2.4. Frothers

Collectors and modifiers give selective flotation of Cu minerals from non-Cu minerals. *Frothers* create the strong but short-lived froth, which holds the floated Cu minerals at the top of the cell. Frothers produce a froth which is strong enough in the flotation cell to support the floated Cu minerals, but breaks down quickly once the forth and its attached minerals overflow the cell.

Branched-chain alcohols are the most common frothers (Woodcock et al., 2007). These include natural (pine oil or terpinol) or synthetic (methyl isobutyl carbinol, polyglycols, and proprietary alcohol blends) products (Chevron Phillips, 2008).

Frothers stabilize the froth by absorbing their OH^- polar end in water, while their branched (non-polar) chains form a cross-linked network in air. The froth should not be long-lived, so the branched-chain hydrocarbon tails should not be too long. The addition of frothers also decreases bubble size, which provides more relative surface to attach mineral particles to.

4.3. SPECIFIC FLOTATION PROCEDURES FOR CU ORES

Selective flotation of Cu-sulfide minerals (chalcopyrite, chalcocite, bornite) from oxide rock and Fe-sulfide minerals (mainly pyrite) is usually done with xanthate, dithiophosphate, mercaptan, or thionocarbamate collectors; burnt lime (CaO) for pH (OH^- ion) control; and branched-chain alcohol frothers. A representative flowsheet, industrial data, and example reagents are shown in Fig. 4.3 and Tables 4.1–4.3.

TABLE 4.1 Cu contents of Ores Around the World. Industrial Recoveries to Concentrate are also Shown

Country	Mine	% Cu in ore	Recovery of Cu to concentrate, %
Chile	Candelaria	0.54	91
Indonesia	Grasberg open pit	0.9	84
Peru	Cerro Verde	0.4	86
Poland	Lower Silesia	1.64	~90
USA	Bagdad	0.36	85
USA	Bingham Canyon	0.6	84
USA	Sierrita	0.25	82

The flowsheet in Fig. 4.3 shows four sets of flotation cells:

(a) The *rougher—scavenger cells* float the incoming ground-ore slurry under conditions, which give efficient recovery of Cu to the froth with a reasonable froth grade (15—20% Cu);
(b) The *cleaner cells* in which flotation of non-Cu minerals in reground concentrate from the rougher—scavenger cells is depressed with CaO to produce a higher grade Cu concentrate;
(c) The *re-cleaner cells* maximize concentrate grade (% Cu) by giving rock and Fe-sulfide minerals, a final depression under quiescent conditions;
(d) The *cleaner—scavenger cells* scavenge the last traces of Cu to the froth from the cleaner tails by the addition of more collector before the tails are discarded.

The froths from the rougher—scavengers and cleaner—scavengers are ground before being sent to the cleaners (Fig. 4.3). This releases previously unliberated Cu mineral grains.

The rougher—scavenger and cleaner—scavenger cells are designed to maximize Cu recovery to concentrate. The cleaner and re-cleaner cells maximize concentrate grade.

Circuits such as that in Fig. 4.3 give ~90% recovery of Cu-sulfide minerals and ~30% Cu concentrate grade (Tables 4.2 and 4.3).

4.4. FLOTATION CELLS

Figure 4.2 shows a *mechanical* flotation cell. Air bubbles are introduced into the slurry through a rotating agitator at the bottom of the cell. The agitator sheers the air into the 1 mm diameter bubbles needed for Cu mineral attachment. It also disperses the bubbles across the cell.

4.4.1. Column Cells

Many new Cu flotation plants use column cells for cleaner flotation (Tables 4.3 and 4.4; Figs. 4.6 and 4.7). These cells provide separate zones for particle—bubble attachment and

TABLE 4.2 % Cu in Concentrate Feeds at Smelters Around the World Pure chalcopyrite contains 35.4% Cu

Country	Smelter location	% Cu in concentrate
Australia	Mount Isa	25–27
Brazil	Caraiba	32
Canada	Noranda	22
Chile	Caletones (Teniente)	31
China	Guixi	28
Finland	Harjavalta	28
Germany	Hamburg	30–33
India	Tutincorin	30–32
Indonesia	Gresik	30
Iran	Khatoon Abad	27
Japan	Toyo (Niihama)	32
Korea	Onsan	32.5
Mexico	La Caridad	26
Peru	Ilo	28
Philippines	Isabel (Leyte)	25–32
Poland	Lower Silesia	23
Russia	Norisk	26
South Africa	Palabora	30–35
Spain	Huelva	31
Sweden	Ronnskar	30
U.S.A.	Kennecott	26.5
Zambia	Mopani	28–42

for draining of non-attached low-Cu particles from the froth (Finch, Cilliers, & Yianatos, 2007).

Column cells provide a long vertical particle/bubble contact zone and a well-controlled froth-draining zone. They are an excellent tool for maximizing % Cu in a concentrator's final concentrate.

TABLE 4.3a Industrial Flotation Data from Three Copper Concentrators. The Equivalent Crushing/Grinding Data are Given in Table 3.1

Concentrator	Candelaria, Chile	El Soldado, Chile	Los Bronces, Chile
Ore treated per year, tonnes	25,000,000 (2001)	7,700,000 (2010)	20,500,000 (2010)
Concentrate, tonnes/year		67,000	676,000
Ore grade, % Cu	0.9–1.0	0.63	
Sulfide		0.54	1.055
'Oxide'		0.09	0.049
Concentrate grade, % Cu	29–30	22.5	28.4
Tailings grade, % Cu		0.18	0.133
Cu recovery to concentrate, %	95% Cu, 82% Au 87% Mo		
Rougher–scavenger flotation			
Feed	Cyclone overflow	Cyclone overflow	Grinding product
Number of cells	24	32	30
Volume of each cell, m^3	85 and 128	42.5	3 × 127 + 27 × 85
Cell type	Eimco	Wemco	Wemco
Mass% solids in feed		38	30
pH	10.4	10.3	10.3
Collector	SF 323	K dithiophosphate	Dithiophosphate & polysulfate
kg/tonne of ore		0.19	0.035
Oily collector, kg/t of ore		No	0.012
Frother	MIBC	Dowfroth 250: MIBC 1:5	5 Nalfloat: 1 Dowfoth 250
kg/tonne of ore		0.033	0.022
CaO, kg/tonne of ore	0.7	0.65	0.45
Residence time, minutes		33	25–30
Cleaner flotation			

TABLE 4.3a Industrial Flotation Data from Three Copper Concentrators. The Equivalent Crushing/Grinding Data are Given in Table 3.1—cont'd

Concentrator	Candelaria, Chile	El Soldado, Chile	Los Bronces, Chile
Feed	Rougher concentrate ground in 4.27 m × 6.7 m ball mill	Rougher–scavenger concentrate	Reground rougher & scavenger concentrate
Number of cells		1 square + 1 round columns	5
Volume of each cell, m^3		8.8 m^2 square, 16 m^2 round	224 c/u
Cell type		Column	Column
Mass% solids in feed		16	16
pH		10.5	11.3
Reagents		CaO	No
Residence time, minutes			15.8
Cleaner–scavenger flotation			No
Feed		Cleaner tails	Cleaner tails
Number of cells		11 Wemco	21
Volume of each cell, m^3		10 × 28.3 + 1 × 127	42.5
Cell type		Mechanical	Wemco
Mass% solids in feed		18	16
pH		10–10.5	11.3
Reagents		No	No
Residence time, minutes			15

TABLE 4.3b Industrial Flotation Data from Three Copper Concentrators, 2010. The Equivalent Crushing/Grinding Data are Given in Table 3.1

Concentrator	Africa open pit	Africa under ground	Mantos Blancos Chile, open pit
Ore treated per year, tonnes	4,000,000	900,000	4,500,000

(Continued)

TABLE 4.3b Industrial Flotation Data from Three Copper Concentrators, 2010. The Equivalent Crushing/Grinding Data are Given in Table 3.1—cont'd

Concentrator	Africa open pit	Africa under ground	Mantos Blancos Chile, open pit
Concentrate, tonnes/year	35,000	21,000	125,000
Ore grade, % Cu	0.21	0.45	1.13
Sulfide	0.15	Chalcopyrite	0.92
'Oxide'	0.06	–	0.21
Concentrate grade, % Cu	25	17	35.2
Tailings grade, % Cu	0.018	0.04	0.12
Cu recovery to concentrate, %	80	85	89
Rougher–scavenger flotation			
Feed	Cyclone overflow	Cyclone overflow	Cyclone overflow
Number of cells	5	7	11
Volume of each cell, m^3	150	40	43
Cell type	OK150 tank	OK40 tank	Wemco
Mass% solids in feed	30	36	42
pH	9.0	7.9	9.5
Collector	K amyl xanthate Aerophine 3418A	K amyl xanthate Aerophine 3418A	Sodium di-isobutyl dithiophosphate
kg/tonne of ore	0.12 g PAX, 0.08 g Aerophine	0.1 g PAX, 0.03 g Aerophine	0.042
Oily collector, kg/t of ore	No	No	No
Frother	MIBC	Inter-froth 64	MB78-MIBC
kg/tonne of ore	0.06 g	0.06 g	0.02
CaO, kg/tonne of ore	0.5	2	0.808
Residence time, minutes	25	35	5
Cleaner flotation			

TABLE 4.3b Industrial Flotation Data from Three Copper Concentrators, 2010. The Equivalent Crushing/Grinding Data are Given in Table 3.1—cont'd

Concentrator	Africa open pit	Africa under ground	Mantos Blancos Chile, open pit
Feed	Various reground concentrates 80% < 30 μm	Various reground concentrates 80% < 20 μm	Rougher, scavenger, & sand flotation concentrates
Number of cells	7	6	1
Volume of each cell, m³	8	17	86
Cell type	OK8 trough cell	Mechanical	Column
Mass% solids in feed	14	20	15
pH	11.5	11.5	11
Reagents	CaO, 3418A, CMC (depressant)	CaO, 3418A, CMC (depressant)	None
Residence time, minutes	9	12	16.5
Cleaner–scavenger flotation	None		
Feed		Cleaner tails	Cleaner tails
Number of cells		4	6
Volume of each cell, m³		17	3 × 28 + 3 × 14
Cell type		OK 16 mechanical	Wemco
Mass% solids in feed		16	18
pH		11.5	11.5
Reagents		Aerophine 3418A	0.005 kg/tonne
Residence time, minutes		10	16.5
	Re-cleaner float	**Re-leaner float**	**Sand flotation**
Feed	Cleaner concentrate	Cleaner concentrate	Various tailings
Number of cells	5	Various	20
Cell volume, m³	4.25	Various	14
Cell type	OK3HG trough cell	Various	Wemco

(Continued)

TABLE 4.3b Industrial Flotation Data from Three Copper Concentrators, 2010. The Equivalent Crushing/Grinding Data are Given in Table 3.1—cont'd

Concentrator	Africa open pit	Africa under ground	Mantos Blancos Chile, open pit
Mass% solids in feed	10	24	42
pH	11	11	11.5
Reagents	CaO, Aerophine 3418A, MIBC	CaO, 3418A, Inter-froth, CMC	NaHS (4%) + CytecAero 3753
Residence time, minutes	14	16	19

TABLE 4.3c Industrial Flotation Data from Three Copper Concentrators, 2010. The Equivalent Crushing/Grinding Data are Given in Table 3.1

Concentrator	Cerro Verde, Peru	Sierrita, U.S.A.	Ray, U.S.A.
Ore treated per year, tonnes	39,000,000	37,000,000	10,000,000
Concentrate, tonnes/year	750,000	325,000	150,000
Ore grade, % Cu		0.26	0.453
Sulfide	0.6	0.23 Chalcopyrite & chalcocite	Chalcopyrite & chalcocite
'Oxide'	0.02	0.03	0.025
Concentrate grade, % Cu	25	26	28
Tailings grade, % Cu	0.075	0.041	0.035
Cu recovery to concentrate, %	88	84.3	90
Rougher–scavenger flotation			
Feed	Cyclone overflow	Cyclone overflow	Cyclone overflow
Number of cells	40	380	15
Volume of each cell, m^3	160	2.9; 14.2; 29.3	9.4
Cell type	Wemco	Denver & Wemco	Wemco

TABLE 4.3c Industrial Flotation Data from Three Copper Concentrators, 2010. The Equivalent Crushing/Grinding Data are Given in Table 3.1—cont'd

Concentrator	Cerro Verde, Peru	Sierrita, U.S.A.	Ray, U.S.A.
Mass% solids in feed	30	35	35
pH	11.2	~11.8	10.5
Collector	Thionocarbamate Na isopropyl xanthate	Nalco 9740 + K amyl xanthate	Xanthate
kg/tonne of ore	0.01	0.01 N & 0.005 K	0.015
Oily collector, kg/t of ore	0.012	Chevron Phillips MCO, 0.05	0.012
Frother	Molycop H 75	Nalco 9873/Ore Prep X-133 80/20	Pine oil
kg/tonne of ore	0.02	0.018	0.015
CaO, kg/tonne of ore	1.26	1.02	0.92
Residence time, minutes	18	8	27
Cleaner flotation			
Feed	Reground cleaner–scavenger & rougher–scavenger froth & column tail, 80% < 101 μm	Rougher froth	Reground rougher–scavenger & cleaner–scavenger froth
Number of cells	6	12	6
Volume of each cell, m^3	160	8.5	42.5
Cell type	Wemco	Wemco	Denver Agitair
Mass% solids in feed	19	20	14
pH	12.2		11
Reagents	CaO, Molyflow	None	CaO, collector, frother
Residence time, minutes	20	6.2	12
Cleaner–scavenger flotation			
Feed	Cleaner tails	Cleaner + re-cleaner tails	Cleaner tails

(Continued)

TABLE 4.3c Industrial Flotation Data from Three Copper Concentrators, 2010. The Equivalent Crushing/Grinding Data are Given in Table 3.1—cont'd

Concentrator	Cerro Verde, Peru	Sierrita, U.S.A.	Ray, U.S.A.
Number of cells	4	6	4
Volume of each cell, m³	160	14.2	42.5
Cell type	Wemco	Wemco	Denver Agitair
Mass% solids in feed	19	16	11.3
pH	12.2		11.5
Reagents	CaO, Molyflow	None	CaO, xanthate
Residence time, minutes	23	4.2	8
Re-cleaner flotation			No
Feed	Rougher & cleaner froths	Ground cleaner froth 80% < 44 μm	
Number of cells	4	12	
Cell volume, m³	236	54.5	
CELL type	Minovex column cell	Column	
Mass% solids in feed	27	20	
pH	12.3		
Reagents	CaO, Molyflow	None	
Residence time, minutes	36		

4.5. SENSORS, OPERATION, AND CONTROL

Modern flotation plants are equipped with sensors and automatic control systems (Shang, Ryan, & Kennedy, 2009). The principal objectives of the control are to maximize Cu recovery, concentrate grade (% Cu), and ore throughput rate. The principal variables sensed are (Table 4.5):

(a) Ore particle size after grinding and regrinding (Outotec, 2010a);
(b) % Cu, % solids, pH, and mass flowrate of the process streams (especially the input and output streams);
(c) Froth height in the flotation cells.

TABLE 4.4 Column Flotation Cells Around the World. Virtually all are used for cleaning gangue from the final concentrate

Country	Mine	Column cell use	Column diameter × height, m	Number
Australia	Cobar	Cu cleaner	2.3 × ?	1
Brazil	Sossego	Cu cleaners	4.3 × 14	6
Canada	Duck Pond	Cu cleaners	1.2 × 10	2
Chile	Escondida	Cu cleaners	4 × 4 × 13.9	8
China	Dashan copper	Cu cleaners	4.3 × 12	7
Indonesia	Batu Hijau	Cu cleaners	4.3 × 14	2
Mexico	Capstone mining	Cu cleaner	3.1 × 7.6	1
Mongolia	Oyu Tolgoi	Cu cleaners	5.5 × 16	4
Peru	Toquepala	Cu	2.4 × 13.3	3
Philippines	Philex Mining	Cu cleaner	3.1 × 12	1
USA	Bingham Canyon	Cu cleaners	4.3 × 9.6	4
		Scavengers	4.3 × 9.6	4
Venezuela	Auriferas Brisas	Cu cleaners	3.1 × 9	2
Zambia	Nchanga	Cu cleaners	4 × ?	6
		Cu cleaners	2.5 × ?	2

Impeller speeds and air input rates in the flotation cells are also often sensed. The adjustments made on the basis of the sensor readings are:

(a) Water flow rates into hydrocyclone feed sumps to control grinding recycle, and thus flotation feed (ore) particle size;
(b) Flotation reagent (collector, frother, and depressant) and water addition rates throughout the flotation plant;
(c) Slurry level in the flotation cells, by adjusting the underflow valves in each cell.

4.5.1. Continuous Chemical Analysis of Process Streams

Of particular importance in flotation control is continuous measurement of Cu concentration in the solids of the process streams. This is typically done by X-ray fluorescence analysis, which may be accomplished by either sending samples to a central X-ray analysis unit or by using small X-ray units in the process streams themselves. The analyses are often done by fixed-crystal wavelength-dispersive spectrometry (Outotec, 2010b).

These analyses are used to monitor and optimize plant performance by automatically controlling (for example) reagent addition rates, grind size, and flotation cell operation. In modern plants, the control is done by a supervisory computer (Barette, Bom, Taylor, & Lawson, 2009).

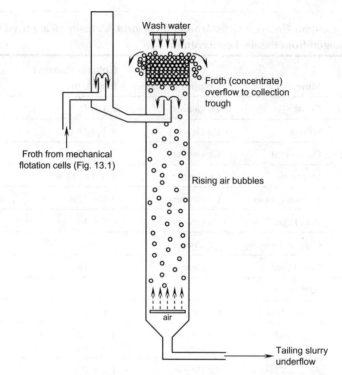

FIGURE 4.6 Sketch of column flotation cell. Air bubbles carry Cu-sulfide particles up to the top of the column where they overflow in concentrate froth. Water spraying on top of the froth layer washes weakly held pyrite and gangue from the froth back into the slurry and eventually out in the underflow tailings slurry. Industrial column cell dimensions are given in Table 4.4.

FIGURE 4.7 Industrial column flotation cells. Their sizes around the world are given in Table 4.4.

TABLE 4.5 Sensors and Their Use in Automatic Flotation Control and Optimization

Sensing Instrument	Purpose	Type of Device	Use in Automatic Control
Process stream particle size analyzer	Senses particle sizes after grind and regrind mills	Laser beam diffraction (Outotec, 2010a)	Controls water addition rates to hydrocyclone feed (which controls particle recycle to ball mills & final grind size)
In-stream X-ray chemical analyzer	Senses Ni content of solids in process streams	X-ray fluorescence analysis of automatically taken process stream samples (Outotec, 2010b)	Controls collector, frother, depressant, & water addition rates throughout the plant. Adjusts valves in flotation cells to vary froth thickness
Slurry level sensor	Indicates slurry surface location	Float level, hydro-static pressure, conductivity	Adjusts underflow valves to maintain prescribed froth layer thicknesses
Slurry mass flow gage and % solids in slurry gage	Determines mass and volumetric flow rates of process streams	Magnetic induction, Doppler effects, ultrasonic energy loss	Measures flows in flotation circuit, permits optimization of recycle streams

Advanced mineralogical techniques are also being more commonly employed to improve the performance and control of flotation circuits (Celik, Can, & Sherazadishvili, 2011; Hamilton, Martin, & Zhou, 2007).

4.5.2. Machine Vision Systems

Several large concentrators have installed machine vision systems on their flotation cells (Metso, 2010). These systems determine froth quality and froth overflow rate. They are used to automatically adjust grind size, reagent addition rates, and froth depth, to maximize Cu recovery in their rougher and scavenger cells and Cu grade in their cleaner and re-cleaner cells (Chadwick, 2010; Supomo et al., 2008).

Freeport's Grasberg concentrator installed 172 machine vision systems. These systems and new process strategies increased Cu-to-concentrate recovery by several percent (Supomo et al., 2008). The Escondida Phase IV concentrator has installed 102 systems (Metso, 2010).

4.6. THE FLOTATION PRODUCTS

4.6.1. Thickening and Dewatering

The product from flotation contains ~75 mass% water, most of which must be removed before the concentrate can be transported and smelted. Most of this *dewatering* is performed by settling in large quiescent *thickeners*. The solids settle under the influence of gravity to the bottom of the thickener from where they are scraped to a central discharge by a slowly rotating rake. Faster settling is encouraged by adding small quantities of organic flocculants, such as polyacrylamides (Ferrera, Arinaitwe, & Pawlik, 2009), to the input slurry. These cause the fine particles to flocculate, which results in faster settling velocities.

The underflows from the thickeners still contain 30—40% water. This is reduced to 10—15% in rotary vacuum filters and then further reduced to 8% in pressure filters or ceramic-disc vacuum filters (sometimes pressurized to 3 bar gauge; Outotec, 2010c). The water content of the concentrate when it is shipped (approximately 8%) is a compromise between the cost of shipping water and the need to minimize concentrate losses as dust during transport.

4.6.2. Tailings

Flotation tailings account for ~98% of the ore fed to the concentrator. They are stored in large *tailings dams* near the mine property. Water is reclaimed from the dams and recycled to the concentrator.

Most concentrators are *zero water discharge* plants, which means that no water can be discharged from the plant but must be internally recycled. This minimizes water consumption and avoids mixing concentrator effluents with the surrounding water table. The pH of the tailings water is close to that required for rougher—scavenger flotation, so this recycle minimizes CaO consumption. Cu flotation tailings typically contain 0.02—0.15% Cu on a dry basis (Table 4.3).

4.7. OTHER FLOTATION SEPARATIONS

Copper flotation consists mainly of separating Cu-sulfide minerals from non-sulfide rock and Fe-sulfide minerals. Many Cu deposits also contain molybdenite. Others contain pentlandite [$(Ni,Fe)_9S_8$], sphalerite (ZnS), or galena (PbS). These can all be separated from Cu minerals by selective flotation. Molybdenite flotation is discussed in Chapter 21. The flotation of sphalerite, galena, and Ni and Cu oxides is discussed by Biswas and Davenport (1994). Pentlandite flotation is described by Crundwell, Moats, Ramachandran, Robinson, and Davenport (2011).

4.7.1. Gold Flotation

Au is present in many Cu-sulfide ores, from just traces to ~1 g/t Au (1×10^{-4}% Au). Au is mostly present in Cu-sulfide ores as metal and metal alloy (e.g., electrum Au, Ag) grains. Some may also be present as tellurides. These grains of Au and Au alloy may be liberated particles, attached to other mineral particles, encapsulated in sulfide mineral particles (e.g., in chalcopyrite $CuFeS_2$ and pyrite FeS_2 particles), or encapsulated in oxide gangue (e.g., andesite) particles.

The Au, Au alloy, and Au mineral grains in Cu—Au ores may be smaller than their companion Cu-sulfide mineral grains, so Au recovery may be increased by finer grinding. Optimum grind size is best determined by controlled in-plant testing.

Fortunately, all of these gold mineral types float efficiently under the same conditions as Cu-sulfide minerals, using xanthate, dithiophosphate and thionocarbamate collectors and conventional frothers such as methyl isobutyl carbinol (Woodcock et al., 2007). Generally speaking, liberated Au metal, Au alloy, and Au mineral particles float with chalcopyrite and other Cu-sulfide minerals and report to the bulk concentrate (Zahn et al., 2007). Au recoveries to the concentrate of 80−85% are typical.

Kendrick, Baum, Thompson, Wilkie, and Gottlieb (2003) suggest that Au recovery to concentrate can be increased ~10% by decreasing the pH of the ore slurry from 10.7 to 9.5. This also appreciably lowers CaO consumption. Special collectors have also been developed for copper ores with significant gold content. These collectors are a blend of monothiophosphates and dithiophosphates, and may be worth considering for Cu−Au concentrators.

4.8. SUMMARY

Copper sulfide ores must be concentrated before they can be economically transported and smelted. The universal technique for this concentration is froth flotation of finely ground ore.

Froth flotation entails attaching fine Cu-sulfide mineral particles to bubbles and floating them out of a water−ore mixture. The flotation is made selective by using reagents, which make the Cu-sulfide minerals water-repellant (hydrophobic) while leaving the other minerals wetted (hydrophilic).

Typical Cu sulfide recoveries to concentrate are 85−90%. Typical concentrate grades are 30% Cu (higher with chalcocite, bornite, and native copper mineralization). Column flotation cleaner cells are particularly effective at giving a high Cu grade in the concentrate.

Modern concentrators are automatically controlled to give maximum Cu recovery, maximum % Cu in the concentrate, and maximum ore throughput rate at minimum cost. Expert control systems help to optimize the performance of flotation plants. On-stream particle size and on-stream X-ray fluorescence analyses are key components of this automatic control.

REFERENCES

Barette, R., Bom, A., Taylor, A., & Lawson, V. (2009). Implementation of expert control systems at Vale Inco Limited's Clarabelle mill. In H. Shang, L. Ryan & S. Kennedy (Eds.), *Process control: Applications in mining and metallurgical plants* (pp. 25−37). Montreal, Canada: CIM.

Biswas, A. K., & Davenport, W. G. (1994). *Extractive metallurgy of copper* (3rd ed.). New York: Elsevier Science Press.

Boldt, J. R., & Queneau, P. (1967). *The winning of nickel*. Toronto: Longmans Canada Ltd.

Celik, I. B., Can, N. M., & Sherazadishvili, J. (2011). Influence of process mineralogy on improving metallurgical performance of a flotation plant. *Mineral Processing and Extractive Metallurgy Review*, 32(1), 30−46.

Chadwick, J. (2010). Grasberg concentrator. *International Mining*, 27(5), 8−20. http://www.infomine.com/publications/docs/InternationalMining/Chadwick2010p.pdf.

Chevron Phillips. (2008). Mining chemicals. www.cpchem.com/bl/specchem/en-us/Pages/MiningChemicals.aspx.

Crundwell, F., Moats, M., Ramachandran, R., Robinson, T., & Davenport, W. G. (2011). *Extractive metallurgy of nickel, cobalt and Pt group metals*. Oxford: Elsevier.

Ferrera, V., Arinaitwe, E., & Pawlik, M. (2009). A role of flocculant conformation in the flocculation process. In C. O. Gomez, J. E. Nesset & S. R. Rao (Eds.), *Advances in mineral processing science and technology: Proceedings of the 7th UBC-MCGill-UA international symposium on fundamentals of mineral processing* (pp. 397−408). Montreal, Canada: CIM.

Finch, J. A., Cilliers, J., & Yianatos, J. (2007). Froth flotation. In M. C. Fuerstenau, G. Jameson & R.-H. Yoon (Eds.), *Froth flotation, a century of innovation* (pp. 681−737). Littleton, CO, USA: SME.

Fuerstenau, M. C., Jameson, G., & Yoon, R.-H. (2007). *Froth flotation, a century of innovation*. Littleton, CO, USA: SME.

Hagihara, H. (1952). Mono- and multiplayer adsorption of aqueous xanthate on galena surfaces. *Journal of Physical Chemistry*, 56, 616−621.

Hamilton, C., Martin, C., & Zhou, J. (2007). Ore characterization and the value of QEMSCAN. In R. del Villar, J. E. Nesset, C. O. Gomez & A. W. Stradling (Eds.), *Copper 2007, Vol. II: Mineral processing* (pp. 137−147). Montreal, Canada: CIM.

Kapusta, J. P. T. (2004). JOM world nonferrous smelters survey, part I: copper. *JOM*, 56(7), 21−27.

Kendrick, M., Baum, W., Thompson, P., Wilkie, G., & Gottlieb, P. (2003). The use of the QemSCAN automated mineral analyzer at the Candelaria concentrator. In C. O. Gomez & C. A. Barahona (Eds.), *Copper−Cobre 2003, Vol. III: Mineral processing* (pp. 415−430). Montreal: CIM.

Klimpel, R. R. (2000). Optimizing the industrial flotation performance of sulfide minerals having some natural floatability. *International Journal of Mineral Processing*, 58, 77−84.

Metso. (2010). *Vision systems*. http://www.metso.com/miningandconstruction/mm_proc.nsf/WebWID/WTB-041208-2256F-8CDAA?OpenDocument&mid=CBE875DB6E40FD7DC22575BD0023C839.

Nagaraj, D. R., & Ravishankar, S. A. (2007). Flotation reagents − a critical overview from an industry perspective. In M. C. Fuerstenau, G. Jameson & R.-H. Yoon (Eds.), *Froth flotation, a century of innovation* (pp. 375−424). Littleton, CO, USA: SME.

Outotec. (2010a). *Outotec PSI® 300*. www.outotec.com/pages/Page.aspx?id=36489&;epslanguage=EN.

Outotec. (2010b). *Outotec courier® 6 SL*. www.outotec.com/pages/Page.aspx?id=36492&;epslanguage=EN.

Outotec. (2010c). *LAROX® automatic pressure filters C series*. www.larox.smartpage.fi/en/pfcseries/pdf/Larox_PF_C_series.pdf.

Shang, H., Ryan, L., & Kennedy, S. (2009). *Process control applications in mining and metallurgical plants*. Montreal, Canada: CIM.

Supomo, A., Yap, E., Zheng, X., Banini, G., Mosher, J., & Partanen, A. (2008). PT freeport Indonesia's mass-pull control strategy for rougher flotation. *Minerals Engineering*, 21, 808−816.

Wark, I. W., & Cox, A. G. (1934). Principles of flotation, III, an experimental study of influence of cyanide, alkalis and copper sulfate on effect of sulfur-bearing collectors and mineral surfaces. *AIME Transactions*, 112, 288.

Woodcock, J. T., Sparrow, G. J., Bruckard, W. J., Johnson, N. W., & Dunne, R. (2007). Plant practice: sulfide minerals and precious metals. In M. C. Fuerstenau, G. Jameson & R.-H. Yoon (Eds.), *Froth flotation, a century of innovation* (pp. 781−817). Littleton, CO, USA: SME.

Zahn, R. P., Baum, W., Magnuson, R., Jensen, D., & Thompson, P. (2007). QEMSCAN characterization of selected ore types for plant optimization. In R. del Villar, J. E. Nesset, C. O. Gomez & A. W. Stradling (Eds.), *Copper 2007, Vol. II: Mineral processing* (pp. 147−156). Montreal, Canada: CIM.

SUGGESTED READING

Brandt, D., Martinez, I., French, T., Slattery, L., & Baker, G. (2010). Better fragmentation through team work at Dos Pobres Mine Safford, Arizona. In *36th conference on explosives and blasting technique, Vol. 1* (pp. 13−22). Solon, OH, USA: ISEE.

Castro, S. H., Henriquez, C., & Beas, E. (1999). Optimization of the phosphate Nokes process at the El Teniente by-product molybdenite plant. In B. A. Hancock & M. R. L. Pon (Eds.), *Copper 99−Cobre 99, Vol. II: Mineral processing/environment, health and safety* (pp. 41−50). Warrendale, PA, USA: TMS.

Fuerstenau, M. C., Jameson, G., & Yoon, R.-H. (2007). *Froth flotation, a century of innovation*. Littleton, CO, USA: SME.

McCaffery, K., Mahon, J., Arif, J., & Burger, B. (2006). Batu Hijau — controlled mine blasting and blending to optimise process production at Batu Hijau. In M. J. Allan (Ed.), *International autogenous and semi-autogenous grinding technology 2006, Vol. 1*. Vancouver: University of British Columbia.

Shang, H., Ryan, L., & Kennedy, S. (2009). *Process control applications in mining and metallurgical plants*. Montreal, Canada: CIM.

Wills, B. A., & Napier-Munn, T. J. (2006). *Mineral processing technology* (7th ed.). Oxford: Elsevier.

Chapter 5

Matte Smelting Fundamentals

5.1. WHY SMELTING?

Beneficiation of copper ores produces concentrates consisting mostly of sulfide minerals, with small amounts of gangue oxides (Al_2O_3, CaO, MgO, and SiO_2). Theoretically, this material could be directly reacted to produce metallic Cu by oxidizing the sulfides to elemental copper and ferrous oxide:

$$CuFeS_2 + 2.5O_2 \rightarrow Cu° + FeO + 2SO_2 \quad (5.1)$$

$$Cu_2S + O_2 \rightarrow 2Cu° + SO_2 \quad (5.2)$$

$$FeS_2 + 2.5O_2 \rightarrow FeO + 2SO_2 \quad (5.3)$$

These reactions are exothermic, meaning they generate heat. As a result, the smelting of copper concentrate should generate (a) molten copper and (b) molten slag containing flux oxides, gangue oxides, and FeO. However, under oxidizing conditions, Cu forms Cu oxide as well as metal:

$$Cu_2S + 1.5O_2 \rightarrow Cu_2O + SO_2 \quad (5.4)$$

When this happens, the Cu_2O dissolves in the slag generated during coppermaking. The large amount of iron in most copper concentrates means that a large amount of slag would be generated. More slag means more lost Cu. As a result, eliminating some of the iron from the concentrate before final coppermaking is a good idea.

Figure 5.1 illustrates, what happens when a mixture of FeO, FeS, and SiO_2 is heated to 1200 °C. The left edge of the diagram represents a solution consisting only of FeS and FeO. In silica-free melts with FeS concentrations above ~31 mass%, a single oxysulfide liquid is formed. However, when silica is added, a liquid-state miscibility gap appears. This gap becomes larger as more silica is added.

Lines a, b, c and d represent the equilibrium compositions of the two liquids. The sulfide-rich melt is known as *matte*. The oxide-rich melt is known as *slag*. Heating a sulfide concentrate to this temperature and oxidizing some of its Fe to generate a molten matte and slag,

$$CuFeS_2 + O_2 + SiO_2 \rightarrow (Cu, Fe, S)\,(matte) + FeO \cdot SiO_2\,(slag) + SO_2 \quad (5.5)$$

is known as *matte smelting*. It accomplishes the partial removal of Fe needed to make final coppermaking successful. Matte smelting is now performed on nearly all Cu–Fe–S and Cu–S concentrates. This chapter introduces the fundamentals of matte smelting and the influence of process variables. Following chapters describe current smelting technology.

FIGURE 5.1 Simplified partial phase diagram for the Fe–O–S–SiO$_2$ system showing liquid–liquid (slag–matte) immiscibility caused by SiO$_2$ (Yazawa & Kameda, 1953). The heavy arrow shows that adding SiO$_2$ to an oxysulfide liquid causes it to split into FeS-rich matte and FeS-lean slag. The compositions of points A and B (SiO$_2$ saturation) and the behavior of Cu are detailed in Table 5.1.

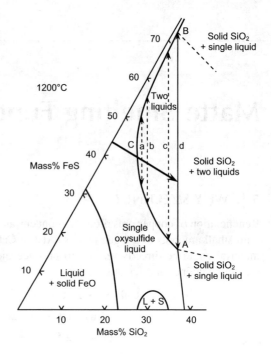

5.2. MATTE AND SLAG

5.2.1. Slag

Slag is a solution of molten oxides. These oxides include FeO from Fe oxidation, SiO$_2$ from flux, and oxide impurities from concentrate. Oxides commonly found in slags include ferrous oxide (FeO), ferric oxide (Fe$_2$O$_3$), silica (SiO$_2$), alumina (Al$_2$O$_3$), calcia (CaO), and magnesia (MgO). As Fig. 5.1 shows, small amounts of sulfides can also be dissolved in FeO–SiO$_2$ slags. Small amounts of calcia and alumina in slags decrease this sulfide solubility (Table 5.1).

The molecular structure of molten slag is described by dividing its oxides into three groups — acidic, basic, and neutral. The best-known acidic oxides are silica and alumina. When these oxides melt, they polymerize, forming long polyions such as those shown in Fig. 5.2. These polyions give acidic slags high viscosities, making them difficult to handle, and increasing the amount of entrained copper or matte. Acidic slags also have low solubilities for other acidic oxides. This can cause difficulty in coppermaking because impurities, which form acidic oxides (e.g., As$_2$O$_3$, Bi$_2$O$_3$, Sb$_2$O$_3$) are not readily removed in acidic slag. Instead, they will remain in matte or copper.

Adding basic oxides such as calcia and magnesia to acidic slags breaks the polyions into smaller structural units. As a result, basic slags have low viscosities and high solubilities for acidic oxides. Up to a certain limit, adding basic oxides also lowers the melting point of a slag. Coppermaking slags generally contain small amounts of basic oxides.

Neutral oxides such as FeO and Cu$_2$O react less strongly with polyions in a molten slag. Nevertheless, they have much the same effect. FeO and Cu$_2$O have low melting points, so they tend to lower a slag's melting point and viscosity.

TABLE 5.1 Compositions of Immiscible Liquids in the SiO_2-Saturated Fe–O–S System, 1200 °C (Yazawa & Kameda, 1953). Points A (slag) and B (matte) correspond to A and B in Fig. 5.1

System	Phase	Composition (mass%)					
		FeO	FeS	SiO_2	CaO	Al_2O_3	Cu_2S
FeS–FeO–SiO_2	"A" Slag	54.82	17.90	27.28			
	"B" Matte	27.42	72.42	0.16			
FeS–FeO–SiO_2 + CaO	Slag	46.72	8.84	37.80	6.64		
	Matte	28.46	69.39	2.15			
FeS–FeO–SiO_2 + Al_2O_3	Slag	50.05	7.66	36.35		5.94	
	Matte	27.54	72.15	0.31			
Cu_2S–FeS–FeO–SiO_2	Slag	57.73	7.59	33.83			0.85
	Matte	14.92	54.69	0.25			30.14

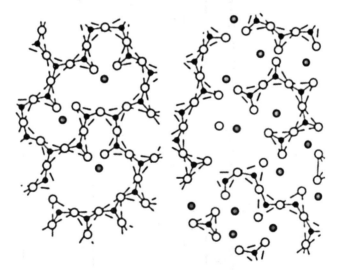

FIGURE 5.2 Impact of basic oxides on the structure of silica polyions in molten slags. Adding basic oxides like CaO and MgO breaks up the polyions, lowering the melting point, and viscosity of the slag. ● = Si; ○ = O; ◉ = Ca^{++} or Mg^{++}.

The slags produced in industrial matte smelting consist primarily of FeO, Fe_2O_3, and SiO_2, with small amounts of Al_2O_3, CaO, and MgO (Table 5.2). They are often referred to as *fayalite* slags; fayalite (Fe_2SiO_4) often precipitates from these slags on cooling. Figure 5.3 shows the composition limits for the liquid region in the FeO–Fe_2O_3–SiO_2 system at 1200 °C and 1250 °C. Along the top line, the slag is saturated with solid silica.

TABLE 5.2 Compositions of Industrial Concentrates, Fluxes, Mattes, Slags, and Dusts for Various Matte-Smelting Processes (Ramachandran et al., 2003)

Smelter/process	Concentrate				Flux			Matte				Slag		
	Cu	Fe	S	SiO$_2$	H$_2$O	SiO$_2$	CaO	Fe	Cu	Fe	S	Cu	SiO$_2$	Total Fe	CaO
La Caridad Outokumpu flash	26	26	35	6	8	90	0.1	0.4	65	10	21	2–3	31	43	
Caraiba Outokumpu flash	32	25	33	7	8	97	2	0	62	12	22	2.3	31	42	
Huelva Outokumpu flash	31	23	31	7	7–8	94–97	0.2	0.9	62	13	20	1.5	28	42	2
Hayden Inco flash	28	25	31	6	10	92	0.4	1.3	59	14	23	1.2	34	36	2
Potrerillos Teniente	30	25	32	4	8	86	0.2	0.5	71	5	21	8	25	38	
Horne Noranda	22	30	30	7	5–10	75	2	2	70	3	20	6.0	30	36	1

Chapter | 5 Matte Smelting Fundamentals

Process															
Mount Isa Isasmelt	25–27	24–26	27–29	11–13	7	87	0.3	1	60	15	23	2.6	33	41	1.5
Zhongtiaoshan Ausmelt	23–26	26–29	29–32	5–13	7–12				60	15	23	0.7	32	36	5
Glogow 2 OK flash direct-to-copper	28	3.3	11	17		92–95		1–2	>95		0.5	14.3	31	6.4	14
Gresik Mitsubishi	30	24	29	10		90	0	<2	68	8	21	0.7	34	36	5
Naoshima Mitsubishi	34	21	24	17	12	85			68	8	21	0.7	36	40	5
SUMZ Vanyukov	15–20	35	33	5	8–10	90	1.5	2.0	45	30	25	0.7	32	30	3
Pasar Electric	25–32	22–30	25–33	6–10	10	>95	<1.5	<2	55	24	23	<0.7	36	38	2
Palabora Reverberatory	30–35	20–30	20–30	3–5	6–7	60–80	0–15	0–15	40–50	25–30	20–27	0.5–0.8	30–40	30–40	0–10

FIGURE 5.3 Liquidus surface in the FeO–Fe$_2$O$_3$–SiO$_2$ system at 1200 °C and 1250 °C (Muan, 1955). Copper smelting processes typically operate near magnetite saturation (line CD).

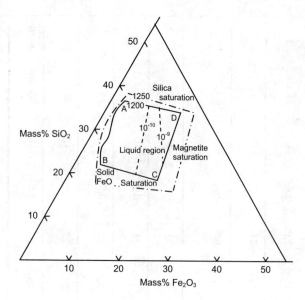

Along the bottom boundary line, the slag is saturated with solid FeO. The boundary at right marks the compositions at which dissolved FeO and Fe$_2$O$_3$ react to form solid magnetite:

$$\text{FeO} + \text{Fe}_2\text{O}_3 \rightarrow \text{Fe}_3\text{O}_4(s) \tag{5.6}$$

Extensive oxidation and lower smelting temperatures encourage the formation of Fe$_2$O$_3$ in the slag. Avoiding these conditions minimizes magnetite precipitation.

Along the left-hand boundary, the slag is saturated either with metallic iron or solid fayalite (Fe$_2$SiO$_4$). Under the oxidizing conditions of industrial copper smelting, this never occurs. Table 5.2 lists the compositions of some smelter slags, including their Cu content. Controlling the amount of Cu dissolved in smelting slag is an important part of smelter strategy (Chapter 12).

Many measurements have been made of the viscosities of molten slags (Vadász et al., 2001; Vartiainen, 1998). These have been used to develop a model, which calculates viscosities as a function of temperature and composition (Utigard & Warczok, 1995). The model relies on calculation of a viscosity ratio (VR). VR is the ratio of A (an equivalent mass% in the slag of acidic oxides) to B (an equivalent mass% of basic oxides):

$$VR = A/B \tag{5.7}$$

$$A = (\%\text{SiO}_2) + 1.5(\%\text{Cr}_2\text{O}_3) + 1.2(\%\text{ZrO}_2) + 1.8(\%\text{Al}_2\text{O}_3) \tag{5.8}$$

$$B = 1.2(\%\text{FeO}) + 0.5(\%\text{Fe}_2\text{O}_3 + \%\text{PbO}) + 0.8(\%\text{MgO}) + 0.7(\%\text{CaO}) \\ + 2.3(\%\text{Na}_2\text{O} + \%\text{K}_2\text{O}) + 0.7(\%\text{Cu}_2\text{O}) + 1.6(\%\text{CaF}_2) \tag{5.9}$$

Utigard and Warczok related VR to viscosity by regression analysis against their existing database, obtaining:

$$\log \mu (\text{kg/m} - \text{s}) = -0.49 - 5.1\sqrt{VR} + \frac{-3660 + 12,080\sqrt{VR}}{T(K)} \tag{5.10}$$

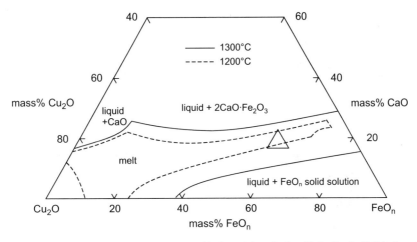

FIGURE 5.4 Liquidus lines at 1200 °C and 1300 °C for calcium ferrite (CaO—Cu$_2$O—FeO$_n$) slags in air, 1 atmosphere pressure (Goto & Hayashi, 1998). The triangle represents Mitsubishi converting furnace slag if all its Cu exists as Cu$_2$O. Slag compositions inside the solid line are fully liquid at 1300 °C. Slag compositions inside the dashed lines are fully liquid at 1200 °C.

Figure 5.4 shows the effect of temperature and composition on the viscosity of FeO—Fe$_2$O$_3$—SiO$_2$. Numerous other models have been developed for viscosity prediction in slags (Kondratiev, Jak, & Hayes, 2002); however, Eqs. 5.7—5.10 appear to be easier to use in practice.

The specific gravity of smelting slags ranges between 3.3 and 4.1. It increases with increasing iron content and increases slightly with increasing temperature (Utigard, 1994; Vadász, Tomášek, & Havlik, 2001).

Slag electrical conductivity is strongly temperature-dependent, ranging at smelting and converting temperatures between 5 and 20 ohm^{-1} cm^{-1} (Hejja, Eric, & Howat, 1994; Ziólek & Bogacz, 1987). It increases with Cu and iron oxide content and with basicity.

The surface tension of smelting slags is 0.35—0.45 N/m (Vadász et al., 2001). It increases with basicity, but is not strongly influenced by temperature.

5.2.2. Calcium Ferrite and Olivine Slags

The ferrous silicate melts discussed in the previous section have been the basis for the slags used in most primary copper smelting and converting technologies. However, ferrous silicate slags have limitations. The most important is the need to control magnetite formation (Eq. 5.6). In bath smelting or converting, this is a particular problem because the magnetite crust that forms during the initial reaction can act as a process barrier. The second is the relatively small window of compositions where the slag is molten at smelting temperatures (Fig. 5.3). Ferrous silicate slags are relatively acidic, which makes it more difficult to use them to remove impurities such as As, Bi, and Sb that form acidic oxides.

The development of the Mitsubishi converter (Chapter 9) was accompanied by the adoption of CaO-based (rather than SiO$_2$-based) slag, better known as *calcium ferrite* (Jahanshahi & Sun, 2003). Early in the development of the process, it was found that blowing O$_2$-rich blast onto the surface of fayalite slag produced a crust of solid magnetite. This made further converting impossible. CaO, on the other hand, reacts with magnetite,

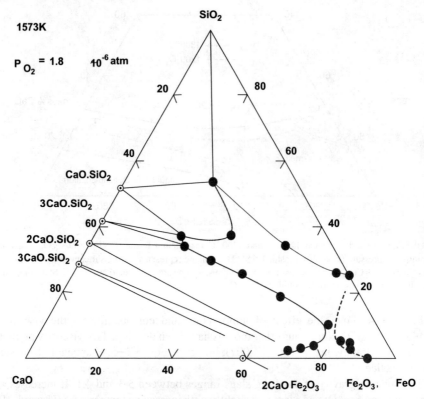

FIGURE 5.5 Liquidus line for the CaO–SiO$_2$–(FeO$_x$) system at 1300 °C (Tsukihashi & Kimura, 2000). The minimal solubility of silica in calcium ferrite slags is apparent, as is the wider solubility range of olivine (ferrous calcium silicate) slags.

molten Cu, and O$_2$ to form a molten Cu$_2$O–CaO–Fe$_3$O$_4$ slag (Fig. 5.4). The slag typically contains 14–16% Cu, 40–55% Fe (mostly Fe^{3+}), and 15–20% CaO.

Calcium ferrite slag has a low viscosity (0.025–0.04 kg/m s), and specific gravities are 3.1–3.8, less than that of comparable fayalite slags (Vadász et al., 2001). Surface tensions are 0.58–0.60 N/m. This improves settling rates after smelting. The CaO in the slag lowers the activity of Fe$_2$O$_3$, and this prevents magnetite formation. The activity coefficient of Cu$_2$O is also higher in calcium ferrite slags. This means that the solubility of Cu$_2$O is reduced, which means lower copper losses. Calcium ferrite slags are more basic than fayalite slags, which lower the activity coefficient of impurity oxides such as As$_2$O$_3$ and Sb$_2$O$_3$. The result enhances the removal of these impurities. Calcium ferrite slags are also less likely to engage in foaming.

The advantages of calcium ferrite slag have led to its adoption for bath converting furnaces (Chapter 9). However, there are drawbacks to the use of this slag, especially in smelting. Figure 5.5 shows the CaO–FeO$_x$–SiO$_2$ phase diagram at 1300 °C (Tsukihashi & Kimura, 2000). As it shows, the solubility of silica in CaO–FeO$_x$ melts is low. This makes smelting of concentrates with significant levels of silica difficult (Vartiainen, Kojo, & Acuña, 2003). The other significant difficulty with the use of calcium ferrite slags is their aggressive behavior toward the 'mag–chrome' (MgO–Cr$_2$O$_3$) refractories used in copper smelting and converting furnaces (Fahey, Swinbourne, Yan, & Osborne,

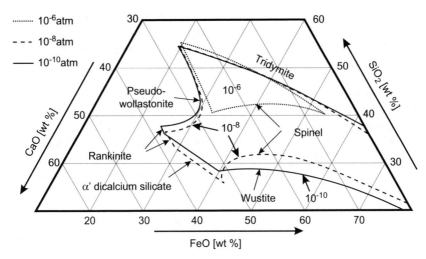

FIGURE 5.6 Liquidus isotherms in the CaO–FeO$_x$–SiO$_2$ system at 1250 °C and varying oxygen potential (Jak et al., 2000).

2004). The frequent lining repair and replacement that occurs as a result increases downtime and expense. As a result, extensive research has been conducted over the past decade on a more suitable type of slag. Figure 5.6 shows the liquid region at 1250 °C for slags in the CaO–FeO$_x$–SiO$_2$ system at different oxygen potentials (Jak, Zhao, & Hayes, 2000). Slags in this region have been called *ferrous calcium silicate (FCS)* or *olivine* ((Fe,Ca)$_2$SiO$_4$) slags. Their properties represent a compromise between those of calcium ferrite and fayalite slags. As a result, they are a possible alternative to fayalite slags for copper smelting. They have also been chosen for the Ausmelt C3 converting process (Wood, Matusewicz, & Reuter, 2009).

5.2.3. Matte (Sundström, Eksteen, & Georgalli, 2008)

As Fig. 5.1 shows, immiscibility between matte and fayalite slag increases with increasing silica content (Yazawa, 1956). A high sulfur/iron ratio also increases the completeness of separation, as do the presence of calcia and alumina in the slag (Table 5.1).

There is some silica and oxygen solubility in matte, but Li and Rankin (1994) demonstrated that increasing Cu$_2$S in matte decreases these solubilities "dramatically". As a result, the typical industrial matte contains only about one percent oxygen (Table 5.2).

The structure of molten mattes is uncertain. Shimpo, Goto, Ogawa, and Asakura (1986) treat them as molten salts, but Zhang (1997) modeled their electrical conductance by treating them as an intermetallic alloy. Their specific gravity is higher than that of slags, so they form the bottom layer in smelting furnaces. As Fig. 5.5 shows, their melting points are lower than the 1200 °C of most slags.

The viscosities of mattes are much lower than those of slags. Data obtained in the 1960s showed that the values of dynamic viscosity of melts in the FeS–Cu$_2$S system ranged between 0.002–0.004 N-s/m^2, with a minimum at about 65 wt.-% FeS (Sundström et al., 2008). Temperature has little impact on these values. As a result, the choice of operating temperature in smelting furnaces is driven primarily by the characteristics of the molten slag.

The surface tension of molten Cu_2S-FeS mattes ranges from 0.35 to 0.40 N/m, increasing slightly with Cu_2S content. Temperature has little effect (Kucharski, Ip, & Toguri, 1994; Sundström et al., 2008). The effect of temperature on matte density is also small; values range from 3.97 g/cm^2 for pure Cu_2S to 5.27 for pure FeS, with a roughly linear relationship to mole fraction.

Measurements of interfacial tension between molten mattes and slags were reviewed by Nakamura and Toguri (1991). Interfacial tension increases from near zero in low-Cu mattes to about 0.20 N/m for high-Cu mattes (~70 mass% Cu_2S). The interfacial tension is lowered by small additions of CaF_2, and is also reduced in mattes with higher Fe/SiO_2 ratios (Li, Huang, & Chen, 1989). Higher interfacial tension is a driving force for improved separation of matte and slag after smelting, and thus favors production of higher-grade mattes. However, higher-grade mattes are also associated with higher dissolved-copper losses in slag.

The electrical conductance of molten matte depends on concentration (Sundström et al., 2008). For pure Cu_2S, the conductance at coppermaking temperatures is less than 100 $ohm^{-1} cm^{-1}$. The conductance of pure molten FeS in the same temperature range is 1500–1600 $ohm^{-1} cm^{-1}$. Excess sulfur levels above those needed for strict stoichiometry also cause substantial increases in conductance. For typical smelter mattes, conductances are in the 350–450 $ohm^{-1} cm^{-1}$ range.

5.3. REACTIONS DURING MATTE SMELTING

The primary purpose of matte smelting is to turn the sulfide minerals in solid copper concentrate into three products: molten matte, molten slag, and offgas. This is done by reacting them with O_2. The oxygen is almost always fed as oxygen-enriched air. The initial reaction takes the form:

$$CuFeS_2 + O_2 \rightarrow (Cu, Fe, S)(matte) + FeO\,(slag) + SO_2 \quad (5.11)$$

The stoichiometry varies, depending on the levels of chalcopyrite and other Cu−Fe sulfide minerals in the concentrate and on the degree of oxidation of the Fe.

As will be seen, smelting strategy involves a series of trade-offs. The most significant is that between matte grade (mass% Cu) and recovery. Inputting a large amount of O_2 will oxidize more of the Fe in the concentrate, so less Fe sulfide ends up in the matte. This generates a higher matte grade. On the other hand, using too much oxygen encourages oxidation of Cu, as shown previously:

$$Cu_2S + 1.5O_2 \rightarrow Cu_2O + SO_2 \quad (5.4)$$

The Cu oxide generated by this reaction dissolves in the slag, which is undesirable. As a result, adding the correct amount of O_2 needed to produce an acceptable matte grade without generating a slag too high in Cu is a key part of smelter strategy.

A second set of reactions important in smelter operation involves the FeO content of the slag. If the activity of FeO in the slag is too high, it will react with Cu_2S in the matte:

$$FeO\,(slag) + Cu_2S\,(matte) \rightarrow FeS\,(matte) + Cu_2O\,(slag) \quad (5.12)$$

This reaction is not thermodynamically favored ($K_{eq} \sim 10^{-4}$ at 1200 °C). However, a high activity of FeO in the slag and a low activity of FeS in the matte generate higher activities of Cu_2O in the slag. (This occurs if too much of the iron in the concentrate is oxidized.) This again gives too much Cu in the slag. In addition, FeO reacts with O_2 to form solid magnetite if the FeO activity is too high:

$$3FeO + 0.5O_2 \rightarrow Fe_3O_4\,(s) \tag{5.13}$$

As a result, lowering the activity of FeO in the slag is important. It is done by adding silica as a flux:

$$FeO + SiO_2 \rightarrow FeO - SiO_2\,(slag) \tag{5.14}$$

However, again there is a trade-off. Flux costs money and the energy required to heat and melt, and it also costs more as more silica flux is used. In addition, as Fig. 5.4 shows, the viscosities of smelting slags increase as the silica concentration increases. This makes slag handling more difficult, and also reduces the rate at which matte particles settle through the slag layer. If the matte particles settle too slowly, they will remain entrained in the slag when it is tapped. This increases Cu losses. As a result, the correct levels of FeO and SiO_2 in the slag require another balancing act.

5.4. THE SMELTING PROCESS: GENERAL CONSIDERATIONS

While industrial matte smelting equipment and procedures vary, all smelting processes have a common sequence of events. The sequence includes:

(a) Contacting particles of concentrate and flux with an O_2-containing gas in a hot furnace. This causes the sulfide minerals in the particles to rapidly oxidize (Eq. 4.11). The reactions are exothermic, and the energy they generate heats and melts the products.

In flash smelting, the contact time between concentrate particles and the gas is less than a second, so ensuring good reaction kinetics is essential (Jorgensen & Koh, 2001). This is difficult, as the percentage of volatiles in copper concentrate is much smaller than in coal, which makes stable combustion of the concentrate more unlikely (Caffery, Shook, Grace, Samarasekara, & Meadowcroft, 2000; González, Richards, & Rivera, 2006). As a result, good mixing is essential. Nearly all smelters mix concentrate with the gas prior to injecting it into the smelting furnace. Decreasing the particle size of the concentrate increases flame speeds, but this requires a finer grind, which increases both processing cost and dust losses. The use of oxygen-enriched air instead of air also improves reaction kinetics, and is increasingly popular (Chapter 6). Nevertheless, about 5% of the concentrate fed to a flash furnace winds up in the dust collector (Asaki, Taniguchi, & Hayashi, 2001). The use of computational fluid dynamics (CFD) has become an increasingly important tool in the modeling of flash furnace processing (Li, Mei, & Xiao, 2002; Li, Peng, Han, Mei, & Xiao, 2004).

Use of oxygen-enriched air or oxygen also makes the process more autothermal. Because less nitrogen is fed to the furnace, less heat is removed in the offgas. This means that more of the heat generated by the reactions goes into the matte and slag. As a result, less (or no) hydrocarbon fuel combustion is required to ensure the proper final slag and matte temperature (~1250 °C). This is especially important in the melting of high-grade concentrates, which have less FeS to oxidize and thus generate less heat (Caffery et al., 2000).

Another method for contacting concentrate and O_2 is used in submerged tuyere *bath smelting* furnaces (Chapter 7 and 9). In these furnaces, concentrate is blown into a mixture of molten matte and slag, and the oxidation process takes place indirectly. The limiting factor in bath-smelting kinetics appears to be the rate at which silica flux dissolves in the matte, where it reacts with dissolved FeS and oxygen to generate slag (Asaki et al., 2001). Because of this, the particle size of the concentrate is of less importance, and the finer grind used by some flash smelters is not needed. The rapid dissolution of concentrate particles in the bath also means that less dust is generated.

(b) *Letting the matte settle through the slag layer into the matte layer below the slag.* Most smelting furnaces provide a quiet settling region for this purpose. During settling, FeS in the matte reacts with dissolved Cu_2O in the slag by the reverse of Reaction (5.12):

$$FeS\ (matte) + Cu_2O\ (slag) \rightarrow FeO\ (slag) + Cu_2S\ (matte) \quad (5.15)$$

This further reduces the amount of Cu in the slag. The importance of low slag viscosity in encouraging settling has already been mentioned. Keeping the slag layer still also helps. Here too a trade-off is at work. Higher matte and slag temperatures encourage reaction (5.15) to go to completion and decrease viscosity, but they cost more in terms of energy and refractory wear.

(c) *Periodically tapping the matte and slag through separate tap holes*. Feeding of smelting furnaces and withdrawing of offgas is continuous. Removal of matte and slag is, however, done intermittently, when the layers of the two liquids have grown deep enough. The location of tap holes is designed to minimize tapping matte with slag.

5.5. SMELTING PRODUCTS: MATTE, SLAG AND OFFGAS

5.5.1. Matte

In addition to slag compositions, Table 5.2 shows the composition of mattes tapped from various smelters. The most important characteristic of a matte is its grade (mass% Cu), which typically ranges between 45 and 75% Cu (56–94% Cu_2S equivalent). At higher levels, the activity of Cu_2S in the matte rises rapidly, pushing Reaction (5.12) to the right. Figure 5.7 shows what happens as a result.

The rapidly increasing concentration of Cu in slag when the matte grade rises above 65% is a feature of many smelter operators prefer to avoid (Henao, Yamaguchi, & Ueda, 2006). However, producing higher-grade mattes increases heat generation, reducing fuel costs. It also decreases the amount of sulfur and iron to be removed during subsequent converting (decreasing converting requirements), and increases SO_2 concentration in the offgas (decreasing gas-treatment costs). In addition, almost all copper producers now recover Cu from smelting and converting slags (Chapter 12). As a result, production of higher-grade mattes has become more popular.

Most of the rest of the matte consists of iron sulfide (FeS). Table 5.3 shows the distribution of other elements in copper concentrates between matte, slag, and offgas. Precious metals report almost entirely to the matte (Henao et al., 2006), as do most Ni, Se, and Te.

5.5.2. Slag

As Table 5.2 shows, the slag tapped from the furnace consists mostly of FeO and SiO_2, with a small amount of ferric oxide. Small amounts of Al_2O_3, CaO, and MgO are also

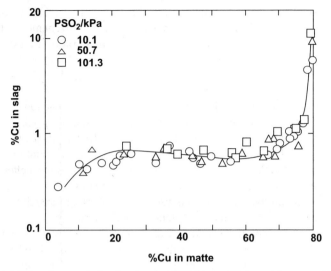

FIGURE 5.7 Solubility of copper in fayalite slags at 1300 °C in equilibrium with mattes of increasing grade (Roghani et al., 2000).

TABLE 5.3 Estimated Distribution of Impurities During Flash Furnace Production of 55% Cu Matte (Steinhauser et al., 1984)

	Matte	Slag	Volatilized[a]
Copper	99	1	0
Alkali/alkaline-earth elements, aluminum, titanium	0	100	0
Ag, Au, Pt-group elements	99	1	0
Antimony	30	30	40
Arsenic	10	10	80
Bismuth	15	5	80
Cobalt	40	55	5
Lead	20	10	70
Nickel	50	45	5
Selenium	75	5	20
Zinc	15	45	40

[a]Not including solid dust from the furnace.

present, as is a small percentage of dissolved sulfur (typically <1%). Cu contents range from less than 1 to as high as 7%. Higher Cu levels are acceptable if facilities are available for recovering Cu from smelter slag. SiO_2/Fe mass ratios are usually 0.7–0.8.

5.5.3. Offgas

The offgas from smelting contains SO_2 generated by the smelting reactions, N_2 from the air used for oxidizing the concentrate, and small amounts of CO_2, H_2O, and volatilized impurity compounds. The strength of the offgas is usually 10–60 vol.-% SO_2. The strength depends on the type of O_2-containing gas used for smelting, the amount of air allowed to leak into the furnace and the grade of matte produced. Vol.-% SO_2 in smelter offgases has risen in recent years. This is due to increased use of oxygen in smelting, which reduces the amounts of nitrogen and hydrocarbon combustion gases passing through the furnace.

Smelter offgases may also contain substantial levels of dust (up to 0.3 kg/Nm^3). This dust comes from (a) small particles of unreacted concentrate or flux, (b) droplets of matte/slag that did not settle into the slag layer in the furnace, and (c) volatilized elements in the concentrate such as arsenic, antimony, bismuth, and lead, which have either solidified as the gas cools or reacted to form non-volatile compounds. The dust generally contains 20–40 mass% Cu, making it potentially valuable. It is nearly always recycled to the smelting furnace, but it may be treated hydrometallurgically to recover Cu and remove deleterious impurities from the smelting circuit (Chapter 21).

5.6. SUMMARY

Matte smelting is the most common way of smelting Cu–Fe–S concentrates. It entails heating, oxidizing (almost always with oxygen-enriched air), and fluxing the concentrate at high temperatures, 1250 °C. The products are:

(a) Molten Cu–Fe–S matte, 45–75% Cu, which is sent to oxidation converting to molten metallic copper
(b) Molten Fe silicate slag, which is treated to recover Cu and then sold or stockpiled; and
(c) SO_2-bearing offgas, which is cooled, cleaned, and sent to sulfuric acidmaking.

Matte smelting oxidizes most, but not all, of the Fe and S in its input concentrates. Total oxidation of Fe and S would produce molten Cu, but would also result in large Cu_2O losses in slag. The expense of reducing this Cu_2O and settling the resulting copper almost always overwhelms the advantage of direct-to-copper smelting.

The next four chapters describe industrial techniques for matte smelting and converting.

REFERENCES

Asaki, Z., Taniguchi, T., & Hayashi, M. (2001). Kinetics of the reactions in the smelting process of the Mitsubishi Process. *JOM, 53*(5), 25–27.
Caffery, G. A., Shook, A. A., Grace, J. R., Samarasekara, I. V., & Meadowcroft, T. R. (2000). Comparisons between sulfide flash smelting and coal combustion – with implications for the flash smelting of high-grade concentrate. *Metallurgical and Materials Transactions B, 31B*, 1005–1012.
Fahey, N. P., Swinbourne, D. R., Yan, S., & Osborne, J. M. (2004). The solubility of Cr_2O_3 in calcium ferrite slags at 1573K. *Metallurgical and Materials Transactions B, 35B*, 197–202.

González, O., Richards, J. F., & Rivera, J.de D. R. (2006). Measurement of flame speed in copper concentrate clouds. *Journal of the Chilean Chemical Society, 51*, 869–874.

Goto, M., & Hayashi, M. (1998). *The Mitsubishi Continuous Process — A description and detailed comparison between commercial practice and metallurgical theory*. Tokyo: Mitsubishi Materials Corporation.

Hejja, A. A., Eric, R. H., & Howat, D. D. (1994). Electrical conductivity, viscosity and liquidus temperature of slags in electric smelting of copper–nickel concentrates. In G. W. Warren (Ed.), *EPD congress 1994* (pp. 621–640). Warrendale, PA: TMS.

Henao, H. M., Yamaguchi, K., & Ueda, S. (2006). Distribution of precious metals (Au, Pt, Pd, Rh and Ru) between copper matte and iron-silicate slag at 1573K. In F. Kongoli & R. G. Reddy (Eds.), *Sohn international symposium, Vol. 1: Thermo and physicochemical principles* (pp. 723–729). Warrendale, PA: TMS.

Jahanshahi, S., & Sun, S. (2003). Some aspects of calcium ferrite slags. In F. Kongoli, K. Itagaki, C. Yamauchi & H. Y. Sohn (Eds.), *Yazawa international symposium, Vol. 1: Materials processing fundamentals and new technologies* (pp. 227–244). Warrendale, PA: TMS.

Jak, E., Zhao, B., & Hayes, P. (2000). Phase equilibria in the system $FeO-Fe_2O_3-Al_2O_3-CaO-SiO_2$, with applications to non-ferrous smelting slags. In S. Seetharaman & D. Sichen (Eds.), *6th int. conf. slags, fluxes, molten salts*. Stockholm: KTH. Paper 20.

Jorgensen, F. R. A., & Koh, P. T. L. (2001). Combustion in flash smelting furnaces. *JOM, 53*(5), 16–20.

Kondratiev, A., Jak, E., & Hayes, P. C. (2002). Predicting slag viscosities in metallurgical systems. *Journal of Metals, 54*(11), 41–45.

Kucharski, M., Ip, S. W., & Toguri, J. M. (1994). The surface tension and density of Cu_2S, FeS, Ni_3S_2 and their mixtures. *Canadian Metallurgical Quarterly, 33*, 197–203.

Li, H., & Rankin, J. W. (1994). Thermodynamics and phase relations of the $Fe-O-S-SiO_2$ (sat) system at 1200 °C and the effect of copper. *Metallurgical and Materials Transactions B, 25B*, 79–89.

Li, J., Huang, K., & Chen, X. (1989). Interfacial tension between slag and copper matte. *Acta Metallurgica Sinica (English Edition) Series B, 2*, 386–391.

Li, X., Mei, C., & Xiao, T. (2002). Numerical simulation analysis of Guixi copper flash smelting furnace. *Rare Metals, 21*, 260–265.

Li, X., Peng, S., Han, X., Mei, C., & Xiao, T. (2004). Influence of operation parameters on flash smelting furnace based on CFD. *Journal of University of Science and Technology Beijing, 11*, 115–119.

Muan, A. (1955). Phase equilibria in the system $FeO-Fe_2O_3-SiO_2$. *Transactions of AIME, 205*, 965–976.

Nakamura, T., & Toguri, J. M. (1991). Interfacial phenomena in copper smelting processes. In C. Díaz, C. Landolt, A. A. Luraschi & C. J. Newman (Eds.) *Copper 91–Cobre 91, Vol. IV: Pyrometallurgy of copper* (pp. 537–551). New York, NY: Pergamon Press.

Ramachandran, V., Díaz, C., Eltringham, T., Jiang, C. Y., Lehner, T., Mackey, P. J., Newman, C. J., & Tarasov, A. (2003). Primary copper production — a survey of operating world copper smelters. In C. Díaz, J. Kapusta & C. Newman (Eds.) *Copper 2003–Cobre 2003, Vol. IV, Book 1: Pyrometallurgy of copper, The Hermann Schwarze symposium on copper pyrometallurgy* (pp. 3–106). Montreal: CIM.

Roghani, G., Takeda, Y., & Itagaki, K. (2000). Phase equilibrium and minor element distribution between FeO_x-SiO_2-MgO-based slag and Cu_2S-FeS matte at 1573 K under high partial pressure of SO_2. *Metallurgical and Materials Transactions B, 31B*, 705–712.

Shimpo, R., Goto, S., Ogawa, O., & Asakura, I. (1986). A study on the equilibrium between copper matte and slag. *Canadian Metallurgical Quarterly, 25*, 113–121.

Steinhauser, J., Vartiainen, A., & Wuth, W. (1984). Volatilization and distribution of impurities in modern pyrometallurgical copper processing from complex concentrates. *Journal of Metals, 36*(1), 54–61.

Sundström, A. W., Eksteen, J. J., & Georgalli, G. A. (2008). A review of the physical properties of base metal mattes. *Journal of the South African Institute of Mining and Metallurgy, 108*, 431–448.

Tsukihashi, F., & Kimura, H. (2000). Phase diagram for the $CaO-SiO_2-FeO_x$ system at low oxygen partial pressure. In S. Seetharaman & D. Sichen (Eds.), *6th int. conf. slags, fluxes, molten salts*. KTH. Paper 22.

Utigard, T. A. (1994). Density of copper/nickel sulfide smelting and converting slags. *Scandinavian Journal of Metallurgy, 23*, 37–41.

Utigard, T. A., & Warczok, A. (1995). Density and viscosity of copper/nickel sulfide smelting and converting slags. In W. J. Chen, C. Díaz, A. Luraschi & P. J. Mackey (Eds.) *Copper 95—Cobre 95, Vol. IV: Pyrometallurgy of copper* (pp. 423—437). Montreal: CIM.

Vadász, P., Tomášek, K., & Havlik, M. (2001). Physical properties of $FeO-Fe_2O_3-SiO_2-CaO$ melt systems. *Archives of Metallurgy, 46*, 279—291.

Vartiainen, A. (1998). Viscosity of iron—silicate slags at copper smelting conditions. In J. A. Asteljoki & R. L. Stephens (Eds.), *Sulfide smelting '98* (pp. 363—371). Warrendale, PA: TMS.

Vartiainen, A., Kojo, I. V., & Acuña, R. C. (2003). Ferrous calcium slags in direct-to-blister flash smelting. In F. Kongoli, C. Yamauchi & H. Y. Sohn (Eds.), *Yazawa international symposium, Vol. 1: Materials processing fundamentals and new technologies* (pp. 277—290). Warrendale, PA: TMS.

Wood, J., Matusewicz, R., & Reuter, M. A. (2009). Recent developments at Ausmelt: C3 continuous converting. In *High temperature processing seminar: Final program and abstract book*. Swinburne University of Technology. http://www.swinburne.edu.au/engineering/htp/seminars/2009/Abstract%20Book.pdf.

Yazawa, A. (1956). Copper smelting. V. Mutual solution between matte and slag produced in the $Cu_2S-FeS-FeO-SiO_2$ system. *Journal of the Mining Institute of Japan, 72*, 305—311.

Yazawa, A., & Kameda, A. (1953). Copper smelting. I. Partial liquidus diagram for $FeS-FeO-SiO_2$ system. *Technology Reports of the Tohoku University, 16*, 40—58.

Zhang, J. (1997). Calculating models of mass action concentrations for mattes ($Cu_2S-FeS-SnS$) involving eutectic. *Acta Metallurgica Sinica (English Letters), 10*, 392—397.

Ziólek, B., & Bogacz, A. (1987). Electrical conductivity of liquid slags from the flash-smelting of copper concentrates. *Archives of Metallurgy, 32*, 631—643.

SUGGESTED READING

Mackey, P. J. (1982). The physical chemistry of copper smelting slags — a review. *Canadian Metallurgical Quarterly, 21*, 221—260.

Nakamura, T., & Toguri, J. M. (1991). Interfacial phenomena in copper smelting processes. In C. Díaz, C. Landolt, A. A. Luraschi & C. J. Newman (Eds.), *Copper 91—Cobre 91, Vol. IV: Pyrometallurgy of copper* (pp. 537—551). New York: Pergamon Press.

Sundström, A. W., Eksteen, J. J., & Georgalli, G. A. (2008). A review of the physical properties of base metal mattes. *Journal of the South African Institute of Mining and Metallurgy, 108*, 431—448.

Utigard, T. A., & Warczok, A. (1995). Density and viscosity of copper/nickel sulfide smelting and converting slags. In W. J. Chen, C. Díaz, A. Luraschi & P. J. Mackey (Eds.), *Copper 95—Cobre 95 proceedings of the third international conference, Vol. IV Pyrometallurgy of copper* (pp. 423—437). Montreal: CIM.

Chapter 6

Flash Smelting

Flash smelting accounts for around half of all Cu matte smelting. It entails blowing oxygen, air, dried Cu−Fe−S concentrate, silica flux, and recycle materials into a 1250 °C hearth furnace. Once in the hot furnace, the sulfide mineral particles of the concentrate react rapidly with the O_2 of the blast. This results in controlled oxidation of the concentrate's Fe and S, a large evolution of heat and melting of the solids.

The process is continuous. When extensive oxygen enrichment of the blast is practiced, it is nearly autothermal. It is perfectly matched to smelting fine particulate concentrates (~100 μm) produced by froth flotation.

The products of flash smelting are:

(a) Molten Cu−Fe−S matte, ~65% Cu, considerably richer in Cu than the input concentrate
(b) Molten iron-silicate slag containing 1−2% Cu
(c) Hot dust-laden offgas containing 30−70 vol.-% SO_2.

Three flash furnaces produce molten copper directly from concentrate (Chapter 10). Two other Outotec flash furnaces produce molten copper from solidified/ground matte. This is flash converting (Chapter 8).

The goals of flash smelting are to produce:

(a) Molten matte at constant composition and temperature for feeding to converters
(b) Slag which, when treated for Cu recovery, contains only a tiny fraction of the Cu input to the flash furnace
(c) Offgas strong enough in SO_2 for its efficient capture as sulfuric acid.

There are two types of flash smelting: the Outotec process (~30 furnaces in operation) and the Inco process (four furnaces in operation). The Outotec process was formerly known as the Outokumpu process.

6.1. OUTOTEC FLASH FURNACE

Figure 6.1 shows a typical 2010-design Outotec flash furnace. It is 26 m long, 8 m wide, and 2 m high (all dimensions inside the refractories). It has a 7 m diameter, 8 m high reaction shaft and a 5 m diameter, 10 m high offgas uptake. It has one concentrate burner and is designed to smelt up to 4500 tonnes of concentrate per day. It has four matte tapholes and four slag tapholes.

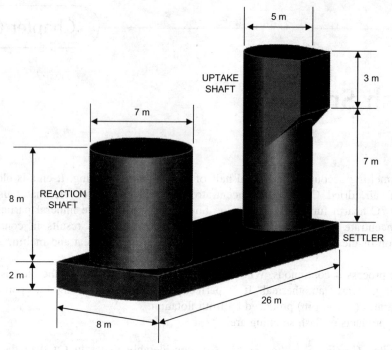

FIGURE 6.1 Schematic of a year 2010 Outotec flash furnace. This furnace is designed to smelt up to 4500 tonnes of concentrate per day. A concentrate burner is shown in Fig. 6.2. It sits atop the reaction shaft.

Outotec flash furnaces vary considerably in size and shape (Table 6.1). However, they all have five main features:

(a) Concentrate burners (usually one, but a few have four), which combine dry particulate feed with O_2-bearing blast and blow them downward into the furnace
(b) A reaction shaft where most of the reaction between O_2 and Cu−Fe−S feed particles takes place
(c) A settler where molten matte and slag droplets collect and form separate layers
(d) Water-cooled copper block tapholes for removing molten matte and slag
(e) An uptake for removing hot SO_2-bearing offgas.

6.1.1. Construction Details (Fagerlund, Lindgren, & Jåfs, 2010)

The interior of an Outotec flash furnace consists of high-purity direct-bonded magnesia−chrome bricks. The bricks are backed by water-cooled copper cooling jackets on the walls and by sheet steel elsewhere. Reaction shaft and uptake refractory is backed by water-cooled copper cooling jackets or by sheet steel, cooled with water on the outside.

The furnace rests on a 2-cm thick steel plate on steel-reinforced concrete pillars. The bottom of the furnace is air cooled by forced convection (Janney, George-Kennedy, & Burton, 2007). Slag line bricks may have eroded, but the furnace can usually continue to operate without them. This is because magnetite-rich slag freezes on cool regions of the furnace walls.

TABLE 6.1a Dimensions and Production Details of Three Outotec Flash Furnaces

Smelter	PASAR Leyte, Philippines	Arubis Hamburg, Germany	Pan Pacific Copper Saganoseki, Japan
Startup date	1983	1972	1973
Size, inside brick, m			
Hearth: w × l × h	7.5 × 20.2 × 2.6	6 × 20 × 3	6.8 × 20.1 × 2.2
Reaction shaft			
Diameter	6.5	6	6.2
Height above settler roof	6.5	7.5	5.9
Gas uptake			
Diameter	2.5	4 × 8	3.1
Height above roof	9.4	10	6.3
Slag layer thickness	0.3–0.6	0.7	0.3
Matte layer thickness	0.65–0.85	0.2–0.5	0.8
Active slag tapholes	2	2	1
Active matte tapholes	4	4	6
Concentrate burners	1	1	1
Feed details tonnes/day			
New concentrate (dry)	2000 (23–29% Cu)	3100 (29% Cu)	3581 (28.4% Cu)
Silica flux	100	350–450	543
Oxygen	440		
Recycle flash furnace dust	95–120	230	215
Converter dust	20–30	15	40
Slag concentrate	75	No	
Reverts		150 (ladle sculls, slimes, various dusts)	107 (converter dust, leach plant residue, gypsum)
Other	36 t/d coal	430 molten converter slag	70 purchased scrap

(Continued)

TABLE 6.1a Dimensions and Production Details of Three Outotec Flash Furnaces—cont'd

Smelter	PASAR Leyte, Philippines	Arubis Hamburg, Germany	Pan Pacific Copper Saganoseki, Japan
Blast details			
Temperature, °C	200	Ambient	Ambient
Vol.-% O_2	45–52	50–65	70–80
Flowrate, thousand Nm^3/h	36	45	27–35
Production details			
Matte, tonnes/day	1080 (54–56%Cu)	1450 (63% Cu)	1632 (64.8% Cu)
Slag, tonnes/day	800–900	2000	2220
Slag, mass% Cu	0.5–0.75	1.5	
Slag mass% SiO_2/mass% Fe	0.87–1	0.82	0.93
Cu recovery, flash slag	Electrodes and coal in flash furnace	Electric furnace	Electric furnace
Cu recovery, converter slag	Solidify/flotation	Recycle to flash furnace	Solidify/flotation
Offgas, thousand Nm^3/h	36.6	50–68	40–45
Vol.-% SO_2, leaving furnace	28–30	30–45	50–60 (calculated)
Dust production, tonnes/day	48–60 flash furnace 28–30 dryer dust	230	215
Matte/slag/offgas temperature, °C	1220/1240/750	1220/1230/1350	1242/1245/1250
Fuel inputs, kg/h			
Hydrocarbon fuel burnt in reaction shaft	0–0.2 m^3/h bunker C oil, coal added to solid feed	None	70 fine coal with feed
Hydrocarbon fuel in settler burners	None, but 1.2–1.5 MW electric power	None	None

TABLE 6.1b Dimensions and Production Details of Three Outotec Flash Furnaces

Smelter	Hibi Kyodo Smelting Co. Tamano, Japan	Sumitomo Toyo, Japan	Rio Tinto Kennecott Utah Copper Magna, UT, USA
Startup date	1972	1971	1995
Size, inside brick, m			
Hearth: w × l × h	7.0 × 19.8 × 2.5	6.7 × 19.9 × 3.3	7.7 × 23.9 × 1.9
Reaction shaft			
Diameter	6.2	6.1	7
Height above settler roof	6.7	6.4	8.1
Gas uptake			
Diameter	2.5	4	5.0
Height above roof	8.7	7.5	11.9
Slag layer thickness	0.4	0.3–0.5	0.4
Matte layer thickness	0.7	0.6–0.7	0.5
Active slag tapholes	2	2	3
Active matte tapholes	4	4	4
Concentrate burners	1	1	1
Feed details tonnes/day			
New concentrate (dry)	2241 (29.2% Cu)	3830 (28.8% Cu)	4200 (25% Cu)
Silica flux	220	564	528
Oxygen		1036	720–1100
Recycle flash furnace dust	178	262	307
Converter dust	20	36	75
Slag concentrate	70	0	240–500
Reverts	70 (gypsum, tankhouse sludge)	64	
Other	20 purchased scrap	43	
Blast details			
Temperature, °C	210	200	Ambient
Vol.-% O_2	66–69	77	82–88
Flowrate, thousand Nm^3/h	25–29	38	27–34

(Continued)

TABLE 6.1b Dimensions and Production Details of Three Outotec Flash Furnaces—cont'd

Smelter	Hibi Kyodo Smelting Co. Tamano, Japan	Sumitomo Toyo, Japan	Rio Tinto Kennecott Utah Copper Magna, UT, USA
Production details			
Matte, tonnes/day	1129 (63.5% Cu)	1895 (64% Cu)	1800 (66.5–74.5% Cu)
Slag, tonnes/day	1326	2477	2100–3100
Slag, mass% Cu	0.74	1	0.5–4
Slag mass% SiO_2/mass% Fe	0.85	1	0.64
Cu recovery, flash slag	Electric furnace	Electric furnace with coal	
Cu recovery, converter slag	Solidify/flotation	Solidify/mill plant/flash furnace	
Offgas, thousand Nm^3/h	27	57	43–60
Vol.-% SO_2, leaving furnace	50 (calculated)	38	30–40
Dust production, tonnes/day	178	210	
Matte/slag/offgas temperature, °C	1222/1254/1230	1220/1250/1300	1260/1320/1370
Fuel inputs, kg/h			
Hydrocarbon fuel burnt in reaction shaft	1200 fine coal with feed	404 oil (including settler) 147 pulverized coal	Occasionally
Hydrocarbon fuel in settler burners	None		

6.1.2. Cooling Jackets

Recent design cooling jackets are solid copper with Monel 400 (Ni–30% Cu) alloy tube embedded inside. The tube is bent into many turns to maximize heat transfer from the solid copper to water flowing in the Monel tube (McKenna, Voermann, Veenstra, & Newman, 2010). The hot face of the cooling jacket is cast in a waffle shape. This provides a jagged face for refractory retention and magnetite-slag deposition. Jackets are typically 0.75 m × 0.75 m × 0.1 m thick with 0.03 m diameter, 0.004 m wall Monel tube.

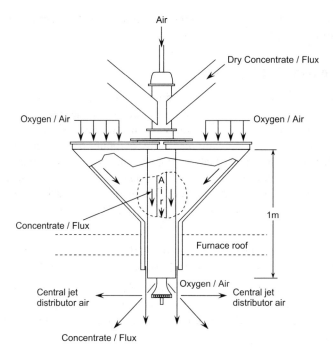

FIGURE 6.2 Central jet distributor Outotec concentrate burner. The main goal of the burner is to create a uniform concentrate-blast suspension 360° around the burner. This type of burner can smelt up to 200 tonnes of feed per hour. Its feed consists mainly of dry Cu−Fe−S concentrate, ~100 μm, silica flux, ~1 mm, recycle dust and recycle crushed reverts, ~1 mm.

6.1.3. Concentrate Burner (Fig. 6.2)

Dry concentrate and O_2-rich blast are combined in the furnace reaction shaft by blowing them through a concentrate burner. Dry flux, recycle dust, and crushed reverts (matte and slag inadvertently frozen during transport around the smelter) are also added through the burner. As of 2010, a concentrate burner consisted of:

(a) An annulus through which O_2-rich blast is blown into the reaction shaft
(b) A central pipe through which concentrate falls into the reaction shaft
(c) A distributor cone at the burner tip, which blows air horizontally through the descending solid feed.

Special attention is paid to uniform distribution of blast and solid feed throughout the reaction shaft. It is achieved by introducing blast and solids vertically and uniformly into quadrants around the burner and by blowing the solids outwards with central jet distributor air.

A variation of this burner design is the Sumitomo-type concentrate burner, which uses an oxygen−fuel burner located at the central jet distributor to accelerate the ignition of the concentrate particles (Nagai, Kawanaka, Yamamoto, & Sasai, 2010).

6.1.4. Supplementary Hydrocarbon Fuel Burners

All Outotec flash furnaces are equipped with hydrocarbon fuel burners atop the reaction shaft and through the settler walls and roof. Shaft-top burners keep the process in thermal balance. Settler burners eliminate cool zones in the furnace. They are also used to adjust slag temperature.

6.1.5. Matte and Slag Tapholes

Molten matte and slag are tapped through single-hole water-cooled copper chill blocks embedded in the furnace walls. The holes are typically 60–80 mm diameter. They are plugged with moist fireclay, which is solidified by the heat of the furnace when the clay is pushed into the hole. They are opened by chipping out the clay and by melting it out with steel oxygen lances. Matte is tapped via copper or refractory-lined steel launders into cast steel ladles for transport to converting. Matte is also granulated to ~0.5 mm granules in a water torrent where Outotec flash converting is used instead of Peirce–Smith converting (Chapter 8). Slag is tapped down water-cooled copper launders into:

(a) An electric settling furnace for Cu settling and recovery, or
(b) Ladles for truck haulage to Cu recovery by slow cooling/grinding/flotation.

Both withdrawals are only partial. Reservoirs of matte and slag (~0.5 m deep each) are maintained in the furnace. Tapping of matte is continuously rotated around its tapholes. This washes out solid buildups on the furnace floor by providing matte flow over the entire hearth.

6.2. PERIPHERAL EQUIPMENT

The Outotec flash furnace is surrounded by:

(a) Concentrate blending equipment
(b) Solids feed dryer
(c) Flash furnace feed bins and feed system
(d) Oxygen plant
(e) Blast preheater (optional)
(f) Heat recovery boiler (Chapter 12)
(g) Dust recovery and recycle system
(h) Gas cleaning system (Chapter 12)
(i) Sulfuric acid plant (Chapter 12)
(j) Secondary gas collection and treatment system (Chapter 12)
(k) Cu-from-slag recovery system (Chapter 11).

6.2.1. Concentrate Blending System

Most flash furnaces smelt several concentrates plus small amounts of miscellaneous materials, such as precipitate copper. They also smelt recycle dusts, sludges, slag flotation concentrate, and reverts. These materials are blended to give constant composition feed to the flash furnace. Constant composition feed is the surest way to ensure smooth flash furnace operation and continuous attainment of target compositions and temperatures.

Two techniques are used:

(a) Bin-onto-belt blending by which individual feed materials are dropped from holding bins at controlled rates onto a moving conveyor belt

(b) Bedding, where layers of individual feed materials are placed on long (occasionally circular) shaped piles, then reclaimed as vertical slices of blend (FLSmidth, 2010).

The blended feed is sent to a dryer. Flux may be included in the blending or added just before the dryer.

6.2.2. Solids Feed Dryer

Concentrate and flux are always dried prior to flash smelting to ensure even flow through the concentrate burner. Steam and rotary dryers are used (Chen & Mansikkaviita, 2006; Hoshi, Toda, Motomura, Takahashi, & Hirai, 2010; Talja et al., 2010). The water contents of moist and dry feed are typically 6—14% and 0.2% H_2O.

Rotary dryers evaporate water by passing hot gas from natural gas or oil combustion through the moist feed. The temperature of the drying gas is kept below ~500 °C (by adding nitrogen, recycle combustion gas, or air) to avoid spontaneous oxidation of the concentrate. Some smelters recycle anode furnace offgas (Chapter 12) or warm air from acid plant SO_3 coolers (Chapter 12) to their rotary dryers to decrease fuel consumption. In addition, steam from the flash furnace heat recovery boiler can also be used to indirectly pre-heat rotary kiln air to decrease fuel consumption (Hoshi et al., 2010).

Steam drying is being widely adopted in new and existing Outotec flash smelters (Talja et al., 2010). Steam dryers rotate hot, steam-heated stainless steel coils through the moist feed. Modern designs have a drum that rotates in the same direction as the steam coils to minimize tube wear and improve water evaporation efficiency. Dryer capacities are as high as 200 t/h wet feed.

Steam drying has the advantages of:

(a) Efficient use of flash furnace heat recovery boiler steam
(b) Little SO_2, dust, and offgas evolution because hydrocarbon combustion is not used
(c) Low risk of concentrate ignition because steam drying is done at a lower temperature (~200 °C) than combustion-gas drying (~500 °C).
(d) Approximately 14% less energy consumption than a similar-capacity rotary dryer (Chen & Mansikkaviita, 2006)

6.2.3. Bin and Feed System

Dried feed is blown up from the dryer by a pneumatic lift system (Walker, Coleman, & Money, 2010). It is caught in acrylic bags and dropped into bins above the flash furnace reaction shaft. It is fed from these bins onto drag or screw conveyors and sometimes into a loss-in-weight feeder. From the loss-in-weight feeder it flows on an air slide to the concentrate burner (Jones et al., 2007; Reed, Jones, Snowdon, & Coleman, 2010; Outotec, 2010). The rate of feeding is adjusted by varying the speed of the conveyors below the bins (Contreras, Bonaño, Fernández-Gil, & Palacios, 2003). The use of the loss-in-weight feeder provides an accurate measurement of the mass flow of feed to the furnace. Older feed systems were based on volumetric flow.

Bin design is critical for controlled feeding of the flash furnace. Fine dry flash furnace feed tends to hang up on the bin walls or flood into the concentrate burner. This is avoided by mass flow bins (Dudley, 2009; Marinelli & Carson, 1992) that are steep enough and smooth enough to give even flow throughout the bin.

6.2.4. Oxygen Plant

The principal oxygen plant in an Outotec flash smelter is usually a cryogenic liquefaction/distillation unit producing 200—1100 tonnes oxygen per day. It delivers 92—99.95 vol.-% O_2 industrial oxygen gas (0.8—2 bar gauge) to the flash furnace (Deneys, Dray, Mahoney, & Barry, 2007).

Some smelters also use molecular sieve oxygen plants (vacuum pressure swing absorption (VPSA) or pressure swing absorption (PSA)) to supplement their liquefaction/distillation oxygen. Molecular sieve plants are small (<200 tonnes oxygen/day) units that produce oxygen with a purity of 90—93 vol.-% O_2: Ar (5%) and N_2 are the main impurities. They are suitable for incremental additions to the output of the main oxygen plant.

Oxygen-enriched blast is prepared by mixing industrial oxygen and air as they flow to the concentrate burner. The oxygen is added through a diffuser (holed pipe) protruding into the air duct. The diffuser is located about six duct diameters ahead of the concentrate burner to ensure good mixing.

The rates at which oxygen and air flow into the concentrate burner are important flash furnace control parameters. They are measured by orifice or mass flow flowmeters and are adjusted by butterfly valves.

6.2.5. Blast Heater (optional)

A few Outotec flash furnaces use heated blast. The blast is typically heated to 100—450 °C using hydrocarbon-fired shell-and-tube heat exchangers. Hot blast ensures rapid Cu—Fe—S concentrate ignition in the flash furnace. It also provides energy for smelting. Modern, highly oxygen-enriched flash furnaces use ambient ($\sim$30 °C) blast. Concentrate ignition is rapid with this blast at all temperatures.

6.2.6. Heat Recovery Boiler (Köster, 2010)

Offgas leaves an Outotec flash furnace at about 1300 °C. Its sensible heat is recovered as steam in a horizontal heat recovery boiler (Chapter 12).

6.2.7. Dust Recovery and Recycle System

Outotec flash furnace offgases typically contain 0.1—0.2 kg of dust per Nm^3 of offgas. The dust consists of mainly oxides (e.g. Cu_2O, Fe_3O_4). It becomes sulfated by SO_2 and O_2 in the heat recovery boiler and electrostatic precipitators. This sulfated dust is collected and recycled to the flash furnace for Cu recovery. In many flash smelters, air or industrial oxygen are deliberately fed into the offgas to sulfate the dust (Davenport, Jones, King, & Partelpoeg, 2001). Controlling the oxidation and sulfation behavior of the dust minimizes accretion formation in the uptake and heat recovery boiler (Miettinen, Ahokainen, & Eklund, 2010).

About 70% of the dust leaving the uptake drops out in the heat recovery boiler. The remainder is caught in electrostatic precipitators (Ryan, Smith, Corsi, & Whiteus, 1999) where the particles are charged in a high voltage electrical field, caught on a charged wire or plate and periodically collected as dust clumps by rapping the wires and plates. Electrostatic precipitator exit gas contains ~0.1 g of dust per Nm^3 of gas (Conde, Taylor, & Sarma, 1999; Parker, 1997).

The collected dust contains approximately 25% Cu. It is almost always recycled to the flash furnace for Cu recovery. It is removed from the boilers and precipitators by drag-and-screw conveyors, transported pneumatically to a dust bin above the flash furnace, and combined with the dried feed just before it enters the concentrate burner.

6.3. FLASH FURNACE OPERATION

Table 6.1 indicates that Outotec flash furnaces:

(a) Smelt up to 4200 tonnes per day of new concentrate
(b) Typically produce ~65% Cu matte (depending on the matte converting process)
(c) Use 45–88% O_2 blast, sometimes slightly heated
(d) Burn hydrocarbon fuel to some extent mostly in the reaction shaft.

This section describes how the furnaces operate.

6.3.1. Startup and Shutdown

Operation of an Outotec flash furnace is begun by heating the furnace to its operating temperature with hydrocarbon burners or hot air blowers (Severin, 1998). The heating is carried out gently and evenly over a week or two to prevent uneven expansion and spalling of the refractories. Adjustable springs attached to fixed-position I-beams keep the walls and hearth under constant pressure during the heating. During initial heatup, paper is inserted between newly laid hearth bricks to burn out and compensate for brick expansion. Concentrate feeding is begun as soon as the furnace is at its target temperature. Full production is attained in a day or so. Shutdown consists of:

(a) Overheating the furnace for 7–10 days to melt solid buildups
(b) Starting hydrocarbon burners
(c) Stopping the concentrate burner
(d) Draining the furnace with hydrocarbon burners on
(e) Turning off the hydrocarbon burners
(f) Turning off the cooling water
(g) Allowing the furnace to cool at its natural rate.

6.3.2. Steady-state Operation

Steady-state operation of a flash furnace entails:

1. Feeding solids and blast at a constant rate
2. Drawing SO_2-rich gas from the gas uptake at a constant rate
3. Tapping matte from the furnace on a scheduled basis or as needed by the converter(s)
4. Tapping slag from the furnace on a scheduled basis or when it reaches a prescribed level in the furnace.

The next section describes how steady-state operation is attained.

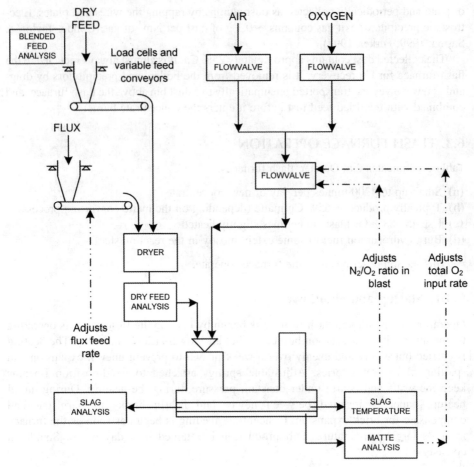

FIGURE 6.3 Example control system for Outotec flash furnace. The three loops control slag temperature, slag composition, and matte composition. Slag temperature may also be controlled by adjusting reaction shaft hydrocarbon burner combustion rate. It is fine-tuned by adjusting settler burner combustion rates. Matte grades ± 1.5% Cu and temperatures ± 20 °C are obtained.

6.4. CONTROL (FIG. 6.3)

The Outotec flash furnace operator must smelt concentrate at a steady, specified rate (Fig. 6.3) while:

(a) Producing matte of specified Cu grade
(b) Producing slag of specified SiO_2 content
(c) Producing slag at specified temperature
(d) Maintaining a protective coating of magnetite-rich slag on the furnace interior.

6.4.1. Concentrate Throughput Rate and Matte Grade Controls

Basic Outotec flash furnace strategy is to charge dried concentrate mix to the furnace at a prescribed rate and to base all other controls on this rate. Having chosen the concentrate feed rate, the flash furnace operator must next select the grade (% Cu) of the product matte, i.e. the extent of Fe and S oxidation. It is selected as a compromise between:

1. Maximizing SO_2 evolution in the flash furnace (where it is captured efficiently) and:
2. Keeping enough Fe and S in the matte so that subsequent converting can operate autothermally while melting the required amount of Cu scrap and smelter recycle materials.

Physically, matte grade is set by adjusting the ratio of oxygen input rate in the blast to concentrate feed rate until the target matte composition is obtained. A large ratio results in extensive Fe and S oxidation and high-grade (% Cu) matte. A small ratio gives the opposite. Physically, the ratio is controlled by adjusting the rates at which air and oxygen enter the furnace, after setting a constant concentrate feed rate.

6.4.2. Slag Composition Control

The iron oxide formed by concentrate oxidation is fluxed with SiO_2 to form liquid slag. The amount of SiO_2 is based upon the slag having a low solubility for Cu and sufficient fluidity for easy tapping and a clean matte-slag separation. An SiO_2/Fe mass ratio of 0.7–1.0 is used. It is controlled by adjusting the rate at which flux is fed to the solids feed dryer.

6.4.3. Temperature Control

Matte and slag temperatures are measured as matte and slag flow from the furnace. Disposable thermocouple probes and optical pyrometers are used. Matte and slag temperatures are controlled by adjusting (a) the rate at which N_2 coolant enters the furnace (mainly in air) or (b) hydrocarbon burner combustion rates. Slag temperature is adjusted somewhat independently of matte temperature by adjusting settler hydrocarbon burner combustion rates.

Matte and slag temperatures are typically 1220 °C and 1250 °C, respectively. They are chosen for rapid matte/slag separation and easy tapping. They are also high enough to keep matte and slag molten during transport to their destinations. Excessive temperatures are avoided to minimize refractory and cooling jacket wear.

6.4.4. Reaction Shaft and Hearth Control (Davenport et al., 2001)

Long flash furnace campaign lives require that magnetite-rich slag be deposited in a controlled manner on the furnace walls and hearth. Magnetite slag deposition is encouraged by (Chapter 6):

(a) Highly oxidizing conditions in the furnace
(b) Low operating temperature
(c) Low SiO_2 concentration in slag

It is discouraged by reversing these conditions and by adding coke or coal to the furnace.

Campaigns up to 11 years in duration have been achieved for flash smelting furnaces (Janney et al., 2007).

6.5. IMPURITY BEHAVIOR

Flash furnace concentrates inevitably contain impurities from their original ore. They must be separated from Cu during smelting and refining. Table 6.2 shows that this is partially accomplished during flash smelting, as portions of the impurities report to the slag and offgas while almost all the Cu reports to the matte. Important exceptions to this are gold, silver, and platinum group metals. They accompany Cu through to electrorefining where they are recovered as byproducts (Chapter 21). Most Ni also follows Cu.

Industrial impurity distribution is complicated by recycle of:

(a) Flash furnace and converter dusts
(b) Flash furnace slag concentrate
(c) Converter slag concentrate and (occasionally) molten converter slag
(d) Solid reverts from around the smelter
(e) Acid plant sludges

Nevertheless, Table 6.2 provides guidance as to how impurities distribute themselves during flash smelting.

6.5.1. Non-recycle of Impurities in Dust

Impurities are also found in flash furnace dust (Table 6.2). This dust is usually recycled to the flash furnace for Cu recovery, so it is not usually an escape route for impurities.

TABLE 6.2 Distribution of Elements in Outotec Flash Smelting (Davenport et al., 2001)

Element	% to matte	% to slag	% to offgas[a]
Cu	97	2	1
Ag	90–95	2–5	3–8
Au	95	2	3
As	15–40	5–25	35–80
Bi	30–75	5–30	15–65
Cd	20–40	5–35	25–60
Co	45–55	45–55	0–5
Ni	70–80	20–25	0–5
Pb	45–80	15–20	5–40
Sb	60–70	5–35	5–25
Se	85	5–15	0–5
Te	60–80	10–30	0–10
Zn	30–50	50–60	5–15

[a] Collected as precipitated solids during gas cleaning.

However, some flash smelters, including Chuquicamata (EcoMetales, 2010), Kennecott (Gabb, Howe, Purdie, & Woerner, 1995; Kaur, Nexhip, Wilson, & George-Kennedy, 2010), Saganoseki (Larouche, 2001, pp. 118–119) and Kosaka (Maeda, Inoue, Hoshikawa, & Shirasawa, 1998), recover Cu from some of their dust hydrometallurgically rather than by recycle to the flash furnace (Chapter 21). This allows the impurities to leave the smelter in leach plant residues. It is particularly effective for removing As, Bi, and Cd from the smelter. Lead is removed to a lesser extent.

6.5.2. Other Industrial Methods of Controlling Impurities

Flash smelters that treat several concentrates blend their high- and low-impurity concentrates to ensure that impurity levels in their product anodes are low enough for efficient electrorefining. Flash smelters that are dedicated to treating high-impurity concentrates control anode impurity levels by (a) minimizing dust recycle, or (b) modifying converting and fire refining to increase impurity removal (Larouche, Harris, & Wraith, 2003).

6.6. OUTOTEC FLASH SMELTING RECENT DEVELOPMENTS AND FUTURE TRENDS

Outotec flash furnace smelting capacity and campaign life has increased in recent years through enhancements in the design of the feed system, concentrate burner, and furnace cooling elements. Advances in process control and dust sulfation techniques have also contributed. The future development foreseen by Outotec is use of its flash furnace for direct-to-copper smelting (Chapter 10) of lower grade concentrates and further use of continuous flash converting (Chapter 8). Both have the advantages of improved SO_2 capture and an SO_2-free workplace because they eliminate batch Peirce–Smith converting (Ahokainen & Eklund, 2007; Kojo & Lahtinen, 2007).

6.7. INCO FLASH SMELTING

Inco flash smelting blows industrial oxygen, dried Cu–Fe–S concentrate, SiO_2 flux, and recycle materials horizontally into a hot ($\sim 1250\ °C$) furnace. Once in the furnace, the oxygen reacts with the concentrate to generate:

(a) Molten matte, 55–60% Cu
(b) Molten slag, 1–2% Cu
(c) Offgas, 60–75 vol.-% SO_2.

The matte is tapped into ladles and sent to converting. The slag is tapped into ladles and sent to stockpile, with or without Cu-from-slag removal (Chapter 11). The offgas is water-quenched, cleaned of dust, and sent to a sulfuric acid plant. The Inco flash furnace is also used to recover Cu from molten recycle converter slag. The slag is poured into the furnace via a steel chute and water-cooled door (Fig. 6.4).

At the start of 2010 there were four Inco flash furnaces in operation: at Almalyk, Uzbekistan (Ushakov et al. 1975); Hayden, Arizona (Marczeski & Aldrich, 1986), and Sudbury, Ontario (two furnaces; Carr, Humphris, & Longo, 1997; Humphris, Liu, & Javor, 1997; Molino et al., 1997; Zanini, Wong, & Cooke, 2009).

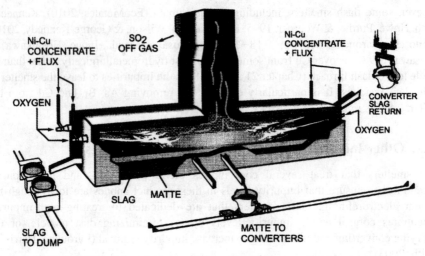

FIGURE 6.4 Inco flash furnace at Sudbury, Ontario (Queneau & Marcuson, 1996).

The Almalyk and Hayden furnaces smelt Cu—Fe—S concentrates. The Sudbury furnaces smelt Ni—Cu—Co—Fe—S concentrates to produce ~45% Ni + Cu + Co matte and ~1% Ni + Cu + Co slag.

6.7.1. Furnace Details

The Inco flash furnace is made of high-quality MgO and MgO—Cr_2O_3 brick (Fig. 6.4). Its main components are:

(a) Concentrate burners, two at each end of the furnace
(b) End- and sidewall water-cooled copper cooling jackets
(c) A central offgas uptake
(d) Sidewall tapholes for removing matte
(e) An endwall taphole for removing slag
(f) An endwall chute for charging molten converter slag

6.7.2. Concentrate Burner

Inco concentrate burners are typically ~0.25 m diameter, 1 m long, 1 cm thick stainless steel pipes, water-cooled with an internal ceramic sleeve. They are fed with industrial oxygen, blown in horizontally, and dry feed, dropped in from above through an angled tube (Fig. 6.5) or flexible metal hose.

The diameter of the burner barrel gives an oxygen/feed entry velocity of ~40 m/s. This velocity creates a concentrate/oxygen flame that reaches the central uptake. The burners are typically angled 7° down and 7° in so that the flame plays on the slag surface rather than on the roof and walls.

6.7.3. Water Cooling

Inco furnace side and endwalls are fitted with water-cooled copper fingers, plates, and shelves to maintain the integrity of the furnace structure (McKenna et al., 2010). As in

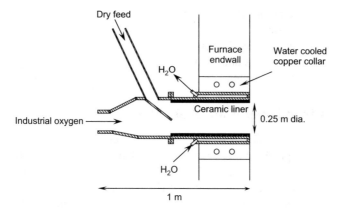

FIGURE 6.5 Details of Inco flash furnace concentrate burner.

the Outotec furnace, water cooling causes magnetite-rich slag to deposit on the furnace walls. This protects the bricks and water-cooled copper, extending furnace life. An Inco furnace can be operated for five or six years without reconstruction. Repairs are done annually and while the smelter is stopped for non-furnace reasons, such as acid plant maintenance. The end of a furnace campaign is usually caused by excessive refractory erosion, especially at the slag line.

6.7.4. Matte and Slag Tapholes

The Inco furnace is equipped with four sidewall matte tapholes and an endwall slag taphole. Each consists of a single-hole refractory block fitted in a water-cooled copper plate. Tapping and plugging are similar to Outotec flash furnace tapping and plugging (Section 6.1.5).

Reservoirs of slag (~ 0.5 m deep) and matte (0.2–1.0 m deep) are always kept in the furnace. The matte level is kept well below the slag taphole. This avoids accidental tapping of matte with slag. Matte tapping is rotated continuously between the four tapholes to keep matte flowing across the entire floor. This minimizes localized magnetite buildups on the furnace floor.

6.7.5. Gas Uptake

The central offgas uptake is brick or castable refractory backed by copper cooling jackets or stainless steel. A small amount of oxygen is injected into the uptake. It burns the elemental sulfur that is present in Inco furnace offgas. This prevents the sulfur from precipitating in downstream cooling and dust collection equipment.

6.7.6. Auxiliary Equipment

The auxiliary equipment required for Inco flash smelting is:

(a) Oxygen plant (Section 6.2.4)
(b) Concentrate blending system (Section 6.2.1)
(c) Solids feed dryer

(d) Four concentrate burner feed systems
(e) Offgas cooling system
(f) Dust recovery and recycle system
(g) Secondary gas collection and treatment system (Chapter 12)
(h) Sulfuric acid plant (Chapter 12).

6.7.7. Solids Feed Dryer (Carr et al., 1997)

Inco furnace solid feed (concentrate, flux, reverts, recycle sludges) is dried to give an even flow of solids through the concentrate burners. This is essential to create a uniform oxygen-concentrate suspension in the furnace. Natural gas-fired fluid bed dryers are used (Liu, Vahed, Yotingco, Macnamara, & Warner, 2007). The dryer feed is typically 10% H_2O; the dried product, ~0.2% H_2O. The fluidization gas is combustion gas plus air at ~330 °C. A fixed bed of 1—10 mm gravel distributes the fluidization gas across the bed.

The fluidized dried feed particles and their surrounding gas are drawn at ~90 °C through acrylic or aramid fabric tubes stretched on wire frameworks. About 99.99% of the concentrate is caught on the outside of the fabric. It is removed by reverse nitrogen or air pulsation and dropped into a large storage bin. From there it is sent by conveyors or air slides to four dry charge bins above the furnace, one for each concentrate burner.

6.7.8. Concentrate Burner Feed System

The solid charge to the Inco flash furnace consists mainly of dried feed blend from the fluid bed dryer. It drops from its dry charge bins onto drag or screw conveyors, which then drop it into the concentrate burners. The feed system may also include separate touch-up flux, touch-up revert, and converter dust bins. Material is drawn from these bins as needed to control flash furnace product temperature and slag composition. The rate of withdrawal from each bin is controlled by adjusting the speed of the drag or screw conveyor beneath the bin.

6.7.9. Offgas Cooling and Dust Recovery Systems (Humphris et al., 1997; Liu, Baird, Kenny, Macnamara, & Vahed, 2006)

Inco offgas is cooled and cleaned of dust in (a) a water-spray evaporation cooler where the offgas is cooled from ~1230 °C to 80 °C and where 90% of the entrained dust is removed as sludge, or (b) scrubbers and wet electrostatic precipitators. The equipment is stainless steel to minimize corrosion. The offgas (60—75 vol.-% SO_2) is pulled through the equipment by fans, which push the gas onwards to a sulfuric acid plant for SO_2 treatment.

Solids from the cooler and dust removal equipment contain ~35% Cu. The Cu is recovered by neutralizing and dewatering the sludge then recycling it through the concentrate dryer and flash furnace.

6.8. INCO FLASH FURNACE SUMMARY

The Inco flash furnace uses industrial oxygen (no air) to smelt Cu—Fe—S and Ni—Cu—Co—Fe—S concentrates. It produces high Cu and high Ni—Cu—Co mattes. It introduces dry feed and industrial oxygen through four horizontal burners and removes

SO$_2$ offgas through a central gas uptake. The offgas is water-quenched and sent to a sulfuric acid plant to capture its SO$_2$.

Very little nitrogen enters the Inco furnace, so its blast and offgas handling systems are small. Also, the strong offgas is ideal for SO$_2$ capture.

Inco flash furnace slag can contain less than 1% Cu, which can be discarded without Cu-recovery treatment. This gives it a cost advantage over most other modern smelting techniques. If desired, converter slag can also be recycled through the furnace for Cu recovery. However, this procedure upsets an otherwise steady process.

6.9. INCO VS. OUTOTEC FLASH SMELTING

There are many more Outotec flash furnaces than Inco flash furnaces. Outotec advantages include:

(a) Higher concentrate throughput rates (double that of the Inco flash furnace)
(b) Single concentrate burner in place of four Inco burners
(c) Recovery of offgas heat in a heat recovery boiler as useful steam
(d) Dry dust recycle
(e) Engineering and operational support

6.10. SUMMARY

Outotec flash smelting accounts for around half of all Cu matte smelting. It is also used in three locations for direct-to-copper smelting and in two locations for continuous converting.

It blows a well-dispersed mixture of oxygen, air, dried concentrate, flux, and particulate recycle materials into a hot reaction shaft. Smelting reactions are extremely fast under these conditions. Outotec flash furnaces smelt ~4000 tonnes of new concentrate per day.

Modern Outotec flash furnaces operate with high oxygen blast and very little hydrocarbon fuel. Most of the energy for heating and melting comes from Fe and S oxidation. This operation also gives strong SO$_2$ offgas from which SO$_2$ can be captured efficiently as sulfuric acid.

Outotec flash furnaces are operated under automatic control to give constant temperature, constant composition products at a rapid rate and with minimum energy consumption. Matte and slag compositions are controlled by adjusting the ratios of O$_2$ input rate to concentrate feed rate, and of flux input to concentrate input. Product temperatures are controlled by adjusting the N$_2$/O$_2$ ratio of the input blast and the hydrocarbon fuel combustion rate.

Wide adoption of Outotec flash smelting is due to its efficient capture of SO$_2$, its rapid production rate, and its small energy requirement. Its only limitation is its inability to smelt scrap.

REFERENCES

Ahokainen, T., & Eklund, K. (2007). Flexibility of the Outokumpu flash smelting for low and high grade concentrates − evaluation by CFD modeling. In A. E. M. Warner, C. J. Newman, A. Vahed, D. B. George, P. J. Mackey & A. Warczok (Eds.), *Cu2007, Vol. III, Book 1: The Carlos Díaz symposium on pyrometallurgy* (pp. 343−356). Montreal: CIM.

Carr, H., Humphris, M. J., & Longo, A. (1997). The smelting of bulk Cu−Ni concentrates at the Inco Copper Cliff smelter. In C. Diaz, I. Holubec & C. G. Tan (Eds.), *Nickel−Cobalt 97, Vol. III: Pyrometallurgical operations, environment, vessel integrity in high-intensity smelting and converting processes* (pp. 5−16). Montreal: CIM.

Chen, S. L., & Mansikkaviita, H. (2006). The beneficial effects of feeding dry copper concentrate to smelting furnaces and development of the dryers. In R. T. Jones (Ed.), *Southern African pyrometallurgy 2006* (pp. 265−272). Johannesburg: South African Institute of Mining and Metallurgy. http://www.saimm.co.za/Conferences/Pyro2006/265_Kumera.pdf.

Conde, C. C., Taylor, B., & Sarma, S. (1999). Philippines associated smelting electrostatic precipitator upgrade. In D. B. George, W. J. Chen, P. J. Mackey & A. J. Weddick (Eds.), *Copper 99−Cobre 99, Vol.: Smelting operations and advances* (pp. 685−693). Warrendale, PA: TMS.

Contreras, J., Bonaño, J. L., Fernández−Gil, R., & Palacios, M. (2003). Recent operating advances at the Huelva smelter. In C. Díaz, J. Kapusta & C. Newman (Eds.), *Copper 2003−Cobre 2003, Vol. IV, Book 1: Pyrometallurgy of copper: The Herman Schwarze symposium on copper pyrometallurgy* (pp. 225−238). Montreal: CIM.

Davenport, W. G., Jones, D. M., King, M. J., & Partelpoeg, E. H. (2001). *Flash smelting, analysis, control and optimization.* Warrendale, PA: TMS.

Deneys, A., Dray, J., Mahoney, W., & Barry, M. (2007). Industrial gases for copper production. In A. E. M. Warner, C. J. Newman, A. Vahed, D. B. George, P. J. Mackey & A. Warczok (Eds.), *Cu2007 − Vol. III, Book 2: The Carlos Díaz symposium on pyrometallurgy* (pp. 271−288). Montreal: CIM.

Dudley, L. V. (2009). Secrets of the one-dimensional convergence hopper. In J. Liu, J. Peacey, M. Barati, S. Kashani-Nejad & B. Davis (Eds.), *Pyrometallurgy of nickel and cobalt 2009* (pp. 577−589). Montreal: CIM.

EcoMetales. (2010). http://www.EcoMetales.cl.

Fagerlund, K., Lindgren, M., & Jåfs, M. (2010). Modern flash smelting cooling systems. In *Copper 2010, Vol. 2: Pyrometallurgy I* (pp. 699−711). Clausthal-Zellerfeld, Germany: GDMB.

FLSmidth, (16 November 2010). *Stacker and reclaimer systems for cement plants.* http://www.flsmidth.com/~/media/Brochures/Brochures%20for%20crushers%20and%20raw%20material%20stores/stacker%20and%20reclaimer.ashx.

Gabb, P. J., Howe, D. L., Purdie, D. J., & Woerner, H. J. (1995). The Kennecott smelter hydrometallurgical impurities process. In W. C. Cooper, D. B. Dreisinger, J. E. Dutrizac, H. Hein & G. Ugarte (Eds.), *Copper 95−Cobre 95, Vol. III: Electrorefining and hydrometallurgy of copper* (pp. 591−606). Montreal: CIM.

Hoshi, M., Toda, K., Motomura, T., Takahashi, M., & Hirai, Y. (2010). Dryer fuel reduction and recent operation of the flash smelting furnace at Saganoseki smelter & refinery after the SPI project. In *Copper 2010, Vol. 2: Pyrometallurgy I* (pp. 779−791). Clausthal-Zellerfeld, Germany: GDMB.

Humphris, M. J., Liu, J., & Javor, F. (1997). Gas cleaning and acid plant operations at the Inco Copper Cliff smelter. In C. Diaz, I. Holubec & C. G. Tan (Eds.), *Nickel−Cobalt 97, Vol. III: Pyrometallurgical operations, environment, vessel integrity in high-intensity smelting and converting processes* (pp. 321−335). Montreal: CIM.

Janney, D., George-Kennedy, D., & Burton, R. (2007). Kennecott Utah Copper smelter rebuild: over 11 million tonnes of concentrate smelted. In A. E. M. Warner, C. J. Newman, A. Vahed, D. B. George, P. J. Mackey & A. Warczok (Eds.), *Cu2007, Vol. III Book 2: The Carlos Díaz symposium on pyrometallurgy* (pp. 431−443). Montreal: CIM.

Jones, C. U., Reed, M. E., Fallas, D. E., Cockburn, P. A., Snowden, B., Walker, P. E., & Sims, R. C. (2007). Transforming flash furnace feed and burner stability: optimizing furnace performance and productivity. In A. E. M. Warner, C. J. Newman, A. Vahed, D. B. George, P. J. Mackey & A. Warczok (Eds.), *Cu2007, Vol. III, Book 1: The Carlos Díaz symposium on pyrometallurgy* (pp. 357−376). Montreal: CIM.

Kaur, R., Nexhip, C., Wilson, M., & George-Kennedy, D. (2010). Minor element deportment at the Kennecott Utah copper smelter. In *Copper 2010, Vol. 6: Economics, process control, automatization and optimization* (pp. 2415−2432). Clausthal-Zellerfeld, Germany: GDMB.

Kojo, I. V., & Lahtinen, M. (2007). Outokumpu blister smelting processes − clean technology standards. In A. E. M. Warner, C. J. Newman, A. Vahed, D. B. George, P. J. Mackey & A. Warczok (Eds.), *Cu2007, Vol. III, Book 1: The Carlos Díaz symposium on pyrometallurgy* (pp. 183−190). Montreal: CIM.

Köster, S. (2010). Waste heat boilers for copper smelting applications. In *Copper 2010, Vol. 2: Pyrometallurgy I* (pp. 879−891). Clausthal-Zellerfeld, Germany: GDMB.

Larouche, P. (2001). *Minor elements in copper smelting and electrorefining, M.Eng thesis*. Canada: McGill University. http://digitool.library.mcgill.ca:8881/R/?func=dbin-jump-full&object_id=33978.

Larouche, P., Harris, R., & Wraith, A. (2003). Removal technologies for minor elements in copper smelting. In C. Díaz, J. Kapusta & C. Newman (Eds.), *Copper 2003−Cobre 2003, Vol. IV, Book 1: Pyrometallurgy of copper, The Herman Schwarze symposium on copper pyrometallurgy* (pp. 385−404). Montreal: CIM.

Liu, J., Baird, M. H. I., Kenny, P., Macnamara, B., & Vahed, A. (2006). Inco flash furnace froth column modifications. In F. Kongoli & R. G. Reddy (Eds.), *Sohn international symposium, Vol. 8: International symposium on sulfide smelting 2006* (pp. 291−302). Warrendale, PA: TMS.

Liu, J., Vahed, A., Yotingco, R., Macnamara, B., & Warner, A. E. M. (2007). Study of elemental sulfur formation in the fluid bed dryer baghouse. In A. E. M. Warner, C. J. Newman, A. Vahed, D. B. George, P. J. Mackey & A. Warczok (Eds.), *Cu2007, Vol. III, Book 2: The Carlos Díaz symposium on pyrometallurgy* (pp. 289−299). Montreal: CIM.

McKenna, K., Newman, C., Voermann, N., Veenstra, R., King, M., & Bryant, J. (2010). Extending copper smelting and converting furnace campaign life through technology. In *Copper 2010, Vol. 3: Pyrometallurgy II* (pp. 971−986). Clausthal-Zellerfeld, Germany: GDMB.

McKenna, K., Voermann, N., Veenstra, R., & Newman, C. (2010). High intensity cast cooling element design and fabrication considerations. In *Copper 2010, Vol. 3: Pyrometallurgy II* (pp. 987−1002). Clausthal-Zellerfeld, Germany: GDMB.

Maeda, Y., Inoue, H., Hoshikawa, Y., & Shirasawa, T. (1998). Current operation in Kosaka smelter. In J. A. Asteljoki & R. L. Stephens (Eds.), *Sulfide smelting '98: Current and future practices* (pp. 305−314). Warrendale, PA: TMS.

Marczeski, W. D., & Aldrich, T. L. (1986). *Retrofitting Hayden plant to flash smelting*. Paper A86−65. Warrendale, PA: TMS.

Marinelli, J., & Carson, J. W. (1992). Solve solids flow problems in bins, hoppers, and feeders. *Chemical Engineering Progress, 88*(5), 22−28.

Miettinen, E., Ahokainen, T., & Eklund, K. (2010). Management of copper flash smelting off-gas line gas flow and oxygen potential. In *Copper 2010, Vol. 3: Pyrometallurgy II* (pp. 1003−1012). Clausthal-Zellerfeld, Germany: GDMB.

Molino, L., Diaz, C. M., Doyle, C., Hrepic, J., Slayer, R., Carr, H., & Baird, M. H. I. (1997). Recent design improvements to the Inco flash furnace uptake. In C. Diaz, I. Holubec & C. G. Tan (Eds.), *Nickel−Cobalt 97, Vol. III: Pyrometallurgical operations, environment, vessel integrity in high-intensity smelting and converting processes* (pp. 527−537). Montreal: CIM.

Nagai, K., Kawanaka, K., Yamamoto, K., & Sasai, S. (2010). Development of Sumitomo concentrate burner. In *Copper 2010, Vol. 3: Pyrometallurgy II* (pp. 1025−1034). Clausthal-Zellerfeld, Germany: GDMB.

Outotec. (2010). *Minerals and metals, copper, flash smelting and flash converting*. http://www.outotec.com/pages/Page____38082.aspx?epslanguage=EN.

Parker, K. R. (1997). *Applied electrostatic precipitation*. London, U.K: Chapman & Hall.

Queneau, P. E., & Marcuson, S. W. (1996). Oxygen pyrometallurgy at Copper Cliff – a half century of progress. *JOM, 48*(1), 14−21.

Reed, M. E., Jones, C. U., Snowdon, B., & Coleman, M. (2010). Clyde-WorleyParsons' flash furnace feed system: the development cycle. In *Copper 2010, Vol. 3: Pyrometallurgy II* (pp. 1079−1094). Clausthal-Zellerfeld, Germany: GDMB.

Ryan, P., Smith, N., Corsi, C., & Whiteus, T. (1999). Agglomeration of ESP dusts for recycling to plant smelting furnaces. In D. B. George, W. J. Chen, P. J. Mackey & A. J. Weddick (Eds.), *Copper 99−Cobre 99, Vol. V: Smelting operations and advances* (pp. 561−571). Warrendale, PA: TMS.

Severin, N. W. (1998). Refractory dryout, how can we improve it? *Canadian Ceramics, 67*(2), 21−23.

Talja, J., Chen, S., & Mansikkaviita, H. (2010). Kumera technology for copper smelters. In *Copper 2010, Vol. 3: Pyrometallurgy II* (pp. 1183−1197). Clausthal-Zellerfeld, Germany: GDMB.

Ushakov, K. I., Bochkarev, L. M., Ivanov, A. V., Shurchov, V. P., Sedlov, M. V., & Zubarev, V. I. (1975). Assimilation of the oxygen flash smelting process at the Almalyk plant. *Tsvetnye Metally, 16*(2), 5−9, (English Translation).

Walker, P., Coleman, M., & Money, G. (2010). Energy efficient and reliable pneumatic conveying solutions at Aurubis. In *Copper 2010, Vol. 6: Economics, process control, automatization and optimization* (pp. 2517−2526). Clausthal-Zellerfeld, Germany: GDMB.

Zanini, M., Wong, D., & Cooke, D. (2009). Quality control improvements in Bessemer matte production at the Vale Inco Copper Cliff smelter. In J. Liu, J. Peacey, M. Barati, S. Kashani-Nejad & B. Davis (Eds.), *Pyrometallurgy of nickel and cobalt 2009* (pp. 351−370). Montréal: CIM.

SUGGESTED READING

Davenport, W. G., Jones, D. M., King, M. J., & Partelpoeg, E. H. (2001). *Flash smelting, analysis, control and optimization*. Warrendale, PA: TMS.

Outotec. (2010). *Minerals and metals, copper, flash smelting and flash converting*. http://www.outotec.com/pages/Page____38082.aspx?epslanguage=EN.

Chapter 7

Submerged Tuyere Smelting: Noranda, Teniente, and Vanyukov

Teniente and Noranda smelting use large, 4–5 m diameter × 14–26 m long, cylindrical furnaces. The furnaces always contain layers of molten matte (70–74% Cu) and slag. Oxygen for concentrate oxidation is provided by blowing enriched air through submerged tuyeres into the molten matte layer.

Cu–Fe–S concentrate is (a) dried and blown into the furnace through 4–6 dedicated submerged tuyeres, or (b) thrown moist (~8% H_2O) with flux, recycle materials, and scrap onto the slag surface through an endwall.

The products of the processes are super high-grade molten matte (72–74% Cu, 3–7% Fe, ~20% S), slag (6–10% Cu, 1200–1250 °C), and offgas with 20–26 vol.-% SO_2 (before dilution). The matte is sent to Peirce–Smith converters for coppermaking, the slag is sent to a Cu recovery process, and the offgas is sent to cooling, dust recovery, and a sulfuric acid plant.

All or most of the energy for heating and melting the charge comes from Fe and S oxidation:

$$\underset{\substack{\text{injected concentrate} \\ \text{in matte/slag bath}}}{CuFeS_2(s)} + \underset{\text{in injected oxygen-air blast}}{O_2(g)} \rightarrow \underset{\text{molten matte}}{Cu\text{–}Fe\text{–}S(l)} + \underset{\text{in offgas}}{SO_2(g)} + heat \quad (7.1)$$

Natural gas, coal or coke may be burnt to supplement this heat.

In 2010, there were three Noranda furnaces and seven Teniente furnaces operating around the world (ICSG, 2010). Operating data for two Noranda and two Teniente furnaces are given in Tables 7.2 and 7.4.

This chapter describes operation and control of (a) Noranda submerged-tuyere smelting (Sections 7.1–7.4), (b) Teniente submerged-tuyere smelting (Sections 7.5–7.10), and (c) Vanyukov submerged-tuyere smelting (Section 7.11). It then discusses submerged-tuyere smelting as a whole.

7.1. NORANDA PROCESS (PREVOST, LETOURNEAU, PEREZ, LIND, & LAVOIE, 2007; ZAPATA, 2007)

The Noranda furnace is a 0.1 m thick horizontal molybdenum steel barrel lined inside with about 0.5 m of magnesia–chrome refractory. Industrial furnaces are ~5 m diameter and 21–26 m long (Table 7.1). They have 54–66 submerged tuyeres (5 or 6 cm diameter) along the length of the furnace (Fig. 7.1).

TABLE 7.1 Teniente and Noranda Smelting Furnaces Around the World (ICSG, 2010)

Smelting furnace type	Location	Length × diameter, m	Annual Cu production, kilotonnes per year
Teniente	Chuquicamata, Chile	23 × 5	~300 from Teniente furnace
	Potrerillos, Chile	22 × 5	190
	Paipote, Chile	14 × 4	090
	Las Ventanas, Chile	15 × 4	130
	Teniente, Chile	22 × 5	400 from 2 Teniente furnaces
	La Caridad, Mexico	21 × 4.5	~140 from Teniente furnace
Noranda	Noranda, Canada	21 × 5	220
	Altonorte, Chile	26 × 5	390
	Daye, China		150 from Noranda furnace

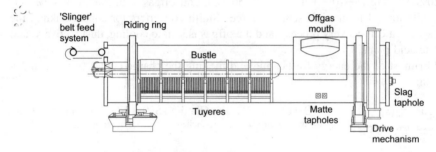

FIGURE 7.1 Noranda smelting furnace.

Noranda smelting entails:

(a) Continuously blowing oxygen-enriched air *blast* (40 vol.-% O_2, 1.6 bar gauge) through submerged tuyeres into the molten matte
(b) Continuously blowing dried concentrate and blast through five special submerged tuyeres into the matte layer (Antofagasta, Chile) and/or continuously feeding moist concentrate through an endwall onto the slag bath surface (Rouyn-Noranda, Canada)
(c) Continuously drawing offgas through a large mouth and hood at the top of the furnace
(d) Intermittently tapping matte and slag through submerged tapholes.

Flux, reverts, scrap, and coal/coke are also thrown onto the top of the slag. Molten converter slag is also occasionally recycled through the Noranda furnace mouth.

The blast tuyeres are periodically cleared by breaking blockages with a steel bar. This ensures an even flow of blast. A Gaspé puncher is used (Heath and Sherwood, 2011; Fig. 7.2).

TABLE 7.2 Operating Details of Two Noranda Smelting Furnaces (Yves Prevost, Personal Communication, 2010; Zapata, 2007)

Smelter	Noranda, Canada	Altonorte, Chile
Startup date	1973	2002
Furnace details		
Length × diameter, m	21.3 × 5.1	26.4 × 5.3
Slag layer thickness, m	0.3–0.6	0.5
Matte layer thickness, m	0.9–1.15	1.2
Active slag tapholes	1	1
Active matte tapholes	1	2
Auxiliary burners	0	2 (natural gas)
Campaign life, years	2.5	2.5
Tuyere details		
Tuyeres (total)	54	61
Diameter, cm	5.4	6.4
Active air blast tuyeres	44	47
Concentrate injection tuyeres	0	5
Type of charge	100% to top of bath	95% thru tuyeres, 5% to top of bath
Feed, t/day (dry basis)		
New concentrate	1900–2200 (~28% Cu)	3000 (37% Cu)
Silica flux	200–300	200
Slag concentrate	80–260	250 (37% Cu)
Recycle dust	50–75 + converter dust	40 Noranda
Reverts	140–440	120–140
Other	0–64 liquid converter slag	Solid converter slag
Tuyere blast details		
Vol.-% O_2	38–42	37–40
Flowrate per tuyere, Nm^3/minute	18.4–27.3	20–22
Products, tonnes/day		
Matte, tonnes/day	700–900 (71–72% Cu; 3–5% Fe)	1350 (74% Cu, 3.5% Fe)
Slag, tonnes/day	1300–1500 (3–4% Cu)	1350 (6% Cu)
Mass% SiO_2/mass% Fe	0.6–0.8	0.7
Cu recovery, Noranda slag	Solidification/flotation	Solidification/flotation
Cu recovery, converter slag	Molten, to Noranda converter or Noranda smelting furnace	Solid to Noranda furnace, 15–20% Cu
Offgas, thousand Nm^3/hour	125–140 (before dilution)	72
Vol.-% SO_2, leaving furnace (wet)	15–20 (before dilution)	20–25
Dust production, t/day	70–100	40
Matte/slag/offgas T, °C	1210/1230/1300	1230/1220/1200
Hydrocarbon consumption, kg/tonne of charge	1–2 coal if needed; 349 oxygen	5–10 metallurgical coke in solid charge

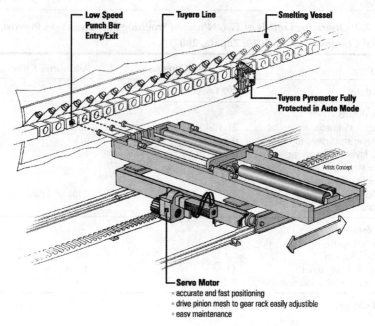

FIGURE 7.2 Sketch of automatic *Gaspé* tuyere puncher for clearing Noranda and Teniente blast input tuyeres (Heath and Sherwood, 2011). It automatically positions itself exactly in front of the tuyeres. It then pushes the three punch-rods through tuyeres to clear frozen blockages.

The furnace is equipped with a rotation mechanism. It is used to correctly position the tuyere tips in the molten matte layer and to roll the tuyeres above the liquids during maintenance and repair. It also automatically rolls the tuyeres above the liquids in the event of a power failure or other emergency.

7.2. REACTION MECHANISMS

The reaction mechanisms in the Noranda furnace are:

(a) Sulfide concentrate is blown into the matte layer or thrown by slinger belt onto the slag surface
(b) The concentrate particles enter the matte layer where they melt and are oxidized by tuyere O_2
(c) Fe oxides from the oxidation react with SiO_2 flux to form slag, which rises to the top of the bath
(d) SO_2 from the oxidation rises through the bath and leaves the furnace along with N_2 from the tuyere blast and CO_2 and water vapor from hydrocarbon combustion.

Other parts of the charge, such as scrap, sludges, and recycle materials, melt and undergo oxidation and slagging. Oxides rise to the slag layer while copper and precious metals (mostly from scrap) descend and join the matte layer.

7.2.1. Separation of Matte and Slag

Matte and slag are intimately mixed in the tuyere region. They are allowed to separate in a quiet tuyere-free zone at the taphole end of the furnace (Fig. 7.1). Matte falls,

TABLE 7.3 Impurity Distribution During Noranda Smelting (Harris, 1999)

Element	% to matte	% to slag	% to offgas
As	8	12	80
Bi	9	12	79
Ni	77	22	1
Pb	13	13	74
Sb	15	31	54
Zn	6	84	10

SO_2/N_2 gas rises, and slag forms a top layer dilute enough in Cu for tapping from the furnace. It contains ~6% Cu, 30% dissolved and 70% as entrained matte. It is tapped from the furnace, and sent for Cu recovery by solidification/comminution/flotation (Chapter 11).

7.2.2. Choice of Matte Grade

The Noranda process was initially conceived as a direct-to-copper smelting process. The furnace at Noranda produced molten copper from 1973 to 1975. It was switched to high-grade matte production to (a) lower impurity levels in the smelter's anode copper and (b) increase smelting rate. Noranda furnaces now produce super-high (70–74% Cu), low-sulfur (~20% S) matte. This maximizes SO_2 production in the continuous smelting furnace and minimizes SO_2 production in discontinuous Peirce–Smith converters, maximizing SO_2 capture as sulfuric acid.

7.2.3. Impurity Behavior

Table 7.3 describes impurity behavior during Noranda smelting. It shows that harmful impurities report mainly to slag and offgas. Most As, Bi, Pb, and Sb can be removed from the Cu circuit by not recycling offgas solids (recovered dust) to smelting or converting.

7.2.4. Scrap and Residue Smelting

The feed to the Noranda furnace at Noranda, Quebec includes 10+% end-of-use scrap (Coursol, Mackey, Prevost, & Zamalloa, 2007). The scrap includes precious metal and copper-containing slags, ashes, residues (up to 14% moisture), precious metal catalysts, wire cables, jewelry, telephone scrap, automobile parts, and precious metal computer and electronic scrap. However, recycling of beryllium (a potential carcinogen) and radioactive materials are strictly forbidden.

Tuyere-blast stirring in the Noranda furnace rapidly melts these materials and causes their precious metals and Cu to be rapidly absorbed in matte. Also, the high temperature and intensity of smelting cause potentially harmful organic compounds to be oxidized completely to CO_2 and H_2O.

7.3. OPERATION AND CONTROL

Noranda smelting is started by heating the furnace with hydrocarbon burners. Molten matte is then poured in through the furnace mouth (tuyeres elevated). Once a meter or so of molten matte is in place, the tuyere blast is started and the tuyeres are rolled into the molten matte to begin oxidation and heat generation. Concentrate and flux feeding is then started and normal smelting is begun. About a week is taken to heat the furnace, provide the molten matte, and attain full production.

The initial molten matte is prepared by melting matte pieces or high-Cu concentrate in a converter or unused furnace in the smelter. Smelting is terminated by inserting hydrocarbon burners into the furnace, stopping smelting, and pouring slag then matte out the furnace mouth.

7.3.1. Control (Zapata, 2007)

Once steady operation has been reached, the furnace is controlled to:

(a) Smelt concentrates, scrap, and other metal-bearing solids at the company's prescribed rates
(b) Produce matte and slag of prescribed composition and temperature
(c) Maintain prescribed depths of matte and slag in the furnace.

Matte composition is controlled by adjusting the ratio of oxygen input in the blast to the rate of solid feed input. Increasing this ratio increases matte grade (by increasing Fe and S oxidation) and vice versa. It is often altered by adjusting solid feed input rate at a constant O_2-in-blast injection rate. This gives constant-rate SO_2 delivery to the sulfuric acid plant.

Matte/slag temperature (1200−1250 °C) is controlled by altering the ratio of hydrocarbon combustion rate to solid feed input rate. The ratio is increased to raise temperature and vice versa. It is altered by adjusting coal/coke feed rate and natural gas combustion rate. Matte/slag temperature may also be controlled by adjusting the N_2/O_2 ratio of the blast.

Slag composition is controlled by adjusting the ratio of flux input rate to solid feed input rate. The target SiO_2/Fe ratio is ~0.7.

In addition, the mix of metal-bearing solid feed to the furnace is controlled to keep impurity levels in the matte below pre-set values. This is done to avoid excessive impurity levels in the anodes produced by the smelter.

Feed rates and O_2 input rate are monitored continuously. Matte samples are taken every hour (analyses being returned 15 min later), and slag samples every two hours. Bath temperature is monitored continuously with optical pyrometers (Fig. 7.3) sighted through two tuyeres.

Matte and slag depths are monitored hourly with a 2.5 cm diameter vertical steel bar. This is done to (a) ensure that there is enough matte above the tuyeres for efficient O_2 utilization, and (b) produce an even blast flow by maintaining a constant static liquid pressure at the tuyere tips. The depths are adjusted by altering matte and slag tapping frequency.

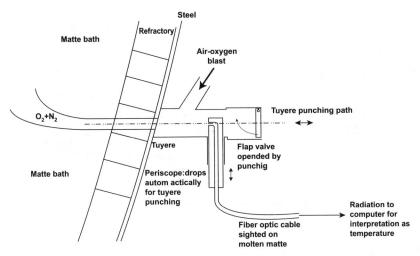

FIGURE 7.3 Schematic of in-tuyere optical pyrometer. Its main component is an optical fiber, which transmits a color image of in-furnace matte to a distant computer for temperature interpretation. Its other key component is a periscope optical fiber holder, which automatically lowers when the Gaspé puncher moves into position to punch this tuyere. Industrially, ball valves (Heath and Sherwood, 2011) are used instead of simple flap valves to prevent blowback.

7.4. PRODUCTION RATE ENHANCEMENT

The smelting rate of the furnace near Antofagasta, Chile has more than doubled since 2003. Most of the increase has been due to increased furnace blowing availability, now over 90%. This in turn is the result of improved control over matte and slag compositions and temperatures, and better taphole dependability, which improves control over matte and slag depths (Zapata, 2007).

7.5. TENIENTE SMELTING

Teniente smelting shares many features with Noranda smelting (Carrasco, Bobadilla, Duarte, Araneda, & Rubilar, 2007; Moyano, Rojas, Caballero, Font, Rosario, & Jara, 2010). It:

(a) Uses a cylindrical furnace with submerged tuyeres (Fig. 7.4)
(b) Blows oxygen-enriched air through the tuyeres into molten matte
(c) Feeds dry concentrate through dedicated tuyeres
(d) (Sometimes) charges moist concentrate onto its matte/slag surface
(e) Produces high-Cu, low Fe, low S matte, which it sends to Peirce—Smith converting for coppermaking (i.e. Fe and S oxidation).

Table 7.4 gives operating details of two Teniente smelting furnaces.

7.5.1. Seed Matte

Teniente smelting evolved from smelting concentrates in Peirce—Smith converters (Chapter 8). Early Teniente smelting always included molten matte from another smelting furnace in the charge. This is now rare.

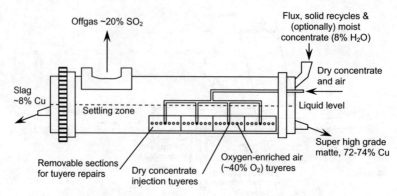

FIGURE 7.4 Schematic of Teniente smelting furnace, ~20 m long.

Teniente furnaces have proven to be successful stand-alone smelting units, and molten matte is no longer needed. This has permitted shutdown of the reverberatory furnaces that formerly supplied Teniente furnaces with matte.

7.6. PROCESS DESCRIPTION

Teniente furnaces are 4—5 m diameter and 14—23 m long. The furnace barrels are steel, ~0.05 m thick, lined with about 0.5 m of magnesia—chrome refractory. The furnaces have 35—50 tuyeres (5 or 6 cm diameter) along 65% of their length. The remaining 35% of the furnace length is a quiet settling zone.

All Teniente furnaces blow dry concentrate into the furnace through four or six dedicated tuyeres (Moyano et al., 2010; Rosales, Font, Fuentes, Moyano, Rojas, Caballero, & Mackay, 2010). Flux, recycle materials and (sometimes) moist concentrate are charged onto the matte/slag surface. Reactions are similar to those in the Noranda furnace. The principal products of the process are molten matte (72—74% Cu, 3—7% Fe, ~21% S), molten Fe-silicate slag (6—10% Cu), and offgas (~25 vol.-% SO_2).

7.7. OPERATION (MOYANO ET AL., 2010)

Teniente smelting is begun by:

(a) Preheating the furnace with hydrocarbon burners
(b) Charging molten matte to the furnace (with tuyeres elevated)
(c) Blowing oxygen-enriched air through the tuyeres
(d) Rotating the tuyeres into the molten matte
(e) Starting normal feeding of concentrate, flux, and recycles.

Feed rates are then gradually increased until full production is attained. Startup to full production takes about one week.

The initial charge of matte comes from another furnace in the smelter — a reverberatory, flash, or electric slag cleaning furnace. In smelters without another furnace, the initial molten matte is prepared by melting matte pieces or high-grade concentrate in a converter or other unused furnace.

Chapter | 7 Submerged Tuyere Smelting: Noranda, Teniente, and Vanyukov

TABLE 7.4 Mechanical and Operating Details of Teniente Smelting Furnaces at Caletones, Chile and Chuquicamata, Chile (2010). Data Courtesy of Codelco

Smelter	Codelco Caletones, Chile	Codelco Chuquicamata, Chile
Startup date	1989	2003
Furnace details, inside brick		
Length × diameter, m	21.1 × 4.2	22 × 5
Slag layer thickness, m	0.45–0.75	0.9–1.0
Matte layer thickness. m	0.9–1.2	1.0–1.2
Active slag tapholes	1	1
Active matte tapholes	2	1
Temperature measurement	0	
Auxiliary burners	0	0
Time between tuyere line repairs	³⁄₄ year	1.5 years
Time between complete re-lines	1.5 years	1.5 years
Refractory consumption	13,500 kg/campaign	1.8 kg per tonne of Cu
Tuyere details		
Tuyeres (total)	48	55
Diameter, cm	6.35	6.35
Active air blast tuyeres	43 (operating 34)	49
Concentrate injection tuyeres	5	6
Feed, tonnes/day (dry basis)		
Dry concentrate through tuyeres	2100	2200–2500
%Cu in dry concentrate	26–28	30–33
Quartz flux	160–220	150–200
%SiO$_2$ in flux	94.5–98.4	90
Other	264–300 solid reverts	400–600 solid reverts
	0–75 molten slags	0–50 Teniente or flash furnace dust
Tuyere blast details		
Flowrate per tuyere	20–27 Nm^3 per minute	18–22 Nm^3 per minute
Vol.-% O$_2$	35–38	39–42
Oxygen consumption per tonne per tonne of concentrate	310–330 kg industrial oxygen, 95.5% O$_2$	210 kg
Products, tonnes/day		
Matte, tonnes/day	801–909 (72–74% Cu)	990–1150 (72–74% Cu)
Slag, tonnes/day	1600–1900, 6–10% Cu	1550–1800 (6–10% Cu)
Mass% SiO$_2$/mass% Fe	0.57	0.66–0.68
Cu-from-slag recovery method	Slag cleaning furnace	Electric slag cleaning fce
Cu recovery, converter slag	Recycle to Teniente fce	Electric slag cleaning fce
Vol.-% SO$_2$, leaving furnace	25.5–26.4	25
Matte/slag/offgas temps, °C	1220/1240/1250	1220/1240/1250
Hydrocarbon fuel	0	0

7.8. CONTROL (MORROW & GAJAREDO, 2009; MOYANO ET AL., 2010)

Steady operation of a Teniente furnace consists of:

(a) Continuous injection of dried concentrate and air through four to six dedicated submerged tuyeres
(b) Continuous blowing of oxygen-enriched air (~40 vol.-% O_2) through submerged tuyeres
(c) Continuous surface charging of flux and solid recycle materials onto the bath surface
(d) Continuous withdrawal of offgas
(e) Intermittent tapping of matte and slag through submerged tapholes
(f) Occasional recycling of molten converter matte through the furnace mouth.

The operation is controlled to (a) produce matte and slag of specified compositions and temperature and (b) protect the furnace refractories, while smelting solid feed at a specified or maximum rate.

7.8.1. Temperature Control

Liquid temperature is measured by optical pyrometers through two of the tuyeres or onto the slag tapping stream. Slag temperature (~1240 °C) is controlled by adjusting revert feed rates and blast oxygen enrichment level. Reducing the enrichment means that more N_2 is input, which acts as a coolant. Temperature is typically controlled within about ± 20 °C.

7.8.2. Slag and Matte Composition Control

Matte and slag compositions are measured by on-site X-ray analysis. Results are available 15–120 minutes after a sample is taken.

Slag composition is controlled by adjusting flux feed rate. It is controlled to an SiO_2/Fe ratio of 0.65. This, plus good temperature control gives a slag Fe_3O_4 content of 15–20%, which maintains a protective (but not excessive) layer of solid magnetite on the furnace refractory.

Matte grade is controlled by adjusting the ratio of total O_2 input to concentrate feed rate. This ratio controls the degree of Fe and S oxidation, hence matte composition.

7.8.3. Matte and Slag Depth Control

Matte and slag depths are measured frequently by inserting a steel bar vertically from above. Matte depth is controlled to give ~0.5 m of matte above the tuyeres. This ensures efficient use of tuyere O_2.

Heights of matte and slag above the tuyeres are also controlled to be as constant as possible. This gives a constant static pressure of liquid above the tuyeres, hence a constant flow of blast. The heights are kept constant by adjusting matte and slag tapping frequencies.

7.9. IMPURITY DISTRIBUTION

Table 7.5 shows impurity behavior during Teniente smelting. As with Noranda smelting, As, Bi, Pb, Sb, and Zn are largely removed in slag and offgas. Se is removed less efficiently.

TABLE 7.5 Impurity Distribution During Teniente Smelting (Harris, 1999)

Element	% to matte	% to slag	% to offgas
As	6	7	87
Bi	23	40	37
Ni	80	19	1
Pb	22	25	53
Sb	19	30	51
Se	58	39	1
Zn	11	85	4

Teniente impurity removal appears to be slightly less effective than Noranda impurity removal (Table 7.3). This may, however, be due to differences in furnace feeds and measurement techniques.

7.10. DISCUSSION

7.10.1. Super-high Matte Grade and SO_2 Capture Efficiency

Noranda and Teniente smelting oxidize most of the Fe and S in their concentrate feed. This is shown by the super high-Cu grade (72–74% Cu) of their product matte. Extensive S oxidation is advantageous because continuous smelting furnaces capture SO_2 more efficiently than discontinuous batch converters.

Teniente and Noranda smelting gain this SO_2 advantage from the violent stirring created by submerged injection of blast. The stirring dissolves and suspends magnetite in slag, preventing excessive deposition on the furnace refractories even under the highly oxidizing conditions of super high-grade matte production.

7.10.2. Campaign Life and Hot Tuyere Repairing

The campaign lives of Teniente and Noranda furnaces are one to two years. Refractory wear in the tuyere region is often the limiting factor.

Most Teniente furnaces mount their tuyeres in four detachable panels (Carrasco et al., 2007; Guzman, Gomez, & Silva, 2007). These panels can be detached and replaced without cooling the furnace. This significantly improves furnace availability.

7.10.3. Furnace Cooling

Chapters 6 and 9 show that flash and vertical lance furnaces need to be cooled by many copper water-jackets and sprays. Teniente and Noranda furnaces use very little water cooling due to their simple barrel design and submerged oxidation reactions. This cooling simplicity is a significant advantage.

7.10.4. Offgas Heat Recovery

Teniente and Noranda furnaces cool their offgases by water evaporation rather than by heat recovery boilers.

Improved smelting control and increased waste heat boiler reliability may make heat recovery boilers economic for future Teniente and Noranda furnace retrofits. The steam will be especially valuable for steam drying of tuyere-injection concentrates and heap leach lixiviants.

7.11. VANYUKOV SUBMERGED-TUYERE SMELTING

Vanyukov smelting is performed in stationary, submerged-tuyere furnaces (Fig. 7.5). It was developed in the former Soviet Union. It is still used by one Kazakh smelter (Balkash) and two Russian smelters (Norilsk Copper, Sredneuralsk), usually in a smelter complex along with other types of furnaces, e.g. reverberatory, electric and blast furnaces (ICSG, 2010). Operating data for Vanyukov furnaces is provided in Table 7.6.

The process entails (Tarasov & Paretsky, 2003):

(a) Charging moist concentrate (up to 8% H_2O), reverts, flux, and occasionally lump coal through two roof ports (Fig. 7.5)
(b) Blowing oxygen-enriched air (50–95 vol.-% O_2, 9 bar gauge) into the molten slag layer through tuyeres on both sides of the furnace. The tuyeres are located ~0.5 m below the slag surface.

Smelting is continuous. The furnace always contains layers of molten matte and slag. The smelting reactions are similar to those in Noranda and Teniente smelting furnaces. Matte and slag are tapped intermittently through tapholes at opposite ends of the furnace. Weirs are provided to give quiet matte/slag separation near the slag taphole.

Matte grade is 45–74% Cu, and slag Cu content is 0.7–2%. The offgas strength is 25–40 vol.-% SO_2, depending upon blast oxygen enrichment and hydrocarbon combustion rate.

Unlike the rotatable Noranda and Teniente furnaces, the Vanyukov furnace is stationary. The advantages of this are a directly connected gas collection system and no moving parts. The disadvantage is that the Vanyukov furnace cannot lift its tuyeres above the slag for maintenance and repair or in a blower emergency. This may be the reason that the process is not more widespread.

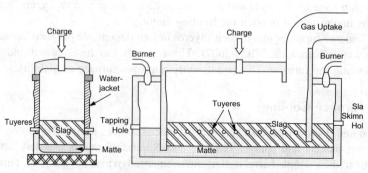

FIGURE 7.5 End view and side view of Vanyukov matte smelting furnace.

TABLE 7.6 Operating Details of Vanyukov Furnaces at Norilsk, Siberia and Sredneuralsky, Urals, 2003 (Ramachandran et al., 2003; Tarasov & Paretsky, 2003)

Smelter	Norilsk Copper	Sredneuralsky
Number of Vanyukov furnaces	3	1
Copper production, tonnes per year		35,000
Furnace volume, inside brick, m^3	36; 39; 6	35
Campaign life, years	5; 5; 4	5
Feed details		
Concentrate composition		
%Cu	19–23	13–15
%Fe	36–40	26–30
%S	28–34	35–39
Flux as % of concentrate	18/17/–	13
Fuel type	Natural gas	Natural gas
Blast details		
Number of operating tuyeres	~16	~16
Tuyere diameter, cm	~2	~2
Vol.-% O$_2$ in blast	75; 75; 94	65
Blast temperature	Ambient	Ambient
Production details		
Operating temperature, °C	1320; 1320; 1250	1280
Matte composition		
%Cu	65; 55; 74	45
%Fe	8; 13; 3	30
%S	24; 24; 22	25
% Cu recovery to matte	97.3	96.2
Slag composition		
%Cu	1.2; 0.9; 2.0	0.7
%Fe	48; 45; 25	30
%SiO$_2$	29; 29; 35	32
%CaO	2; 2; 8	3
Offgas production, thousand Nm^3/hour	75; 50; 6	35
Vol.-% SO$_2$ in offgas	40; 35; –	25
Temperature, °C	1300; 1300; 1300	1290
Dust production, % of solid charge	0.5; 0.7; 0.6	0.9
Fuel (gas) requirement	2–3% of total heat requirement	1–2% of total heat requirement

7.12. SUMMARY

Teniente and Noranda smelting are submerged-tuyere smelting processes. They oxidize Fe and S by blowing oxygen-enriched air through submerged tuyeres into a matte-slag bath. The principal product is super high-grade matte, 72–74% Cu.

Both use horizontal refractory-lined cylindrical furnaces with a horizontal line of submerged tuyeres. The furnaces are rotatable so that their tuyeres can be rolled out of the liquids when blowing must be interrupted.

Concentrate feed is dried and blown into the matte-slag bath through dedicated tuyeres or charged moist onto the bath surface. Tuyere injection is increasing due to its even concentrate and heat distributions, high thermal efficiency and small dust evolution.

Submerged blowing of blast causes violent stirring of the matte/slag bath. This results in rapid melting and oxidation of the furnace charge. It also prevents excessive deposition of solid magnetite in the furnace even under highly oxidizing conditions. The violent stirring also permits extensive smelting of scrap and reverts.

Teniente and Noranda smelting furnaces account for about 15% of world copper smelting (Diaz, 2010). They are the dominant smelting method in Chile and are used in several other countries. Installation rate has slowed appreciably during the last decade.

A third submerged tuyere copper smelting technology is Vanyukov smelting. It uses stationary furnaces and is used in Russia and Kazakhstan.

REFERENCES

Carrasco, C., Bobadilla, J., Duarte, G., Araneda, J., & Rubilar, S. (2007). Evolution of the Teniente converter — Caletones smelter. In A. E. M. Warmer, C. J. Newman, A. Vahed, D. G. George, P. J. Mackey & A. Warczok (Eds.), *Copper—Cobre 2007, Vol. III Book 2: The Carlos Diaz symposium of pyrometallurgy* (pp. 325—337). Montreal: CIM.

Coursol, P., Mackey, P. J., Prevost, Y., & Zamalloa, M. (2007). Noranda process reactor at Xstrata copper — impact of minor slag components (CaO, Al_2O_3, MgO, ZnO) on the optimum $\%Fe/SiO_2$ in slag and operating temperature. In A. E. M. Warmer, C. J. Newman, A. Vahed, D. G. George, P. J. Mackey & A. Warczok (Eds.), *Copper—Cobre 2007, Vol. III Book 2: The Carlos Diaz symposium of pyrometallurgy* (pp. 79—93). Montreal: CIM.

Diaz, C. M. (2010). Copper sulphide smelting: past achievements and current challenges. *Copper 2010, Vol. 7: Plenary, recycling, mineral processing, posters.* Clausthal-Zellerfeld, Germany: GDMB. 2543—2562.

Guzman, G., Gomez, Z., & Silva, A. (2007). Automatic flux feeding into El Teniente converter. In A. E. M. Warmer, C. J. Newman, A. Vahed, D. G. George, P. J. Mackey & A. Warczok (Eds.), *Copper—Cobre 2007, Vol. III Book 2: The Carlos Diaz symposium on pyrometallurgy* (pp. 301—313). Montreal: CIM.

Harris, C. (1999). Bath smelting in the Noranda process reactor and the El Teniente process converter compared. In D. B. George, W. J. Chen, P. J. Mackey & A. J. Weddick (Eds.), *Copper 99—Cobre 99, Vol. V: Smelting operations and advances* (pp. 305—318). Warrendale, PA: TMS.

Heath and Sherwood. (2010). *Gaspé tuyere puncher.* www.heathandsherwood64.com/tuyere.html#GaspePuncher.

Heath and Sherwood. (2011). *Smelter tuyere line management products.* www.heathandsherwood64.com/tuyere.html.

ICSG. (2010). *Directory of copper mines and plants.* Lisbon: International Copper Study Group.

Morrow, A. B., & Gajaredo, M. (2009). Control of copper Teniente smelting units using integrated advanced control technologies. In H. Chang, L. Ryan & S. Kennedy (Eds.), *Process control applications in mining and metallurgical plants* (pp. 187—198). Montreal: CIM.

Moyano, A., Rojas, F., Caballero, C., Font, J., Rosales, M., & Jara, H. (2010). The Teniente converter: a high smelting rate and versatile reactor. In *Copper 2010, Vol. 3: Pyrometallurgy II* (pp. 1013—1023). Clausthal-Zellerfeld, Germany: GDMB.

Prevost, Y., Letourneau, P., Perez, H., Lind, P., & Lavoie, A. (2007). Improving flexibility: recent developments at the Horne smelter. In A. E. M. Warmer, C. J. Newman, A. Vahed, D. G. George, P. J. Mackey & A. Warczok (Eds.), *Copper—Cobre 2007, Vol. III Book 1: The Carlos Diaz symposium on pyrometallurgy* (pp. 167—179). Montreal: CIM.

Ramachandran, V., Lehner, T., Diaz, C., Mackey, P. J., Eltringham, T., Newman, C. J., et al. (2003). Primary copper production — a survey of operating world copper smelters. In C. Diaz, J. Kapusta & C. Newman (Eds.), *Copper 2003—Cobre 2003, Vol. IV, Book 1: Pyrometallurgy of copper, The Hermann Schwarze symposium on copper pyrometallurgy.* Montreal: CIM. 3—119.

Rosales, M., Font, J., Fuentes, R., Moyano, A., Rojas, F., Caballero, C., & Mackay, R. (2010). A fluid-dynamic review of the Teniente Converter. In *Copper 2010, Vol. III: Pyrometallurgy II* (pp. 1095−1113). Clausthal-Zellerfeld, Germany: GMDB.

Tarasov, A. V., & Paretsky, V. M. (2003). Development of autogenous copper smelting processes in Russia and CIS countries. In C. Diaz, J. Kapusta & C. Newman (Eds.), *Copper 2003−Cobre 2003, Vol. IV, Book 1: Pyrometallurgy of copper, The Hermann Schwarze symposium on copper pyrometallurgy* (pp. 173−187). Montreal: CIM.

Zapata, R. (2007). Continuous reactor Altonorte smelter. In A. E. M. Warmer, C. J. Newman, A. Vahed, D. G. George, P. J. Mackey & A. Warczok (Eds.), *Copper−Cobre 2007, Vol. III Book 1: The Carlos Diaz symposium on pyrometallurgy* (pp. 141−153). Montreal: CIM.

SUGGESTED READING

Diaz, C. M. (2010). Copper sulphide smelting: past achievements and current challenges. *Copper 2010, Vol. 7: Plenary, recycling, mineral processing, posters* (pp. 2543−2562). Clausthal-Zellerfeld, Germany: GDMB.

Jak, E., & Tsymbulov, L. (2010). Development of the continuous copper converting using two-zone Vaniukov converter. *Copper 2010, Vol. 2: Pyrometallurgy I* (pp. 793−810). Clausthal-Zellerfeld, Germany: GDMB.

Kozhakhmetov, S. M., Kvyatkovskiy, S. A., Ospanov, E. A., Abisheva, Z. S., & Zagorodnyaya, A. N. (2010). Processing of high-silicon copper sulfide concentrates by Vanyukov smelting. *Copper 2010, Vol. 2: Pyrometallurgy I* (pp. 893−905). Clausthal-Zellerfeld, Germany: GDMB.

Morrow, A. B., & Gajaredo, M. (2009). Control of copper Teniente smelting units using integrated advanced control technologies. In H. Chang, L. Ryan & S. Kennedy (Eds.), *Process control applications in mining and metallurgical plants* (pp. 187−198). Montreal: CIM.

Chapter 8

Converting of Copper Matte

Converting is oxidation of molten Cu−Fe−S matte to form molten *blister* copper (99% Cu). It entails oxidizing Fe and S from the matte with oxygen-enriched air or air *blast*. It is mostly done in the Peirce−Smith converter, which blows the blast into molten matte through submerged tuyeres (Figs. 8.1 and 8.2). Several other processes are also used or are under development,

The main raw material for converting is molten Cu−Fe−S matte from smelting. Other raw materials include silica flux, air, and industrial oxygen. Several Cu-bearing materials are recycled to the converter, such as solidified Cu-bearing reverts and copper scrap.

The products of converting (Table 8.4) are:

(a) Molten blister copper, which is sent to fire refining and electrorefining
(b) Molten iron silicate slag, which is sent to Cu recovery, then discard
(c) SO_2-bearing offgas, which is sent to cooling, dust removal, and H_2SO_4 manufacture.

The heat for converting is supplied entirely by Fe and S oxidation, i.e. the process is autothermal.

8.1. CHEMISTRY

The overall converting process may be described by the schematic reaction:

$$\underset{\substack{\text{molten} \\ \text{matte}}}{\text{Cu}-\text{Fe}-\text{S}} + \underset{\substack{\text{in air and} \\ \text{oxygen}}}{O_2} + \underset{\text{in flux}}{SiO_2} \rightarrow Cu°(l) + \underset{\substack{\text{molten slag with} \\ \text{some solid Fe}_3O_4}}{\left\{ \begin{array}{c} 2FeO \cdot SiO_2 \\ Fe_3O_4 \end{array} \right\}} + SO_2 \quad (8.1)$$

Converting takes place in two stages. The first stage is the *slag-forming* stage when Fe and S are oxidized to FeO, Fe_3O_4, and SO_2 by reactions like:

$$FeS + 1.5O_2 \rightarrow FeO + SO_2 \quad (8.2)$$
$$3FeS + 5O_2 \rightarrow Fe_3O_4 + 3SO_2 \quad (8.3)$$

The melting points of FeO and Fe_3O_4 are 1377 °C and 1597 °C, so silica flux is added to form a liquid slag with FeO and Fe_3O_4. The slag-forming stage is finished when the Fe in the matte has been lowered to about 1%. The principal product of the slag-forming stage is impure molten Cu_2S (*white metal*) at ~1200 °C (Table 8.4).

The second stage is *coppermaking* when the sulfur in Cu_2S is oxidized to SO_2. Copper is not appreciably oxidized until it is almost devoid of S. Thus, the blister copper product of converting is low in both S and O (0.001−0.03% S, 0.1−0.8% O). Nevertheless, if this

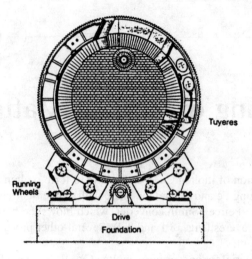
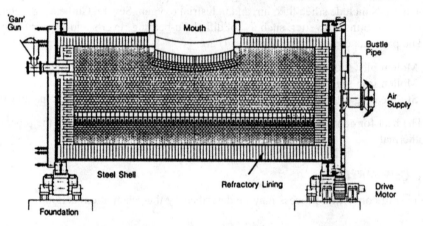

FIGURE 8.1 Peirce–Smith Converter. Note the blast supply, rotating seal, and bustle pipe connected to the tuyeres. Flux and Cu-bearing reverts are added from bins above the offgas uptake or by air gun (Garr Gun) through an end of the converter. The converter rotates around its long axis in order to lift its tuyeres out of the liquids when it is not blowing and submerge them correctly in the matte during blowing, Fig. 8.2. (Converter drawing courtesy Harbison–Walker Refractories, Pittsburgh, PA).

copper were cast, the S and O would form SO_2 bubbles or blisters, which give blister copper its name.

Industrially, matte is charged to the converter in several steps with each step followed by oxidation of FeS from the charge. Slag is poured from the converter after each oxidation step and a new matte addition is made (Fig. 8.3). In this way, the amount of Cu in the converter gradually increases until there is sufficient (100–250 tonnes Cu as molten Cu_2S) for a final coppermaking blow. At this point, the Fe in the matte is reduced by oxidation to about 1%, a final slag is removed, and the resulting white metal is oxidized to molten blister copper. The converting process is terminated when copper oxide begins to appear in samples of the molten copper.

The molten copper is poured from the converter into ladles and transported by crane to a fire-refining furnace for S and O removal and casting of anodes. A start-to-finish converting cycle typically lasts 6–12 h (Tables 8.1 and 8.2).

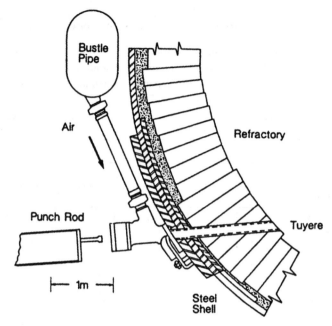

FIGURE 8.2 Details of Peirce–Smith converter tuyere (from Vogt, Mackey, & Balfour, 1979). The tuyeres are nearly horizontal during blowing. 'Blast' pressure is typically 1.2 bar (gauge) at the tuyere entrance. Reprinted by permission of TMS.

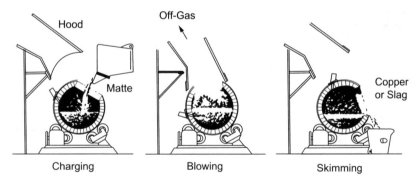

FIGURE 8.3 Positions of a Peirce–Smith converter for charging, blowing, and skimming (Boldt & Queneau, 1967 courtesy Vale). SO_2 gas escapes the system unless the hooding is tight.

8.1.1. Coppermaking Reactions

Blowing air and oxygen into molten white metal creates a turbulent Cu_2S–copper mixture. The products of oxidation in this mixture are SO_2, molten copper, and copper oxide. The molten copper is dense and fluid. It quickly sinks below the tuyeres. The most probable coppermaking reactions are:

$$Cu_2S + 1.5O_2 \rightarrow Cu_2O + SO_2 \tag{8.4}$$

$$Cu_2S + 2Cu_2O \rightarrow 6Cu(l) + SO_2 \tag{8.5}$$

TABLE 8.1 Dimensions and Production Details of Three Peirce–Smith Converter Operations

Smelter	Hibi Kyodo Smelting Co., Tamano, Japan	Sumitomo Toyo, Japan	Pan Pacific Copper Saganoseki, Japan
Number of converters			
Total	3	4	3
Hot	1	3	2
Blowing at one time	1	2	2
Converter details			
Diameter × length, inside shell, m	3.96 × 13.8	4.2 × 11.9	4.2 × 11.5
Number of tuyeres			
Total	60	58 (No. 1–3) 60 (No. 4)	56
Active	60	46	
Tuyere diameter, mm	40		49.5
Usual blast rate per converter			
Slag blow, Nm^3/minute	730	690	650
Copper blow, Nm^3/minute	750	690	650
Usual vol.-% O_2-in-blast			
Slag blow	27.5	25–26	22.3
Copper blow	22.1	21	22.3
SO_2 in offgas, vol.-%	7–10	10–15	7–10
Production details (per converter)			
Inputs, tonnes/cycle			
Molten matte	220 (64% Cu)	227 (64% Cu)	(68% Cu)
Source	Outotec flash furnace + ESCF	Outotec flash furnace	Outotec flash furnace + ESCF
Other inputs, tonnes			
Slag blow	30 ladle skulls + 10 secondaries	0	
Copper blow	80	62	
Outputs, tonnes/cycle			
Blister copper	200	204	

TABLE 8.1 Dimensions and Production Details of Three Peirce–Smith Converter Operations—cont'd

Smelter	Hibi Kyodo Smelting Co., Tamano, Japan	Sumitomo Toyo, Japan	Pan Pacific Copper Saganoseki, Japan
Slag	50	54	
Slag average mass %Cu	8.7	4.8	5.9
Slag mass%SiO$_2$/mass%Fe	0.47	0.42	0.47
Cycle time			
Usual converter cycle time, hours	4.7	12.8	4.83
Slag blow, hours	1.2	0.9	
Copper blow, hours	3.5	3.4	
Campaign details			
Time between tuyere line repairs, days	180	120	210
Copper produced between tuyere line repairs, tonnes	90,000	63,000	
Time between complete converter re-lines, years	—	—	
Refractory consumption, kg/tonne Cu	0.97	1.1	1.7

though some copper may be made directly by:

$$Cu_2S + O_2 \rightarrow 2Cu(l) + SO_2 \qquad (8.6)$$

In principle, there are three sequential steps in coppermaking, as indicated on the Cu–S phase diagram (Fig. 8.4).

(a) The first blowing of air and oxygen into the Cu$_2$S removes S as SO$_2$ to give S-deficient white metal, but no metallic copper. The reaction for this step is:

$$Cu_2S + xO_2 \rightarrow Cu_2S_{1-x} + xSO_2 \qquad (8.7)$$

It takes place until the S is lowered to 19.6% (Fig. 8.4, point b, 1200 °C).

(b) Subsequent blowing of air and oxygen causes a second liquid phase, metallic copper (1% S, point c), to appear. It appears because the average composition of the liquids is now in the liquid-liquid immiscibility region. The molten copper phase is dense and sinks to the bottom of the converter (Fig. 8.5). Further blowing oxidizes

TABLE 8.2 Dimensions and Production Details of Three Peirce–Smith Converter Operations

Smelter	Aurubis Hamburg, Germany	Boliden Rönnskär, Sweden	CODELCO Caletones, Chile
Number of converters			
Total	3	3	4
Hot	2	2	3
Blowing at one time	2	1	—
Converter details			
Diameter × length, inside shell, m	4.6 × 12.2	4.5 × 13.4	Three 4.5 × 10.6 One 4.0 × 10.6
Number of tuyeres			
Total	62	60	48
Active	60	54	46
Tuyere diameter, mm	60	50	63.5
Usual blast rate per converter			
Slag blow, Nm^3/minute	700	750	Only copper blow
Copper blow, Nm^3/minute	700–800	850	600
Usual volume% O_2-in-blast			
Slag blow	23	24	None
Copper blow	23	24	21
SO_2 in offgas, volume%	8–14	5–12	15
Production details (per converter)			
Inputs, tonnes/cycle			
Molten matte	300 (63% Cu)	350 (55% Cu)	200 (74.3% Cu)
Source	Outotec flash furnace + ESCF	Flash- electric- Kaldo furnace	Teniente and slag cleaning furnaces
Other inputs, tonnes			
Slag blow	15 ladle skulls + 10 secondaries	20 ladle skulls + 10 secondaries	None
Copper blow	75 Cu scrap	90–120 Cu scrap	35 reverts
Outputs, tonnes/cycle			
Blister copper	260–280	290–310	145
Slag	70	150–160	30

TABLE 8.2 Dimensions and Production Details of Three Peirce−Smith Converter Operations—cont'd

Smelter	Aurubis Hamburg, Germany	Boliden Rönnskär, Sweden	CODELCO Caletones, Chile
Slag average mass %Cu	4	5	25
Slag mass%SiO$_2$/mass%Fe	0.55	0.6	—
Cycle time			
Usual converter cycle time, hours	9.5	8	7−7.5
Slag blow, hours	2	2	None
Copper blow, hours	5	4.5	5
Campaign details			
Time between tuyere line repairs, days	60	80	30 tuyere line (180 tuyere line + body)
Copper produced between tuyere line repairs, tonnes	55,000	51,000	11,200
Time between complete converter re- lines, years	1	0.7	2.0
Refractory consumption, kg/tonne Cu	1.9	4.7	4.5

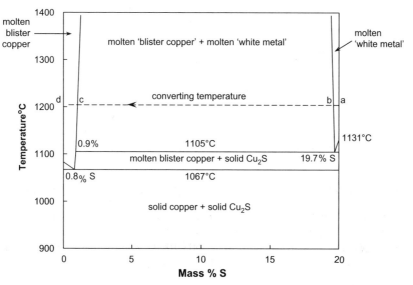

FIGURE 8.4 Cu−S equilibrium phase diagram showing coppermaking reaction path (a, b, c, d, 1200 °C) (Sharma & Chang, 1980).

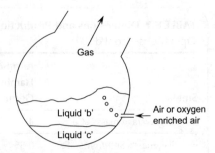

FIGURE 8.5 Sketch of Peirce−Smith converter and its two immiscible liquids during the coppermaking stage of converting (after Peretti, 1948). In practice, the liquid 'b' region is a $Cu_2S-Cu-Cu_2O$−gas foam/emulsion from which metallic copper 'c' descends and SO_2 and N_2 ascend. The immiscibility of copper and Cu_2S is due to their different structures; copper is metallic while Cu_2S is a semiconductor.

additional S from the Cu_2S and the amount of molten copper increases at the expense of the white metal, according to reaction (8.6). As long as the combined average composition of the system is in the immiscibility range, the converter contains both white metal (19.7% S) and molten copper (1% S). Only the proportions change.

(c) Eventually the white metal becomes so S-deficient that the sulfide phase disappears and only molten copper (1% S) remains. Further blowing removes most of the remaining S (point d). Great care is taken during this period to ensure that the copper is not overoxidized to Cu_2O. This care is necessary because Cu_2S is no longer available to reduce Cu_2O back to Cu by reaction (8.5).

Step (a) is very brief, because very little S oxidation is required. Step (c) is also brief. Its beginning is marked by a change in the converter flame color from clear to green when metallic copper begins to be oxidized in front of the tuyeres. This tells the converter operator that the copper blow is nearly finished.

8.1.2. Elimination of Impurities During Converting

The principal elements removed from matte during converting are Fe and S. However, many other impurities are partially removed as vapor or in slag. Table 8.3 shows some distributions. The outstanding feature of the data is that impurity retention in the product blister copper increases significantly with increasing matte grade (%Cu in matte). This is because high-Cu mattes have less blast blown through them, and they form less slag.

The table also shows that significant amounts of impurities report to the offgas. They are eventually collected during gas cleaning. They contain sufficient Cu to be recycled to the smelting furnace. However, such recycle returns all impurities to the circuit as well. For this reason, some smelters treat the dusts for impurity removal before they are recycled (Larouche, 2001, pp. 118−119). Bismuth, in particular, is removed because it causes brittleness in the final copper anodes and it can be a valuable byproduct.

8.2. INDUSTRIAL PEIRCE−SMITH CONVERTING OPERATIONS

Modern industrial Peirce−Smith converters are typically 4.5 m diameter by 12 m long. They consist of a 5 cm steel shell lined with ~0.5 m of magnesite−chrome refractory brick. Converters of these dimensions treat 600−1000 tonnes of matte per day to produce 500−900 tonnes of copper per day. A smelter has two to five converters depending on its overall smelting capacity.

Chapter | 8 Converting of Copper Matte

TABLE 8.3 Distribution of Impurity Elements during Peirce–Smith Converting of Low and High-Grade Mattes (Mendoza & Luraschi, 1993; Vogt et al., 1979)

	54% Cu matte feed			70% Cu matte feed		
	distribution %			distribution %		
Element	To blister copper	To converter slag	To converter offgas	To blister copper	To converter slag	To converter offgas
As	28	13	58	50	32	18
Bi	13	17	67	55	23	22
Pb	4	48	46	5	49	46
Sb	29	7	64	59	26	15
Se	72	6	21	70	5	25
Zn	11	86	3	8	79	13

Oxygen-enriched air or air is blown into a converter at 600–850 Nm^3/min and 1.2 bar gauge. It is blown through a single line of 50 mm diameter tuyeres, 40–60 per converter. It enters the matte 0.5–1 m below its surface, nearly horizontal (Lehner et al., 1993).

The flowrate per tuyere is about 14 Nm^3/minute at a velocity of 100–130 m/s. Blowing rates above about 17 Nm^3/minute/tuyere cause slopping of matte and slag from the converter (Johnson, Themelis, & Eltringham, 1979). High blowing rates without slopping are favored by deep tuyere submergence in the matte (Richards, Legeard, Bustos, Brimacombe, & Jorgensen, 1986).

About half of the operating Peirce–Smith converters enrich their air blast with industrial oxygen, up to ~29 vol.-% O_2-in-blast.

TABLE 8.4 Representative Analyses of Converter Raw Materials and Products (Mass%). Industrial Surveys, Johnson et al. (1979), Pannell (1987), and Lehner et al. (1993)

	Cu	Fe	S	O	As	Bi	Pb	Sb	Zn	Au	Ag
Matte	45–75	3–30	20–23	1–3	0–0.5	0–0.1	0–1	0–0.5	0–1	0–0.003	0–0.3
White metal (Cu_2S)	79	~1	~20	<1							
Blister copper	~99	0.001–0.3	0.001–0.3	0.1–0.8	0–0.2	0–0.03	0–0.5	0–0.1	0	0–0.004	0–0.5

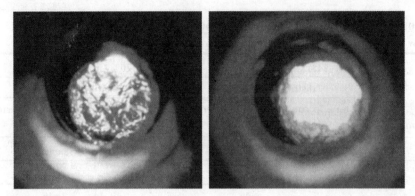

FIGURE 8.6 Photograph showing buildup of accretion at the interior end of a Peirce–Smith converter tuyere (Bustos et al., 1984). Left, tuyere is nearly blocked; right, the accretion has dislodged spontaneously. Bustos et al. (1988) report that accretion 'tubes' are formed in front of the tuyeres. They also indicate that tuyere blockage is discouraged by high matte temperature and oxygen-enrichment of the blast. This is particularly important near the end of the slag blow and the start of the copper blow. Clear tuyere conditions at the beginning of the copper blow often give 'free blowing' conditions (without punching) during most or all of the copper blow. (Photograph courtesy of Dr. Alejandro Bustos, Air Liquide).

8.2.1. Tuyeres and Offgas Collection

Peirce–Smith tuyeres are carbon steel or stainless steel pipes embedded in the converter refractory (Fig. 8.2). They are joined to a distribution *bustle pipe*, which is affixed the length of the converter and connected through a rotatable seal to a blast supply flue. The blast air is pressurized by electric or steam driven blowers. Industrial oxygen is added to the supply flue just before it connects to the converter.

Steady flow of blast requires periodic clearing (*punching*) of the tuyeres to remove matte accretions, which buildup at their tips, especially during the slag blow (Fig. 8.6; Bustos, Richards, Gray, & Brimacombe, 1984; Bustos, Brimacombe, & Richards, 1988). Punching is done by ramming a steel bar completely through the tuyere. It is usually done with a Gaspé mobile carriage puncher (Fig. 8.7), which runs on rails behind the converter. The puncher is sometimes automatically positioned and operated (Marinigh, 2009). An alternative to the Gaspé puncher is the Kennecott 4B-5 system, which consists of a pneumatic cylinder actuated punch rod contained in a housing bolted to each tuyere body (Pasca, Ushakov, & Welker, 2005).

Peirce–Smith converter offgas is collected by a steel hood (usually water cooled), which fits as snugly as possible over the converter mouth. This is shown as the primary hood in Fig. 8.8 (Mori, Nagai, Morita, & Nakano, 2009). The gas then passes through a recovery heat boiler or water-spray cooler, electrostatic precipitators and a sulfuric acid plant. Peirce–Smith converter offgases contain ~8 vol.-% SO_2 (slag blow) to ~10 vol.-% SO_2 (copper blow) after cooling and dust removal (Tables 8.1 and 8.2).

Also shown in Fig. 8.8 is the secondary hood and fugitive gas-collection system. This system collects dilute SO_2 gas not captured by the primary hood. Primary and secondary hood system designs vary considerably. All Peirce–Smith converters have a primary gas capture system, but some do not have secondary gas capture systems.

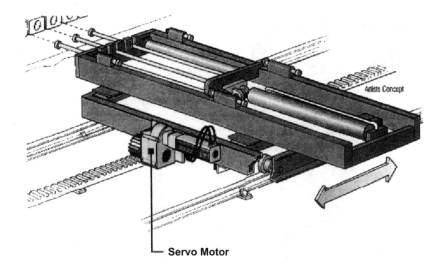

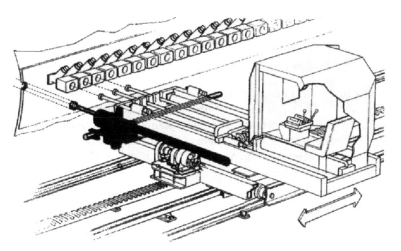

FIGURE 8.7 Gaspé mobile carriage punchers. Two versions are shown. The top one requires an operator who sits on the puncher. A side mounted drill is shown (dark). This drill is used to create new tuyere holes in the refractory after installation of new brick. It is not used during normal operations. The puncher shown at bottom is an automatic type that uses non-contact infrared distance measuring for positioning. It does not require an operator (Marinigh, 2009).

8.2.2. Temperature Control

All heat required for maintaining the converter liquids at their specified temperatures results from Fe and S oxidation, i.e. from reactions like:

$$FeS + 1.5O_2 \rightarrow FeO + SO_2 + heat \tag{8.2}$$

$$Cu_2S + O_2 \rightarrow 2Cu(l) + SO_2 + heat \tag{8.6}$$

Converter temperature can be raised by oxygen enrichment, which reduces the rate at which N_2 coolant enters the converter, and can be lowered by increasing the amount of revert and scrap copper coolant added.

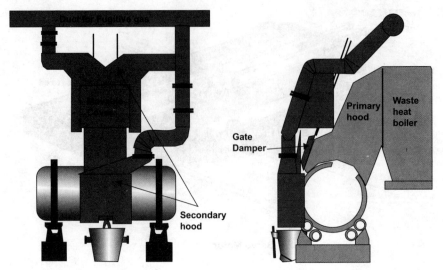

FIGURE 8.8 Offgas-collection systems for a Peirce—Smith converter at the Sumitomo Toyo smelter (Mori et al., 2009).

8.2.3. Choice of Temperature

Representative liquid temperatures during converting are 1200 °C for the input matte and blister copper, and 1220 °C for the skimmed slag. The high temperature during the middle of the cycle is designed to give rapid slag formation and fluid slag with a minimum of entrained matte. It also discourages tuyere blockage (Bustos, Brimacombe, Richards, Vahed, & Pelletier, 1987). An upper limit of about 1250 °C is imposed to prevent excessive refractory wear.

8.2.4. Temperature Measurement

Converter liquid temperature is measured by means of an optical pyrometer sighted downwards through the converter mouth or a two-wavelength optical pyrometer periscope sighted through a tuyere (Marinigh, 2009). The tuyere pyrometer (Fig. 8.9) is preferred because it sights directly on the matte rather than through a dust-laden atmosphere.

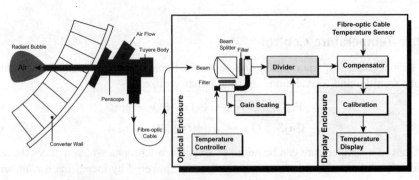

FIGURE 8.9 Tuyere pyrometer schematic (Marinigh, 2009).

8.2.5. Slag and Flux Control

The chief objective of creating a slag in the converter is to liquify newly formed solid FeO and Fe_3O_4, so they can be poured from the converter. SiO_2-bearing flux (e.g. quartz, quartzite, sand) is added for this purpose. A common indicator of slag composition is the mass ratio of SiO_2 to total Fe. Enough SiO_2 flux is added to produce an SiO_2/Fe ratio of ~0.5. Acceptable Fe_3O_4 levels are typically 12−18% (Eltringham, 1993). Some smelters use Au- and Ag-bearing siliceous material as converter flux. The Au and Ag dissolve in the matte and proceed with copper to the electrorefinery where they are profitably recovered. These smelters tend to maximize flux input. Most smelters, however, use just enough flux to obtain an appropriately fluid slag. This minimizes flux cost, slag handling, and the expense of recovering Cu from the slag.

8.2.6. Slag Formation Rate

Flux is added through chutes above the converter mouth or via a high-pressure air gun (*Garr Gun*) at one end of the converter. It is added at a rate that matches the rate of Fe oxidation (usually after an initial several-minute delay while the converter heats up). The flux is commonly crushed to 1−5 cm diameter. Sand (0.1 cm) is used in some smelters. Rapid reaction between O_2, matte, and flux to form liquid slag is encouraged by:

(a) High operating temperature
(b) Steady input of small and evenly sized flux (Schonewille, O'Connell, & Toguri, 1993)
(c) Deep tuyere placement in the matte (to avoid overoxidation of the slag)
(d) The vigorous mixing provided by the Peirce−Smith converter
(e) Reactive flux.

Casley, Middlin, and White (1976) and Schonewille et al. (1993) report that the most reactive fluxes are those with a high percentage of quartz (rather than tridymite or feldspar).

8.2.7. End Point Determinations

8.2.7.1. Slag Blow

The slag-forming stage is terminated and slag is poured from the converter when there is about 1% Fe left in the matte. Further blowing causes excessive Cu and solid magnetite in slag. The blowing is terminated when:

(a) Metallic copper begins to appear in matte samples or when X-ray fluorescence shows 76−79% Cu in matte (Mitarai, Akagi, & Masatoshi, 1993)
(b) The converter flame turns green from Cu vapor in the converter offgas
(c) PbS vapor (from Pb in the matte feed) concentration decreases and PbO vapor concentration increases (Persson, Wendt, & J. Demetrio, 1999).

8.2.7.2. Copper Blow

The coppermaking stage is terminated the instant that copper oxide begins to appear in copper samples. Copper oxide attacks converter refractory, so it is avoided as much as possible. The copper blow is ended and metallic copper is poured from the converter when:

(a) Copper oxide begins to appear in the samples
(b) SO_2 concentration in the offgas falls because S is nearly gone from the matte (Shook, Pasca, & Eltringham, 1999)
(c) PbO concentration in the offgas falls and CuOH concentration increases (H from moisture in the air blast; Persson et al., 1999).

8.3. OXYGEN ENRICHMENT OF PEIRCE–SMITH CONVERTER BLAST

Most smelters enrich their converter blast during part or all of the converting cycle. The advantages of O_2 enrichment are:

(a) Oxidation rate is increased for a given blast input rate
(b) SO_2 concentration in offgas is increased, making gas handling and acid making cheaper
(c) The amount of N_2 'coolant' entering the converter per kg of O_2-in-blast is diminished.

The diminished amount of N_2 'coolant' is important because it permits:

(a) Generation of high temperatures even with high-grade mattes, which have less FeS 'fuel'
(b) Rapid heating of the converter and its contents
(c) Increased melting of valuable coolants such as Cu-bearing reverts and copper scrap.

The only disadvantages of high-O_2 blast are the cost of the oxygen, and the high reaction temperature that results at the tuyere tip. This leads to rapid refractory erosion in the tuyere area. This erosion is discouraged by blowing at a high velocity, which promotes tubular accretion formation and pushes the reaction zone away from the tuyere tip (Bustos et al., 1988).

On balance, the advantages of O_2 enrichment outweigh the refractory erosion disadvantages, especially in smelters, which wish to:

(a) Convert high-grade mattes
(b) Maximize converting rate, especially if converting is a production bottleneck
(c) Maximize melting of solids, such as flux, reverts, and scrap.

The present upper practical limit of oxygen enrichment seems to be about 29 vol.-% O_2. Above this level, refractory erosion becomes excessive. This is because strong tubular accretions do not form in front of the tuyeres above 29 vol.-% O_2, causing the O_2-matte reactions to take place flush with the tuyere tip and refractory. Sonic high-pressure blowing is expected to permit higher oxygen levels in the future (Hills, Warner, & Harris, 2007).

8.4. MAXIMIZING CONVERTER PRODUCTIVITY

The production rate of a converter is maximized by:

(a) Charging high-Cu grade (low FeS) matte to the converter (Fig. 8.10)
(b) Blowing the converter blast at its maximum rate (including avoidance of tuyere blockages)
(c) Enriching the blast to its maximum feasible O_2 level

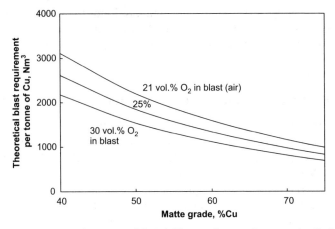

FIGURE 8.10 Theoretical air and oxygen-enriched air blast requirements for converting Cu_2S—FeS mattes to copper. Blast requirement decreases with increasing matte grade and % O_2-in-blast. 100% O_2 efficiency is assumed.

(d) Maximizing O_2 utilization efficiency (deep tuyere submergence)
(e) Minimizing idle time along with efficient crane scheduling for material transfers
(f) Maximizing campaign life (Section 8.4.3).

High-grade matte contains little FeS, so it requires little O_2 (and time) to convert (Fig. 8.10). Rapid blowing of blast, a high % O_2-in-blast, and high O_2 utilization efficiency all lead to rapid oxidation. However, the blast rate should be maintained at a level that prevents bath slopping, which leads to excessive wave motion on the surface and uncontrolled splashing (Kapusta, 2010). This splashing results in rapid accretion growth on the converter mouth, ultimately decreasing availability. Efficient scheduling, high crane availability, and shortening the time for transferring matte, slag, and blister copper through optimizing ladle size also increase converter productivity (Mori et al., 2009).

8.4.1. Maximizing Solids Melting

An important service of the Peirce—Smith converter is melting of valuable solids using the heat from the converting reactions. Typical solids include Cu-bearing revert materials, scrap copper, and Au- and Ag-bearing flux. Cu concentrate is also melted in several smelters. Melting of solids is maximized by (a) maximizing blast O_2 enrichment and (b) blowing the converter at a rapid rate with the tuyeres deep in the matte. This maximizes reaction rate, and thus heat production rate (at an approximately constant heat loss rate from the converter).

The solids are added steadily to avoid excessive cooling of the converter liquids. This is easily done with flux and reverts, which can be crushed and added at controlled rates from storage bins above the converter. Scrap copper, on the other hand, is often large and uneven in shape. It is usually added in batches by crane with the converter in the charging position. This has the disadvantages that blowing must be stopped and the large batch of scrap may excessively cool the converter liquids.

Several converters have conveyor systems, which feed large pieces of copper (scrap anodes and purchased blister copper) at a steady rate during blowing (Fukushima, Baba,

Kurokawa, & Yamagiwa, 1988, Maruyama, Saito, & Kato, 1998). This avoids excessive cooling and maximizes the converter's scrap melting capability. Up to 30% of a converter's blister copper product comes from copper scrap (Fukushima et al., 1988; Pannell, 1987).

8.4.2. Smelting Concentrates in the Converter

Most smelters melt scrap copper and solid reverts in the Peirce—Smith converter. Several smelters also smelt dried concentrates in their converters by injecting the concentrates through the tuyeres (Godbehere, Cloutier, Carissimi, & Vos, 1993; Kawai, Nishiwaki, & Hayashi, 2005; Mast, Arrian, & Benavides, 1999). The technology is well-proven and has two advantages: (a) it can increase smelter capacity without major investment in a larger smelting furnace, and (b) it can lengthen the converting blow and improve removal of impurities, especially bismuth and antimony (Godbehere et al., 1993).

8.4.3. Maximizing Campaign Life

Converters produce 50 000—90 000 tonnes of blister copper before they must be taken out of service for tuyere-refractory replacement. The replacement takes about two weeks and it is done many times before the converter must be completely relined (*shelled*). Copper production per tuyere-refractory replacement period (campaign life) has continued to increase due to:

(a) Improved, higher quality refractories
(b) Higher-grade matte feeds (requiring less blowing per tonne of Cu)
(c) Better temperature measurement and control.

The most durable refractories in 2010 are burned or direct bonded chrome—magnesite bricks. Industrial evidence suggests that oxygen enrichment up to 25% O_2 enhances converter productivity without shortening campaign life. This is especially true if converter-blowing rates are high (Verney, 1987). Enrichment above this level should be tracked to determine the optimum from the points of view of converter productivity and campaign life.

8.5. RECENT IMPROVEMENTS IN PEIRCE—SMITH CONVERTING

8.5.1. Shrouded Blast Injection

ALSI (Air Liquide Shrouded Injector) technology has been successfully demonstrated in Peirce—Smith converters (Pagador, Wachgama, Khuankla, & Kapusta, 2009). The objectives of the ALSI process are to:

(a) Oxidize matte using 30—60% O_2 blast, thereby increasing converter productivity and its ability to melt solids
(b) Eliminate the need to punch the converter (Section 8.2.1)
(c) Minimize refractory wear in the tuyere area.

The tuyere used to achieve these objectives consists of two concentric pipes: the inner pipe for oxygen-enriched air blast (30—60% O_2) and the annulus for nitrogen 'coolant'. The purpose of the nitrogen is (a) to cool the circumference of the tuyere tip and (b) to protect the refractory around the tuyere by building up an accretion of solidified matte and slag.

The blast and nitrogen are blown in at high pressure, 4–8 bar gauge. This prevents the accretion from bridging across the tuyere and eliminates the need for punching.

ALSI technology has been successfully applied to Peirce–Smith converters at Hoboken, Belgium, Xstrata Nickel in Falconbridge, Canada, and Thai Copper in Rayong, Thailand.

8.5.2. Scrap Injection

Pneumatic injection of electronic scrap (<12 mm) at 1 tonne per hour into a converter through a tuyere has been successfully implemented on a commercial scale (Coleman & Money, 2009). This avoids charging through the mouth, allowing particulate (soot) and dioxins to be captured by the primary gas handling system.

8.5.3. Converter Shell Design

The design of the steel converter shell has been improved through the use of curved disk heads integrated with the shell (Chen, Mansikkaviita, Rytkonen, & Kylmakorpi, 2009). This design eliminates the gap between the head and the shell, minimizing the risk of a molten metal leak.

8.6. ALTERNATIVES TO PEIRCE–SMITH CONVERTING

Peirce–Smith converting has been used for over 100 years, and accounts for over 90% of Cu matte converting. Around 250 converters are in operation worldwide in the combined copper, nickel, and platinum-group metals industries (Price, Harris, Hills, Boyd, & Wraith, 2009). This is due to its simplicity and high chemical efficiency. Peirce–Smith converting has problems, however:

(a) It leaks SO_2-bearing gas into the workplace during charging and pouring
(b) It leaks air into its offgas between its mouth and gas-collection hood, producing a relatively weak SO_2 gas
(c) It operates batchwise, giving uneven flow of SO_2 offgas into the sulfuric acid plant.

These deficiencies are addressed by several different alternative converters:

(a) The Hoboken or siphon converter, a Peirce–Smith converter with an improved gas-collection system (four units operating, 2010)
(b) Flash converting, which oxidizes solidified crushed matte in a small Outotec flash furnace (two units operating, 2010)
(c) Noranda continuous converting, which uses submerged tuyeres to blow oxygen-enriched air into matte in a Noranda-type furnace (one unit operating, 2010).
(d) The Mitsubishi top-blown converter, which blows oxygen-enriched blast onto the molten matte surface via vertical lances (four units operating, 2010)
(e) Ausmelt TSL (top submerged lance) batch converting (two units operating for matte converting, 2010)
(f) Isasmelt (top submerged lance) batch converting (one unit operating for matte converting, 2010).

Hoboken converting, flash converting, and Noranda converting are discussed next. Mitsubishi, Ausmelt, and Isasmelt converting are discussed in Chapter 9.

8.6.1. Hoboken Converter

The Hoboken converter collects its offgas through an axial flue at one end of the converter (Binegar & Tittes, 1993; Pasca et al., 2005). A gooseneck is provided to allow the offgas (but not the liquids) to enter the flue. The offgas is collected efficiently. Considerable care must be taken to prevent buildup of splash and dust in the gooseneck.

In 2010, four Hoboken converters were operating at the Freeport McMoRan smelter in Miami, Arizona. Three Hoboken converters were commissioned in 2004 at the Thai Copper Industries smelter in Thailand (Pagador et al., 2009), but the smelter has since shut down for other reasons.

8.6.2. Flash Converting

Flash converting (Table 8.5) uses a small Outotec flash furnace to convert solidified/crushed matte (50 μm) to molten metallic copper (Davenport, Jones, King, & Partelpoeg, 2001; Kojo, Lahtinen, & Miettinen, 2009). Flash converting entails (Fig. 8.11):

(a) Tapping molten 70% Cu matte from a smelting furnace
(b) Granulating the molten matte to ~0.5 mm granules in a water torrent
(c) Crushing the matte granules to 50 μm followed by drying
(d) Continuously feeding the dry crushed matte to the flash converter with 80 vol.-% O_2 blast and CaO flux
(e) Continuously collecting offgas
(f) Periodically tapping molten blister copper and molten calcium ferrite slag.

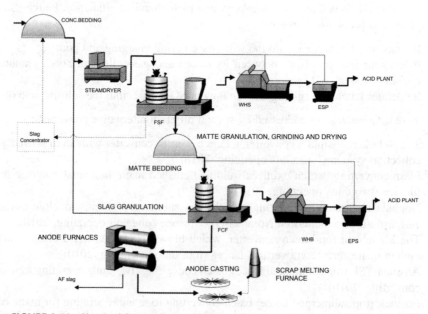

FIGURE 8.11 Sketch of Outotec flash smelting and flash converting process (Kojo et al., 2009).

TABLE 8.5 Physical and Operating Details of Kennecott's Flash Converter, 2010

Smelter	Rio Tinto Kennecott Utah Copper, Magna, UT, USA
Flash converter startup date	1995
Size, inside brick, m	
Hearth: w × l × h	6.5 × 18.75 × 3
Reaction shaft	
Diameter	4.25
Height above settler roof	6.5
Gas uptake	
Diameter	3
Height above settler roof	8.7
Slag layer thickness, m	0.3
Copper layer thickness, m	0.46
Active copper tapholes	6 tapholes + 2 drain holes
Active slag tapholes	3
Particulate matte burners	1
Feed, tonnes/day	
Granulated/crushed matte (dry)	1265−1725
Cu in matte, mass%	67−72.5
CaO flux	24−41
Recycle flash converter dust	82−200
Blast	
Blast temperature, °C	Ambient
Volume% O_2	82−88
Input rate, thousand Nm^3/h	8−13
Oxygen input rate, tonnes/day	287−425
Products	
Copper, tonnes/day	1100−1500
%S in copper	0.06−0.4
%O in copper	0.15−0.45
Slag, tonnes/day	500
%Cu in slag	16.5−25.5
%CaO/%Fe	0.2−0.38
Offgas, thousand Nm^3/h	22−46
Vol.-% SO_2 in offgas	30−40
Copper/slag/offgas temperatures, °C	1200/1260/1290
Fuel inputs (natural gas)	
Hydrocarbon fuel burnt in reaction shaft	130−260 Nm^3/h
Hydrocarbon fuel into settler burners	180 Nm^3/h

The first flash converting furnace was commissioned in 1995 at Rio Tinto Kennecott Utah Copper in Magna, UT, USA. Kennecott helped Outotec develop the process known as Kennecott−Outotec Flash Converting (Kojo et al., 2009). It is simply referred to as flash converting in this text. In 2010, it was in use at Kennecott and at the Yanggu Xiangguang Copper smelter in China.

The uniqueness of the process is its use of particulate solid matte feed. Preparing this feed involves extra processing, but it is the only way that a flash furnace can be used for converting.

A benefit of the solid matte feed is that it unlinks smelting and converting. A stockpile of crushed matte can be built while the converting furnace is being repaired, and then depleted while the smelting furnace is being repaired.

8.6.2.1. Chemistry

Flash converting is represented by the (unbalanced) reaction:

$$\underset{\text{solidified matte}}{Cu-Fe-S} + \underset{\text{in oxygen air blast}}{O_2} \rightarrow Cu^{\circ}(l) + \underset{\text{in molten calcium ferrite slag}}{Fe_3O_4} + SO_2 \quad (8.8)$$

Exactly enough O_2 is supplied to make metallic copper rather than Cu_2S or Cu_2O. The products of the process are:

(a) Molten copper (0.2% S, 0.3% O)
(b) Molten calcium ferrite slag (~16% CaO) containing ~20% Cu
(c) Sulfated dust, ~0.1 tonnes per tonne of matte feed
(d) 30–40 vol-% SO_2 offgas.

The molten copper is periodically tapped and sent forward to pyro- and electro-refining. The slag is periodically tapped, water-granulated, and sent back to the smelting furnace. The offgas is collected continuously, cooled in a waste heat boiler, cleaned of its dust, and sent to a sulfuric acid plant. The dust is recycled to the flash converter and flash smelting furnace.

8.6.2.2. Choice of Calcium Ferrite Slag

The flash converter uses calcium ferrite slag. This slag is fluid and shows little tendency to foam (Chapter 5). It also absorbs some impurities (As, Bi, Sb, but not Pb) better than SiO_2 slag. It is, however, somewhat corrosive and poorly amenable to controlled deposition of solid magnetite on the converter walls and floor.

8.6.2.3. No Matte Layer

There is no matte layer in the flash converter. This is shown by the 0.2% S content of its blister copper — far below the 1% S that would be in equilibrium with Cu_2S matte. The layer is avoided by adjusting the ratio of O_2 input to matte feed rate to slightly favor Cu_2O formation rather than Cu_2S formation. The matte layer is avoided to minimize the possibility of SO_2 formation (and slag foaming) by the reactions under the slag (Davenport et al., 2001):

$$\underset{\text{in slag}}{2Cu_2O} + \underset{\text{in matte}}{Cu_2S} \rightarrow 6Cu + SO_2 \quad (8.5)$$

$$\underset{\text{in slag}}{2CuO} + \underset{\text{in matte}}{Cu_2S} \rightarrow 4Cu + SO_2 \quad (8.9)$$

$$\underset{\text{in slag}}{2Fe_3O_4} + \underset{\text{in matte}}{Cu_2S} \rightarrow 2Cu + 6FeO + SO_2 \quad (8.10)$$

8.6.2.4. Productivity

Kennecott's flash converter treats ~1700 tonnes of 70% Cu matte and produces ~1300 tonnes of blister copper per day. It is roughly equivalent to three large Peirce–Smith converters.

8.6.2.5. Flash Converting Summary

Flash converting is an extension of the successful Outotec flash smelting process. It was first implemented at Kennecott Utah Copper in 1995. Campaign life has steadily increased to around five years through improvements in the cooling systems and improved process control (Janney, George-Kennedy, & Burton, 2007; McKenna et al., 2010). A new Outotec flash converting furnace was commissioned in 2007 at the Yanggu Xiangguang copper smelter in China.

The process has the disadvantages that (a) it requires granulated and crushed matte, which requires extra energy, and (b) it is not well adapted to melting scrap copper. On the other hand, it has a simple, efficient matte oxidation system, and efficiently collects offgas and dust.

8.6.3. Submerged-Tuyere Noranda Continuous Converting

Noranda continuous converting developed from Noranda submerged-tuyere smelting. It uses a rotary furnace (Fig. 8.12) with:

(a) A large mouth for charging molten matte and large pieces of scrap
(b) An endwall slinger and hole for feeding flux, reverts, and coke
(c) A second large mouth for drawing offgas into a hood and acid plant
(d) Tuyeres for injecting oxygen-enriched air into the molten matte (Fig. 8.2)
(e) Tapholes for separately tapping molten matte and slag
(f) A rolling mechanism for correctly positioning the tuyere tips in the molten matte.

The converter operates continuously and always contains molten copper, molten matte (mainly Cu_2S), and molten slag. It blows oxygen-enriched air continuously through its tuyeres and continuously collects ~20% SO_2 offgas. It taps copper and slag intermittently. It presently operates at the Xstrata Horne (formerly Noranda Horne) smelter in Quebec (Table 8.6).

8.6.3.1. Chemical Reactions

Noranda converting controls matte and O_2 input rates to always have matte (mainly Cu_2S) in the furnace. It is this matte phase that is continuously oxidized by tuyere-injected O_2. The constant presence of this matte is confirmed by the high S content (~1.3%) in the molten copper product.

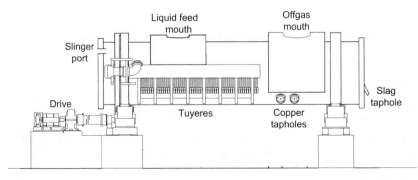

FIGURE 8.12 Sketch of Noranda continuous submerged-tuyere converter. The furnace is 19.8 m long and 4.6 m diameter. It converts matte from a Noranda smelting furnace.

TABLE 8.6 Physical and Operating Details of Noranda Continuous Submerged Tuyere Converting (Ramachandran et al., 2003)

Smelter	Xstrata Copper Horne Smelter
Noranda converter startup date	1997
Noranda converter details	
Shape	Horizontal rotating cylinder
Diameter × length, inside, m	4.6 × 19.8
Tuyeres	42
Diameter, mm	63.5
Feeds, tonnes/day	
Molten matte from Noranda smelting furnace	540
Silica flux	43
'Coolants', e.g. solid matte, smelting furnace slag concentrate, internal and external reverts	135
Blast	
Volume% O_2	24–27
Total input rate, thousand Nm^3/hour	30
Products	
Copper, tonnes/day	330
% Cu/% S	98/1.3
Slag, tonnes/day	241
%Cu in slag	11
Mass% SiO_2/mass% Fe	1.38
Cu-from-slag recovery method	Solidification/flotation
Offgas leaving furnace, thousand Nm^3/hour	30
Volume% SO_2	20

8.6.3.2. Reaction Mechanisms

Reactions in the Noranda continuous converter are as follows:

(a) A ladle of molten ~70% Cu matte (5–10% Fe, ~22% S) is poured into the furnace, mixing with the molten matte layer between copper and slag.
(b) This matte is oxidized by O_2 in the tuyere blast by the reactions:

$$\underset{\text{in molten matte}}{3FeS} + \underset{\text{in tuyere blast}}{5O_2} \rightarrow Fe_3O_4 + 3SO_2 \quad (8.11)$$

$$3Fe_3O_4 + FeS \rightarrow 10FeO + SO_2 \quad (8.12)$$

$$2FeO + \underset{\text{flux}}{SiO_2} \rightarrow \underset{\text{molten slag}}{2FeO \cdot SiO_2} \quad (8.13)$$

then (Prevost, Lapointe, Levac, & Beaudoin, 1999):

$$\underset{\text{in molten matte}}{Cu_2S} + \underset{\text{in tuyere blast}}{O_2} \rightarrow 2Cu°(l) + SO_2 \quad (8.14)$$

(c) As the matte phase is continuously consumed, drops of molten slag rise and drops of molten copper fall below the tuyeres to the molten copper layer.

(d) The matte layer is replenished with Cu, Fe, and S by the next ladle of matte feed.

Slag, matte, gas, and copper are intimately mixed in emulsion form in the converter's tuyere zone, so the above reaction scheme is an oversimplification. Nevertheless, the concept of slag formation, copper formation, matte consumption, and intermittent matte replenishment is probably correct.

8.6.3.3. Silicate Slag

Noranda continuous converting uses SiO_2 slag rather than the CaO slag used for Mitsubishi and flash converting. This is because (a) the Cu_2S layer in the Noranda converter tends to reduce magnetite so that magnetite solubility in CaO-base slag is not critical and (b) SiO_2 slag is cheaper, less corrosive, and more easily controlled than CaO slag.

8.6.3.4. Control

The critical control parameters in Noranda continuous converting are (a) matte temperature and (b) matte layer position and thickness (to ensure that tuyere O_2 blows into the matte rather than the slag or copper).

Matte temperature is measured continuously with a Noranda tuyere two-wavelength optical pyrometer (Prevost et al., 1999). It is adjusted by increasing or decreasing the rate at which solid coolants (solid matte, slag concentrate, reverts, etc.) are charged to the converter. Natural gas combustion rate and coke addition rate are also used to control temperature. Matte layer thickness is controlled by adjusting the ratio of O_2 input to matte feed rate. A high ratio decreases matte mass (hence matte layer thickness), a low ratio the opposite. Matte layer position is controlled by adjusting the amount of copper below the matte. It is altered by adjusting the frequency at which copper is tapped from the furnace. Blowing of O_2 into the slag is avoided. It tends to overoxidize the slag, precipitate magnetite, and cause slag foaming. It is avoided by controlling copper and matte layer thicknesses as described above.

8.6.3.5. Noranda Converting Summary

The Noranda continuous converter is a compact, highly productive, submerged tuyere converting process. It charges its matte via ladle through a large mouth, which is also used for charging large pieces of scrap copper. It produces 1.3% S molten copper, which is sent to a desulfurizing furnace prior to pyro- and electrorefining.

8.6.4. Recent Developments in Peirce–Smith Converting Alternatives

Three continuous converting processes are presently under development as an alternative to batch Peirce–Smith converting. They are:

(a) Codelco–Chile Continuous Converting (CCC) process (Moyano, Caballero, & Font, 2007, Moyano, Caballero, Toro, Morales, & Font, 2009). A Teniente Converter (Chapter 7) is used to convert liquid matte or white metal to blister copper. An olivine-type slag (CaO–FeO$_x$–SiO$_2$, Chapter 5) is used.

(b) Ausmelt Continuous Copper Converting (C3) technology (Wood, Matusewicz, & Reuter, 2009). The top submerged lance (TSL) C3 process is based on the commercial use of Ausmelt (Chapter 9) for batch converting of mattes at the Zhong Tiao Shan (ZTS) Non-Ferrous Metals Corporation facility in Houma, China and the Anglo Platinum Waterval smelter in Rustenburg, South Africa.

(c) Isaconvert™ uses a TSL furnace to produce blister copper from solid matte (Nikolic, Edwards, Burrows, & Alvear, 2009). It is an extension of the successful Isasmelt process (Chapter 9) and is based on the commercial use of an Isasmelt furnace that batch converts copper matte at Aurubis Precious Metals in Hoboken, Belgium.

Ausmelt C3 and Codelco–Chile Continuous Converting have been validated on a commercial scale. Isaconvert has performed successfully at a pilot scale.

8.7. SUMMARY

Converting is the second half of the smelting/converting sequence by which most Cu–Fe-sulfide concentrates are made into metallic copper. The process oxidizes the Fe and S from molten smelting furnace matte with oxygen-enriched air or air to produce molten metallic copper. It is most often carried out in the cylindrical Peirce–Smith converter. The products of the process are:

(a) Molten blister copper (99% Cu, 0.02% S and 0.6% O), which is sent forward to fire refining for final S and O removal, then anode casting
(b) Molten Fe-silicate slag (4–8% Cu), which is sent to Cu recovery, then discard
(c) SO$_2$-bearing offgas, which is treated for heat, dust, and SO$_2$ capture.

All of the heat for converting comes from Fe and S oxidation. Peirce–Smith converting is a batch process. It produces SO$_2$ intermittently and captures it somewhat inefficiently. Alternatives are:

(a) Hoboken batch converting
(b) Kennecott–Outotec continuous flash converting
(c) Noranda continuous submerged tuyere converting
(d) Mitsubishi continuous downward lance converting
(e) Ausmelt top submerged lance (TSL) batch converting
(f) Isasmelt top submerged lance (TSL) batch converting

Ausmelt converting, Isasmelt converting, and Mitsubishi continuous downward lance converting are described in Chapter 9. Three other continuous converting technologies presently under development include:

(a) Codelco—Chile Continuous Converting (CCC)
(b) Ausmelt Continuous Copper Converting (C3)
(c) Isaconvert

REFERENCES

Binegar, A. H., & Tittes, A. F. (1993). Cyprus Miami Mining Corporation siphon converter operation, past and present. In J. D. McCain & J. M. Floyd (Eds.), *Converting, fire refining and casting* (pp. 297—310). Warrendale, PA: TMS.

Boldt, J. R., & Queneau, P. (1967). *The winning of nickel*. Toronto: Longmans Canada Ltd.

Bustos, A. A., Brimacombe, J. K., & Richards, G. G. (1988). Accretion growth at the tuyeres of a Peirce—Smith copper converter. *Canadian Metallurgical Quarterly, 27*, 7—21.

Bustos, A. A., Brimacombe, J. K., Richards, G. G., Vahed, A., & Pelletier, A. (1987). Developments of punchless operation of Peirce—Smith converters. In C. Díaz, C. Landolt & A. Luraschi (Eds.), *Copper 87, Vol. 4: Pyrometallurgy of copper* (pp. 347—373). Lira 140, Santiago, Chile: Alfabeta Impresores.

Bustos, A. A., Richards, G. G., Gray, N. B., & Brimacombe, J. K. (1984). Injection phenomena in nonferrous processes. *Metallurgical Transactions B, 15B*, 77—79.

Casley, G. E., Middlin, J., & White, D. (1976). Recent developments in reverberatory furnace and converter practice at the Mount Isa Mines copper smelter. In J. C. Yannopoulos & J. C. Agarwal (Eds.), *Extractive metallurgy of copper, Vol. 1: Pyrometallurgy and electrolytic refining* (pp. 117—138). Warrendale, PA: TMS.

Chen, S., Mansikkaviita, H., Rytkonen, M., & Kylmakorpi, I. (2009). Continuous improvement in Peirce Smith converter design — Kumera's approach. In J. Kapusta & T. Warner (Eds.), *International Peirce—Smith converting centennial* (pp. 315—319). Warrendale, PA: TMS.

Coleman, M. E., & Money, G. (2009). Increasing capacity and productivity in the metals markets through pneumatic conveying and process injection technologies. In J. Kapusta & T. Warner (Eds.), *International Peirce—Smith converting centennial* (pp. 217—230). Warrendale, PA: TMS.

Davenport, W. G., Jones, D. M., King, M. J., & Partelpoeg, E. H. (2001). *Flash smelting: analysis, control and optimization*. Warrendale, PA: TMS.

Eltringham, G. A. (1993). Developments in converter fluxing. In J. D. McCain & J. M. Floyd (Eds.), *Converting, fire refining and casting* (pp. 323—331). Warrendale, PA: TMS.

Fukushima, K., Baba, K., Kurokawa, H., & Yamagiwa, M. (1988). Development of automation systems for copper converters and anode casting wheel at Toyo smelter. In E. H. Partelpoeg & D. C. Himmesoete (Eds.), *Process control and automation in extractive metallurgy* (pp. 113—130). Warrendale, PA: TMS.

Godbehere, P. W., Cloutier, J. P., Carissimi, E., & Vos, R. A. (1993). Recent developments and future operating strategies at the Horne Smelter. Paper presented at TMS Annual Meeting, Denver Colorado, February 1993.

Hills, I. E., Warner, A. E. M., & Harris, C. L. (2007). Review of high pressure tuyere injection. In A. E. M. Warner, C. J. Newman, A. Vahed, D. B. George, P. J. Mackey & A. Warczok (Eds.), *Cu2007, Vol. III Book 1: The Carlos Díaz symposium on pyrometallurgy* (pp. 471—481). Montreal: CIM.

Janney, D., George-Kennedy, D., & Burton, R. (2007). Kennecott Utah Copper smelter rebuild: over 11 million tonnes of concentrate smelted. In A. E. M. Warner, C. J. Newman, A. Vahed, D. B. George, P. J. Mackey & A. Warczok (Eds.), *Cu2007, Vol. III Book 2: The Carlos Díaz symposium on pyrometallurgy* (pp. 431—443). Montreal: CIM.

Johnson, R. E., Themelis, N. J., & Eltringham, G. A. (1979). A survey of worldwide copper converter practices. In R. E. Johnson (Ed.), *Copper and nickel converters* (pp. 1—32). Warrendale, PA: TMS.

Kapusta, J. (2010). Gas injection phenomena in converters — an update on buoyancy power and bath slopping. In *Copper 2010, Vol. 2: Pyrometallurgy I* (pp. 839—855). Clausthal-Zellerfeld, Germany: GDMB.

Kawai, T., Nishiwaki, M., & Hayashi, S. (2005). Copper concentrate smelting in Peirce—Smith converters at Onahama smelter. In A. Ross, T. Warner & K. Scholey (Eds.), *Converter and fire refining practices* (pp. 119—123). Warrendale, PA: TMS.

Kojo, I., Lahtinen, M., & Miettinen, E. (2009). Flash converting — sustainable technology now and in the future. In J. Kapusta & T. Warner (Eds.), *International Peirce—Smith converting centennial* (pp. 383—395). Warrendale, PA: TMS. http://www.outotec.com/38209.epibrw.

Larouche, P. (2001). *Minor elements in copper smelting and electrorefining*. Canada: M.Eng thesis, McGill University. http://digitool.library.mcgill.ca:8881/R/?func=dbin-jump-full&object_id=33978.

Lehner, T., Ishikawa, O., Smith, T., Floyd, J., Mackey, P., & Landolt, C. (1993). The 1993 survey of worldwide copper and nickel converter practices. In J. D. McCain & J. M. Floyd (Eds.), *Converting, fire refining and casting* (pp. 1—8). Warrendale, PA: TMS.

McKenna, K., Newman, C., Voermann, N., Veenstra, R., King, M., & Bryant, J. (2010). Extending copper smelting and converting furnace campaign life through technology. *Copper 2010, Vol. 3: Pyrometallurgy II*. Clausthal-Zellerfeld, Germany: GDMB. 971—986.

Marinigh, M. J. (2009). Technology and operational improvements in tuyere punching, silencing, pyrometry and refractory drilling equipment. In J. Kapusta & T. Warner (Eds.), *International Peirce—Smith converting centennial* (pp. 199—215). Warrendale, PA: TMS.

Maruyama, T., Saito, T., & Kato, M. (1998). Improvements of the converter's operation at Tamano smelter. In J. A. Asteljoki & R. L. Stephens (Eds.), *Sulfide smelting '98* (pp. 219—227). Warrendale, PA: TMS.

Mast, E. D., Arrian, V. J., & Benavides, V. J. (1999). Concentrate injection and oxygen enrichment in Peirce—Smith converters at Noranda's Altonorte smelter. In D. B. George, W. J. Chen, P. J. Mackey & A. J. Weddick (Eds.), *Copper 99—Cobre 99, Vol. V: Smelting operations and advances* (pp. 433—445). Warrendale, PA: TMS.

Mendoza, H., & Luraschi, A. (1993). Impurity elimination in copper converting. In J. D. McCain & J. M. Floyd (Eds.), *Converting, fire refining and casting* (pp. 191—202). Warrendale, PA: TMS.

Mitarai, T., Akagi, S., & Masatoshi, M. (1993). Development of the technique to determine the end point of the slag—making stage in copper converter. In J. D. McCain & J. M. Floyd (Eds.), *Converting, fire refining and casting* (pp. 169—180). Warrendale, PA: TMS.

Mori, K., Nagai, K., Morita, K., & Nakano, O. (2009). Recent operation and improvement at the Sumitomo Toyo Peirce—Smith converters. In J. Kapusta & T. Warner (Eds.), *International Peirce—Smith converting centennial* (pp. 151—160). Warrendale, PA: TMS.

Moyano, A., Caballero, C., & Font, J. (2007). Pilot—scale evaluation for the Codelco continuous converting process. In A. E. M. Warner, C. J. Newman, A. Vahed, D. B. George, P. J. Mackey & A. Warczok (Eds.), *Cu2007, Vol. III Book 2: The Carlos Díaz symposium on pyrometallurgy* (pp. 49—61). Montreal: CIM.

Moyano, A., Caballero, C., Toro, C., Morales, P., & Font, J. (2009). The validation of the Codelco—Chile continuous converting process. In J. Kapusta & T. Warner (Eds.), *International Peirce—Smith converting centennial* (pp. 349—360). Warrendale, PA: TMS.

Nikolic, S., Edwards, J. S., Burrows, A. S., & Alvear, G. R. F. (2009). ISACONVERT™ — TSL continuous copper converting update. In J. Kapusta & T. Warner (Eds.), *International Peirce—Smith converting centennial* (pp. 407—414). Warrendale, PA: TMS.

Pagador, R. U., Wachgama, N., Khuankla, C., & Kapusta, J. (2009). Operation of the Air Liquide shrouded injection (ALSI™) technology in a Hoboken siphon converter. In J. Kapusta & T. Warner (Eds.), *International Peirce—Smith converting centennial* (pp. 367—381). Warrendale, PA: TMS.

Pannell, D. G. (1987). A survey of world copper smelters. In J. C. Taylor & H. R. Traulsen (Eds.), *World survey of nonferrous smelters* (pp. 3—118). Warrendale, PA: TMS.

Pasca, O., Ushakov, V., & Welker, E. (2005). Hoboken converter performance improvements at the Phelps Dodge Miami smelter. In A. Ross, T. Warner & K. Scholey (Eds.), *Converter and fire refining practices* (pp. 167—176). Warrendale, PA: TMS.

Peretti, E. A. (1948). An analysis of the converting of copper matte. *Discussion of the Faraday Society, 4*, 179—184.

Persson, W., Wendt, W., & Demetrio, S. (1999). Use of optical on-line production control in copper smelters. In D. B. George, W. J. Chen, P. J. Mackey & A. J. Weddick (Eds.), *Copper 99—Cobre 99, Vol. V: Smelting operations and advances* (pp. 491—503). Warrendale, PA: TMS.

Prevost, Y., Lapointe, R., Levac, C. A., & Beaudoin, D. (1999). First year of operation of the Noranda continuous converter. In D. B. George, W. J. Chen, P. J. Mackey & A. J. Weddick (Eds.),

Copper 99—Cobre 99, Vol. V: Smelting operations and advances (pp. 269—282). Warrendale, PA: TMS.
Price, T., Harris, C., Hills, I. E., Boyd, W., & Wraith, A. (2009). Peirce—Smith converting — another 100 years? In J. Kapusta & T. Warner (Eds.), *International Peirce—Smith converting centennial* (pp. 181—197) Warrendale, PA: TMS.
Ramachandran, V., Díaz, C., Eltringham, T., Jiang, C. Y., Lehner, T., Mackay, P. J., Newman, C. J., & Tarasov, A. (2003). Primary copper production — a survey of operating world copper smelters. In C. Díaz, J. Kapusta & C. Newman (Eds.), *Copper 2003—Cobre 2003, Vol. IV: Pyrometallurgy of copper, Book 1: The Herman Schwarze symposium on copper pyrometallurgy* (pp. 3—119). Montreal: CIM.
Richards, G. G., Legeard, K. J., Bustos, A. A., Brimacombe, J. K., & Jorgensen, D. (1986). Bath slopping and splashing in the copper converter. In D. R. Gaskell, J. P. Hager, J. E. Hoffmann & P. J. Mackey (Eds.), *The Reinhardt Schuhmann international symposium on innovative technology and reactor design in extraction metallurgy* (pp. 385—402). Warrendale, PA: TMS.
Schonewille, R. H., O'Connell, G. J., & Toguri, J. M. (1993). A quantitative method for silica flux evaluation. *Metallurgical Transactions B, 24B*, 63—73.
Sharma, R. C., & Chang, Y. A. (1980). A thermodynamic analysis of the copper sulfur system. *Metallurgical Transactions B, 11B*, 575—583.
Shook, A. A., Pasca, O., & Eltringham, G. A. (1999). Online SO_2 analysis of copper converter off-gas. In D. B. George, W. J. Chen, P. J. Mackey & A. J. Weddick (Eds.), *Copper 99—Cobre 99, Vol. V: Smelting operations and advances* (pp. 465—475). Warrendale, PA: TMS.
Verney, L. R. (1987). Peirce—Smith copper converter operations and economics. In C. Díaz, C. Landolt & A. Luraschi (Eds.), *Copper 87, Vol. 4: Pyrometallurgy of copper* (pp. 375—391). Santiago, Chile: Alfabeta Impresores.
Vogt, J. A., Mackey, P. J., & Balfour, G. C. (1979). Current converter practice at the Horne smelter. In R. E. Johnson (Ed.), *Copper and nickel converters* (pp. 357—390). Warrendale, PA: TMS.
Wood, J., Matusewicz, R., & Reuter, M. A. (2009). Ausmelt C3 converting. In J. Kapusta & T. Warner (Eds.), *International Peirce—Smith converting centennial* (pp. 397—406). Warrendale, PA: TMS.

SUGGESTED READING

Kapusta, J., & Warner, T., Eds. (2009). *International Peirce—Smith converting centennial*. Warrendale, PA: TMS.
Lehner, T., Ishikawa, O., Smith, T., Floyd, J., Mackey, P., & Landolt, C. (1993). The 1993 survey of worldwide copper and nickel converter practice. In J. D. McCain & J. M. Floyd (Eds.), *Converting, fire refining and casting* (pp. 1—58). Warrendale, PA: TMS.
Ramachandran, V., Díaz, C., Eltringham, T., Jiang, C. Y., Lehner, T., Mackay, P. J., et al. (2003). Primary copper production — a survey of operating world copper smelters. In C. Díaz, J. Kapusta & C. Newman (Eds.), *Copper 2003—Cobre 2003, Vol. IV: Pyrometallurgy of copper, Book 1: The Herman Schwarze symposium on copper pyrometallurgy* (pp. 3—119). Montreal: CIM.
Ross, A., Warner, T., & Scholey, K. (2005). *Converter and fire refining practices*. Warrendale, PA: TMS.

Chapter 9

Bath Matte Smelting: Ausmelt/Isasmelt and Mitsubishi

Chapter 6 describes flash smelting, the predominant worldwide technology for producing copper mattes. The advantages of flash smelting are well-known and the technology is well established. However, flash smelting also has disadvantages. The biggest is its use of fine, dry concentrate particles as feed. Fine particles react faster, which is desirable. However, they also settle less quickly. As a result, flash furnaces generate considerable quantities of dust. To reduce this dust generation, a large settling area is built into flash furnaces. This increases the size of the vessel, and thus its cost.

In 1971, researchers at the Australian Commonwealth Scientific and Industrial Research Organization (CSIRO) began investigating the use of top-lancing technology for injecting coal into tin slags to improve reduction kinetics (Arthur, 2006; Collis, 2002, pp. 447–450; Floyd, 2002). This research led to the development of technology suitable for a variety of pyrometallurgical applications (Isasmelt, 2009; Outotec, 2010), including smelting and converting of sulfide concentrates. This technology is known generically as Top Submerged Lance (TSL) technology. It is now licensed by two separate organizations, Isasmelt and Ausmelt (the latter now owned by Outotec, Oy). The technology has found commercial application worldwide. It has become a significant factor in copper smelting, particularly in the last decade. The tonnage of new smelting capacity comprised of TSL furnaces is now roughly equal to that of other smelting furnaces (Ausmelt, 2005; see Fig. 9.1), and may likely be greater in the future.

Both Ausmelt and Isasmelt smelting are based on the technology developed at CSIRO in the 1970s. Their furnaces (Fig. 9.2, Tables 9.1 and 9.2) and operating procedures are similar. Because of this, they are described together throughout this chapter.

9.1. BASIC OPERATIONS

Ausmelt/Isasmelt copper smelting entails dropping moist solid feed into a tall cylindrical furnace while blowing oxygen-enriched air through a vertical lance into the mixture of molten matte and slag known as the *bath* (Matusewicz, Hughes, & Hoang, 2007). The products of the process are a matte/slag mixture and strong SO_2 offgas. The matte/slag mixture is tapped periodically (Isasmelt) or continuously (Ausmelt) into a fuel-fired (Isasmelt only) or electric settling furnace for separation. The choice of separating furnace is decided largely by the local cost of electricity. The settled matte (~60% Cu) is sent to conventional converting. The slag (0.7% Cu) is discarded.

The offgas (~25% SO_2) is drawn from the top of the smelting furnace through a vertical flue. It is passed through a recovery heat boiler, gas cleaning and on to a sulfuric

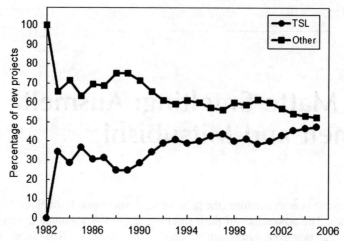

FIGURE 9.1 Percentage of new smelting capacity using TSL (Ausmelt/Isasmelt) and 'Other' (mostly flash smelting) technology (Ausmelt, 2005).

acid plant. A small amount of oxygen is added above the bath line to ensure that sulfur leaves the furnace as SO_2 rather than S_2. This prevents sulfur condensation in the gas cleaning system. In Ausmelt furnaces, this is done using the outermost pipe in the lance assembly, known as a *shroud*. Isasmelt furnaces accomplish this by injecting air in the freeboard area of the furnace.

Most of the energy for smelting comes from oxidizing the concentrate charge. Additional energy is provided by combusting (a) oil, gas, or coal fines blown through the vertical lance and (b) coal fines in the solid charge. TSL furnaces are versatile, and can use a variety of fuels, depending on cost.

9.2. FEED MATERIALS

Ausmelt/Isasmelt feed is moist concentrate, flux and recycle materials, sometimes pelletized (Tables 9.1 and 9.2). Drying of the feed is not necessary because the smelting reactions take place in the matte/slag bath rather than above it. Moist feed also decreases dust evolution. About one percent of the feed is captured as dust.

Oxygen enrichment of the air-blown into an Ausmelt/Isasmelt furnace is standard practice. The blast typically contains 50–65 vol.-% O_2, a percentage that has been steadily increasing in practice. Oxygen levels higher than this tend to cause excessive lance wear.

Because of (a) this upper limit on O_2 enrichment and (b) the presence of moisture in the solid feed, autothermal operation is usually not achieved. Instead, hydrocarbon fuel is added. Ausmelt/Isasmelt furnaces are designed to use natural gas, oil, or coal. A cool lance tip is important for reducing lance wear. As a result, coal is often added to the feed as a partial substitute for flammable fuel oil and natural gas (Arthur, 2006).

9.3. THE TSL FURNACE AND LANCES

Figure 9.2 shows a typical TSL furnace. It is a vertically aligned steel barrel, 3.5–5 m in diameter and 12–16 m high. The largest of these process well over 5000 tonnes of

Chapter | 9 Bath Matte Smelting: Ausmelt/Isasmelt and Mitsubishi   157

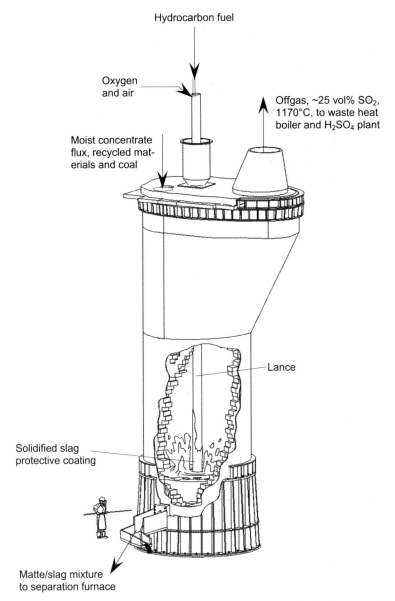

FIGURE 9.2 Cutaway view of Isasmelt furnace, 2001. A furnace is typically ~3.5 m diameter and 12 m high. It smelts up to 3000 tonnes of new concentrate per day. The outside of the furnace is often water-cooled with copper cooling blocks. The main product of the furnace is a mixture of molten matte and slag, which is sent to an electric or gas-fired matte/slag separation furnace.

concentrate per day. It is lined inside with chrome−magnesite refractory, sometimes backed with copper water-cooling blocks in Ausmelt furnaces (Table 9.1 and 9.2). Its roof consists of water-cooled copper slabs or steel panels (Binegar, 1995; Voltura, 2004). Tapping is done through one or more water-cooled tapping blocks (Isasmelt) or a weir system (Ausmelt; Reuter & Matusewicz, 2010). The weir system allows continuous operation of the furnace, which improves operation of the subsequent settling furnace.

TABLE 9.1 Operating Data for Isasmelt Smelting Furnaces Burrows, 2011

Plant	Mount Isa, Australia	Kunming, China	Miami, USA	Ilo, Peru	Sterlite, India	Mufulira, Zambia	Ust-Kamenogorsk, Kazakhstan[b]
Capacity (t/a concentrate)	1 000 000	800 000	750 000	1 200 000	1 200 000	650 000	290 000
Furnace details							
Diameter, inside brick (m)	3.75	4.4	3.75	4.4	4.4	4.4	3.6
Molten bath depth (m)	1–2	1–2	1–2	1–2	1–2	1–2	1–2
Number of tapholes	2	2	2	2	2	1	1
Furnace Lining	Refractory only	Refractory only	Water-cooled copper staves behind slag line refractory	Refractory only	Water-cooled copper staves behind slag line refractory	Refractory only	Refractory only
Lance details							
Nominal bore (mm)	450	350	400	450	400	300	250
Vol.-% O$_2$ in lance air	60–65	45–50	45–50	60–67	80–90	60–70	50–55

Chapter | 9 Bath Matte Smelting: Ausmelt/Isasmelt and Mitsubishi

Hydrocarbon fuel	Natural gas	Diesel	Natural gas	#2 Fuel Oil	Furnace oil	Diesel	Diesel
Feed details							
Solid fuel	Coal or coke breeze	Coal	Nil	Coal	Pet Coke	Coal	Coal
Feed mixing	Pelletizer	Multiple Pelletizers	Paddle mixer	Paddle Mixer	Conveyed direct to furnace	Conveyed direct to furnace	Shaft Mixer
% Cu in concentrate	22–26	18–22	25–29	25–29	26–31	28–32	22–26
% H_2O in concentrate	9	9	9	9	8	9	9
Matte/slag product							
Destination	RHF (2)	EF	EF	RHF (2)	RHF (2)	EF	EF
Temperature (°C)	1190	1180	1185	1185	1185	11	1180
Matte grade (% Cu)[a]	60–63	52–55	55–60	60–65	60–65	60–65	55–60
Slag SiO_2:Fe[a]	0.85	0.85	0.65	0.75	0.75	0.85	0.8

RHF = Rotary Holding Furnace, EF = Electric Furnace.
[a] After Settling.
[b] Startup in 2011.

TABLE 9.2 Operating Data for Ausmelt Smelting Furnaces (2010)

Smelter	Zhong Tiao Shan, China	Tongling China	RCC Russia	Jinjian China	Huludao China	NCS Namibia	Daye China	YTCL China	Xinjiang Wuxin China
Startup date	1999	2003	2006	2008	2010	2008	2010	2011	Under Design
Furnace details									
Diameter, m	4.4	4.4	5	4	4.4	4.4	5	4.4	5
Height, m	~12	~14	~14	~15	~12	~12	~18	~16	~16
Bath depth, m	~1.2	~1.5	~1.5	~1.5	~1.5	~1.5	~1.5	~1.5	~1.5
Number of tapholes	Weir	Weir	Weir	Weir	Weir	Taphole	Weir	Weir	Weir
Standby burners	Yes	Yes	Yes	Yes	Yes	Yes	Yes	Yes	Yes
Furnace cooling system	Partial	Partial	Partial	Partial	Partial	Partial	Partial	Partial	Partial
Lance details									
Construction material	Mixed steel	Mixed steel	Mixed steel	Mixed steel	Mixed steel	Mixed steel	Mixed steel	Mixed steel	Mixed steel
Lance pipe diameter, mm	~300	~400	~500	~300	~400	~300	~500	~400	~400
Construction material	Mixed steel	Mixed steel	Mixed steel	Mixed steel	Mixed steel	Mixed steel	Mixed steel	Mixed steel	Mixed steel
Tip penetration into bath, m	~0.3	~0.3	~0.3	~0.3	~0.3	~0.3	~0.3	~0.3	~0.3
Overall length, m	16	21	22	19	20	15	25	23	23
Capacity (t/a concentrate)	200 000	700 000	550 000	480 000	500 000	140 000	1 500 000	450 000	550 000

Type of charge	Moist, agglomerated	Moist, agglomerated	Moist, agglomerated	Moist, agglomerated	Moist, agglomerated	Moist, agglomerated	moist, agglomerated	moist, agglomerated	moist, agglomerated
% H_2O in charge	~10	~10	~10	~10	~10	~10	~10	~10	~10
% Cu (dry basis)	17–22	25	14–23	26	22	25	20	22	19
Lance inputs									
Vol.-% O_2 in blast	40	40	60	50	40	30	60	40	45
Hydrocarbon fuel	Coal	Coal	Natural gas	Coal	Oil (LFO)	Oil	Coal	Coal	Coal
Products									
Matte/slag mixture destination	Electric furnace	Electric furnace	Rotary furnace	Electric furnace	Electric furnace	Reverberatory furnace	Electric furnace	Electric furnace	Electric furnace
Offgas destination	Acid plant	Acid plant	Acid plant	Acid plant	Acid plant	Spray cooler	Acid plant	Acid plant	Acid plant
Liquids/offgas temperatures, °C	(1200–1300)/1180	1180/1100	1180/1100	1180/1100	1180/1100	1180/1100	1180/1100	1180/1100	1180/1100
Liquid products after settling									
Matte, approx. t/d	250	1120	710	800	700	240	1750	540	600
Approx. Grade, % Cu	60	50	40	50	50	50	55	60	56
%Cu, approx.	0.6	0.6	1	0.6	0.6	1	0.6	0.6	0.6
%SiO_2/%Fe, approx.	0.8	0.7	0.7	0.7	0.7	0.7	0.7	0.7	0.7
Destination	Granulation/discard	Granulation/concentrator	Mill/concentrator	Granulation/discard	Granulation/discard	Granulation/concentrator	Granulation/discard	Granulation/discard	Granulation/discard

FIGURE 9.3 Sketch of Isasmelt lance tip. Vanes for swirling blast down the lance tip are shown. The swirling gives rapid heat transfer from lance to blast, causing a protective layer of slag to freeze on the outside of the lance tip (Solnordal & Gray, 1996).

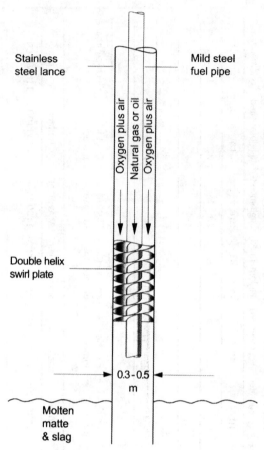

Figure 9.3 shows a TSL lance. It consists of a stainless steel outer pipe (up to 0.5 m diameter) for oxygen-enriched air and a carbon-steel inner pipe for oil or natural gas. The outer pipe is normally immersed about 0.3 m into the furnace slag. The inner pipe ends about 1 m above the slag surface. The lance pressure is low enough to allow use of a single-stage blower, rather than a compressor (Alvear, Hunt, & Zhang, 2006; Bill et al., 2002). The Ausmelt lance assembly includes an additional outer pipe for oxygen addition, which does not extend into the bath.

The lance tip is cooled by swirling the enriched air blast in the annulus between the pipes (Bill et al., 2002). The swirled gas extracts heat from the outside pipe and causes a protective slag layer to freeze on the pipe surfaces. Helical swirl vanes in the annulus are used to create this effect.

Despite this, the immersed lance tip slowly erodes away. The lance is lowered to compensate for this erosion. It is removed and replaced when ~1 m has eroded. Better operational practice has extended lance replacement time from one week to as many as four in some cases (Arthur, 2006; Bill et al., 2002; Reuter & Matusewicz, 2010; Walqui, Noriega, Partington, & Alvear, 2006). The used lances are refurbished for re-use by welding a new 1 m section of outside pipe to the bottom tip, a process that takes less than an hour. Lance wear is minimized by avoiding excessive immersion and limiting the oxygen content of the blast.

9.4. SMELTING MECHANISMS

TSL smelting is different from flash smelting in that the smelting reactions take place primarily in the bath rather than above the melt. As a result, the reaction sequence is different. It is commonly believed that dissolved magnetite in the slag serves as a catalyst of sorts for the overall process. Alvear et al. (2010) propose this reaction sequence:

$$6CuFeS_2 + 18Fe_3O_4 \rightarrow 54FeO + 7SO_2 + 3Cu_2S + 2FeS + 4FeO \quad (9.1)$$

$$54FeO + 9O_2 \rightarrow 18Fe_3O_4 \quad (9.2)$$

As a result, the presence of about 5% magnetite in the slag is important (Binegar, 1995). This requires a low bath temperature (1150–1200 °C) and low silica/iron ratio (0.65–0.85 for Isasmelt, 0.6–0.7 for Ausmelt). The low bath temperature also reduces refractory wear.

9.4.1. Impurity Elimination

Bill et al. (2002) report impurity elimination into slag and offgas during Isasmelt smelting as: 91% As, 85% Cd, 76% Bi, 68% Zn, 61% Sb, 60% Tl, 44% Pb, and 29% Te. Arsenic retention in the matte is less than half of that obtained using other smelting technology (Alvear et al., 2006), making TSL smelting technology especially suitable for high-arsenic concentrates. Elimination of As, Bi, Pb, and Sb is encouraged by decreasing oxygen enrichment. Offgas solids are recycled to the smelting furnace for Cu recovery, so they are not usually an escape route for impurities.

9.5. STARTUP AND SHUTDOWN

Smelting is started by:

(a) Preheating the furnace
(b) Slowly charging about 2 m of solid slag pieces
(c) Melting the slag, using the lance as an oxy-fuel burner
(d) Immersing the smelting lance in the molten slag
(e) Beginning normal concentrate smelting.

At least one day is required. Smelting is terminated by stopping the solid feed, draining the furnace, and turning off the lance. The furnace is then allowed to cool at its natural rate.

Steady operation consists of continuous feeding of solid charge through the roof feed port and continuous blowing of oxygen-enriched air into the molten bath. The furnace is computer controlled to give a specified concentrate smelting rate while producing matte and slag of desired composition and temperature. Matte/slag temperature is sensed by thermocouples embedded in the furnace walls (MacLeod, Harlen, Cockerell, & Crisafulli, 1997). It is controlled by adjusting the rate at which fossil fuel is supplied through the lance.

Matte and slag compositions are determined by X-ray fluorescence analysis of samples from each matte/slag tap. The compositions are controlled by adjusting the O_2/concentrate and flux/concentrate ratios in the furnace input.

9.6. CURRENT INSTALLATIONS

The number of TSL furnaces has expanded dramatically over the past decade. Figure 9.4 shows the location of TSL primary copper smelting furnaces in 2011. The figure does not include:

- Secondary copper smelting furnaces using Ausmelt and Isasmelt technology (see Chapter 19)
- Two Ausmelt converters in China (Yunnan and Houma)
- Numerous TSL furnaces smelting other nonferrous metals.

9.7. COPPER CONVERTING USING TSL TECHNOLOGY

TSL smelting is the outgrowth of technology originally designed for use in tin slag reduction. Both Ausmelt and Isasmelt have been active since then in developing uses for their furnace beyond sulfide matte smelting (Isasmelt, 2009; Outotec, 2010).

One of these is matte converting, which has been commercialized over the past decade (Wood, Matusewicz, & Reuter, 2009). The Ausmelt furnace for converting is similar to that used for smelting (Mounsey, Li, & Floyd, 1999). In fact, smelting and converting were originally performed in the same furnace (Mounsey, Floyd, & Baldock, 1998).

Because the lance tip can be easily replaced, the upper limit for enrichment of Peirce—Smith converters is no longer a concern. As a result, higher oxygen enrichment levels (30—40%) are used (Reuter & Matusewicz, 2010). This allows autothermal operation of the process, although some coal is added to reduce the copper oxide content of the slag to about 15% Cu. A disadvantage of this process compared with Peirce—Smith converting is that use of scrap and reverts is limited to smaller-size material (Rapkoch & Cerna, 2006). However, converting in TSL furnaces has numerous advantages over Peirce—Smith, especially in continuous operation (Outotec, 2010):

(a) Higher availability (no 'dead time' while the vessel is being filled or emptied)
(b) Lower fugitive emissions, since the vessel can be completely sealed
(c) Higher SO_2 concentration in the offgas (~15%), which makes acid-plant operation easier
(d) More stable operation, since conditions remain constant

FIGURE 9.4 Location of Ausmelt (circles) and Isasmelt (squares) primary smelting furnaces, 2011.

TSL converting was originally developed as a batch process, but efforts have since been made to link it to continuous smelting and settling (Matusewicz et al., 2007; Nikolic, Edwards, Burrows, & Alvear, 2009; Wood et al., 2009). This results in what is effectively the TSL equivalent of the Mitsubishi process. Continuous converting allows the system to be sealed, eliminating fugitive vapors.

TSL technology is also useful for recovering copper from non-sulfide materials, particularly slags and sludges (Hughes, 2000; see Chapter 19). Its ability to control air and fuel inputs means that conditions can be changed from oxidizing to reducing without transferring material to a second furnace. This is particularly effective for smelting Cu/Ni hydrometallurgical residues.

9.8. THE MITSUBISHI PROCESS

9.8.1. Introduction

The most recent Ausmelt processing operations feature (a) a high-intensity smelting furnace that produces a mixture of matte and slag, (b) a settling furnace (usually electric) that separates matte from slag, and (c) a converting furnace — usually a Peirce—Smith but increasingly an Ausmelt C3 converter. The smelting and settling furnace are designed to operate on a semi-continuous basis, leading to the possibility of a continuous process that inputs concentrate, flux, and fuel at one end and outputs blister copper, offgas, and slag for disposal.

However, this has been done before. In a sense, the Ausmelt process is a high-intensity version of a process first developed in the 1980s that continues to operate at four smelters today. The Mitsubishi process operates continuously, using three connected furnaces to smelt, settle, and convert copper concentrates. Although this process no longer appears to be the future of pyrometallurgical copper processing, it is worth introducing as the forerunner of bath smelting and converting technology.

Continuous copper concentrate smelting and converting has several advantages:

(a) Ability to smelt all concentrates, including $CuFeS_2$ concentrates
(b) Elimination of Peirce—Smith converting with its SO_2 collection and air infiltration difficulties
(c) Continuous production of high SO_2-strength offgas, albeit from two sources
(d) Relatively simple Cu-from-slag recovery
(e) Minimal materials handling.

9.8.2. The Mitsubishi Process

Figure 9.5 shows the three furnaces of the Mitsubishi process (Ajima et al., 1999; Mitsubishi Materials Corporation, 2004). The S (smelting) furnace blows oxygen-enriched air, dried concentrates, SiO_2 flux, and recycles into the furnace liquids via vertical lances. It oxidizes the Fe and S of the concentrate to produce ~68% Cu matte and Fe-silicate slag. Its matte and slag flow together into the electric slag-cleaning furnace.

The CL (slag-cleaning) furnace separates the smelting furnace's matte and slag. Its matte flows continuously to the converting furnace. Its slag (0.7—0.9% Cu) flows continuously to water granulation and sale or stockpile.

The C (converting) furnace blows oxygen-enriched air, $CaCO_3$ flux, and granulated converter slag coolant into the matte via vertical lances. It oxidizes the Fe and S in the

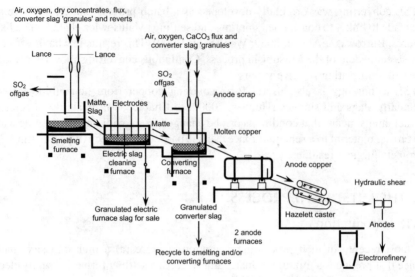

FIGURE 9.5 Mitsubishi process flowsheet and vertical layout at Gresik, Indonesia (Ajima et al., 1999). Note the continuous gravity flow of liquids between furnaces. The smelting furnace is about 15 m higher than the Hazelett caster. The smelting and converting furnaces each have 9 or 10 rotating lances.

matte to produce molten copper. The copper continuously exits the furnace into one of two holding furnaces for subsequent fire- and electrorefining. The slag (14% Cu) flows continuously into a water-granulation system. The resulting slag granules are recycled to the smelting furnace for Cu recovery and the converting furnace for temperature control.

A major advantage of the process is its effectiveness in capturing SO_2. It produces two continuous strong SO_2 streams, which are combined to make excellent feed gas for sulfuric acid or liquid SO_2 manufacture. The absence of crane-and-ladle transport of molten material also minimizes workplace emissions.

9.8.3. Smelting Furnace Details

Figure 9.6 shows details of the Mitsubishi smelting furnace. Solid particulate feed and oxidizing gas are introduced through nine vertical lances in two rows across the top of the furnace. Each lance consists of two concentric pipes inserted through the furnace roof. The diameter of the inside pipe is 5 cm; the diameter of the outside pipe, 10 cm. Dried feed is air-blown from bins through the central pipe; oxygen-enriched air (55 vol.-% O_2) is blown through the annulus between the pipes. The outside pipes are continuously rotated (7–8 rpm) to prevent them from sticking to their water-cooled roof collars.

The outside pipes extend downward to ~0.7 m above the molten bath, the inside pipes to the furnace roof or just above. The outside pipes are high chromium steel (27% Cr); they burn back ~0.4 m per day and are periodically slipped downward to maintain their specified tip positions. New 3 m sections are welded to the top of the shortened pipes to maintain continuous operation. The inside pipes are 304 stainless steel. They do not wear back because their tips are high above the reaction zone.

Concentrate/flux/recycle feed meets oxidizing gas at the exit of the inside pipe. The mixture jets onto the molten bath to form a matte–slag–gas foam/emulsion in which

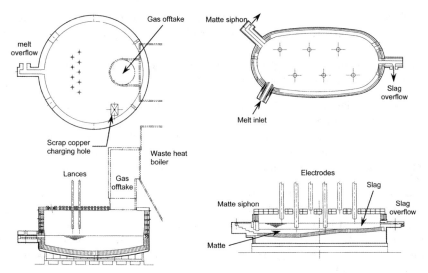

FIGURE 9.6 Details of Mitsubishi smelting and electric slag-cleaning furnaces (Goto & Hayashi, 1998). Note particularly the siphon by which matte underflows the slag-cleaning furnace. Most of the refractories are fused cast and direct-bonded magnesite chrome ($MgO-Cr_2O_3$).

liquids, solids, and gas react with each other to form matte and slag. These continuously overflow together through a taphole and down a sloped launder into the electric slag-cleaning furnace.

The offgas (25–30 vol.-% SO_2) from the oxidation reactions is drawn up a large uptake. It passes through recovery heat boilers, electrostatic precipitators, and a wet gas cleaning system before being pushed into a sulfuric acid or liquid SO_2 plant.

The velocities of solids and gas leaving Mitsubishi smelting furnace lances are 130–150 m/s (Goto & Hayashi, 1998). Times of flight of the particles across the 0.7 m distance between lance tip and melt surface are, therefore, on the order of $10^{-3}-10^{-2}$ s. The temperature rise of the gas/solid jet during this time is calculated to be ~50 °C. This is not enough to cause ignition of the concentrate. Consequently, melting and oxidation of concentrate particles occurs entirely after entry into the gas–slag–matte foam/emulsion beneath the lances (Asaki, Taniguchi, & Hayashi, 2001).

Industrial evidence indicates that the smelting furnace contains mainly matte (1.2–1.5 m deep) with a gas/slag/matte foam/emulsion beneath the lances (Shibasaki & Hayashi, 1991). Away from the lances, SO_2/N_2 gas disengages from the foam and the matte and slag begin to separate. Newly formed slag (~0.05 m thick and containing some entrained matte) flows toward the taphole where it overflows. Matte also continuously overflows as new matte is made under the lances.

9.8.4. Electric Slag-Cleaning Furnace Details

The electric slag-cleaning furnace (3600 kW) is elliptical with three or six graphite electrodes in one or two rows along the long axis. It accepts matte and slag from the smelting furnace and separates them into two layers — a bottom layer of matte 0.5–0.8 m thick, and a top layer of slag ~0.5 m thick. The residence times of the liquids are 1–2 h. These times, plus electromagnetic stirring in the furnace, allow the slag and matte to approach equilibrium. Passage of electricity through the slag ensures that the slag is hot and fluid. This, in turn, creates conditions for efficient settling of matte droplets from the slag.

The slag is the main route of Cu loss from the smelter. It is important, therefore, that the total amount of Cu-in-slag be minimized. This is done by:

(a) Maximizing slag residence time in the electric furnace (to maximize matte settling)
(b) Keeping the slag hot, fluid, and quiescent
(c) Minimizing slag mass (per tonne of Cu) by smelting high-Cu grade concentrates and minimizing fluxing
(d) Optimizing slag composition to minimize slag viscosity and density.

Matte continuously underflows from the bottom layer of the electric furnace to the converting furnace. A siphon-and-launder system is used. Slag continuously overflows through a taphole. It is granulated in flowing water and sold or stockpiled.

The purpose of the electrodes and electrical power is to keep the slag hot and fluid. Heat is obtained by resistance to electric current flow between the graphite electrodes in the slag, selectively heating the slag to 1250 °C.

Only a tiny amount of offgas is generated in the electric furnace. It is collected from the slag taphole hood, drawn through an electrostatic precipitator, and vented to atmosphere.

9.8.5. Converting Furnace Details

The converting furnace continuously receives matte from the electric slag-cleaning furnace. It blows oxygen-enriched air blast (30–35 vol.-% O_2), $CaCO_3$ flux and converter slag granules onto the surface of the matte. It also receives considerable copper scrap, including scrap anodes.

It produces molten copper (~0.7% S), (b) molten slag (14% Cu), and (c) SO_2-bearing offgas, 25–30 vol.-% SO_2.

The molten copper continuously departs the furnace through a siphon and launder into one of two anode furnaces. The slag continuously overflows into a water-granulation system. The offgas is drawn through a large uptake, recovery heat boiler, electrostatic precipitators, and a wet gas cleaning system before being pushed into a sulfuric acid or liquid SO_2 plant. The smelting and converting offgases are combined prior to entering the electrostatic precipitators.

The oxygen-enriched air and solids are introduced into the furnace through 10 lances like the smelting furnace lances, at a speed of ~120 m/s. The tips of the outside pipes are 0.3–0.8 m above the bath; the inside pipe tips are above or just through the roof. The outer pipes are continuously rotated to prevent them from sticking to their roof collars. They are also slowly slipped downwards as they burn back.

9.8.5.1. Converting Furnace Slag

A unique feature of Mitsubishi converting is its CaO-based calcium ferrite slag (see Chapter 5). Early in the development of the process, it was found that blowing strong O_2 blast onto the surface of SiO_2-based slag made a crust of solid magnetite. This made further converting impossible. CaO, on the other hand, reacts with solid magnetite, molten Cu and O_2 to form a liquid $CaO-Cu_2O-FeO_n$ slag. The slag typically contains

12–16% Cu (~60% as Cu_2O, balance Cu), 40–55% Fe (~70% as Fe^{3+}, balance Fe^{2+}), and 15–20% CaO. Its viscosity is ~10^{-1} kg/m s (Wright, Zhang, Sun, & Jahanshahi, 2000).

9.8.5.2. Converting Furnace Copper

Mitsubishi-process blister copper contains more sulfur than Peirce–Smith converter copper, ~0.7% vs. ~0.02%. This is the result of operating with excess O_2 addition to avoid a permanent matte layer in the furnace. This prevents slag foaming. The only disadvantage of the higher S level is a longer oxidation period in the anode furnace. The S content can be lowered in the Mitsubishi converter by supplying more O_2, but this increases the amount of Cu in slag.

9.8.6. Optimum Matte Grade

The Cu grade of the matte produced by the smelting furnace (and flowing into the converting furnace) is a balance between:

(a) The amount of Cu lost in the discard slag from the electric furnace;
(b) The amount of granulated high-Cu slag which must be recycled back from the converting furnace to the smelting furnace; and
(c) The amount of coal, which must be added to the smelting furnace and the amount of coolant, which must be added to the converting furnace.

The optimum matte grade for the Mitsubishi process is 68% Cu. With this matte grade, the discard slag from the electric slag-cleaning furnace contains 0.7–0.9% Cu and converter slag recycle is 0.1–0.3 tonnes per tonne of concentrate feed (Tables 9.3 and 9.4).

9.8.7. Process Control in Mitsubishi Smelting/Converting (Goto & Hayashi, 1998)

Bath temperature is the most important control parameter in the Mitsubishi process. Maintaining an optimum temperature ensures good slag and matte fluidity while at the same time minimizing refractory erosion. Typical operating temperatures at the Naoshima smelter are shown in Table 9.5.

Table 9.5 shows that the electric slag-cleaning furnace slag is operated hotter than the smelting furnace matte/slag. This is done to avoid precipitation of solid Fe_3O_4 and/or SiO_2. Precipitation of Fe_3O_4 caused formation of a muddy layer between molten matte and molten slag. This layer prevents good matte/slag separation, and results in high matte entrainment in the discard slag, hence high Cu-in-slag losses. Precipitation of SiO_2 produces SiO_2 rich floating solids. These solids (~10% Cu) do not settle into the matte layer and also result in high Cu-in-slag losses.

Process control at Naoshima combines quantitative and qualitative process information (Tables 9.6 and 9.7). The quantitative information is continuously inputted into an expert type computer control system. The overall state of the process is best represented by the following quantitative variables:

(a) Smelting furnace matte/slag outlet temperature
(b) Converting furnace slag temperature

TABLE 9.3 Physical Details of Three Mitsubishi Coppermaking Systems, 2001

Smelter	Mitsubishi Materials Corp. Naoshima, Japan	PT Smelting Co., Gresik, Indonesia	LG Nikko Onsan, Korea
Furnace commissioning date	1991	1998	1998
Copper production rate, tonnes/day	900–1000	~750	800
Smelting furnace			
Shape	Circular	Circular	Circular
Diameter × height (inside brick), m	10 × 4	10.1 × 4	10 × 4
Number of lances	9	9	9
Outside pipe diameter, cm	ID: 10	10	8 × 10 cm, 2 × 9 cm
Inside pipe diameter, cm	ID: 5	5	5
Lance rotations per min	7.7	6.5	7.8
Slag layer thickness, m	0.1	0.1	0.1
Matte layer thickness, m	1.4	1.4	1.4
Liquids, offgas temperature, °C	1225–1240, 1225–1240	1240, 1240	1240, 1250
Electric slag-cleaning furnace			
Power rating, kW	3600	2100(No.1); 1500 (No.2)	3600
Shape	Elliptical	Elliptical	Elliptical
Width × length × height, m	6.0 × 12.5 × 2.0	5.9 × 12.5 × 2.0	6.0 × 12.5 × 2.0
Electrodes			
Material	Graphite	Graphite	Graphite
Number	6	6	6
Diameter, m	0.4	0.4	0.4
Immersion in slag, m	0.5	2–0.3	5–0.6
Voltage between electrodes, V	90–120	90–120	80–100
Current between electrodes, kA	5–12.0	6.5(No.1); 5 (No.2)	5–7.0
Applied power, kW	2400	3000	1700

TABLE 9.3 Physical Details of Three Mitsubishi Coppermaking Systems, 2001—cont'd

Smelter	Mitsubishi Materials Corp. Naoshima, Japan	PT Smelting Co., Gresik, Indonesia	LG Nikko Onsan, Korea
Slag layer thickness, m	0.6	0.45	0.60
Matte layer thickness, m	0.45	0.65	0.45
Matte, slag, offgas temperatures, °C	1210, 1250, —	1230, 1270, —	1220, 1250, 1000
Estimated slag residence time, hours	~1	1–2	1–2
Converting furnace			
Shape	Circular	Circular	Circular
Diameter × height, m	8.0 × 3.6	9.0 × 3.7	8.1 × 3.6
Number of lances	10	10	10
Outside pipe diameter, cm	ID: 10	10, 9	8 × 10 cm, 2 × 8.2 cm
Inside pipe diameter, cm	ID: 5	5	5
Lance rotations per min	7.7	6.5	7.8
Slag layer thickness, m	0.12	0.15	0.13
Copper layer thickness, m	0.97	1.1	0.96
Copper, slag, offgas temperature, °C	1220, 1235, 1235	1220, 1235, 1230	1220, 1240, 1250

(c) Matte grade
(d) %SiO_2 in electric slag-cleaning furnace slag
(e) %CaO in electric slag-cleaning furnace slag
(f) %Cu in converting furnace slag
(g) %CaO in converting furnace slag.

Table 9.6 shows how each parameter is controlled.

Control of the Mitsubishi process is a challenge for its operators. The difficulty lies in the fact that (a) there are many process variables to control and (b) they all interact with each other. When one variable goes out of control the rest of the process is affected. For example, the solution to the problem of high viscosity slags, floating solids, and muddy layers has typically been to increase furnace temperature. Unfortunately, this increases refractory erosion, ultimately decreasing smelter availability.

However, recent development of sophisticated continuous melt-temperature devices coupled with the use of an expert-type computer control system has helped minimize process excursions. This has greatly increased Mitsubishi process availability.

TABLE 9.4 Operating Data for Three Mitsubishi Coppermaking Systems, 2001

Smelter	Naoshima, Japan	Gresik, Indonesia	Onsan, Korea
Smelting furnace			
Inputs, tonnes/day			
Concentrate	2300 (34% Cu)	2000–2300 (31.7% Cu)	2109 (33.2% Cu)
Silica flux	340	300–400	386 (82% SiO2)
Limestone flux	42	35	52
Granulated converting furnace slag	240	160–180	96
Smelting furnace dust	67	60	60
Converting furnace dust	61	60	60
Other	5 compressed copper scrap	40 sludge from wastewater treatment plant	14 reverts
Blast			
Vol.-% O_2	56	50–55	45–55
Input rate, Nm^3/min	540	600–650	600
Oxygen, tonnes/day	500		450 (99% O2)
Hydrocarbon Fuel (coal, tonnes/day)	63	90	140
Output offgas			
Vol.-% SO_2 (entering boiler)	34	30	31
Flowrate, Nm^3/min	560	600–650	570
Dust tonnes/day	67	60	60
Electric slag-cleaning furnace			
Inputs	Matte and slag from smelting furnace		
Outputs			
Matte, tonnes/day	1400	1270	1018
Matte %Cu	68	68	68.8
Slag, tonnes/day	1300	1450	1331
Slag %Cu	0.7	0.7	0.8
Slag %SiO$_2$/%Fe	0.9	0.8	0.9
Offgas, Nm^3/min (ppm SO_2)			500 (<150 ppm)
kWh consumption/tonne of slag	42	55	30

TABLE 9.4 Operating Data for Three Mitsubishi Coppermaking Systems, 2001—cont'd

Smelter	Naoshima, Japan	Gresik, Indonesia	Onsan, Korea
Converting furnace (autothermal)			
Inputs, tonnes/day			
Matte from electric furnace	1400	1270	1018
Limestone flux	50	30	69
Granulated converting furnace slag	360	160–180	246
Anode scrap copper	120	95–100	78
Purchased scrap copper	34	0	44
Blast			
Vol.-% O_2	35	25–28	32–35
Input rate, Nm^3/min	490	460–470	430
Oxygen, t/d	180		133 (99% O_2)
Outputs			
Copper, tonnes/day	900–1000	850–900	820
Copper, %Cu	98.4	98.5	98.5
Copper, %O	0.3	0.3	—
Copper, %S	0.7	0.7	0.9
Slag, t/d	600	160–180	360
Slag, %Cu	14	14	15
Slag, mass% CaO/mass% Fe	0.4	0.4	0.34
Offgas, Nm^3/min	480	450	410
Vol.-% SO_2	31	25	24
Dust, t/d	61	60	60

TABLE 9.5 Maximum, Minimum, and Optimum Operating Temperatures in the Mitsubishi Smelter at Naoshima

	Minimum (°C)	Maximum (°C)	Optimum (°C)
Smelting furnace — matte/slag outlet	1200	1240	1220
Electric slag-cleaning furnace — slag outlet	1220	1260	1240
Converting furnace — slag outlet	1210	1250	1230

TABLE 9.6 Quantitative Process Parameters in Mitsubishi Smelting/Converting

Information	Adjustable or Result	Comments
Smelting furnace		
Melt temperature	Result	Continuous measurement by submerged K-type thermocouple at smelting furnace melt outlet.
Feedrate	Adjustable	Continuous monitoring of blended concentrate, silica, $CaCO_3$, coal, converter slag granules, Cu scrap, and dust input rates.
Blowing rate	Adjustable	Continuous monitoring of oxygen and air input rates.
Matte composition	Result	Hourly analysis by X-ray fluorescence analyzer (Cu and Pb). Samples are taken at electric slag-cleaning furnace outlet.
Slag composition	Result	Hourly analysis by X-ray fluorescence analyzer (Cu, Fe, SiO_2, CaO, and Al_2O_3). Samples are taken at electric slag-cleaning furnace outlet.
Converting furnace		
Slag temperature	Result	Hourly manual measurement by disposable K-type thermocouples.
Copper temperature	Result	Automatic measurement by submerged glass fiber optical pyrometer — every 15 min.
Feedrate	Adjustable	Continuous monitoring of $CaCO_3$, recycled converter slag granules and Cu scrap input rates.
Matte flowrate	Result	Continuous measurement by infrared scanner at electric slag-cleaning furnace matte siphon.
Blowing rate	Adjustable	Continuous measurement of oxygen and air input rates.
Slag composition	Result	Hourly sample and analysis (Cu, SiO_2, CaO, S, and Pb) by X-ray fluorescence analyzer.

9.9. THE MITSUBISHI PROCESS IN THE 2000s

Improvements in the Mitsubishi Process since 2000 have focused on two concerns: refractory life and process control. Refractory life is a challenge in the S furnace (due to the turbulence created by the injection lances), and in the C furnace (due to turbulence and the more corrosive nature of the calcium ferrite slag). Increasing the S-furnace lance diameter from 0.060 to 0.069 m reduced penetration depth and improved refractory life (Cho, Lee, Chang, Seo, & Kim, 2006; Taniguchi, Matsutani, & Sato, 2006), as did improved distribution control for the solid feed to create even penetration depth. C-furnace refractory life has been extended by adjusting the refractory in front of the water-cooled copper jacket at the bath line (Goto, Konda, Muto, Yamashiro, & Matsutano, 2002).

TABLE 9.7 Qualitative Process Parameters in Mitsubishi Smelting/Converting

Information	Adjustable or Result	Comments
Smelting furnace		
Melt flowrate and melt surface condition	Result	Visual inspection by remote camera and at the furnace.
Converting furnace		
Slag flowrate and slag surface condition	Result	Visual inspection by remote camera and at the furnace.
Matte inflow rate	Result	Visual inspection by remote camera and at the furnace.

One of the challenges to successful operation of the C furnace is magnetite control. Some magnetite is necessary to provide a protective coating for the refractories, but too much causes accretions to build up in the furnace, eventually blocking the siphon through which molten blister copper is recovered (Hasegawa & Sato, 2003; Taniguchi et al., 2006). As a result, careful control of the slag chemistry is needed (Tanaka, Iida, & Takeda, 2003). Increasing the CaO content of the slag reduces accretions, but in turn requires better heat removal through the water jacket at the slag line. Plant operations at Naoshima now control the Fe/CaO ratio separately from the %Cu in the C-furnace slag; maintaining 2.2 < Fe/CaO < 2.45 seems to work best.

The Mitsubishi Process currently operates at three copper smelting facilities: Birla in India, Naoshima in Japan, Onsan in Korea, and Gresik in Indonesia. The Gresik facility is the most recent of the three (1999), and there is no indication of new installations in the near future.

9.10. SUMMARY

Ausmelt and Isasmelt smelting is done in vertically aligned cylindrical furnaces 3.5–5 m diameter and 12–16 m high. The smelting entails:

(a) Dropping moist concentrate, flux, and recycle materials into a molten matte/slag bath in a hot furnace
(b) Blowing oxygen-enriched air through a vertical lance into the matte/slag bath.

Most of the energy for smelting is obtained from oxidizing the Fe and S in the concentrate.

The principal product of the furnace is a matte/slag mixture. It is tapped into a hydrocarbon-fired or electric settling furnace. The products after settling are 60% Cu matte and 0.7% Cu slag. The main advantages of the process are:

(a) Its small footprint, which makes it easy to retrofit into existing smelters
(b) Its ability to process a variety of feed materials, and
(c) Its small evolution of dust.

The Mitsubishi Process also performs smelting and converting in a molten bath. It uses three interconnected furnaces to continuously smelt copper concentrates, settle, and separate matte from discard slag, and convert the matte to blister copper. When performing properly, continuous processing solves many of the problems created by more traditional copper smelting technology, most notably the elimination of ladle transfer of molten matte, the fugitive emissions associated with Peirce–Smith converting, and the uneven SO_2 offgas content that results from batch processing of copper concentrates.

REFERENCES

Ajima, S., Koichi, K., Kanamori, K., Igarashi, T., Muto, T., & Hayashi, S. (1999). Copper smelting and refining in Indonesia. In D. B. George, W. J. Chen, P. J. Mackey & A. J. Weddick (Eds.), *Copper 99—Cobre 99, Vol. V: Smelting operations and advances* (pp. 57—69). Warrendale, PA: TMS.

Alvear, G. R., Arthur, P., & Partington, P. (2010). Feasibility to profitability with Copper ISASMELT™. In *Copper 2010, Vol. 2: Pyrometallurgy I* (pp. 615—630). Clausthal-Zellerfeld, Germany: GDMB.

Alvear, G. R., Hunt, S. P., & Zhang, B. (2006). Copper Isasmelt — dealing with impurities. In F. Kongoli & R. G. Reddy (Eds.), *Sohn international symposium, Vol. 8: International symposium on sulfide smelting 2006* (pp. 673—685). Warrendale, PA: TMS.

Arthur, P. S. (2006). ISASMELT™ — 6,000,000 TPA and rising. In F. Kongoli & R. G. Reddy (Eds.), *Sohn international symposium, Vol. 8: International symposium on sulfide smelting* (pp. 275—290). Warrendale, PA: TMS. www.isasmelt.com/downloads/Arthur%20-%20ISASMELT%20-%206000000tpa%20and%20Rising%20060326.pdf.

Asaki, Z., Taniguchi, T., & Hayashi, M. (2001). Kinetics of the reactions in the smelting furnace of the Mitsubishi process. *JOM, 53*(5), 25—27.

Ausmelt. (2005). *Ausmelt TSL technology: Capability & experience in copper smelting & converting*. Ausmelt Global Environmental Technologies. 22 September 2005.

Bill, J. L., Briffa, T. E., Burrows, A. S., Fountain, C. R., Retallick, D., Tuppurainen, J. M. I., et al. (2002). Isasmelt — Mount Isa copper smelter performance update. In R. L. Stephens & H. Y. Sohn (Eds.), *Sulfide smelting 2002* (pp. 181—193). Warrendale, PA: TMS.

Binegar, A. H. (1995). Cyprus Isasmelt start-up and operating experience. In W. J. Chen, C. Díaz, A. Luraschi & P. J. Mackey (Eds.), *Copper 95—Cobre 95, Vol. IV: Pyrometallurgy of copper* (pp. 117—132). Montreal: CIM.

Burrows, A. (2011). Personal communication.

Cho, H. Y., Lee, K. H., Chang, J. S., Seo, W. C., & Kim, J. Y. (2006). Expansion and modernization of Onsan Smelter. In F. Kongoli & R. G. Reddy (Eds.), *Sohn International Symposium, Vol. 8: International symposium on sulfide smelting* (pp. 97—110). Warrendale, PA: TMS.

Collis, B. (2002). *Fields of discovery: Australia's CSIRO*. Crows Nest, Australia: Allen & Unwin.

Floyd, J. M. (2002). Converting an idea into a worldwide business commercializing smelting technology. *Metallurgical and Materials Transactions B, 36B*, 557—575.

Goto, M., & Hayashi, M. (1998). *The Mitsubishi continuous process — a description and detailed comparison between commercial practice and metallurgical theory*. Tokyo: Mitsubishi Materials Corporation.

Goto, M., Konda, K., Muto, T., Yamashiro, A., & Matsutano, A. (2002). Gresik copper smelter and refinery — current operation and expansion plan. In R. L. Stephens & H. Y. Sohn (Eds.), *Sulfide smelting 2002* (pp. 61—72). Warrandale, PA: TMS.

Hasegawa, N., & Sato, H. (2003). Control of magnetite behavior in the Mitsubishi process at Naoshima. In F. Kongoli, K. Itagaki, C. Yamauchi & H. Y. Sohn (Eds.), *Yazawa international symposium, Vol. II: High-temperature metals production* (pp. 509—519). Warrendale, PA: TMS.

Hughes, S. (2000). Applying Ausmelt technology to recover Cu, Ni, and Co from slags. *Journal of Metals, 52*(8), 30—33.

Isasmelt. (2009). *IsaSmelt installations*. http://www.isasmelt.com/index.cfm?action=dsp_content&contentID=28.

MacLeod, I. M., Harlen, W., Cockerell, R. A., & Crisafulli, S. (1997). Advanced temperature control for a copper Isasmelt plant. In *ECC 97*. http://www.cds.caltech.edu/conferences/related/ECC97/proceeds/251_500/ECC316.PDF Paper TH-M-H-6.

Matusewicz, R., Hughes, S., & Hoang, J. (2007). The Ausmelt continuous copper converting (C3) process. In A. E. M. Warner, C. J. Newman, A. Vahed, D. B. George, P. J. Mackey & A. Warczok (Eds.), *Cu2007 — Vol. III Book 2: The Carlos Díaz symposium on pyrometallurgy* (pp. 29–46). Montreal: CIM.

Mitsubishi Materials Corporation. (2004). *The Mitsubishi process: Copper smelting for the 21st century*. http://www.mmc.co.jp/sren/MI_Brochure.pdf.

Mounsey, E. N., Floyd, J. M., & Baldock, B. R. (1998). Copper converting at Bindura Nickel Corporation using Ausmelt technology. In J. A. Asteljöki & R. L. Stephens (Eds.), *Sulfide smelting '98* (pp. 287–301). Warrendale, PA: TMS.

Mounsey, E. N., Li, H., & Floyd, J. W. (1999). The design of the Ausmelt technology smelter at Zhong Tiao Shan's Houma smelter, People's Republic of China. In D. B. George, W. J. Chen, P. J. Mackey & A. J. Weddick (Eds.), *Copper 99—Cobre 99, Vol. V: Smelting operations and advances* (pp. 357–370). Warrendale, PA: TMS.

Nikolic, S., Edwards, J. S., Burrows, A. S., & Alvear, G. R. F. (2009). ISACONVERT™ – TSL continuous copper converting update. In J. Kapusta & T. Warner (Eds.), *International Peirce—Smith centennial* (pp. 407–414). Warrendale, PA: TMS.

Outotec. (2010). *Ausmelt smelting and converting*. http://www.outotec.com/pages/Page____39927.aspx?epslanguage=EN.

Rapkoch, J. M., & Cerna, M. (2006). Isasmelt copper converting: a review of possibilities and challenges. In F. Kongoli & R. G. Reddy (Eds.), *Sohn international symposium, Vol. 8: International symposium on sulfide smelting 2006* (pp. 251–259). Warrendale, PA: TMS.

Reuter, M.A., & Matusewicz, R. (2010). Personal communication.

Shibasaki, T., & Hayashi, M. (1991). Top blowing injection smelting and converting — the Mitsubishi process. In T. Lehner, P. J. Koros & V. Ramachandran (Eds.), *Injection in process metallurgy* (pp. 199–213). Warrendale, PA: TMS.

Solnordal, C. B., & Gray, N. B. (1996). Heat transfer and pressure drop considerations in the design of Sirosmelt lances. *Metallurgical and Materials Transactions B, 27B*(4), 221–230.

Tanaka, F., Iida, O., & Takeda, Y. (2003). Thermodynamic fundamentals of calcium ferrite slag and their application to Mitsubishi continuous copper converter. In F. Kongoli, K. Itagaki, C. Yamauchi & H. Y. Sohn (Eds.), *Yazawa international symposium, Vol. II: High-temperature metals production* (pp. 495–508). Warrendale, PA: TMS.

Taniguchi, T., Matsutani, T., & Sato, H. (2006). Technological innovations in the Mitsubishi process to achieve four years campaign. In F. Kongoli & R. G. Reddy (Eds.), *Sohn international symposium, Vol. 8: International symposium on sulfide smelting* (pp. 261–273). Warrendale, PA: TMS.

Voltura, S. A. (2004). Continuous improvements at the Phelps Dodge Miami mining smelter. *Minerals & Metallurgical Processing, 21*, 158–163.

Walqui, H., Noriega, C., Partington, P., & Alvear, G. (2006). SPCC's 1200 TPA copper Isasmelt. In F. Kongoli & R. G. Reddy (Eds.), *Sohn international symposium, Vol. 8: International symposium on sulfide smelting 2006* (pp. 199–209). Warrendale, PA: TMS.

Wood, J., Matusewicz, R., & Reuter, M. A. (2009). Ausmelt C3 converting. In J. Kapusta & T. Warner (Eds.), *International Peirce—Smith centennial* (pp. 397–406). Warrendale, PA: TMS.

Wright, S., Zhang, L., Sun, S., & Jahanshahi, S. (2000). Viscosity of calcium ferrite slags and calcium aluminosilicate slags containing spinel particles. In S. Seetharaman & D. Sichen (Eds.), *Sixth international conference on molten slags, fluxes and salts*. Stockholm, Sweden: Division of Metallurgy, KTH. paper number 059.

SUGGESTED READING

Binegar, A. H. (1995). Cyprus Isasmelt start-up and operating experience. In W. J. Chen, C. Díaz, A. Luraschi & P. J. Mackey (Eds.), *Copper 95—Cobre 95, Vol. IV: Pyrometallurgy of copper* (pp. 117–132). Montreal: CIM.

Goto, M., & Hayashi, M. (1998). *The Mitsubishi continuous process — a description and detailed comparison between commercial practice and metallurgical theory.* Tokyo, Japan: Mitsubishi Materials Corporation.

Mitsubishi Materials Corporation. (2004). *The Mitsubishi process: Copper smelting for the 21st century.* http://www.mmc.co.jp/sren/MI_Brochure.pdf.

Mounsey, E. N., Li, H., & Floyd, J. W. (1999). The design of the Ausmelt technology smelter at Zhong Tiao Shan's Houma smelter, People's Republic of China. In D. B. George, W. J. Chen, P. J. Mackey & A. J. Weddick (Eds.), *Copper 99—Cobre 99, Vol. V: Smelting operations and advances* (pp. 357—370). Warrendale, PA: TMS.

Chapter 10

Direct-To-Copper Flash Smelting

Previous chapters show that coppermaking from sulfide concentrates entails two major steps: *smelting* and *converting*. They also show that smelting and converting are part of the same chemical process: oxidation of Fe and S from a Cu–Fe–S phase. It has long been the goal of metallurgical and chemical engineers to combine these two steps into one *continuous direct-to-copper* smelting process. In 2011 this has been accomplished in three smelters, all using flash furnaces:

- Olympic Dam, Australia (Ranasinghe, Russell, Muthuraman, & Dryga, 2010);
- Głogów II, Poland (Byszynski, Garycki, Gostynski, Stodulski, & Urbanowski, 2010);
- Chingola, Zambia (Taskinen & Kojo, 2009).

The Głogów smelting complex is planning a second direct-to-copper flash furnace. It will replace their existing blast furnaces, which are environmentally difficult.

Figure 10.1 shows the interior of the new Zambian direct-to-copper flash furnace. Figure 10.2 shows a flowsheet for the Olympic Dam direct-to-copper flash furnace smelter.

10.1. ADVANTAGES AND DISADVANTAGES

The principal advantages of combining smelting and converting are:

(a) Isolation of SO_2 emission to a single, continuous, SO_2-rich gas stream, ideal for capturing sulfur in sulfuric acid;
(b) Minimization of energy consumption;
(c) Minimization of capital and operating costs.

This chapter describes direct-to-copper smelting. The above three advantages have been achieved. Current disadvantages of the process are that (a) about 25% of the Cu entering a direct-to-copper smelting furnace ends up dissolved in the slag (compared with <10% in traditional Peirce–Smith converting; see Chapter 8) and (b) the cost of recovering this Cu will probably restrict future expansion of direct-to-copper smelting to low-Fe chalcocite and bornite concentrates rather than high-Fe chalcopyrite concentrates; smelting of low-Fe concentrates generates less slag and therefore less copper loss.

Tables 10.1a-c describe the concentrates currently being smelted in direct-to-copper furnaces.

10.2. THE IDEAL DIRECT-TO-COPPER PROCESS

Figure 10.3 is a sketch of the ideal direct-to-copper process. The principal inputs to the process are concentrate, oxygen, air, flux, and recycles. The principal outputs are molten copper, low-Cu slag, and high-SO_2 offgas.

FIGURE 10.1 New Chingola direct-to-copper flash furnace. The photo shows the top of the reaction shaft and the bottom tip of the concentrate burner. The diameter of the reaction shaft is ~6 m. Photograph by W.G. Davenport.

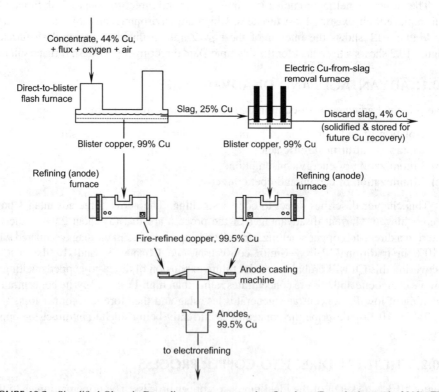

FIGURE 10.2 Simplified Olympic Dam direct-to-copper smelter flowsheet (Ranasinghe et al., 2010). The Outokumpu flash smelting furnace, electric Cu-from-molten slag recovery furnace, fire refining, and anode casting systems are notable. The slag is molten until it is solidified for storage (right). The copper is molten until it is cast as anodes (bottom).

TABLE 10.1a Elements and Equivalent Minerals in Olympic Dam Direct-to-Copper Flash Furnace Concentrate Feed (Ranasinghe et al., 2010; Solnordal, Jorgensen, Koh, & Hunt, 2006)

Element	Cu	Fe	S	SiO_2		
Mass %	44	20	25	5		
Minerals	Bornite Cu_5FeS_4	Digenite $Cu_{1.8}S$	Chalcopyrite $CuFeS_2$	Quartz SiO_2	Pyrite FeS_2	Hematite Fe_2O_3
Mass %	28–45	25–35	11–27	2–6	0.2–2	8–12

TABLE 10.1b Elements and Equivalent Minerals in Głogów Direct-to-Copper Flash Furnace Concentrate Feed (Czernecki et al., 2010; Dobrzanski & Kozminski, 2003)

Element	Cu	Fe	S	'Free' carbon	Pb	As	Other
Mass %	17–31	2–7	8–11	5–8	1.5–3	0.05–4	42–50
Minerals	In chalcocite Cu_2S and bornite Cu_5FeS_4	In bornite Cu_5FeS_4 and hematite Fe_2O_3	Mainly in chalcocite and bornite	Carbon	In galena, PbS	In arsenopyrite, $FeAsS_2$, enargite, Cu_3AsS_4, tennantite	20% SiO_2 6% Al_2O_3 6% CaO^a 4% MgO K_2O, Na_2O

^{a}CaO, MgO, K_2O, and Na_2O are present as carbonates.

TABLE 10.1c Elements and Equivalent Minerals in Chingola Direct-to-Copper Flash Furnace Concentrate Feed (Syamujulu)

Element	Cu	Fe	S	SiO_2	Other
Mass %	39 mainly in chalcocite, bornite & chalcopyrite	8 mainly in bornite & chalcopyrite (minor in pyrite)	19 mainly in chalcocite bornite & chalcopyrite	18 mainly in feldspar and clay	10% Al_2O_3 (mainly in clay) 2% CaO 2% MgO

The process is continuous and autothermal. With highly oxygen-enriched (low-N_2) blast, there is enough excess reaction heat to melt all the Cu-bearing recycle materials from the smelter and adjacent refinery, including scrap anodes.

The remainder of this chapter indicates how close the process has come to this ideality.

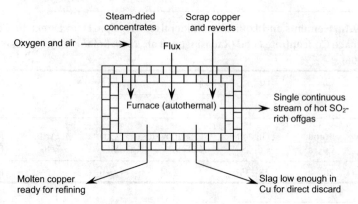

FIGURE 10.3 Ideal single-furnace coppermaking process. Ideally the copper is low in impurities, the slag is discardable without Cu-recovery treatment, and the offgas is strong enough in SO_2 for sulfuric acid manufacture. Little or no fossil fuel is required.

TABLE 10.2 Compositions of Slags from Direct-to-Copper Flash Smelting

Smelter								
	Cu	Fe	SiO_2	CaO	MgO	Al_2O_3	K_2O	Na_2O
Element/compound				Mass %				
Olympic Dam, Australia	23–28	33	18	0	0	3.5		
Głogów, Poland	12–15	6	31	14	6	9	Yes	Yes
Chingola, Zambia	17–20	17–29	28–32	5	3–7	5–7		

10.3. INDUSTRIAL SINGLE FURNACE DIRECT-TO-COPPER SMELTING

In 2011, single furnace direct-to-copper smelting is done by only one process: Outotec (formerly Outokumpu) flash smelting (Fig. 10.1). For several years the Noranda submerged-tuyere process (Chapter 7) also produced copper directly. This plant now produces high-grade (72–74% Cu) matte. The change was made to avoid slag foaming (Section 10.7.1) and to increase impurity elimination.

The products of direct-to-copper flash smelting (Tables 10.2 and 10.3) are impure molten copper (99% Cu, 1300 °C), molten slag (12–28% Cu, ~1310 °C), and offgas (16–41 vol.-% SO_2, 1250–1400 °C). As with conventional matte flash smelting, the temperature in the smelting furnace is controlled by adjusting (a) the degree of oxygen enrichment of the blast (which changes the amount of N_2 coolant entering the furnace) and (b) the rate at which fossil fuel is burnt in the furnace.

The O_2 content of industrial direct-to-copper flash furnace blast is 65–95 vol.-% O_2, depending on the concentrate composition. Considerable fossil fuel is burnt in the reaction shaft and in settler burners (Table 10.3).

TABLE 10.3 Details of Chingola and Glogów Direct-to-Copper Outokumpu Flash Furnaces (2010)

Smelter	Konkola Copper Mines, Chingola, Zambia	KGHM Polska Miedz Glogów, Poland
Startup date	2008	1978 modernized 2010
Size, inside brick, m		
Hearth: w × l × h	7.4 × 24 × 2.4	8.9 × 25.6 × 3.8
Reaction shaft		
Diameter	~6	7.55
Height, above settler roof	6	9.1
Gas uptake		
Diameter	4	6.7
Height above settler roof	8	12.3
Slag layer thickness, m	0.7–1.35	0.5
Copper layer thickness, m	0–0.7	0.7
Active slag tapholes	3	5
Active copper tapholes	8	13
Concentrate burners	1	1
Settler burners	10	Normally none
Feed details, tonnes/day		
New concentrate (dry)	1440–2600, 31–38% Cu	2000 (27% Cu)
Oxygen	400–700	
Silica flux		Self-fluxing
Recycle flash furnace dust	0-300	250
Other	80–130 CaO in flux	115 desulfurizing dust
Blast details		
Blast temperature, °C	Ambient	Ambient
Vol.-% O_2	95	65
Flowrate, thousand Nm^3/hour	22	32
Production details		
Copper production, tonnes/day	~1000	420

(Continued)

TABLE 10.3 Details of Chingola and Glogów Direct-to-Copper Outokumpu Flash Furnaces (2010)—cont'd

Smelter	Konkola Copper Mines, Chingola, Zambia	KGHM Polska Miedz Glogów, Poland
Composition	99.2% Cu, 0.5% O; 0.3% S	98.7% Cu; 0.01% Fe, 0.48% O; 0.31% Pb, 0.05% S
Temperature, °C	1280–1300	1300 maximum
Slag production, tonnes/day	900–1400	1060
Mass% Cu	17–20	14
Mass% SiO_2/mass%Fe	0.9	3.95
Temperature, °C	1300–1320	1310 maximum
Cu-from-slag recovery method	Electric furnace	Electric furnace
Offgas, thousand Nm^3/hour	26–46	47
Volume% SO_2, leaving furnace	32–41	16
Temperature, °C	1400	1250
Dust production, tonnes/day	Boiler 150, ESP 80	250
Hydrocarbon fuel inputs, kg/hour		
Burnt in reaction shaft	Oil, 150–650	Oil, 80
Burnt in settler burners	Oil, 3000	0

10.4. CHEMISTRY

Direct-to-copper smelting takes place by the schematic (unbalanced) reaction:

$$\underset{\text{in concentrate}}{Cu_2S, Cu_5FeS_4(s)} + \underset{\text{in oxygen-enriched blast}}{O_2(g)} + \underset{\text{in flux}}{SiO_2(s)} \xrightarrow{\sim 1300\,°C}$$

$$\underset{\text{impure molten copper}}{Cu°(l)} + \underset{\text{slag}}{FeO, Fe_3O_4, SiO_2(l)} + \underset{\text{in offgas}}{SO_2(g)} \quad (10.1)$$

Just enough O_2 is supplied to produce metallic copper rather than Cu_2S or Cu_2O.

In practice, the flash furnace reaction shaft product is a mixture of over-oxidized (oxide) and under-oxidized (sulfide) materials. Individual particles may be over-oxidized on the outside and under-oxidized on the inside. The over- and under-oxidized components react to give metallic copper, as shown in the reactions below:

$$\underset{\text{over-oxidized Cu}}{2Cu_2O(s)} + \underset{\text{under-oxidized Cu sulfide}}{Cu_2S(s)} \xrightarrow{\sim 1300\,°C} \underset{\text{impure molten copper}}{6Cu°(l)} + \underset{\text{in offgas}}{SO_2(g)} \quad (10.2)$$

and

$$\underset{\text{over-oxidized FeO}}{2Fe_3O_4(s)} + \underset{\substack{\text{under-oxidized} \\ \text{Cu sulfide}}}{Cu_2S(s)} \xrightarrow{\sim 1300\ ^\circ C} \underset{\substack{\text{impure molten} \\ \text{copper}}}{2Cu^\circ(l)} + \underset{\text{in silicate slag}}{6FeO(l)} + \underset{\text{in offgas}}{SO_2(g)} \quad (10.3)$$

Industrially, the overall extent of Reaction (10.1) is controlled by (a) monitoring the Cu content of the product slag and the S content of the product copper and (b) adjusting the ratio of oxygen in the blast to concentrate fed to the furnace, based on these measured Cu-in-slag and S-in-copper values. An increasing Cu content in the slag is reversed by decreasing the O_2/concentrate ratio and vice versa. The copper slag content is kept between 12 and 28%.

10.5. EFFECT OF SLAG COMPOSITION ON % Cu-IN-SLAG

Table 10.2 gives the compositions of direct-to-copper flash furnace slags. The most notable feature is that the concentration of Cu in Głogów slag is half that of Olympic Dam slag. This is fortunate because Głogów produces about 2.5 tonnes of slag per tonne of copper (Table 10.3), while Olympic Dam produces about 1.4 t slag per tonne of copper. The net result is that both smelters send about 25% of the copper in the feed to slag.

The process at Głogów produces a lower copper slag because its composition is essentially a Ca—Al—Mg—K—Na silicate slag (Table 10.3), which has a relatively low solubility for Cu oxide (Taskinen & Kojo, 2009).

Conventional smelting/converting smelters produce ~2.5 t slag per tonne of copper production, about the same as Głogów. However, the slags contain much less Cu (1—2% for smelting; 4—8% for converting, some it entrained rather than dissolved). Cu removal from these slags is quicker and cheaper than from the direct-to-copper smelting slags.

10.6. INDUSTRIAL DETAILS

Operating details of two direct-to-copper flash furnaces are given in Table 10.3. The furnaces are similar to conventional flash furnaces. Differences are:

(a) The hearths are deeply bowl-shaped to prevent molten copper from contacting the furnace sidewalls
(b) The hearths are more radically arched and compressed to prevent their refractory from being floated by the dense (7.8 t/m^3) molten copper layer (Brillo & Egry, 2003; Hunt, 1999)
(c) The furnace walls are extensively water-cooled and the hearth extensively air-cooled to prevent metallic copper from seeping too far into the refractories
(d) The bottom refractories are monolithic to prevent molten copper from seeping under the bricks, solidifying, and lifting them
(e) The copper tapholes are designed to prevent the out-flowing molten copper from enlarging the taphole to the point where molten copper contacts cooling water

At Olympic Dam, molten copper passes through magnesite—chrome brick (inside) and a silicon carbide insert and graphite insert (outside) (Hunt, Day, Shaw, Montgomerie, & West, 1999). The graphite insert is replaced after ~1200 t tapped copper and the silicon carbide insert is replaced after ~2400 t. Excessive copper flow (the result of an excessive taphole diameter) initiates earlier replacement.

10.7. CONTROL

The compositions of the industrial furnace products are controlled by adjusting the ratios of oxygen in the blast to concentrate feed rate and flux/concentrate feed rate. The temperatures of the products are controlled by adjusting the oxygen-enrichment level of the blast (as represented by the N_2/O_2 ratio) and the rate at which fossil fuel is burnt in the furnace.

10.7.1. Target: No Matte Layer to Avoid Foaming

Industrial direct-to-copper furnaces are operated with O_2/concentrate ratios that are high enough to avoid forming a Cu_2S layer. This is done to avoid the possibility of foaming slag out the top of the furnace (Hunt et al., 1999). A molten Cu_2S layer, once built up between the molten copper and molten slag layers, has the potential to react with slag by reactions such as (10.2) and (10.3), which can produce SO_2 *beneath the slag layer.*

Foaming is particularly favored if the input O_2/concentrate ratio is subsequently increased to shrink or remove an existing Cu_2S layer. This results in a highly oxidized slag, full of dissolved Fe_3O_4 and Cu_2O, which has great potential for producing SO_2 beneath the slag layer.

The foaming problem is avoided by ensuring that the O_2/concentrate ratio is always at or above its set point, never below. This may lead to high Cu-in-slag levels but it avoids the potentially serious operational problems caused by foaming (Hunt et al., 1999). S-in-copper below ~1% S ensures that a Cu_2S layer does not form.

10.7.2. High % Cu-in-Slag from No-Matte-Layer Strategy

An unfortunate side-effect of the above no-matte-layer strategy is a high Cu content in the slag, mainly as dissolved Cu_2O. It arises because there is no permanent layer of Cu_2S in the furnace to reduce Cu_2O to metallic copper, Reaction (10.2).

Simply stated, direct-to-copper smelting is operated in a slightly over-oxidizing mode to prevent the foaming described in Section 10.7.1. The downside of operating this way is that 12–28% Cu reports to the slag (Table 10.2).

10.8. ELECTRIC FURNACE Cu-FROM-SLAG RECOVERY

All direct-to-copper smelters recover Cu from their slag in an electric slag-cleaning furnace. The slag flows from the direct-to-blister furnace directly into an electric furnace where it is settled for about 10 h under a 0.25 m blanket of metallurgical coke (Czernecki et al., 1999). The coke reduces the oxidized Cu from the slag by reactions such as:

$$\underset{\text{dissolved in slag}}{Cu_2O(l)} + \underset{\text{in metallurgical coke}}{C} \xrightarrow{1300\ °C} \underset{\text{impure molten copper}}{2Cu°(l)} + \underset{\text{offgas}}{CO(g)} \quad (10.4)$$

Magnetite (molten and solid) is also reduced:

$$\underset{\text{dissolved in slag and solid}}{Fe_3O_4(l,s)} + \underset{\text{in metallurgical coke}}{C} \xrightarrow{1300\ °C} \underset{\text{dissolved in slag}}{3FeO(l)} + \underset{\text{offgas}}{CO(g)} \quad (10.5)$$

and some FeO is inadvertently reduced to Fe by the reaction:

$$\underset{\text{dissolved in slag}}{FeO(l)} + \underset{\text{in metallurgical coke}}{C} \xrightarrow{1300\ °C} \underset{\text{dissolved in copper}}{Fe(l)} + \underset{\text{offgas}}{CO(g)} \quad (10.6)$$

The resulting Fe joins the newly reduced copper.

10.8.1. Głogów

The Cu content of the Głogów direct-to-copper slag is lowered from ~14% Cu to ~0.6% Cu in an 25 000 kVA electric furnace. The molten product analyzes 70–80% Cu, 5% Fe, and 15–25% Pb, from Pb in the concentrate (Czernecki, Smieszek, Miczkowski, Krawiec, & Gizicki, 2010).

The Pb and Fe are oxidized in a Hoboken converter (Chapter 8), upgrading the metal to >98% Cu, 0.3% Pb. This product is sent to anode-making and electrorefining. The PbO-rich slag is sent to a lead smelter (along with lead- and zinc-rich dusts).

10.8.2. Olympic Dam

Olympic Dam lowers its direct-to-copper slag from 24% to ~4% Cu in its 15 000 kVA electric furnace (Hunt et al., 1999). This Cu content could be lowered further by using more coke and a longer residence time, but the copper product would then contain excessive radioactive ^{216}Pb and ^{210}Po, from the original concentrate (Ranasinghe et al., 2010). Instead, the slag is solidified and stored for future Cu recovery using solidification, comminution, and flotation (Fig. 10.2).

10.9. Cu-IN-SLAG LIMITATION OF DIRECT-TO-COPPER SMELTING

The principal advantage of direct-to-copper smelting is isolation of SO_2 evolution to one furnace. The principal disadvantage of the process is its large amount of Cu-in-slag. Balancing these factors, it appears that direct-to-copper smelting is best suited to concentrates containing covellite (Cu_2S) and bornite (Cu_5FeS_4). Processing these concentrates produces little slag, so Cu recovery from slag is not too costly.

Direct-to-copper smelting will probably not be suitable for most chalcopyrite concentrates. These concentrates produce about 2.5 t slag per tonne of Cu, and the energy requirement and cost of recovering Cu from their slag are considerable. Only about 60% of new Cu in the concentrate would report directly to copper; the other 40% would wind up in the slag.

Davenport, Jones, King, and Partelpoeg (2010) affirm this view. However, Outotec remains dedicated to direct-to-blister flash smelting for normal chalcopyrite concentrates, as well as for high-grade chalcocite and bornite concentrates (Tuominen & Kojo, 2005).

10.10. DIRECT-TO-COPPER IMPURITIES

The compositions of the anode copper produced by the direct-to-copper smelters are given in Table 10.4. The impurity levels of the copper are within the normal range of electrorefining anodes (Chapter 14). The impurity levels could be reduced further by avoiding recycle of the flash furnace dust.

TABLE 10.4 Anode Compositions from Direct-to-Copper Smelters

Impurity	Olympic Dam (2010) ppm in copper	Glogów II (2010) ppm in copper
Ag	188	1800–3300
As	597	500–1900
Au	24	~2
Bi	182	2–94
Fe	11	23–263
Ni	81	680–1620
Pb	241	500–3800
S	12	5–19
Sb	36	20–290
Se	263	118
Te	75	NA

Impurities do not seem, therefore, to be a problem in the three existing direct-to-copper smelters. However, metallic copper is always present in the direct-to-copper furnace, ready to absorb impurities. For this reason, concentrates destined for direct-to-copper smelting should always be carefully tested in a pilot furnace before being accepted by the smelter.

10.11. SUMMARY

Direct-to-copper smelting is the smelting of concentrate directly to molten copper in one furnace. In 2011, it is practiced in three smelters: Olympic Dam (Australia), Glogów II (Poland), and Chingola (Zambia). All of these plants use an Outotec flash furnace.

The main advantage of the process is its restriction of SO_2 evolution to a single, continuous source of high SO_2-strength gas. In principal, the energy, operating, and capital costs of producing metallic copper are also minimized by the single-furnace process.

Metallic copper is obtained in a flash furnace by setting the ratio of oxygen input in the blast to concentrate feed rate to the point where all the Fe and S in the input concentrate are oxidized. The ratio must be precisely controlled to avoid the production of Cu_2S or Cu_2O. Avoiding formation of a molten Cu_2S layer in the furnace is critical. Reactions between Cu_2S layers and oxidizing slag may cause rapid SO_2 evolution and slag foaming.

Direct-to-copper flash smelting has proven effective for SO_2 capture. However, about 25% of the Cu in the input concentrate is oxidized, ending up as copper oxide dissolved in slag. This copper oxide must be reduced back to metallic copper, usually with coke. The expense of recovering Cu from slag will probably restrict future direct-to-copper smelting to concentrates that produce little slag. Chalcopyrite concentrates will probably continue to be treated by multi-furnace processes, either by conventional smelting and converting or by continuous multi-furnace processing (Chapter 9).

REFERENCES

Brillo, J., & Egry, I. (2003). Density determination of liquid copper, nickel and their alloys. *International Journal of Thermophysics, 24*, 1155−1170.

Byszynski, L., Garycki, L., Gostynski, Z., Stodulski, T., & Urbanowski, J. (2010). Present and future modernization of metallurgical production lines of the Głogów copper smelter. *Copper 2010, Vol. 2: Pyrometallurgy I* (pp. 631−647). Clausthal-Zellerfeld, Germany: GDMB.

Czernecki, J., Smieszek, Z., Miczkowski, Z., Dobrzanski, J., Bas, W., Szwancyber, G., et al. (1999). The process flash slag cleaning in electric furnace at the Głogów II copper smelter. In F. Niemenen (Ed.), *9th international flash smelting congress* (pp. 375−389). Espoo, Finland: Outokumpu Technology.

Czernecki, J., Smieszek, Z., Miczkowski, Z., Krawiec, G., & Gizicki, S. (2010). Problems of lead and arsenic removal from copper production in a one-stage flash-smelting process. *Copper 2010, Vol. 2: Pyrometallurgy I* (pp. 669−683). Clausthal-Zellerfeld, Germany: GDMB.

Davenport, W. G., Jones, D. M., King, M. J., & Partelpoeg, E. H. (2010). *Flash smelting, analysis, control and optimization* (2nd ed.). New York, USA: Wiley.

Dobrzanski, J., & Kozminski, W. (2003). Copper smelting in KGHM Polska Miedz S.A. In C. Díaz, J. Kapusta & C. Newman (Eds.), *Copper/Cobre 2003, Vol. IV, Book 1: Pyrometallurgy of copper* (pp. 239−252). Montreal, Canada: CIM.

Hunt, A. G., Day, S. K., Shaw, R. G., Montgomerie, D., & West, R. C. (1999). Start up and operation of the #2 direct-to-copper flash furnace at Olympic Dam. In F. Niemenen (Ed.), *9th international flash smelting congress* (pp. 375−389). Espoo, Finland: Outokumpu Technology.

Hunt, A. G., Day, S. K., Shaw, R. G., & West, R. C. (1999). Developments in direct-to-copper smelting at Olympic Dam. In D. B. George, W. J. Chen, P. J. Mackey & A. J. Weddick (Eds.), *Copper 99−Cobre 99, Vol. V: smelting operations and advances* (pp. 239−253). Warrendale, USA: TMS.

Ranasinghe, D. J., Russell, R., Muthuraman, R., & Dryga, Z. (2010). Process optimization by means of heat and mass balance based modelling at Olympic Dam. *Copper 2010, Vol. 3: Pyrometallurgy II* (pp. 1063−1078). Clausthal-Zellerfeld, Germany: GDMB.

Solnordal, C. B., Jorgensen, R. R. A., Koh, P. T. L., & Hunt, A. (2006). CFD modeling of the flow and reactions in the Olympic Dam flash furnace smelter reaction shaft. *Applied Mathematical Modelling, 30*, 1310−1325.

Syamujulu, M. (2007). Opportunities, problems and survival strategies from recent developments in the copper concentrate treatment and smelting practices at Vedanta's Konkola copper mines in the Zambian copperbelt. In A. E. M. Warmer, C. J. Newman, A. Vahed, D. G. George, P. J. Mackey & A. Warczok (Eds.), *Copper/Cobre 2007, Vol. III, Book 1: The Carlos Diaz symposium of pyrometallurgy* (pp. 155−166). Montreal, Canada: CIM.

Taskinen, P., & Kojo, I. (2009). Fluxing options in the direct-to-blister copper smelting. In M. Sánchez, R. Parra, G. Riveros & C. Díaz (Eds.), *MOLTEN 2009* (pp. 1139−1151). Santiago, Chile: Gecamin.

Tuominen, J., & Kojo, I. (2005). Blister flash smelting − efficient and flexible low-cost continuous copper process. In A. Ross, T. Warner & K. Scholey (Eds.), *Converter and fire refining practices* (pp. 271−282). Warrendale, PA, USA: TMS.

SUGGESTED READING

Gostynski, Z., & Haze, D. (2007). Flash smelting furnace of the KGHM Głogów copper plant − technological and process challenges as a driving force of its continuous modernization. In A. E. M. Warmer, C. J. Newman, A. Vahed, D. G. George, P. J. Mackey & A. Warczok (Eds.), *Copper/Cobre 2007, Vol. III, Book 2: The Carlos Diaz symposium of pyrometallurgy* (pp. 233−243). Montreal, Canada: CIM.

Kojo, I., Lahtinen, M., & Miettinen, E. (2009). Flash converting — sustainable technology now and in the future. In J. Kapusta & T. Warner (Eds.), *International Peirce—Smith converting centennial* (pp. 383—395). Warrendale, PA, USA: TMS.

Kucharski, M. (2007). Modeling of the direct-to-copper flash smelting process. In A. E. M. Warmer, C. J. Newman, A. Vahed, D. G. George, P. J. Mackey & A. Warczok (Eds.), *Copper/Cobre 2007, Vol. III Book 2: The Carlos Diaz symposium of pyrometallurgy* (pp. 159—171). Montreal, Canada: CIM.

Syamujulu, M., & Beene, G. (2003). The smelting characteristics of KCM concentrates. In C. Diaz, J. Kapusta & C. Newman (Eds.), *Copper/Cobre 2003, Vol. IV, Book 1: Pyrometallurgy of copper* (pp. 325—337). Montreal, Canada: CIM.

Chapter 11

Copper Loss in Slag

Pyrometallurgical production of molten copper generates two slags, from smelting and converting. Smelting furnace slag typically contains 1−2% Cu (Table 5.2). The percentage increases as matte grade increases. Converter slag contains 4−8% Cu (Table 8.2). Its percentage increases as converting proceeds, and the grade of the remaining matte increases.

Multiplying these percentages by the mass of each slag shows that a significant fraction of the Cu in the original concentrate is present in these slags. This fraction is increased by the production of higher-grade mattes in the smelting furnace. Because of this, the value of the copper in these slags is usually too high to justify the old practice of simply discarding them.

This chapter discusses the nature of copper in smelting and converting slags. It also describes strategies for minimizing the amount of copper lost from their disposal. The main strategies include:

(a) Minimizing the mass of slag generated
(b) Minimizing the percentage of Cu in the slags
(c) Processing the slags to recover as much Cu as possible.

Slag processing can be divided into two types. The first is pyrometallurgical reduction and settling, performed in an electric or fuel-fired slag-cleaning furnace. The second is minerals processing of solidified slag, including crushing, grinding, and froth flotation, to recover Cu from the slag.

11.1. COPPER IN SLAGS

The Cu in smelting and converting slags is present either as (a) dissolved Cu, present mostly as Cu^+ ions, or as (b) entrained droplets of matte. The dissolved Cu is associated either with O^{2-} ions (i.e., Cu_2O), or with S^{2-} ions (Cu_2S). Cu_2O becomes the dominant form of dissolved Cu at matte grades above 70% Cu (Acuna & Sherrington, 2005; Barnett, 1979; Takeda, 2003), due to the increased activity of Cu_2S in the matte. Higher Cu_2S activity pushes the reaction,

$$Cu_2S(matte) + FeO(slag) \rightarrow Cu_2O(slag) + FeS(matte) \quad (11.1)$$

to the right. The solubility of sulfur in slags is also lower in contact with higher-grade mattes (Roghani, Takeda, & Itagaki, 2000). As a result, dissolved Cu in converter slags is present mostly as Cu_2O (Takeda, 2003). Conversely, the dissolved Cu in smelting slags is present mostly as Cu_2S. This is due to the lower matte grades and oxygen potentials in the smelting furnace.

TABLE 11.1 Matte Settling Rates (mm/h) for $\rho_{matte} - \rho_{slag} = 1\ g/cm^3$ (Ip & Toguri, 2000)

Slag viscosity (Pa-s)	Matte particle diameter		
	2 μm	20 μm	40 μm
0.05	0.157	15.690	62.578
0.08	0.098	9.806	39.224
0.1	0.078	7.845	31.379
0.12	0.065	6.537	26.149

There are several sources of entrained matte in slags. The most obvious are droplets of matte that have failed to settle completely through the slag layer during smelting. The Hadamard−Rybczynski formula predicts the rate at which matte droplets will settle through molten slag (Warczok, Riveros, Degel, Kunze, & Oterdoom, 2007a):

$$V = (\rho_M - \rho_S)gr^2/3\eta_S \qquad (11.2)$$

In this expression V is the settling rate of the matte droplets (m/s), g the gravitational constant (9.8 m/s^2), ρ_M the matte density (3900−5200 kg/m^3), ρ_S the slag density (3300−3700 kg/m^3), η_S the slag kinematic viscosity, and r the radius (m) of the settling matte droplet.

The expression is most accurate for systems with Reynolds numbers below 10 (droplet sizes below ~1 mm). Larger matte droplets settle at slower rates than predicted by Eq. (11.2). However, it is the settling rates of the smallest droplets that are of greatest concern.

Table 11.1 (Ip & Toguri, 2000) shows just how long matte droplets can take to settle. Besides droplet size, the biggest influences on settling rate are temperature and slag silica content. Higher temperatures and lower silica levels decrease slag viscosities, increasing settling rate. A more reducing environment also encourages settling, by decreasing the $Fe_3O_4(s)$ content of the slag.

In addition, matte grade has an impact on settling rates. Low-grade mattes have lower densities than high-grade mattes and therefore settle at slower rates (Fagerlund & Jalkanen, 1999).

Matte droplets can become suspended in smelter slags by several other mechanisms. Some are carried upwards from the molten matte layer by gas bubbles generated by the reaction (Poggi, Minto, & Davenport, 1969):

$$3Fe_3O_4(slag) + FeS(matte) \rightarrow 10FeO(slag) + SO_2 \qquad (11.3)$$

Still others appear by precipitation from the slag in colder areas of the smelting furnace (Barnett, 1979). Converter slag returned to a smelting furnace also contains suspended matte droplets, which may not have time to completely settle. As a result, entrained matte can represent from 50% to 90% of the copper contained in smelter slag (Acuna & Sherrington, 2005; Imriš et al., 2000). Figure 11.1 shows the solubility of oxidic and sulfidic copper in smelter slag, calculated using thermodynamic modeling

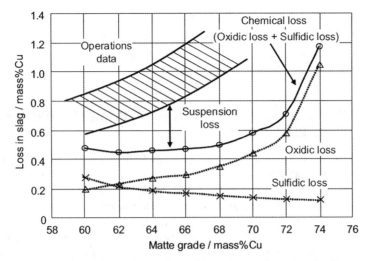

FIGURE 11.1 Thermodynamically predicted and actual loss of copper in smelter slag (Furuta et al., 2006). The difference is mechanically entrained matte (suspension loss).

(Furuta, Tanaka, Hamamoto, & Inada, 2006) and compared with actual plant data. The difference between the two is entrained matte.

11.2. DECREASING COPPER IN SLAG I: MINIMIZING SLAG GENERATION

It seems logical to suggest that decreasing the amount of Cu lost in smelting and converting slags could be accomplished by decreasing slag production. However, methods to decrease slag mass may do more harm than good. Possibilities include the following:

(a) *Maximizing concentrate grades.* The less gangue in the concentrate, the less silica required to flux it, and the less overall slag generated. However, increasing concentrate grades may come at the expense of decreasing Cu recoveries in the concentrator.
(b) *Adding less flux.* Adding less flux would decrease slag mass (desirable) and decrease its viscosity, making settling easier (also desirable). However, it would also increase the activity of FeO in the slag, leading to more dissolved Cu_2O by Reaction (11.1) (undesirable) and more magnetite (also undesirable; Mariscal & Herrera, 2009). The precipitation of solid magnetite raises slag viscosities and leads to higher slag copper contents as a result (Imris, 2003; Kunze, Degel, & Warczok, 2003).

11.3. DECREASING COPPER IN SLAG II: MINIMIZING COPPER CONCENTRATION IN SLAG

The concentration of copper in smelting slag is minimized by:

(a) Maximizing slag fluidity, principally by avoiding excessive Fe_3O_4 (s) in the slag and by keeping the slag hot
(b) Providing enough SiO_2 to form distinct matte and slag phases
(c) Providing a large quiet zone in the smelting furnace

(d) Avoiding an excessively thick layer of slag
(e) Avoiding tapping of matte with slag.

Metallurgical coke or coal may also be added to the smelting furnace to reduce Fe_3O_4 (s) to FeO (ℓ) (Furuta et al., 2006).

11.4. DECREASING COPPER IN SLAG III: PYROMETALLURGICAL SLAG SETTLING/REDUCTION

Conditions that encourage suspended matte droplets to settle to a matte layer are low viscosity slag, low turbulence, a long residence time, and a thin slag layer. These conditions are often difficult to obtain in a smelting vessel, particularly the necessary residence time. As desired matte grades increase, the problem gets worse, because the increase in concentrate feed rate decreases settling time further (Furuta et al., 2006). As a result, Cu producers have since the 1960s constructed separate furnaces specifically for *cleaning* smelting and converting slags. These furnaces have two purposes: (a) allowing suspended matte droplets to finish settling to the molten matte layer, and (b) facilitating the reduction of dissolved Cu oxide to suspended Cu sulfide drops.

Inputs to these furnaces vary considerably. Slag-cleaning furnaces associated with bath-smelting units like the Ausmelt/Isasmelt or Mitsubishi smelting furnace accept an unseparated mixture of slag and matte and are required to do all the settling. Others accept converter slag and plant reverts in addition to smelter slag, requiring more emphasis on reduction (Moreno, Sánchez, Worczok, & Riveros, 2003). Most commonly, these furnaces are fed only smelting furnace slag and are used primarily as a final settling furnace.

Figure 11.2 illustrates a typical electric slag-cleaning furnace (Barnett, 1979; Higashi, Suenaga, & Akagi, 1993; Kucharski, 1987). Heat is generated by passing electric current through the slag layer. AC power is used, supplied through three carbon electrodes. This method of supplying heat generates the least amount of turbulence, which improves settling rates. Most furnaces use self-baking Søderberg electrodes (Kunze et al., 2003), but a few use pre-baked electrodes to avoid the need for baking control (Furuta et al., 2006).

The furnace sidewalls are typically cooled by external water jackets to minimize refractory erosion. However, embedded cooling elements are more effective, especially for slag-cleaning furnaces processing high-copper slags (Kunze et al., 2003; Tuominen, Anjala, & Björklund, 2007).

Table 11.2 compares the operating characteristics of seven electric furnaces. The choice of operating temperature represents a balance between the enhanced reaction kinetics of higher temperatures and the reduced refractory wear and lower copper losses of lower temperatures. Required capacities are set by the size of the smelting operation and the choice of input slags. Settling times are usually on the order of one to five hours. Typical energy use is 15–70 kWh per tonne of slag, depending upon furnace inputs, target % Cu, temperature, and residence time. Energy use is higher when the charge contains significant levels of reverts (Kunze et al., 2003). In addition to providing a reducing atmosphere, adding coke helps reduce electrode loss.

While some electric slag-cleaning furnaces process only smelting furnace slag, others are fed a variety of materials. Several furnace operators input converter slag or solid reverts in addition to smelting slag. When this is done, a reducing agent is often required to reduce Cu oxide in the slag to Cu metal or Cu sulfide. Coal or coke is often added for

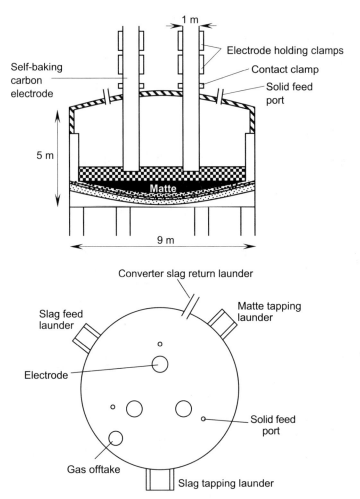

FIGURE 11.2 Electric slag-cleaning furnace. A furnace of this size cleans 1000–1500 tonnes of slag per day.

this reduction (Bergh, Yianatos, & Chacana, 2000). Pyrite may also be added if additional sulfur is needed to form matte (Ponce & Sánchez, 1999):

$$C + Cu_2O(slag) \rightarrow CO + 2Cu(l) \qquad (11.4)$$

$$C + Cu_2O + FeS_2 \rightarrow Cu_2S + FeS + CO \qquad (11.5)$$

Carbon additions also reduce solid magnetite in the slag to liquid FeO (Mackay, Cordero, Esteban, & Alvear, 2000):

$$C + Fe_3O_4(s) \rightarrow CO + 3FeO(slag) \qquad (11.6)$$

This decreases slag viscosity and improves settling rates (Furuta et al., 2006).

Ferrosilicon is occasionally used as a reducing agent (Shimpo & Toguri, 2000), especially in the Mitsubishi slag-cleaning furnace. Recent initiatives in slag-cleaning furnace practice have involved lance injection of solid reductants or gaseous reducing agents such as methane, to improve reduction kinetics (An, Li, & Grimsey, 1998; Sallee & Ushakov, 1999).

TABLE 11.2 Details of Electric Slag-cleaning Furnaces

Smelter	Caraiba Metais Dias d'Avila, Brazil	Norddeutsche Affinerie Hamburg, Germany	KGHM Polska Miedz, Poland	Konkola Copper Nchanga, Zambia	Codelco Las Ventanas Chile	Mexicana de Cobre Mexico Furnace 1	Mexicana de Cobre Mexico Furnace 2
Slag details, tonnes/day							
Smelting furnace slag	880 OK flash furnace	1600 OK flash furnace	1100, flash furnace	1600, flash furnace	1000 Teniente furnace	900 OK flash furnace	740 Teniente furnace
% Cu	1.7	1–1.5	13.86	17	8–12	1.5–2.5	5
Converter slag	65–70	0	52		120	113	184
% Cu			18.76		30	8	8
Products							
Slag, % Cu	0.7	0.6–0.8	0.64	3	1.3	1.26	1.3
Matte, % Cu	65–70	65–70	Cu–Pb–Fe alloy, 77% Cu	99.28	70.5	70.3	70.5
Furnace details							
Shape	Circular	Circular	Circular	Circular	Circular	Circular	Circular
Diameter, m	11	10.2	14.5	13.5	10	10	10
Power rating, MW	2–4	2–3	25	20	4.5	1.5–4.5	1.5–4.5
Electrodes	3	3	3	5	3	3	3
Material	Self-baking	Self-baking	Self-baking	Self-baking	Self-baking	Self-baking	Self-baking
Diameter, m	1	1	1.32	1.1	0.86	0.9	0.9
Operating details							
Slag residence time, hours	2–3	5	8	4	1.5–2.5	0.25–1	0.25–1
Energy use, kWh/tonne of slag	70	40–50	384	150	140	57	69
Reductant, kg/tonne of slag	Coke, 8.3	Coke, 4–5	Coke, 46	Coke, 21	Coke, 13	7.17 Coke	7.32 Coke
Slag layer thickness, m	0.97–1.4	1.5–1.8	<1	1–2	0.8–1.4	0.8–1.5	0.8–1.5
Matte layer thickness, m	0–0.45	0–0.4	0.4–1.0 (alloy)	1.2	0.1–0.5	0–0.2	0–0.2

Table 11.2 includes data for the electric slag-cleaning furnace used by the direct-to-blister smelter at Głogów. This furnace is unusual in several regards, including the production of an alloy rather than matte, due to the lack of sulfide in the charged slag (Czernecki et al., 2006). This operation is discussed in detail in Chapter 10.

Recent research at the University of Chile has highlighted a number of potential improvements to the design and operation of electric furnaces for slag cleaning (Montenegro et al., 2003; Warczok et al., 2007a, Warczok, Riveros, Degel, Kunze, & Oterdoom, 2007b, Warczok, Riveros, & Parada, 2007c). These include:

(a) The use of rectangular rather than circular furnace shape, particularly advantageous for continuous operation
(b) Replacing AC furnace power supply with DC, coupled with a magnetic field to enhance stirring and reduction kinetics
(c) Replacing coke with calcium carbide as a reductant, which improves kinetics and requires less furnace energy

Fuel-fired slag cleaning furnaces are used at a few smelters (Table 11.3). The foremost is the Teniente slag-cleaning furnace, which is similar in design to a rotary fire-refining furnace (Chapter 13; Demetrio et al., 2000; Mackay et al., 2000). It features injection of powdered coke and air into molten slag. In 2000, it began operating on a semi-continuous basis, generating slag with 0.75–0.85% Cu. Semi-continuous operation improved productivity by over 20%, while reducing coke consumption by half. The slag-cleaning furnaces used by bath smelters are usually electric, but fuel-fired furnaces can be used for this purpose as well (Mariscal & Herrera, 2009).

11.5. DECREASING COPPER IN SLAG IV: SLAG MINERALS PROCESSING

Several options are available for recovering Cu from converter slags. Pyrometallurgical cleaning in electric furnaces is quite common. Molten converter slag is also recycled to reverberatory smelting furnaces and Inco flash furnaces. Outotec and Teniente smelting furnaces occasionally accept some molten converter slag (Warczok, Riveros, Mackay, Cordero, & Alvera, 2001).

Cu is also removed from converter slags by slow solidification, crushing/grinding, and froth flotation. It relies on the fact that, as converter slags cool, much of their dissolved Cu exsolves from solution by the reaction (Jalkanen, Poijärvi, & Pajari, 2002):

$$Cu_2O + 3FeO \rightarrow 2Cu(l) + Fe_3O_4 \qquad (11.7)$$

Reaction (11.7) is increasingly favored at low temperatures and can decrease the dissolved Cu content of converter slag to well below 0.5% (Bérubé, Choquette, & Godbehere, 1987; Imriš et al., 2000). After the slag has solidified, the exsolved copper and suspended matte particles respond well to froth flotation. As a result, converter slags have long been crushed, ground and concentrated in the same manner as sulfide ores (Demetrio et al., 2000). The key to successful minerals processing of converter slags is ensuring that the precipitated grains of matte and metallic Cu are large enough to be liberated by crushing and grinding. This is accomplished by cooling the slag slowly to less than 1000 °C, then naturally to ambient temperature. Once this is done, the same minerals processing equipment and reagents that are used to recover Cu from ore can be used to recover Cu from slag (Table 11.4). Some smelting slags are also treated this way (Sarrafi, Rahmati, Hassani, & Shirazi, 2004). The processing of 'historic' slags from old

TABLE 11.3 Details of Teniente Rotary Hydrocarbon-fired Slag Settling Furnace at Caletones, Chile

Smelter	Caletones, Chile
Slag details	
Smelting furnace slag, tonnes/day	3000
% Cu	6–8
Converting furnace slag, tonnes/day	0
% Cu	
Products	
Slag, % Cu	1
Matte, % Cu	72
Matte destination	Peirce–Smith converters Teniente smelting furnace
% Cu recovery	85%
Furnace details	
Number of slag-cleaning furnaces	4
Shape	Horizontal cylinder
Diameter inside refractory, m	4.6
Length inside refractory, m	3 × 10.7; 1 × 12.7
Number of reducing tuyeres	4
Tuyere diameter, cm	6.35
Operating details	
Slag residence time, hours	2
Reductant	Coal, oil, or natural gas
kg per tonne of slag	6
Slag layer thickness, m	1.4
Matte layer thickness, m	0.4
Fuel	Bunker C fuel oil
kg per tonne of slag	8.8

TABLE 11.4 Details of Four Slag Flotation Plants

Smelter	Noranda, Québec	Saganoseki, Japan	Toyo, Japan	PASAR, Philippines
Slag inputs, tonnes/day				
Smelting furnace slag	1700	0	0	0
% Cu	6 (average)			
Converter slag	300	450	450	370
% Cu		8.33	6.5	10–15[a]
Products				
Slag concentrate, % Cu	42	21.8	28	29–33
Slag tailings, % Cu		0.65	0.4	0.5–0.6
Cu recovery, %	90–95	95	95	97–98
Operating details				
Solidification method	Ladle cooling with or without water sprays	~150 kg ingots on moving slag conveyor	~150 kg ingots on moving slag conveyor	
Cooling description		Cooled on slag conveyor	1 h in air then immersion in H$_2$O	
Crushing/grinding				
Equipment	80% semi autogenous grinding, 20% crushing & ball milling	jaw crusher; cone crusher (twice); ball mill (twice)	gyratory crusher; cone crusher (twice); ball mill	jaw crusher; cone crusher; ball mills (primary and regrind)

(Continued)

TABLE 11.4 Details of Four Slag Flotation Plants—cont'd

Smelter	Noranda, Québec	Saganoseki, Japan	Toyo, Japan	PASAR, Philippines
Particle size after grinding	78%–44 μm	40–50%–44 μm	90%–44 μm	65–75%–45 μm
Flotation				
Machinery	Mechanically agitated cells	Mechanically agitated cells	Mechanically agitated cells	Mechanical agitator Agitair 48, Jameson cell (Fig. 3.12)[b]
Flotation residence time	60 min		30 min (rougher + scavenger)	
Flotation reagents				
Promoter				NH_4 & Na dibutyl dithiophosphate
Collector	Thionocarbamate, SIPX	Na isopropyl xanthate, UZ200	Thionocarbamate, PAX	a) Danafloat 245, Penfloat TM3 b) K amyl xanthate
Frother	Propylene glycol	Pine oil, MF550	Pine oil	Pine oil NF 183
CaO+	No	No		Yes
pH	8–9	7–8	7–8	8.5–9.5
Energy use kWh/tonne slag	32.5	58.4	64	40

[a] Non-magnetic 'white metal' (Cu_2S) pieces are isolated magnetically after crushing, leaving 5–6.5% Cu in the ball mill feed slag.
[b] Switching to all Jameson cell.

smelting operations by comminution and flotation is occasionally a source of additional copper for current producers (Aydoğan, Canbazoğlu, Akdemire, & Gürkan, 2000).

Slag flotation is becoming less popular for processing smelter slags, for two reasons (Acuna & Sherrington, 2005). The first is the trend toward production of higher-grade mattes. A higher fraction of the copper in these mattes is present as dissolved copper, rather than entrained matte. This copper is not recovered by flotation, but must be recovered in a settling furnace. The second is the increased throughput at many mills. This means that comminution and flotation equipment are less likely to be available for processing slags.

11.6. SUMMARY

Cu smelters produce two slags: smelting furnace slag with 1−2% Cu and converter slag with 4−8% Cu. Discard of these slags would waste considerable Cu, so they are almost always treated for Cu recovery.

Cu is present in molten slags as (a) entrained droplets of matte or metal and (b) dissolved Cu^+. The entrained droplets are recovered by settling in a slag-cleaning furnace, usually electric. The dissolved Cu^+ is recovered by hydrocarbon reduction and settling of matte.

A second method of recovering Cu from slag is slow-cooling/solidification, crushing/ grinding, and froth flotation. Slowly-cooled/solidified slag contains the originally entrained matte, Cu droplets plus matte, and Cu, which precipitate during cooling/ solidification. These Cu-bearing materials are efficiently recovered from the solidified slag by fine grinding and froth flotation.

Electric furnace settling has the advantage that it can be used for recovering Cu from reverts and miscellaneous materials around the smelter. Slag flotation has the advantages of more efficient Cu recovery and the possibility of using existing crushing/grinding/ flotation equipment.

REFERENCES

Acuna, C. M., & Sherrington, M. (2005). Slag cleaning processes: a growing concern. *Materials Science Forum, 475−479*, 2745−2752.

An, X., Li, N., & Grimsey, E. J. (1998). Recovery of copper and cobalt from industrial slag by top-submerged injection of gaseous reductants. In B. Mishra (Ed.), *EPD congress 1998* (pp. 717−732). Warrendale, PA: TMS.

Aydoğan, S., Canbazoğlu, M., Akdemire, Ü., & Gürkan, V. (2000). Processing of Hafik−Madentepe copper slags using conventional and leaching methods. In G. Özbayoglu, C. Hosten, M. Ümit Atalay, C. Hiçyilmaz & A. Ihsan Arol (Eds.), *Mineral processing on the verge of the 21st century* (pp. 529−533). Rotterdam: Balkema.

Barnett, S. C. C. (1979). The methods and economics of slag cleaning. *Mining Magazine, 140*, 408−417.

Bergh, L. G., Yianatos, J. B., & Chacana, P. B. (2000). Intelligent sensor for coal powder rate injection in a slag cleaning furnace. *Minerals Engineering, 13*, 767−771.

Bérubé, M., Choquette, M., & Godbehere, P. W. (1987). Mineralogie des scories cupriferes. *CIM Bulletin, 80*(898), 83−90.

Czernecki, J., Śmieszek, Z., Miczkowski, Z., Kubacz, N., Dobrzański, J., Staszak, J., et al. (2006). The slag cleaning technologies for one-stage flash smelting of KGHM Polska Miedź concentrates. In F. Kongoli & R. G. Reddy (Eds.), *Sohn international symposium, Vol. 8: International symposium on sulfide smelting 2006* (pp. 181−197). Warrendale, PA: TMS.

Demetrio, S., Ahumada, J., Ángel, D. M., Mast, E., Rosas, U., Sanhueza, J., Reyes, P., & Morales, E. (2000). Slag cleaning: the Chilean copper smelter experience. *JOM, 52*(8), 20−25.

Fagerlund, K. O., & Jalkanen, H. (1999). Some aspects on matte settling in copper smelting. In C. Díaz, C. Landolt & T. Utigard (Eds.), *Copper 99−Cobre 99, Vol. VI: Smelting, technology development, process modeling and fundamentals* (pp. 539−551). Warrendale, PA: TMS.

Furuta, S., Tanaka, S., Hamamoto, M., & Inada, H. (2006). Analysis of copper loss in slag in Tamano type flash smelting furnace. In F. Kongoli & R. G. Reddy (Eds.), *Sohn international symposium, Vol. 8: International symposium on sulfide smelting 2006* (pp. 123−133). Warrendale, PA: TMS.

Higashi, M., Suenaga, C., & Akagi, S. (1993). Process analysis of slag cleaning furnace. In H. Henein & T. Oki (Eds.), *First international conference on processing materials for properties* (pp. 369−372). Warrendale, PA: TMS.

Imris, I. (2003). Copper losses in copper smelting slags. In F. Kongoli, C. Yamauchi & H. Y. Sohn (Eds.), *Yazawa international symposium, Vol. 1: Materials processing fundamentals and new technologies* (pp. 359−373). Warrendale, PA: TMS.

Imriš, I., Rebolledo, S., Sanchez, M., Castro, G., Achurra, G., & Hernandez, F. (2000). The copper losses in the slags from the El Teniente process. *Canadian Metallurgical Quarterly, 39*, 281−290.

Ip, S. W., & Toguri, J. M. (2000). Entrainment of matte in smelting and converting operations. In G. Kaiura, C. Pickles, T. Utigard & A. Vahed (Eds.), *J.M. Toguri symposium on fundamentals of metallurgical processing* (pp. 291−302). Montreal: CIM.

Jalkanen, H., Poijärvi, J., & Pajari, H. (2002). Slags of suspension smelting of chalcopyrite ores and copper matte converting. In R. L. Stephens & H. Y. Sohn (Eds.), *Sulfide smelting 2002* (pp. 363−376). Warrendale, PA: TMS.

Kucharski, M. (1987). Effect of thermodynamic and physical properties of flash smelting slags on copper losses during slag cleaning in an electric furnace. *Archives of Metallurgy, 32*, 307−323.

Kunze, J., Degel, R., & Warczok, A. (2003). Current status and new trends in copper slag cleaning. In C. Díaz, J. Kapusta & C. Newman (Eds.), *Copper 2003−Cobre 2003, Vol. 4 Book 1: Pyrometallurgy of copper* (pp. 459−474). Montreal: CIM.

Mackay, S. R., Cordero, C. D., Esteban, S. J., & Alvear, F. G. R. (2000). Continuous improvement of the Teniente slag cleaning process. In S. Seetharaman & D. Sichen (Eds.), *Proceedings of the 6th international conference molten slags, fluxes and salts*. Stockholm. Paper 208.

Mariscal, L., & Herrera, E. (2009). Isasmelt™ slag chemistry and copper losses in the rotary holding furnaces slag at Ilo Smelter. In M. Sánchez, R. Parra, G. Riveros & C. Díaz (Eds.), *Molten 2009* (pp. 1241−1250). Santiago, Chile: Gecamin.

Montenegro, V., Fujisawa, T., Warczok, A., & Riveros, G. (2003). Effect of magnetic field on the rate of slag reduction in an electric furnace. In F. Kongoli, T. Fujisawa, A. Warczok & G. Riveros (Eds.), *Yazawa international symposium, Vol. II: High-temperature metals production* (pp. 199−209). Warrendale, PA: TMS.

Moreno, A., Sánchez, G., Worczok, A., & Riveros, G. (2003). Development of slag cleaning process and operation of an electric furnace at Las Ventanas smelter. In C. Díaz, J. Kapusta & C. Newman (Eds.), *Copper 2003−Cobre 2003, Vol. 4 Book 1: Pyrometallurgy of copper* (pp. 475−492). Montreal: CIM.

Poggi, D., Minto, R., & Davenport, W. G. (1969). Mechanisms of metal entrapment in slags. *Journal of Metals, 21*(11), 40−45.

Ponce, R., & Sánchez, G. (1999). Teniente Converter slag cleaning in an electric furnace at the Las Ventanas smelter. In D. B. George, W. J. Chen, P. J. Mackey & A. J. Weddick (Eds.), *Copper 99−Cobre 99, Vol. V: Smelting operations and advances* (pp. 583−597). Warrendale, PA: TMS.

Roghani, G., Takeda, Y., & Itagaki, K. (2000). Phase equilibrium and minor element distribution between FeO_x−SiO_2−MgO-based slag and Cu_2S−FeS matte at 1573K under high partial pressures of SO_2. *Metallurgical and Materials Transactions B, 31B*, 705−712.

Sallee, J. E., & Ushakov, V. (1999). Electric settling furnace operations at the Cyprus Miami Mining Corporation copper smelter. In D. B. George, W. J. Chen, P. J. Mackey & A. J. Weddick (Eds.), *Copper 99−Cobre 99, Vol. V: Smelting operations and advances* (pp. 629−643). Warrendale, PA: TMS.

Sarrafi, A., Rahmati, B., Hassani, H. R., & Shirazi, H. H. A. (2004). Recovery of copper from reverberatory furnace slag by flotation. *Minerals Engineering, 17*, 457−459.

Shimpo, R., & Toguri, J. M. (2000). Recovery of suspended matte particles from copper smelting slags. In G. Kaiura, C. Pickles, T. Utigard & A. Vahed (Eds.), *J.M. Toguri symposium: Fundamentals of metallurgical processing* (pp. 481−496). Montreal: CIM.

Takeda, Y. (2003). Thermodynamic evaluation of copper loss in slag equilibrated with matte. In F. Kongoli, C. Yamauchi & H. Y. Sohn (Eds.), *Yazawa international symposium, Vol. 1: Materials processing fundamentals and new technologies* (pp. 359−373). Warrendale, PA: TMS.

Tuominen, J., Anjala, Y., & Björklund, P. (2007). Slag cleaning of Outokumpu direct-to-blister flash smelting slags. In A. E. M. Warner, C. J. Newman, A. Vahed, D. B. George, P. J. Mackey & A. Warczok (Eds.), *Cu2007, Vol. III Book 2: The Carlos Díaz symposium on pyrometallurgy* (pp. 339−350). Montreal: CIM.

Warczok, A., Riveros, G., Degel, R., Kunze, J., Oterdoom, H., et al. (2007a). Computer simulation of slag cleaning in an electric furnace. In A. E. M. Warner (Ed.), *Cu2007, Vol. III Book 2: The Carlos Díaz symposium on pyrometallurgy* (pp. 367−378). Montreal: CIM.

Warczok, A., Riveros, G., Degel, R., Kunze, J., Oterdoom, H., et al. (2007b). Slag cleaning in circular and rectangular electric furnaces. In A. E. M. Warner (Ed.), *Cu2007, Vol. III Book 2: The Carlos Díaz symposium on pyrometallurgy* (pp. 403−415). Montreal: CIM.

Warczok, A., Riveros, G., Parada, R., et al. (2007c). Slag reduction and cleaning with calcium carbide. In A. E. M. Warner (Ed.), *Cu2007, Vol. III Book 2: The Carlos Díaz symposium on pyrometallurgy* (pp. 379−390). Montreal: CIM.

Warczok, A., Riveros, G., Mackay, R., Cordero, G., & Alvera, G. (2001). Effect of converting slag recycling into Teniente converter on copper losses. In P. R. Taylor (Ed.), *EPD congress 2000* (pp. 431−444). Warrendale, PA: TMS.

SUGGESTED READING

Acuna, C. M., & Sherrington, M. (2005). Slag cleaning processes: a growing concern. *Materials Science Forum, 475−479*, 2745−2752.

Barnett, S. C. C. (1979). The methods and economics of slag cleaning. *Mining Magazine, 140*, 408−417.

Demetrio, S., Ahumada, J., Ángel, D. M., Mast, E., Rosas, U., Sanhueza, J., et al. (2000). Slag cleaning: the Chilean copper smelter experience. *JOM, 52*(8), 20−25.

Kunze, J., Degel, R., & Warczok, A. (2003). Current status and new trends in copper slag cleaning. In C. Díaz, J. Kapusta & C. Newman (Eds.), *Copper 2003−Cobre 2003, Vol. 4 Book 1: Pyrometallurgy of copper* (pp. 459−474). Montreal: CIM.

Chapter 12

Capture and Fixation of Sulfur

About 85% of the world's primary copper originates in sulfide minerals. Sulfur is, therefore, evolved by most copper extraction processes. The most common form of evolved sulfur is sulfur dioxide (SO_2) gas from smelting and converting.

SO_2 is harmful to fauna and flora and is detrimental to the human respiratory system. It must be prevented from reaching the environment. Regulations for ground level SO_2 concentrations around copper smelters are presented in Table 12.1. Other regulations such as maximum total SO_2 emission (tonnes per year), percent SO_2 capture, and SO_2-in-gas concentration at point-of-emission, also apply in certain locations.

TABLE 12.1 Standards for Maximum SO_2 Concentration at Ground Level Outside the Perimeters of Copper Smelters

Country	Time period	Maximum $SO_2 + SO_3$ (SO_x) concentration (ppm)
Australia (Australian government, 2010)+	Yearly average	0.02
	Daily average	0.08
	1-h average	0.2
Canada (Health Canada, 2010) (maximum acceptable level)	Annual average	0.023
	Daily average	0.115
	1-h average	0.334
Chile (D.S.No 113/02, 2010)	Yearly average	0.031
	Daily average	0.096
European Union (Council directive 1999/30/EC, 2010)	Yearly average	0.008
	Daily average	0.05
	1-h average	0.13
Japan (MOE Japan, 2010)	Daily average	0.04
	Hourly average	0.1
U.S.A. (USEPA, 2010)	1-h standard	0.075

In the past, SO_2 from smelting and converting was vented directly to the atmosphere. This practice is now prohibited in most of the world, so most smelters capture a large fraction of their SO_2. It is almost always made into sulfuric acid, occasionally into liquid SO_2 or gypsum. Copper smelters typically produce 2.5–4.0 tonnes of sulfuric acid per tonne of product copper depending on the S/Cu ratio of their feed materials.

This chapter describes:

(a) Offgases from smelting and converting
(b) Manufacture of sulfuric acid from smelter gases
(c) Recent and future developments in sulfur capture.

12.1. OFFGASES FROM SMELTING AND CONVERTING PROCESSES

Table 12.2 characterizes the offgases from smelting and converting processes. SO_2 strengths in *smelting furnace* gases vary from about 70 vol.-% in Inco flash furnace gases to 1 vol.-% in reverberatory furnace gases. The SO_2 strengths in *converter* gases vary from about 40% in flash converter gases to 8–12 % in Peirce–Smith converter gases.

Offgases from most smelting and converting furnaces are treated for SO_2 removal in sulfuric acid plants. The exception is offgas from reverberatory furnaces. It is too dilute in

TABLE 12.2 Characteristics of Offgases from Smelting and Converting Processes. The Data are for Offgases as they Enter the Gas-Handling System

Furnace	SO_2 concentration vol.-%	Temperature °C	Dust loading kg/Nm^3	Destination
Inco flash furnace	50–75	1270–1300	0.2–0.25	H_2SO_4 occasionally liquid SO_2 plant
Outotec flash furnace	25–60	1270–1350	0.1–0.25	H_2SO_4 plant, occasionally liquid SO_2 plant
Outotec flash converter	35–40	1290	0.2	H_2SO_4 plant
Outotec direct-to-copper	43	1320–1400	0.2	H_2SO_4 plant
Mitsubishi smelting furnace	30–35	1240–1250	0.07	H_2SO_4, occasionally liquid SO_2 plant
Mitsubishi converting furnace	25–30	1230–1250	0.1	H_2SO_4, occasionally liquid SO_2 plant
Noranda process	15–25	1200–1240	0.015–0.02	H_2SO_4 plant
Teniente furnace	12–25	1220–1250	0.015–0.02	H_2SO_4 plant

TABLE 12.2 Characteristics of Offgases from Smelting and Converting Processes. The Data are for Offgases as they Enter the Gas-Handling System—cont'd

Furnace	SO_2 concentration vol.-%	Temperature °C	Dust loading kg/Nm^3	Destination
Isasmelt or Ausmelt furnace	20–25	1150–1220	~0.01	H_2SO_4 plant
Electric furnace	2–5	400–800		H_2SO_4 or liquid SO_2 plant or vented to atmosphere
Reverberatory furnace	1	1250	~0.03	Vented to atmosphere (made into gypsum in one plant)
Peirce–Smith converter	8–15	1200	0.05–0.1	H_2SO_4 plant or vented to atmosphere
Hoboken converter	12	1200		H_2SO_4 plant
Electric slag cleaning furnaces	0.1	800		Vented to atmosphere (occasionally scrubbed with basic solution)
Anode furnaces	<0.1	1000		Vented to atmosphere (occasionally scrubbed with basic solution or used as dryer dilution gas)
Gas collection hoods around the smelter	<0.1	50		Vented to atmosphere (occasionally scrubbed with basic solution)

SO_2 for economic sulfuric acid manufacture. This is the main reason that reverberatory furnaces continue to be shut down.

The offgases from electric slag cleaning furnaces, anode furnaces, and secondary gas collection hoods around the smelter are dilute in SO_2, <0.1%. These gases are usually vented to atmosphere. In densely populated areas, they may be scrubbed with basic solutions before being vented (Inami, Baba, & Ojima, 1990; Shibata & Oda, 1990; Tomita, Suenaga, Okura, & Yasuda, 1990).

12.1.1. Sulfur Capture Efficiencies

Table 12.3 shows the typical sulfur capture efficiency of a modern smelter with an Outotec flash furnace and Peirce–Smith converters. Gaseous emissions of sulfur compounds are ≤1% of the sulfur entering the smelter.

TABLE 12.3 Typical Copper Smelter Sulfur Distribution

Percent of incoming S in:	
Sulfuric acid	96–97
Gypsum (scrubbing)	1–3
Slag	1–2
Dust (to treatment)	0.2–2
Neutralized liquid effluent	0.3–2
Gaseous emissions (dryer, anode furnace and ventilation stacks, and/or acid plant tail gas)	0.2–1

12.2. SULFURIC ACID MANUFACTURE

Sulfuric acid is manufactured from smelter offgas by one of three methods:

(a) Conventional contact type acid plant (most common)
(b) Haldor Topsøe WSA (Wet gas Sulfuric Acid) plant, one unit treating smelter off-gas, >80 reference installations worldwide
(c) Sulfacid, one unit treating electric furnace offgas, ~20 installations worldwide

The traditional contact type acid plant is described below. WSA and Sulfacid are described in Section 12.7.

Figure 12.1 outlines the steps for producing sulfuric acid from SO_2-bearing smelter offgas. The steps are:

(a) Cooling and cleaning the gas
(b) Drying the gas with 93% sulfuric acid
(c) Catalytically oxidizing SO_2 to SO_3
(d) Absorbing this SO_3 into 98.5% H_2SO_4 sulfuric acid.

The strengthened acid from step (d) is then blended with diluted acid from step (b) and sent to market or used for leach operations.

The acid plant tail gas is cleaned of its acid mist and discharged to the atmosphere. Tail gases typically contain less than 0.5% of the sulfur entering the gas treatment system. Several smelters scrub the remaining SO_2, SO_3, and H_2SO_4 mist with Ca/Na carbonate hydroxide solutions before releasing the gas to atmosphere (Hay, Porretta, & Wiggins, 2003; Kubo, Isshiki, Satou, & Kurokawa, 2007; Shibata & Oda, 1990; Tomita et al. 1990). Basic aluminum sulfate solution is also used (Oshima, Igarashi, Nishikawa, & Kawasaki, 1997).

The following sections describe the principal sulfuric acid production steps and their purposes.

12.3. SMELTER OFFGAS TREATMENT

The key to high acid plant availability in any smelter is to efficiently operate and maintain properly designed gas cooling and cleaning systems as described in the next few sections.

Chapter | 12   Capture and Fixation of Sulfur

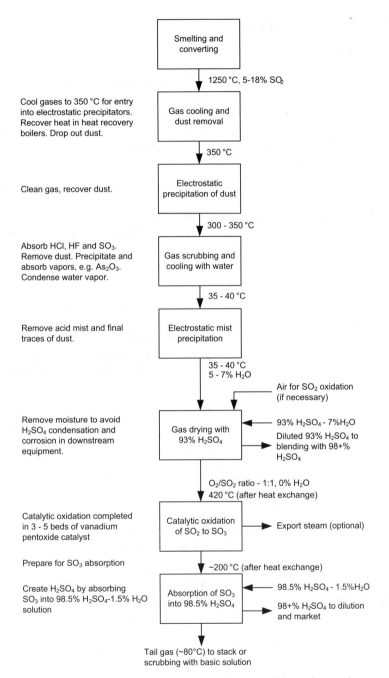

FIGURE 12.1   Flowsheet for producing sulfuric acid from smelting and converting gases.

12.3.1. Gas Cooling and Heat Recovery

The first step in smelter offgas treatment is cooling the gas in preparation for electrostatic precipitation of its dust. Electrostatic precipitators operate at about 350 °C. Above this temperature their steel structures begin to weaken. Below this temperature there is a danger of corrosion by condensation of sulfuric acid from SO_3 and water vapor in the offgas.

Gas cooling is usually done in heat recovery boilers (Fig. 12.2), which not only cool the gas but also recover the heat in a useful form as steam (Köster, 2010; Peippo, Holopainen, & Nokelainen, 1999). The boilers consist of:

(a) A radiation section in which the heat in the gas is transferred to pressurized water flowing through 4 cm diameter tubes in the roof and walls of a large (e.g. 25 m long × 15 m high × 5 m wide) rectangular chamber
(b) A convection section (typically 20 m long × 10 m high × 3 m wide) in which heat is transferred to pressurized water flowing through 4 cm diameter steel tubes suspended in the path of the gas.

The product of the boiler is a water/steam mixture. The water is separated by gravity and re-circulated to the boiler. The steam is often superheated above its dew point and used for generating electricity. It is also used without superheating for concentrate drying and for various heating duties around the smelter and refinery.

Dust falls out of heat recovery boiler gases due to its low velocity in the large boiler chambers. It is collected and usually recycled to the smelting furnace for Cu recovery. It

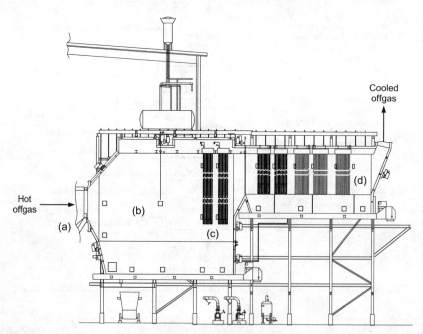

FIGURE 12.2 Heat recovery boiler for the Rönnskär flash furnace (Peippo et al., 1999). Note, left to right, (a) flash furnace gas offtake, (b) radiation section with tubes in the walls, (c) suspended tube baffles in the radiation section to evenly distribute gas flow, (d) convection section with hanging tubes. Note also water tank above radiation section and dust collection conveyors below the radiation and convection sections.

is occasionally treated hydrometallurgically (Larouche, 2001, pp. 118−119). This avoids impurity recycle to the smelting furnace and allows the furnace to smelt more concentrate (Davenport, Jones, King, & Partelpoeg, 2001).

An alternative method of cooling smelter gas is to pass it through sprays of water. Spray cooling (evaporative cooling) lowers the gas temperature to 300−350 °C and avoids the investment in heat recovery equipment, but it wastes the heat in the gases. Dry dust is recovered, since the gases are not completely saturated. It is used primarily for Ausmelt, Isasmelt, Teniente, Noranda, and sometimes Peirce−Smith gases.

Water quenching (Section 12.3.3) can also be used to cool the gases leaving the furnace. All of the Cu-bearing dust ends up in a weak acid solution and must be recovered hydrometallurgically. Water quenching is used primarily for Inco (Humphris, Liu, & Javor, 1997; Liu, Baird, Kenny, Macnamara, & Vahed, 2006) and Peirce−Smith converter offgases (Daum, 2000).

12.3.2. Electrostatic Precipitation of Dust

After cooling, the furnace gases are passed through electrostatic precipitators (Conde, Taylor, & Sarma, 1999; Parker, 1997; Ryan, Smith, Corsi, & Whiteus, 1999) for more dust removal. The dust particles are caught by:

(a) Charging them in the corona of a high voltage electric field
(b) Catching them on a charged plate or wire
(c) Collecting them by neutralizing the charge and shaking the wires or plates.

The precipitators remove 99+% of the dust from their incoming gas (Conde et al., 1999). The dust is usually re-smelted to recover its Cu. About 70% of the dust is recovered in the cooling system and 30% in the electrostatic precipitators.

12.3.3. Water Quenching, Scrubbing, and Cooling

After electrostatic precipitation, the gas is quenched with water and scrubbed to further remove dust and impurities. This quenching (a) removes most of the remaining dust from the gas to avoid fouling of downstream acid plant catalyst and (b) absorbs HCl, HF, SO_3, and vapor impurities such as As_2O_3. The gas is then further cooled to 35−40 °C by direct contact with cool water in a packed tower or by indirect contact with cool water in a heat exchanger.

The gas leaves the cooling section through electrostatic mist precipitators to eliminate fine droplets of liquid remaining in the gas after quenching and cooling. Mist precipitators operate similarly to the electrostatic precipitators described in Section 12.3.2. They must, however, be (a) constructed of acid-resistant materials such as fiber-reinforced plastic, alloy steels, or lead, and (b) periodically turned off and flushed with fresh water to remove collected solids.

12.3.4. Mercury Removal

Smelters treating high-mercury concentrates often use the Boliden Norzink process to remove mercury from the gas leaving the electrostatic mist precipitators (Hasselwander, 2009, pp. 36−43; Hultbom, 2003). In the Boliden Norzink process, the gas passes through a water-based solution of dissolved mercuric chloride ($HgCl_2$) circulating over a packed tower to absorb gaseous elemental mercury and form solid Hg_2Cl_2 (calomel) by the following reaction:

$$HgCl_2(aq) + Hg°(g) \rightarrow Hg_2Cl_2(s) \qquad (12.1)$$

About half of the scrubbing solution is withdrawn and regenerated with chlorine gas by the following reaction:

$$Hg_2Cl_2(s) + Cl_2(g) \rightarrow 2HgCl_2(aq) \qquad (12.2)$$

The overall reaction is:

$$2Hg°(g) + Cl_2(g) \rightarrow Hg_2Cl_2(s) \qquad (12.3)$$

Mercury can be bled from the system as Hg_2Cl_2 (calomel) by solid/liquid separation or as elemental mercury by electrowinning:

$$HgCl_2(aq) + 2e^- \rightarrow Hg°(l) + Cl_2(g) \qquad (12.4)$$

The Boliden Norzink process is efficient, typically removing more than 99.95% of the inlet mercury. This results in a product acid mercury concentration of 0.3−0.5 ppm. This acid will meet most quality specifications. However, if further mercury removal is required, then the gas can be passed through an additional filter that contains either PbS-coated pumice (Dowa process) or Se (Hasselwander, 2009, pp. 36−43; Lossin & Windhager, 1999). Mercury reacts with the Se to form HgSe and with PbS to form HgS. Product acid mercury concentrations below 0.1 mg/L can be achieved with this additional step.

12.3.5. The Quenching Liquid, Acid Plant Blowdown

The water from quenching and direct-contact cooling is passed through water-cooled heat exchangers and used again for quenching/cooling. It becomes acidic (from SO_3 absorption) and impure (from dust and vapor absorption). A bleed stream of this impure solution (acid plant *blowdown*) is continuously withdrawn and replaced with fresh water. The amount of bleed and water replacement is controlled to keep the corrosive H_2SO_4 content of the cooling water below about 10%. The quantity of bleed depends on the amount of SO_3 in the offgas as it enters the water-quench system.

The acid plant blowdown stream is neutralized and either stored or treated for metal recovery (Ante, 2010; Evans, Lawler, Lyne, & Drexler, 1998). Figure 12.3 shows the Atlantic Copper weak acid treatment system.

12.4. GAS DRYING

12.4.1. Drying Tower

The next step in offgas treatment is gas dehydration (drying). It is done to prevent unintentional H_2SO_4 formation and corrosion in downstream ducts, heat exchangers, and catalyst beds. The $H_2O(g)$ is removed by contacting it with 93% (occasionally 96%) acid. H_2O reacts strongly with H_2SO_4 to form hydrated acid molecules. The contacting is done in a counter-current packed tower filled with 5−10 cm ceramic saddles (Fig. 12.4). The sulfuric acid flows down over the saddles. The gas is drawn up by the main acid plant blower(s).

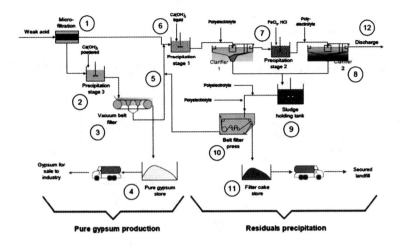

FIGURE 12.3 Acid plant blowdown treatment system at Atlantic Copper (Ante, 2010). Two solids streams are produced — a nearly pure gypsum stream and a heavy metals (residuals) precipitation stream. Clarified effluent is also discharged.

1 Microfiltration: Separation of colored solids for the production of white gypsum
2 Precipitation stage 0: Pure gypsum precipitation takes place by partial neutralization of the weak acid down to a pH smaller than 1
3 Vacuum belt filter (VBF): The gypsum slurry from precipitation stage 0 is dewatered and washed to a residual moisture content below 35%
4 Pure gypsum store
5 Filtrate: Filtrate accumulating from dewatering of the gypsum slurry is routed to precipitation stage 1
6 Precipitation stage 1: Gypsum precipitation, coarse As precipitation and heavy metal precipitation with milk of lime
7 Precipitation stage 2: Final As and heavy metal precipitation as hydroxides
8 Clarifier 1 and 2
9 Sludge holding tank
10 Belt filter press (BFP): Dewatering of dirty gypsum
11 Filter cake store
12 Discharge

The liquid product of gas drying is slightly diluted 93% H_2SO_4. It is strengthened with the 98+% acid produced by subsequent SO_3 absorption (Section 12.5.2). Most of the strengthened acid is recycled to the absorption tower. A portion is sent to storage and then to market.

12.4.1.1. Optimum Absorbing and Composition

The gas product of the drying tower contains typically 50–100 mg H_2O/Nm^3 of offgas. It also contains small droplets of acid mist, which it picked up during its passage up the drying tower. This mist is removed by passing the dry gas through stainless steel or fiber mist eliminator pads or candles.

12.4.2. Main Acid Plant Blowers

The dried gas is drawn into the main acid plant blowers, which push it on to $SO_2 \rightarrow SO_3$ conversion and acidmaking. Two centrifugal blowers, typically 3000–4000 kW each, are

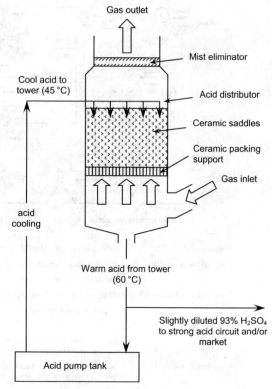

FIGURE 12.4 Drying tower and associated acid circulation and cooling equipment. Acid is pumped around the tubes of the acid—water heat exchanger to the top of the tower, where it is distributed over the packing. It then flows by gravity downward through the packing and returns to the pump tank. The mist eliminator in the top of the tower is a mesh pad. In most SO_3 absorption towers this pad is usually replaced with multiple candle-type mist eliminators.

used. They move 100 000—380 000 Nm^3 of gas per hour (Lüdtke & Müller, 2003). The gas handling system is under a slight vacuum before the blowers (typically <0.07 bar gauge at the smelting furnace) and under pressure (0.3—0.5 bar gauge) after.

12.5. ACID PLANT CHEMICAL REACTIONS

12.5.1. Oxidation of SO_2 to SO_3

The SO_2 in the offgas is oxidized to SO_3 in preparation for absorption in the water component of 98.5% H_2SO_4. The oxidation reaction is:

$$SO_2(g) + 0.5O_2(g) \rightarrow SO_3(g) \quad (12.5)$$

This reaction is often referred to as SO_2 *conversion*. It is very slow without a catalyst, so the gas is always passed through V_2O_5—K_2SO_4 catalyst beds. The volumetric O_2/SO_2 ratio entering the catalyst beds is set at ~1 or above (by adding air, if necessary) to ensure near complete conversion of SO_2 to SO_3.

12.5.1.1. Catalyst Reactions

Typical V_2O_5—K_2SO_4 based catalyst contains 5—10% V_2O_5, 10—20% K_2SO_4, 1—5% Na_2SO_4, and 55—70% SiO_2. It may also contain 5—15% cesium sulfate (Cs_2SO_4) substituted for K_2SO_4. The active components of the catalyst are V_2O_5, K_2SO_4, Na_2SO_4, and Cs_2SO_4 (if present). The inactive material is SiO_2, which acts as a support for the active components (Jensen-Holm, 1986).

V_2O_5–K_2SO_4 catalyst is a supported liquid-phase catalyst (Livbjerg, Jensen, & Villadsen, 1978). At the catalyst operation temperature, ~400 °C, the active catalyst components (V_2O_5, K_2SO_4, Na_2SO_4, Cs_2SO_4) exist as a film of molten salt solution on the solid inactive SiO_2 support. Oxidation of SO_2 to SO_3 in the presence of oxygen takes place by homogeneous reactions in this liquid film. Pores on the surface of the silica substrate provide the large surface area necessary for rapid SO_2 oxidation. The rate of catalytic SO_2 oxidation depends largely upon the rate of mass and heat transfer at the catalyst gas–liquid interface.

One of the most widely cited SO_2 conversion reaction mechanisms is that proposed by Mars and Maessen (1964, 1968). It is based on the experimental observation that during SO_2 conversion, the oxidation state of the vanadium ions in the catalyst changes between the V^{4+} and V^{5+}. This observation suggests that the reaction involves:

(a) Absorption of SO_2, reduction of vanadium ions from V^{5+} to V^{4+}, and formation of SO_3 from SO_2 and O^{2-} ions, i.e.:

$$SO_2 + 2V^{5+} + O^{2-} \rightarrow SO_3 + 2V^{4+} \quad (12.6)$$

and:

(b) Absorption of oxygen, re-oxidation of the vanadium ions, and formation of O^{2-} ions:

$$0.5O_2 + 2V^{4+} \rightarrow 2V^{5+} + O^{2-} \quad (12.7)$$

12.5.1.2. Industrial V_2O_5–K_2SO_4 Catalysts

Catalyst is manufactured by mixing together the active components and substrate to form a paste, which is extruded and baked at ~530 °C into solid cylindrical pellets or rings. Ring-shaped (or *star ring*) catalyst is the most commonly used shape because (a) it gives a small pressure drop in a catalyst bed and (b) its catalytic activity is only slowly affected by dust in the acid plant feed gas. A typical catalyst ring is 10 mm in diameter by 10 mm in length (Fig. 12.5).

12.5.1.3. Catalyst Ignition and Degradation Temperatures

The ignition temperature at which the SO_2 conversion reaction begins with V_2O_5–K_2SO_4 catalyst is ~360 °C. The reaction rate is relatively slow at this ignition temperature. Therefore, the gases entering the catalyst beds are heated to temperatures in the range of 400–440 °C to ensure rapid SO_2 conversion.

Above 650 °C thermal deactivation of the catalyst begins. Several mechanisms for high temperature thermal deactivation have been proposed.

FIGURE 12.5 Photograph of catalyst pieces. The outside diameter of the largest piece (far left) is 20 mm. Pellets are shown far right, rings in the middle, and 'daisy' shapes (modern) left and right of the rings. (Courtesy Haldor Topsoe A/S.)

(a) Silica in the substrate partly dissolves in the catalytic melt. This causes the thickness of the melt film to increase, which in turn blocks the pore structure, preventing gas access to the liquid phase inside the pores.
(b) Sintering of the silica substrate closes pores, restricting gas access to liquid phase inside the pores.

Thermal deactivation proceeds slowly. V_2O_5–K_2SO_4 catalyst can usually be subjected to temperatures of 700–800 °C for short periods without causing significant deactivation. However, long times at these temperatures reduce catalyst activity and decrease SO_2 conversion rate.

12.5.1.4. Cs-promoted Catalyst

Substituting Cs_2SO_4 for K_2SO_4 in the active liquid component of the catalyst lowers the melting point of the liquid, providing higher reaction rates at lower temperatures. The lower melting point of cesium sulfate allows the V^{4+} species to remain in solution at a lower temperature, increasing its low temperature catalytic activity. Cs-promoted catalyst has an ignition temperature of ~320 °C. Its typical operating temperature range is 370–500 °C. Cs-promoted catalyst costs 2–2.5 times that of non Cs-promoted catalyst. Therefore, its use is typically optimized by installing it only in the top half of the first and/ or last catalyst beds.

12.5.1.5. Dust Accumulation in Catalyst Beds

Over time, dust, which inadvertently passes through the gas cleaning section, begins to build up in the catalyst beds. It blocks gas flow through the catalyst and increases the pressure that must be applied to achieve the acid plant's required gas flowrate.

When the pressure drop in the catalyst beds becomes excessive, the acid plant must be shut down and the catalyst screened to remove the accumulated dust. Keeping offgas cleaning apparatus in optimum operating condition is critical to maintaining acid plant availability.

12.5.1.6. $SO_2 \rightarrow SO_3$ Conversion Equilibrium Curve

Oxidation of SO_2 to SO_3 proceeds further toward completion at lower temperatures. Figure 12.6 shows the equilibrium curve for a gas containing 12% SO_2, 12% O_2, balance N_2 at a total pressure of 1.2 bar. The equilibrium curve on the graph represents the maximum attainable conversion of SO_2 to SO_3 at a given temperature. This curve is also shown in Fig. 12.9 with reaction heat-up paths for each catalyst bed.

12.5.1.7. Absorption of SO_3 into H_2SO_4–H_2O Solution

The SO_3 formed by the above-described catalytic oxidation of SO_2 is absorbed into 98.5% H_2SO_4. The process occurs in a packed tower of similar design to a drying tower (Fig. 12.4). In absorption, SO_3 laden gas and sulfuric acid flow counter currently. The overall absorption reaction is:

$$SO_3(g) + H_2O(l) \rightarrow H_2SO_4(l) \quad (12.8)$$

It is not possible to manufacture sulfuric acid by absorbing sulfur trioxide directly into water. Sulfur trioxide reacts with water vapor to form H_2SO_4 vapor. This sulfuric acid

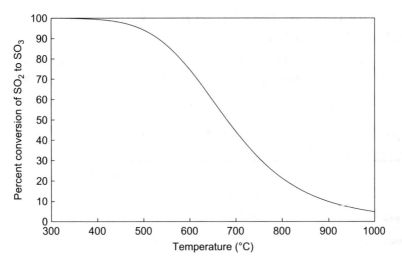

FIGURE 12.6 Equilibrium curve for SO_2 conversion for an initial gas composition of 12 vol.-% SO_2, 12% O_2, and 76% N_2 at a total pressure of 1.2 bar. The curve shows that higher SO_2 conversions are possible at lower temperatures.

vapor condenses as a mist of fine, sub-micron, droplets, which are practically impossible to coalesce. However, the theoretical vapor pressure of water over 98.5% H_2SO_4 is low ($<2 \times 10^{-5}$ bar at 80 °C), avoiding this water vapor problem. The most likely absorption reactions are:

$$SO_3(g) + H_2SO_4(l) \rightarrow H_2S_2O_7(l) \qquad (12.9)$$

followed by:

$$H_2O(l) + H_2S_2O_7(l) \rightarrow 2H_2SO_4(l) \qquad (12.10)$$

Some SO_3 is undoubtedly absorbed directly by water according to Eq. (12.8). Because of the preponderance of H_2SO_4 molecules in the absorbent, however, absorption by Eqs. (12.9) and (12.10) probably predominates. SO_3 absorption is exothermic, so that the strengthened acid must be cooled before it is (a) recycled for further absorption or (b) sent to storage.

12.5.1.8. Optimum Absorbing Acid Composition

The optimum absorbing acid composition is 98–99% H_2SO_4. This is the composition at which the sum of the equilibrium partial pressures of H_2O, SO_3, and H_2SO_4 over sulfuric acid is at its minimum.

Below this optimum, H_2O vapor pressure increases and sulfuric acid mist may form by the reaction of $H_2O(g)$ and SO_3. This mist is difficult to coalesce, so it tends to escape from the acid plant into the environment. Above this optimum, SO_3 and H_2SO_4 partial pressures increase. This also increases the release of sulfur compounds into the environment.

Acid plant flowrates and compositions are controlled to keep the absorbing acid in the 98–99% range before and after SO_3 absorption.

12.6. INDUSTRIAL SULFURIC ACID MANUFACTURE (TABLES 12.4, 12.5, AND 12.6)

Figure 12.7 shows a typical flowsheet for SO_2 conversion and SO_3 absorption. The acid plant is a 3:1 double absorption plant; this means that the gases pass through three catalyst

TABLE 12.4 Operating Details of Three Double Absorption Sulfuric Acid Manufacturing Plants

	Aurubis Hamburg, Germany		
	Line 1	Line 2	Line 3
Start-up date	1972	1972	1991
Manufacturer	Lurgi	Lurgi	Lurgi
Gas source	Outotec flash + Peirce–Smith converters	Outotec flash + Peirce–Smith converters	Outotec flash + Peirce–Smith converters
Single or double absorption	Double	Double	Double
Number of catalyst beds	4	4	5
Intermediate absorption after bed?	2	2	3
Converter design			
Materials of construction	Stainless steel	Stainless steel	Brick lined
First pass diameter, m	8	8	8.5
Others, diameter, m	8	8	8.5
Catalyst bed loading			
Bed 1, liters	75 000	70 000	42 300
Bed 2, liters	58 000	58 000	46 200
Bed 3, liters	64 000	64 000	48 200
Bed 4, liters	64 000	64 000	46 200
Bed 5, liters	–	–	54 200
Catalyst type			
Bed 1	BASF ring + 0.19 m Cs catalyst	Haldor Topsøe ring + 0.22 m Cs catalyst	BASF ring + 0.19 m Cs catalyst
Bed 2	Haldor Topsøe ring catalyst	Haldor Topsøe ring catalyst	BASF ring catalyst
Bed 3	BASF ring catalyst	Haldor Topsøe ring catalyst	BASF ring catalyst

TABLE 12.4 Operating Details of Three Double Absorption Sulfuric Acid Manufacturing Plants—cont'd

	Aurubis Hamburg, Germany		
	Line 1	Line 2	Line 3
Bed 4	BASF Cs ring catalyst	Haldor Topsøe ring catalyst	BASF ring catalyst
Bed 5	—	—	BASF ring catalyst
Gas into converter			
Flowrate, Nm^3/min	1830	1830	1830
SO_2, vol.-%	11	11	11
O_2, vol.-%	>12.1	>12.1	>12.1
CO_2, vol.-%	1830	1830	1830
Design SO_2 conversion	99.7–99.8	99.7–99.8	99.8
H_2SO_4 production rate			
Design, tonnes/day	1265	1265	1265
Actual, tonnes/day	Approximately 3000 total		
Products, %H_2SO_4	94, 96, 98	94, 96, 98	94, 96, 98
Other (oleum, liquid SO_2, etc.)		20% oleum	

TABLE 12.5 Operating Details of Three Double Absorption Sulfuric Acid Manufacturing Plants with Heat Recovery Equipment

	Sumitomo Toyo, Japan Line 1	Sumitomo Toyo, Japan Line 2	Rio Tinto Kennecott Utah Copper Magna, UT, USA
Start-up date	1971	2003	1995
Manufacturer	Sumitomo Chemical Engineering	Sumitomo Chemical Engineering	MECS
Gas source	Outotec flash + Peirce–Smith converters	Outotec flash + Peirce–Smith converters	Outotec flash smelting + Kennecott-Outotec flash converting

(Continued)

TABLE 12.5 Operating Details of Three Double Absorption Sulfuric Acid Manufacturing Plants with Heat Recovery Equipment—cont'd

	Sumitomo Toyo, Japan Line 1	Sumitomo Toyo, Japan Line 2	Rio Tinto Kennecott Utah Copper Magna, UT, USA
Single or double absorption	Double	Double	Double
Number of catalyst beds	5	5	4
Intermediate absorption after bed?	4	4	3
Converter design			
Materials of construction	Stainless steel + carbon steel	Stainless steel	304H stainless steel
First pass diameter, m	12.5	12.3	14.1
Others, diameter, m	12.5	12.3	14.1
Catalyst bed loading			
Bed 1, liters	40 000	45 000	38 500/77 000
Bed 2, liters	50 000	60 000	154 000
Bed 3, liters	65 000	70 000	40 000/152 000
Bed 4, liters	104 000	75 000	288 800
Bed 5, liters	105 000	110 000	
Catalyst type			
Bed 1	Topsøe VK38	Topsøe VK38, Sud Chemie SulfoMax-RR	Haldor Topsøe VK59 Haldor Topsøe VK38
Bed 2	Nihonshokubai 7s Monsanto T-516 Topsøe VK38	Monsanto LP110	Haldor Topsøe VK38
Bed 3	Nihonshokubai R10 Monsanto T-516 Topsøe VK38	Monsanto LP110	Haldor Topsøe VK59 Haldor Topsøe VK48

TABLE 12.5 Operating Details of Three Double Absorption Sulfuric Acid Manufacturing Plants with Heat Recovery Equipment—cont'd

	Sumitomo Toyo, Japan Line 1	Sumitomo Toyo, Japan Line 2	Rio Tinto Kennecott Utah Copper Magna, UT, USA
Bed 4	Topsøe VK48, Nihonshokubai R10	Monsanto LP1150	Haldor Topsøe VK69
Bed 5	Nihonshokubai 7s, R10 BASF O4-111	Monsanto LP1150, BASF O4-111	—
Gas into converter			
Flowrate, Nm^3/min	2917 (maximum)	2833 (maximum)	3965
SO_2, vol.-%	12–13	12–13	14
O_2, vol.-%	12–14	12–14	13.8 (0.8 CO_2)
Heat recovery			
Boiler	After Bed 1, 21.2 t/h steam	After Bed 1, 21.2 t/h steam	After Bed 3
Superheater	None	None	After Bed 1
Economizer	After Bed 4	After Bed 4	None
Acid heat recovery	None	None	HRS
Steam produced (t/t H_2SO_4)	0.23	0.23	0.69
Design SO_2 conversion (%)	99.6	99.6	>99.5
H_2SO_4 production rate			
Design, tonnes/day	2254	2254	3500
Actual, tones/day	1775	1675	—
Products, % H_2SO_4	98, 95	98	94, 98
Other (oleum, liquid SO_2, etc.)	Sodium bisulfite	Oleum	—

TABLE 12.6 Physical and Operating of Two Single Absorption Sulfuric Acid Manufacturing Plants, 2010

Smelter	Incitec Pivot, Mt. Isa, Queensland, Australia	Xstrata Copper, Horne Smelter, Rouyn-Noranda, Canada
Start-up date	1999	1989
Manufacturer	Lurgi	Chemetics
Gas source	Isasmelt, 4 Peirce–Smith converters and sulfur burner	Noranda continuous process reactor and Noranda converter or Peirce–Smith converters
Single or double absorption	Single	Single
Number of catalyst beds	3	3
Converter design		
Materials of construction	304H SS (1 internal Hot heat exchanger)	304H SS (1 internal Hot heat exchanger)
First pass diameter, m	15	14.2
Others, diameter, m	Same	14.2
Catalyst bed loading		
Bed 1, liters	162 300	83 000
Bed 2, liters	183 500	178 000
Bed 3, liters	211 366	216 000
Catalyst type		
Bed 1	$K-V_2O_5$	Topsøe
Bed 2	$K-V_2O_5$	Topsøe
Bed 3	$Cs-K-V_2O_5 + K-V_2O_5$	Topsøe
Gas into converter		
Flowrate, Nm^3/min	6333	3583
SO_2, vol.-%	2 maximum 6 normal operating	6–10.5
O_2, vol.-%	Not measured	11.0 (3.0 CO_2)
Heat recovery		
Boiler	1 after sulfur burner 1 between beds 2 and 3	None
Superheater	2 bundles after bed 3	None
Economizer	2 bundles after bed 3	None
Acid heat recovery	None	None

TABLE 12.6 Physical and Operating of Two Single Absorption Sulfuric Acid Manufacturing Plants, 2010—cont'd

Smelter	Incitec Pivot, Mt. Isa, Queensland, Australia	Xstrata Copper, Horne Smelter, Rouyn-Noranda, Canada
Steam produced (t/t H_2SO_4)	0.8	N/A
Design SO_2 conversion (%)	>98	96 (actual 97.5)
H_2SO_4 production rate		
Design, tonnes/day	3645	2400
Actual, tones/day	3636	2000
Products, %H_2SO_4	98.5	93.5
Other (oleum, liquid SO_2, etc.)	None	None

beds before intermediate absorption and then one catalyst bed before final absorption. Figures 12.9 and 12.10 describe the process thermodynamically. The steps are:

(a) Heating of the incoming gas to the minimum continuous catalyst operating temperature (~410 °C) by heat exchange with the hot gases from SO_2 oxidation

(b) Passing the hot gas through a first bed of catalyst, where partial SO_2 conversion takes place, and where the gases are heated by the heat of the SO_2 reaction

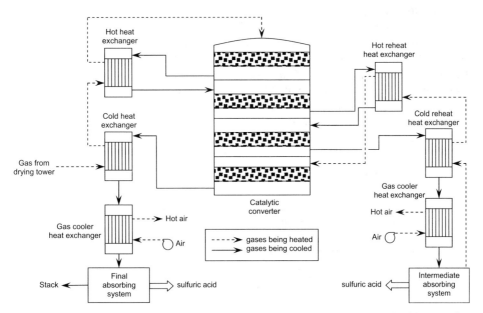

FIGURE 12.7 Flowsheet for a typical 3:1 metallurgical sulfuric acid plant. Catalyst beds with gas cooling between (to avoid overheating the catalyst) are notable. A hydrocarbon fuel-fired heat exchanger (not shown) is used to heat the feed gas during acid plant startup.

(c) Cooling the gas down by heat exchange with cool incoming gas
(d) Passing the cooled gas through a second bed of catalyst, where more SO_2 conversion takes place, and where the gases again become hot
(e) Repeating steps (c) and (d) with a third catalyst bed.

The gas from the third catalyst bed is cooled and its SO_3 absorbed into 98.5% H_2SO_4. The exit gas from this absorption is then passed through a second set of heat exchangers, a fourth catalyst bed, and a second absorption tower. In some plants, initial absorption takes place after the gas passes through two catalyst beds and final absorption after the remaining two catalyst beds.

The above description is for a *double absorption* plant which converts and absorbs >99.7% of the SO_2 entering the acid plant. Single absorption acid plants convert SO_2 to SO_3 in three or four catalyst beds followed by single absorption of SO_3 (Table 12.6). Their conversion of SO_2 to SO_3 is less complete with consequentially lower sulfur capture efficiencies (96—98%).

12.6.1. Catalytic Converter

A catalytic converter typically houses three to five catalyst beds. It is usually made of 304H stainless steel. Figure 12.8 shows the cross-section of a typical catalyst bed.

12.6.2. $SO_2 \rightarrow SO_3$ Conversion Reaction Paths

Figures 12.9 and 12.10 show the schematic steady-state %SO_2 conversion/temperature reaction path for a 12% SO_2, 12% O_2 gas flowing through a double absorption 3:1 sulfuric acid plant. The gas enters the first catalyst bed of the converter at about 410 °C. SO_2 is oxidized to SO_3 in the bed, heating the gas to about 630 °C. About 64% of the input SO_2 is converted to SO_3.

The gas from bed 1 is then cooled to 430 °C in a heat exchanger (Fig. 12.9) and is passed through the second catalyst bed. There, a further 26% of the SO_2 is converted to SO_3 (to a total of 90%) and the gas is heated to about 520 °C by the oxidation reaction. This gas is then cooled to 435 °C in a heat exchanger and is passed through the third catalyst bed. A further 6% of the initial SO_2 is oxidized to SO_3 (to 96% conversion) while the temperature increases to about 456 °C.

At this point, the gas is cooled to ~180 °C and sent to the intermediate absorption tower, where virtually all (99.99%) of its SO_3 is absorbed into 98.5% H_2SO_4.

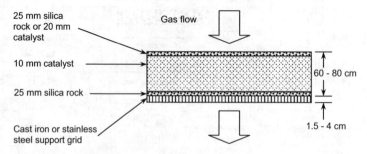

FIGURE 12.8 Catalyst bed showing steel support, catalyst, and silica rock. The bed is typically 8—14 m in diameter. The silica rock on the top of the bed distributes the gas into the catalyst, preventing localized channeling and short-circuiting through the bed.

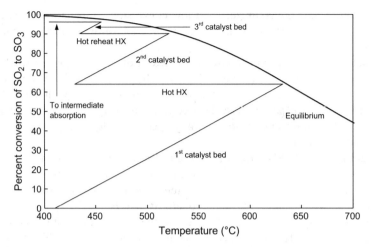

FIGURE 12.9 Equilibrium curve and first through third catalyst bed reaction heat-up paths. The horizontal lines represent cooling between the catalyst beds in the heat exchangers. The feed gas contains 12 vol.-% SO_2, 12% O_2, balance N_2 (1.2 bar gauge overall pressure).

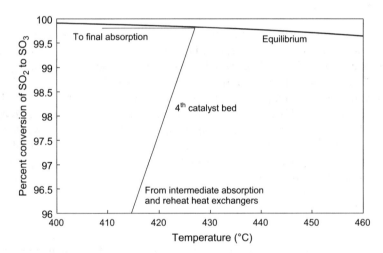

FIGURE 12.10 Equilibrium curve and fourth catalyst bed reaction heat-up path. Almost all of the SO_3 in the gas leaving the third catalyst bed has been absorbed into sulfuric acid in the intermediate absorption tower.

After this absorption, the gas contains about 0.5% SO_2. It is heated to 415 °C and passed through the last catalyst bed in the converter (Fig. 12.10). Here about 90% of its SO_2 is converted to SO_3, leaving only about 0.025% SO_2 in the gas. This gas is again cooled to ~180 °C and sent to the final SO_3 absorption tower. Overall conversion of SO_2 is approximately 99.8%.

12.6.3. Reaction Path Characteristics

Figures 12.9 and 12.10 show some important aspects of SO_2 conversion.

1. Conversion to SO_3 is maximized by a low conversion temperature, consistent with meeting the minimum continuous operating temperature requirement of the catalyst.
2. The maximum catalyst temperature is reached in the first catalyst bed, where most of the SO_2 conversion takes place. This is where a low ignition temperature Cs catalyst can be useful. Catalyst bed temperature increases with increasing SO_2 concentration in the gas because SO_2 conversion energy release has to heat less N_2. Cs catalyst is expensive, so it is only used when low temperature catalysis is clearly advantageous.
3. Conversion of SO_2 to SO_3 after intermediate absorption is very efficient (Fig. 12.10). This is because (a) the gas entering the catalyst contains no SO_3 (driving Reaction (12.4) to the right) and because (b) the temperature of the gas rises only slightly due to the small amount of SO_2 being oxidized to SO_3.
4. Maximum cooling of the gases is required for the gases being sent to SO_3 absorption towers (~440 °C to 180 °C), hence the inclusion of air coolers in Fig. 12.7.
5. Maximum heating of the gases is required for initial heating and for heating after intermediate absorption, hence the passage through several heat exchangers in Fig. 12.7.

12.6.4. Absorption Towers

Double absorption sulfuric acid plants absorb SO_3 twice: after partial SO_2 oxidation and after final oxidation. The absorption is done counter-currently in towers packed with 5–10 cm ceramic saddles, which present a continuous descending film of 98.5% H_2SO_4 into which rising SO_3 absorbs. Typical sulfuric acid irrigation rates, densities, and operating temperatures for absorption towers are shown in Table 12.7.

The strengthened acid is cooled in water-cooled shell-and-tube type heat exchangers. A portion of it is sent for blending with 93% H_2SO_4 from the gas drying tower to produce the grades of acid being sent to market. The remainder is diluted with blended acid and recycled to the absorption towers. These cross-flows of 98+ and 93% H_2SO_4 allow a wide range of acid products to be marketed.

TABLE 12.7 Typical Sulfuric Acid Design Irrigation Rates and Irrigation Densities for Drying and Absorption Towers (Guenkel & Cameron, 2000)

Tower	Sulfuric acid irrigation rate (m^3/tonne of 100% H_2SO_4 produced)	Sulfuric acid irrigation density (m^3/min per m^2 of tower cross-section)	Sulfuric acid temperature (°C) inlet/outlet
Drying tower	0.005	0.2–0.4	45/60
Intermediate absorption tower	0.01	0.6–0.8	80/110
Final absorption tower	0.005	0.4	80/95

12.6.5. Gas to Gas Heat Exchangers and Acid Coolers

Large gas-to-gas heat exchangers are used to transfer heat to and from gases entering and exiting a catalytic converter. The latest heat exchanger designs are radial shell and tube. Older designs were single or double segmental type. Acid plant gas-to-gas heat exchangers typically transfer heat at 10 000—80 000 MJ/h. They must be sized to ensure that a range of gas flowrates and SO_2 concentrations can be processed. This is especially significant for smelters treating offgases generated by batch type Peirce—Smith converters.

The hot acid from SO_3 absorption and gas drying is typically cooled in indirect shell and tube heat exchangers. The water flows through the tubes of the heat exchanger and the acid through the shell. The warm water leaving the heat exchanger is usually cooled in an atmospheric cooling tower before being recycled for further acid cooling.

Anodic protection of the coolers is required to minimize corrosion by the hot sulfuric acid. A non-anodically protected acid cooler has a lifetime on the order of several months whereas anodically protected coolers have lifetimes on the order of 20—30 years.

12.6.6. Grades of Product Acid

Sulfuric acid is sold in grades of 93—98% H_2SO_4 according to market demand. The principal product in cold climates is 93% H_2SO_4 because of its low freezing point, $-30\,°C$ (Davenport & King, 2006).

Oleum, H_2SO_4 into which SO_3 is absorbed, is also sold by several smelters. It is produced by diverting a stream of SO_3-bearing gas and contacting it with $98 + H_2SO_4$ in a small absorption tower.

12.7. ALTERNATIVE SULFURIC ACID MANUFACTURING METHODS

12.7.1. Haldor Topsøe WSA

The WSA (Wet gas Sulfuric Acid) process, developed and commercialized in the early 1980s by Haldor Topsøe, has over 100 installations worldwide and is used to treat wet, low concentration gases (0.2—6.5% SO_2). Sulfur dioxide gas is first cooled and cleaned in a gas cleaning plant. Dust and impurity removal specifications are the same as those used in conventional contact-type acid plants. However, a key difference is the saturated gas temperature, which is typically higher in WSA plants than conventional acid plants. This is because not all of the water entering a WSA plant is absorbed into sulfuric acid. Some of it leaves in the tail gas (Fig. 12.11).

The saturated gas leaves the gas cleaning plant and is preheated in gas-to-gas heat exchangers to the catalyst operating temperature (~410 °C). The SO_2 oxidation step is nearly identical to that in a conventional acid plant (Eq. (11.5)), but the reaction takes place in a wet gas using a specially formulated catalyst. SO_2 oxidation is performed in either two or three beds. Interbed cooling, using an indirect molten salt circulation system, removes surplus heat and produces saturated steam.

When cooled to 290 °C, a substantial amount of the SO_3 undergoes hydration to form sulfuric acid gas by the following reaction:

$$SO_3(g) + H_2O(g) \rightarrow H_2SO_4(g) \qquad (12.11)$$

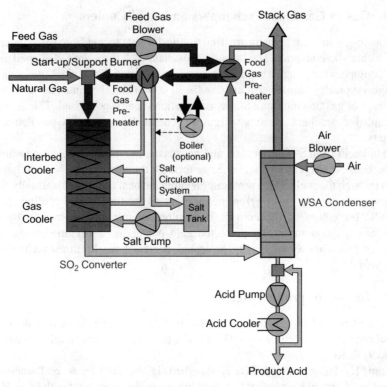

FIGURE 12.11 Haldor Topsøe WSA Process flow schematic. A molten salt mixture is used to transfer heat from the catalyst beds to the cold feed gas entering the plant. Excess heat removed from the catalyst beds is turned into useful steam in a small heat recovery boiler. A direct fired hydrocarbon fuel burner can be used to supply additional heat to the catalyst beds.

The sulfuric acid gas leaves the catalytic converter and passes through a glass tube condenser, where it is indirectly cooled with air. Warm air flows around the tubes, which cools the gas. Concentrated sulfuric acid (~98% H_2SO_4) condenses in the tubes and drains into the bottom of the condenser:

$$H_2SO_4(g) \rightarrow H_2SO_4(l) \tag{12.12}$$

It is subsequently cooled with recycle product sulfuric acid and pumped through a plate heat exchanger for further cooling prior to storage. Clean process gas leaves the WSA condenser at approximately 100 °C and flows through an acid mist eliminator (candle filters or electrostatic mist precipitators) before being exhausted to the atmosphere.

WSA plants are able to achieve greater than 99% SO_2 conversion while producing concentrated (98%) acid (Rosenberg, 2006). WSA plants can treat up to 1 200 000 Nm^3/h of feed gas and produce 1140 t/d of acid. In 2010, a WSA plant is used to treat SO_2 bearing gases generated by an Ausmelt furnace at the Karabashmed smelter in Russia.

12.7.2. Sulfacid®

Sulfacid is presently marketed by Carbon Process and Plant Engineering GmbH. About 20 installations worldwide produce weak sulfuric acid (10–20% H_2SO_4) from very low

TABLE 12.8 Comparison of Conventional, WSA, and Sulfacid Acid Plant Technologies (King & Forzatti, 2009)

Parameter	Conventional	WSA	Sulfacid
Minimum SO_2 concentration	Single Absorption: 4.5[a] Double Absorption: 6	0.6[c]	0.5
Maximum SO_2 concentration	Single Absorption: 10 Double Absorption: 14[b]	6.5	1.0
Product acid, %H_2SO_4	93–98.5	98	10–20
SO_2 removal efficiency (%)	Single Absorption: 98–99 Double Absorption: 99.7–99.95	98–99	65–90

[a]Outotec reports single absorption plants capable of 2 vol.-%.
[b]Without internal gas recycle, for example, Outotec's LUREC™ process (with internal recycle) can treat 18 vol.-% SO_2 gas.
[c]Autothermal at 3 vol.-% SO_2.

concentration gases (<1.0% SO_2). The saturated gas is cleaned (<30 mg solids/dry Nm^3) before entering an activated carbon catalytic reactor where SO_2 reacts with O_2 and H_2O at 30–80 °C (Kruger, 2004) to produce H_2SO_4 by the following reaction:

$$SO_2(g) + 0.5O_2(g) + nH_2O(l) \rightarrow H_2SO_4(g) + (n-1)\ H_2O(l) \quad (12.13)$$

The acid is intermittently washed from the catalyst, which produces weak sulfuric acid. The cleaned gas is discharged to the atmosphere. Sulfacid plants are most economical when their dilute sulfuric acid product can be consumed in nearby operations, such as titanium dioxide manufacture or fertilizer production. Table 12.8 shows a comparison of conventional, WSA, and Sulfacid-type acid plants.

12.8. RECENT AND FUTURE DEVELOPMENTS IN SULFURIC ACID MANUFACTURE

12.8.1. Maximizing Feed Gas SO_2 Concentrations

Since the 1980s, smelting technology has shifted from reverberatory and electric furnace smelting to flash furnace and other intensive smelting processes. Oxygen enrichment of furnace blasts also increased significantly. An important (and desired) effect of these changes has been increased SO_2 strength in the gases that enter smelter sulfuric acid plants (Newman, Collins, & Weddick, 1999).

SO_2 offgases entering the drying tower now average 6–18% SO_2. The low concentrations come from smelters using Peirce–Smith converters. The high concentrations come from direct-to-copper smelting and continuous smelting/converting smelters (Ritschel, Fell, Fries, & Bhambri, 1998).

High SO_2 concentration gases contain little N_2. They heat up more than conventional smelter gas during passage through catalyst beds. This can lead to overheating and degradation of the V_2O_5–K_2SO_4 catalyst (650 °C) and to weakening of the steel catalyst bed support structure (630 °C). These two items limit the maximum strength of sulfuric acid plant feed gas to ~13% SO_2 (with conventional flow schemes) which means that the feed gas entering the acid plant must be diluted with air resulting in a larger gas volume to be treated and higher operating costs associated with the power required for the blower(s). Three approaches have been used to raise permissible SO_2 strength entering a sulfuric acid plant.

(a) Installation of Cs-promoted catalyst in the first-pass catalyst bed. This allows the bed inlet temperature to be operated at ~370 °C, i.e. about 40 °C cooler than conventional catalysts. This allows a larger temperature rise (and more SO_2 conversion) in the first bed without exceeding the bed outlet temperature limit.
(b) Installation of a pre-converter to lower the SO_2 concentration entering the first catalyst bed of the main converter. This approach allowed Olympic Dam to process 18% SO_2 feed gas (Ritschel et al., 1998). Olympic Dam has since constructed a new acid plant designed for a more conventional feed gas SO_2 concentration in the range of 11−12%.
(c) Recycle of SO_3-laden gas from the outlet of the third catalyst bed to the first catalyst bed. This is the LUREC™ process. It was developed by Outotec and installed in 2007 at the Yangqu Xiangguang Copper smelter acid plant in China (Storch, 2009). The use of LUREC allows treatment of 18% SO_2 feed gas at the inlet to the first catalyst bed.

Two other developing technologies are BAYQIK and WSA-DC. They are briefly described below.

12.8.1.1. BAYQIK®

A quasi-isothermal reactor forms the basis of the BAYQIK (Quasi-Isotherme-Katalyse) process developed by Bayer Technology Services (Hands & Connock, 2010, pp. 28−38). The reactor is designed so that the heat of reaction generated by SO_2 conversion to SO_3 can be removed as soon as it is produced. Feed gas SO_2 concentrations up to 50% are proposed. A BAYQIK reactor has been operating at the lead smelter Berzelius Stolberg GmbH since 2009. The reactor acts as a pre-converter ahead of the first catalyst bed, allowing the acid plant to treat 18% SO_2 feed gas.

12.8.1.2. WSA-DC

WSA-DC (Wet gas Sulfuric Acid-Double Condensation) is a modified version of the WSA plant with two stages of acid condensation (Rosenberg, 2009). This allows the wet feed gas SO_2 concentration to be increased to 13−15% SO_2 from the present limit of 6.5%. SO_2 conversion is expected to be greater than 99.95% with a 3:1 catalyst bed configuration. WSA-DC is also expected to be more energy efficient than a conventional contact type acid plant because the heat of SO_2 conversion, SO_3 hydration and acid condensation is recovered and utilized. Less gas pressure drop is also expected which decreases power consumption. Given the success of the WSA process, WSA-DC will likely be implemented on a commercial scale in the near future.

12.8.2. Maximizing Heat Recovery

Heat is generated during SO_2 conversion. In almost all sulfur-burning acid plants and in some metallurgical acid plants this heat is usually recovered into a useful form — steam. The hot gases exiting the catalyst beds are passed through boiler feed water economizers, steam superheaters, and heat recovery boilers (Hashiuchi, Chida, & Nakata, 2007; Kubo et al., 2007). Several metallurgical acid plants capture SO_2 conversion and SO_3 absorption heat, but most remove their excess heat in air coolers (Davenport & King, 2006; Puricelli, Grendel, & Fries, 1998). Some of the warm air produced by the air coolers has been used as concentrate dryer (rotary kiln type) dilution air (Hoshi, Toda, Motomura, Takahashi, & Hirai, 2010). This lowers kiln hydrocarbon fuel consumption.

SO_3 absorption heat can be captured and turned into steam in HRS (Heat Recovery System, MECS) or HEROS (Heat RecOvery System, Outotec) type systems (Jung, Byun, & Shin, 2010; Puricelli et al., 1998). HRS is the most common with ~50 installations worldwide. Low-pressure steam (5—10 bar gauge, ~0.4 t steam/t H_2SO_4) is produced in a kettle tube boiler with hot (~210 °C), concentrated (99.6%) sulfuric acid on the shell side and a steam/water mixture flowing through the tubes. Acid concentration and temperature control are critical in ensuring low corrosion rates are achieved in the HRS system. Thus, HRS is most suited for acid plants that treat feed gas with a steady flow and SO_2 concentration, like those from a flash furnace and flash converter.

12.9. ALTERNATIVE SULFUR PRODUCTS

The SO_2 in Cu smelter gases is almost always captured as sulfuric acid. Other SO_2-capture products have been:

(a) Liquid SO_2
(b) Gypsum
(c) Elemental sulfur (Okura, Ueda, & Yamaguchi, 2006; Platonov, 2006). The processes for making these products are described by King and Forzatti (2009).

12.10. FUTURE IMPROVEMENTS IN SULFUR CAPTURE

Modern smelting processes collect most of their SO_2 at sufficient strength for economic sulfuric acid manufacture. Most reverberatory smelting furnaces have been replaced. Peirce—Smith converting remains a problem for SO_2 collection, especially during charging and skimming when gas leaks into the workplace and at ground level around the smelter. Adoption of continuous converting processes such as Mitsubishi and Kennecott-Outotec flash converting will alleviate this problem.

12.11. SUMMARY

This chapter has shown that most copper is extracted from sulfide minerals so that sulfur, in some form, is a byproduct of most copper extraction processes. The usual byproduct is sulfuric acid, made from the SO_2 produced during smelting and converting. Conventional sulfuric acid production entails:

(a) Cleaning and drying the furnace gases
(b) Catalytically oxidizing their SO_2 to SO_3 (with O_2 in the gas itself or in added air)
(c) Absorbing the resulting SO_3 into a 98.5% H_2SO_4—H_2O sulfuric acid solution.

The process is autothermal when the input gases contain about 4 or more vol.-% SO_2. The double absorption acid plants being installed in the 2000s recover >99.7% of the input SO_2. SO_2 recovery can be increased even further by scrubbing the acid plant tail gas with basic solutions or by adding more catalyst beds with Cs-promoted catalyst.

Some modern smelting processes produce extra-strong SO_2 gases, 13+% SO_2. These strong gases tend to overheat during SO_2 oxidation, causing catalyst degradation and inefficient SO_2 conversion. This problem is leading to the development of catalysts, which have low ignition temperatures and high degradation temperatures. Installation of pre-converters or the recently introduced LUREC process are also used to handle high SO_2 strength feed gas. Two other alternatives are also under development: BAYQIK and WSA-DC.

The Peirce–Smith converter is the major environmental problem remaining in the Cu smelter. It tends to spill SO_2-bearing gas into the workplace and it produces gas discontinuously for the acid plant. Adoption of replacement converting processes began in the 1980s (Mitsubishi converter) and is continuing in the 2000s (Kennecott-Outotec flash converter, Ausmelt/Isasmelt converter, Noranda converter). Replacement has been slow because of the excellent chemical and operating efficiencies of the Peirce-Smith converter.

REFERENCES

Ante, A. (2010). Filsulfor and Gypsulfor: Modern design concepts for weak acid treatment. In *Copper 2010, Vol. 2: Pyrometallurgy I* (pp. 601–614). Clausthal-Zellerfeld, Germany: GDMB.

Australian Government Department of Sustainability, Environment, Water, Population and Communities. (2010). *National standards for criteria air pollutants in Australia.* http://www.environment.gov.au/atmosphere/airquality/publications/standards.html.

Conde, C. C., Taylor, B., & Sarma, S. (1999). Philippines associated smelting electrostatic precipitator upgrade. In D. B. George, W. J. Chen, P. J. Mackey & A. J. Weddick (Eds.), *Copper 99—Cobre 99, Vol. V: Smelting operations and advances* (pp. 685–693). Warrendale, PA: TMS.

Council directive 1999/30/EC. (2010). http://europa.eu/legislation_summaries/environment/air_pollution/l28098_en.htm.

Daum, K. H. (2000). Design of the world's largest metallurgical acid plant. In *Sulphur 2000 preprints* (pp. 325–338). London: British Sulphur.

Davenport, W. G., Jones, D. M., King, M. J., & Partelpoeg, E. H. (2001). *Flash smelting: Analysis, control and optimization.* Warrendale, PA: TMS.

Davenport, W. G., & King, M. J. (2006). *Sulphuric acid manufacture.* Oxford: Elsevier Science Press.

Evans, C. M., Lawler, D. W., Lyne, E. G. C., & Drexler, D. J. (1998). Effluents, emissions and product quality. In *Sulphur 98 Preprints – Vol. 2.* (pp. 217–241). London: British Sulphur.

Guenkel, A. A., & Cameron, G. M. (2000). Packed towers in sulfuric acid plants – review of current industry practice. In *Sulphur 2000 preprints* (pp. 399–417). London: British Sulphur.

Hands, R., & Connock, L. (2010). Achieving lower SO_2 emissions from acid plants. *Sulphur 327*, March–April 2010. London: BCInsight Ltd.

Hashiuchi, F., Chida, H., & Nakata, H. (2007). Recent smelting and acid plant operation at Saganoseki smelter. In A. E. M. Warner, C. J. Newman, A. Vahed, D. B. George, P. J. Mackey & A. Warczok (Eds.), *Cu2007, Vol. III Book 2: The Carlos Díaz symposium on pyrometallurgy* (pp. 613–629). Montreal: CIM.

Hasselwander, K. (2009). Gas cleaning designs for smooth operations. *Sulphur 321*, London, England: BCInsight Ltd. March-April 2009.

Hay, S., Porretta, F., & Wiggins, B. (2003). Design and start-up of acid plant tail gas scrubber. In C. Díaz, J. Kapusta & C. Newman (Eds.), *Copper 2003—Cobre 2003, Vol. IV: Pyrometallurgy of copper, Book 1: The Herman Schwarze symposium on copper pyrometallurgy* (pp. 555–566). Montreal: CIM.

Health Canada. (2010). *Regulations related to health and air quality.* http://www.hc-sc.gc.ca/ewh-semt/air/out-ext/reg-eng.php#a2.

Hoshi, M., Toda, K., Motomura, T., Takahashi, M., & Hirai, Y. (2010). Dryer fuel reduction and recent operation of the flash smelting furnace at Saganoseki Smelter & Refinery after the SPI project. In *Copper 2010, Volume 2: Pyrometallurgy I* (pp. 779–791). Clausthal-Zellerfeld, Germany: GDMB.

Hultbom, K. G. (2003). Industrial proven methods for mercury removal from gases. In M. E. Schlesinger (Ed.), *EPD congress 2003* (pp. 147–156). Warrendale, PA: TMS.

Humphris, M. J., Liu, J., & Javor, F. (1997). Gas cleaning and acid plant operations at the Inco Copper Cliff smelter. In C. Díaz, I. Holubec & C. G. Tan (Eds.), *Nickel-Cobalt 97, Vol. III: Pyrometallurgical operations, environment, vessel integrity in high-intensity smelting and converting processes* (pp. 321–335). Montreal: CIM.

Inami, T., Baba, K., & Ojima, Y. (1990). Clean and high productive operation at the Sumitomo Toyo smelter. Paper presented at the Sixth International Flash Smelting Congress, Brazil, October 14–19, 1990.

Jensen-Holm, H. (1986). *The vanadium catalyzed sulfur dioxide oxidation process.* Denmark: Haldor Topsøe A/S.

Jung, K., Byun, G., & Shin, S. (2010). Profit enhancement through steam selling. In *Copper 2010, Vol. 2: Pyrometallurgy I* (pp. 831–838). Clausthal-Zellerfeld, Germany: GDMB.

King, M. J., & Forzatti, R. J. (2009). Sulphur based by-products from the non-ferrous metals industry. In J. Liu, J. Peacey, M. Barati, S. Kashani-Nejad & B. Davis (Eds.), *Pyrometallurgy of nickel and cobalt 2009* (pp. 137–149). Montréal: CIM.

Köster, S. (2010). Waste heat boilers for copper smelting applications. In *Copper 2010, Vol. 2: Pyrometallurgy I* (pp. 879–891). Clausthal-Zellerfeld, Germany: GDMB.

Kruger, B. (2004). Recovery of SO_2 from low strength off-gases. In *International platinum conference 'platinum surges ahead'* (pp. 59–61). Johannesburg, South Africa: The Southern African Institute of Mining and Metallurgy.

Kubo, N., Isshiki, Y., Satou, H., & Kurokawa, H. (2007). The acid plant expansion and energy saving at Toyo copper smelter. In A. E. M. Warner, C. J. Newman, A. Vahed, D. B. George, P. J. Mackey & A. Warczok (Eds.), *Cu2007, Vol. III Book 2: The Carlos Díaz symposium on pyrometallurgy* (pp. 579–588). Montreal: CIM.

Larouche, P. (2001). *Minor elements in copper smelting and electrorefining.* Canada: M.Eng thesis, McGill University. http://digitool.library.mcgill.ca:8881/R/?func=dbin-jump-full&object_id=33978.

Liu, J., Baird, M. H. I., Kenny, P., Macnamara, B., & Vahed, A. (2006). Inco flash furnace froth column modifications. In F. Kongoli & R. G. Reddy (Eds.), *Sohn international symposium advanced processing of metals and materials, Vol. 8 – International symposium on sulfide smelting 2006* (pp. 291–302). Warrendale, PA: TMS.

Livbjerg, H., Jensen, K., & Villadsen, J. (1978). Supported liquid-phase catalysts. *Catalysis Reviews-Science and Engineering, 17*, 203–272.

Lossin, A., & Windhager, H. (1999). Improving the quality of smelter acid: the example of Norddeutsche Affinerie's "premium quality". In *Sulphur 99* (pp. 209–220). London: British Sulphur Publ.

Lüdtke, P., & Müller, H. (2003). Best available sulfuric acid technology for copper smelters: a state-of-the-art review. In C. Díaz, J. Kapusta & C. Newman (Eds.), *Copper 2003–Cobre 2003, Vol. IV-Pyrometallurgy of copper, Book 1: The Herman Schwarze symposium on copper pyrometallurgy* (pp. 543–554). Montreal: CIM.

Mars, P., & Maessen, J. G. H. (1964). The mechanism of the oxidation of sulfur dioxide on potassium–vanadium oxide catalysts. In W. M. H. Sachtler, G. C. A. Schuit & P. Zweitering (Eds.), 3rd *international congress on catalysis, Vol. 1* (pp. 266–279). Amsterdam: North–Holland.

Mars, P., & Maessen, J. G. H. (1968). The mechanism and the kinetics of sulfur dioxide oxidation on catalysts containing vanadium and alkali oxides. *Journal of Catalysis, 10*, 1–12.

Ministerio Secretaría General de la Presdencia de la República de Chile. (2010). *Norma de calidad primaria de aire para dioxide de azufre (SO_2).* D.S.No 113/02. http://www.temasactuales.com/assets/pdf/gratis/CHLds113-02.pdf.

Ministry of the Environment, Government of Japan. (2010). *Environmental quality standards in Japan — Air quality.* http://www.env.go.jp/en/air/aq/aq.html.

Newman, C. J., Collins, D. N., & Weddick, A. J. (1999). Recent operation and environmental control in the Kennecott smelter. In D. B. George, W. J. Chen, P. J. Mackey & A. J. Weddick (Eds.), *Copper 99—Cobre 99, Vol. V: Smelting operations and advances* (pp. 29—45). Warrendale, PA: TMS.

Okura, T., Ueda, S., & Yamaguchi, K. (2006). Elemental sulfur fixation in smelting gas; is it feasible? In F. Kongoli & R. G. Reddy (Eds.), *Sohn international symposium, Vol. 8: International symposium on sulfide smelting 2006* (pp. 425—431) Warrendale, PA: TMS.

Oshima, E., Igarashi, T., Nishikawa, M., & Kawasaki, M. (1997). Recent operation of the acid plant at Naoshima. In C. Díaz, I. Holubec & C. G. Tan (Eds.), *Nickel—Cobalt 97, Vol. III: Pyrometallurgical operations, environment, vessel integrity in high-intensity smelting and converting processes* (pp. 305—320). Montreal: CIM.

Parker, K. R. (1997). *Applied electrostatic precipitation.* London: Chapman and Hall.

Peippo, R., Holopainen, H., & Nokelainen, J. (1999). Copper smelter waste heat boiler technology for the next millennium. In D. B. George, W. J. Chen, P. J. Mackey & A. J. Weddick (Eds.), *Copper 99—Cobre 99, Vol. V: Smelting operations and advances* (pp. 71—82). Warrendale, PA: TMS.

Platonov, O. I. (2006). Thermal reduction of Vaniukov furnace sulfurous oxygen-bearing off-gas by methane. In F. Kongoli & R. G. Reddy (Eds.), *Sohn international symposium, Vol. 8: International symposium on sulfide smelting 2006* (pp. 509—515). Warrendale, PA: TMS.

Puricelli, S. M., Grendel, R. W., & Fries, R. M. (1998). Pollution to power: a case study of the Kennecott sulfuric acid plant. In J. A. Asteljoki & R. L. Stephens (Eds.), *Sulfide smelting 98* (pp. 451—462). Warrendale, PA: TMS.

Ritschel, P. M., Fell, R. C., Fries, R. M., & Bhambri, N. (1998). Metallurgical sulfuric acid plants for the new millennium. In *Sulphur 98 preprints, Vol 2* (pp. 123—145). London, UK: British Sulphur.

Rosenberg, H. (2006). Topsøe wet gas sulfuric acid (WSA) technology — an attractive alternative for reduction of sulfur emissions from furnaces and converters. In *International platinum conference 'platinum surges ahead'* (pp. 191—198). Johannesburg, South Africa: The Southern African Institute of Mining and Metallurgy.

Rosenberg, H. (2009). WSA-DC — next generation Topsøe WSA technology for stronger SO_2 gases and very high conversion. In *Sulphur and sulfuric acid conference 2009* (pp. 27—35). Johannesburg, South Africa: The Southern African Institute of Mining and Metallurgy.

Ryan, P., Smith, N., Corsi, C., & Whiteus, T. (1999). Agglomeration of ESP dusts for recycling to plant smelting furnaces. In D. B. George, W. J. Chen, P. J. Mackey & A. J. Weddick (Eds.), *Copper 99—Cobre 99, Vol. V: Smelting operations and advances* (pp. 561—571). Warrendale, PA: TMS.

Shibata, T., & Oda, Y. (1990). Environmental protection for SO_2 gas at Tamano smelter. Paper presented at the Sixth International Flash Smelting Congress, Brazil, October 14—19, 1990.

Storch, H. (2009). LUREC™ technology — the economic way to treat strong gases. In J. Liu, J. Peacey, M. Barati, S. Kashani-Nejad & B. Davis (Eds.), *Pyrometallurgy of nickel and cobalt 2009* (pp. 543—551). Montréal: CIM.

Tomita, M., Suenaga, C., Okura, T., & Yasuda, Y. (1990). 20 years of operation of flash furnaces at Saganoseki smelter and refinery. Paper presented at the Sixth International Flash Smelting Congress, Brazil, October 14—19, 1990.

USEPA. (2010). *Fact sheet: Revisions to the primary national ambient air quality standard, monitoring network, and data reporting requirements for sulfur dioxide.* http://www.epa.gov/air/sulfurdioxide/pdfs/20100602fs.pdf.

SUGGESTED READING

Davenport, W. G., & King, M. J. (2006). *Sulfuric acid manufacture.* Oxford: Elsevier Science Press.

Friedman, L. J., & Friedman, S. J. (2007). The metallurgical sulfuric acid plant design, operating & materials considerations 2007 update. In A. E. M. Warner, C. J. Newman, A. Vahed, D. B. George, P. J. Mackey & A. Warczok (Eds.), *Cu2007, Vol. III Book 2: The Carlos Díaz symposium on pyrometallurgy* (pp. 545—566). Montreal: CIM.

Louie, D. K. (2008). *Handbook of sulfuric acid manufacturing* (2nd ed.). Richmond Hill, Canada: DKL Engineering.

Lunt, R. R., & Cunic, J. D. (2000). *Profiles in flue gas desulfurization*. New York: American Institute of Chemical Engineers.

Puricelli, S. M., Grendel, R. W., & Fries, R. M. (1998). Pollution to power: a case study of the Kennecott sulfuric acid plant. In J. A. Asteljoki & R. L. Stephens (Eds.), *Sulfide smelting '98* (pp. 451–462). Warrendale, PA: TMS.

Chapter 13

Fire Refining (S and O Removal) and Anode Casting

Virtually all the molten copper produced by smelting/converting is subsequently electrorefined. It must, therefore, be suitable for casting into thin, strong, smooth anodes for interleaving with cathodes in electrorefining cells (Fig. 14.3). This requires that the copper be fire refined to remove most of its sulfur and oxygen.

The molten blister copper from Peirce−Smith converting contains ∼0.02% S and ∼0.3% O. The copper from single-step smelting and continuous converting contains up to 1% S and 0.2−0.4% O. At these levels, the dissolved sulfur and oxygen would combine during solidification to form bubbles (*blisters*) of SO_2 in newly cast anodes, making them weak and bumpy. In stoichiometric terms, 0.01% dissolved sulfur and 0.01% dissolved oxygen would combine to produce about 2 cm^3 of SO_2 (1083 °C) per cm^3 of copper.

Fire refining removes sulfur and oxygen from liquid blister copper by (a) air-oxidation removal of sulfur as $SO_2(g)$ down to ∼0.003% S, and (b) hydrocarbon-reduction removal of oxygen as $CO(g)$ and $H_2O(g)$ down to ∼0.16% O. Sulfur and oxygen contents at the various stages of fire refining are summarized in Table 13.1.

13.1. INDUSTRIAL METHODS OF FIRE REFINING

Fire refining is mostly carried out in rotary refining furnaces resembling Peirce−Smith converters (Fig. 13.1a) or, less often, in hearth furnaces. It is carried out at about 1200 °C, which provides enough superheat for subsequent casting of anodes. The furnaces are heated by combusting hydrocarbon fuel throughout the process. About 2000−3000 MJ of fuel are consumed per tonne of copper.

TABLE 13.1 Sulfur and Oxygen Contents at Various Stages of Fire Refining

Stage of process	Mass% S	Mass% O
Blister copper[a]	0.01−0.05 (Ramachandran et al., 2003)	0.1−0.8
After oxidation	0.002−0.005	0.3−1
After reduction (poling)	0.001−0.004	0.05−0.2
Cast anodes	0.003 ± 0.002 (Moats et al., 2007)	0.16 ± 0.04 (Moats et al., 2007)

[a]From Peirce−Smith and Hoboken converters. The blister copper from direct-to-copper smelting and continuous converting contains 0.2−1% S and 0.2−0.4% O.

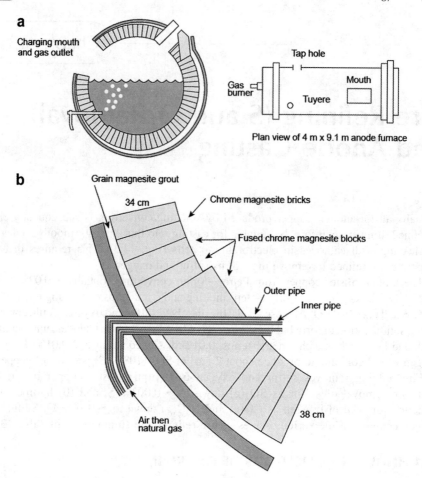

FIGURE 13.1 a. Rotary refining (anode) furnace, end and front views (after McKerrow & Pannell, 1972). The furnaces are typically 3–5 m diameter and 9–14 m long, inside the steel shell. b. Detail of anode furnace tuyere (after McKerrow & Pannell, 1972). Note the two concentric pipes separated by castable refractory, which permit easy replacement of the inside pipe as it wears back. The inside pipe protrudes into the molten copper to prevent seepage of gas back through the refractory wall of the furnace.

13.1.1. Rotary Furnace Refining

Fig. 13.1a shows a rotary refining furnace. Air and hydrocarbon flowrates into refining furnaces are slow, to provide precise control of copper composition. Only ~2 tuyeres are used (Fig. 13.1b, Table 13.2). Gas flowrates are 10–50 Nm^3/min per tuyere at 2–5 bar pressure.

Refining a 250 tonne charge of blister copper (0.03% S) takes up to 3 h: ~1 h for air injection (S removal) and ~2 h for hydrocarbon injection (O removal). High-sulfur copper from direct-to-copper smelting and continuous converting takes up to 5 h (Ramachandran et al., 2003).

A typical sequence in rotary furnace refining is:

(a) Molten copper is delivered by crane and ladle from converters to the anode furnace until 200 or 300 tonnes are accumulated

TABLE 13.2 Details of Two Rotary Anode Furnaces — One Mold-on-Wheel Anode Casting Plant and One Contilanod Anode Casting Plant

Smelter	Norddeutsche Affinerie, Hamburg, Germany	PT Smelting Co. Gresik, Indonesia
Anode production tonnes/year	450 000	300 000
Number of anode furnaces		
Total	2	3
Active	2	3
Furnace dimensions, m		
Diameter × length	4.25 × 10	3.12 × 12.5 (ID)
Tuyeres		
Diameter, cm	0.8, 1, 1.2	
Number per furnace	2	2
Used during oxidation	2	2
Used during reduction	2	2
Reductant	Natural gas	Diesel oil
Production details		
Tap-to-tap duration, h	9	11
Anode production tonnes/cycle	270	400
Oxidation duration, h	0.5	5
Air flowrate, Nm^3/min	6–7	50 air; 5 oxygen
Reduction duration, h	3	2
Reducing gas flowrate Nm^3/min per tuyere	10	15 L/min
Scrap addition, tonnes/cycle	0–10	0–30
Anode casting		
Method	Mold on wheel	Contilanod
Number of wheels, m	1	
Diameter of wheels, m	12.8	
Number of molds per wheel	24	
Casting rate, tonnes/h	75–80	100
Automatic weighing	Yes	Yes
Anode mass, kg	400	400
Variation, kg	±4	±7

(b) The accumulated charge is then desulfurized down to ~0.003% S by blowing air into the molten copper
(c) The copper is deoxidized by blowing gaseous or liquid hydrocarbons into the molten copper bath.

Hydrocarbon blowing is terminated when the oxygen concentration has been lowered to ~0.16% O, as detected with disposable solid electrolyte probes (Heraeus Electro-Nite, 2011) or by examination of copper test blocks. Copper with this oxygen content *sets* flat when it is cast into anodes.

13.1.2. Hearth Furnace Refining

Although the rotary furnace dominates copper fire refining in primary smelters, secondary (scrap) smelters tend to use hearth-refining furnaces (Alarcon, 2005; Hanusch, 2010; Rinnhofer & Zulehner, 2005). They are better for melting scrap and other solid inputs. Sulfur is removed by reaction of the copper solids with an oxidizing flame above the bath and by injecting air through submerged tip steel lances. Deoxidation is also done by injecting hydrocarbons or hydrocarbons plus steam through these steel lances.

Some anode furnaces inject nitrogen into the molten copper through small holes in porous ceramic plugs in the bottom of the furnace (Filzwieser, Wallner, Caulfield, & Rigby, 2003). The injected nitrogen stirs the molten copper and gives it uniform composition and temperature during fire refining and anode casting. This homogenization (a) minimizes solid slag buildups in otherwise cool parts of the anode furnace (Lee, Kim, & Choi, 2003), and (b) produces uniform anode composition throughout an entire *cast* of anodes. Benefit (a) has been crucial in several smelters.

13.2. CHEMISTRY OF FIRE REFINING

Two chemical systems are involved in fire refining: (a) the Cu–O–S system (sulfur removal), and (b) the Cu–C–H–O system (oxygen removal).

13.2.1. Sulfur Removal: the Cu–O–S System

The main reaction for removing sulfur from molten copper with air at 1200 °C is:

$$[S] + O_2(g) \rightarrow SO_2(g) \tag{13.1}$$

while oxygen dissolves in the copper by the reaction:

$$O_2(g) \rightarrow 2[O] \tag{13.2}$$

The equilibrium relationship between gaseous oxygen entering the molten copper and S in the molten copper is, from Eq. (13.1):

$$K = p_{SO_2}/(\%SO_2 \times pSO_2) \tag{13.3}$$

where K is about 10^6 at 1200 °C (Engh, 1992, p. 422).

The large value of this equilibrium constant indicates that even at the end of desulfurization (~0.003% S; p_{O_2} ~0.21 bar), SO_2 formation is strongly favored (i.e. p_{SO_2} > 1 bar) and S is still being eliminated. At the same time, oxygen is still dissolving.

13.2.2. Oxygen Removal: the Cu−C−H−O System

The oxygen concentration in the newly desulfurized molten copper is ~0.3 mass% O. Most of this dissolved O would precipitate as solid Cu_2O inclusions during casting (Brandes & Brook, 1998, p. 1215), so it must be removed to a low level.

Copper oxide precipitation is minimized by removing most of the oxygen from the molten copper with injected gas or liquid hydrocarbons. Representative dissolved oxygen removal reactions are:

$$H_2C(s, l, g) + 2[O] \rightarrow H_2O(g) + CO(g) \tag{13.4}$$

$$CO(g) + [O] \rightarrow CO_2(g) \tag{13.5}$$

$$H_2(g) + [O] \rightarrow H_2O(g) \tag{13.6}$$

13.3. CHOICE OF HYDROCARBON FOR DEOXIDATION

The universal choice for removing S from copper is air. Many different hydrocarbons are used for O removal, but natural gas, oil, liquid petroleum gas, and propane/butane are favored (Ramachandran et al., 2003).

Gas and liquid hydrocarbons are injected into the copper through the same tuyeres used for air injection. Natural gas is blown in directly (sometimes with steam (Fagerlund, 2010). Liquid petroleum gas, propane and butane are blown in after vaporization. Oil is atomized and blown in with steam.

Oxygen removal typically requires ~5 kg of gas or liquid hydrocarbons per tonne of copper (Ramachandran et al., 2003). This is about twice the stoichiometric requirement, assuming that the products of the reaction are CO and H_2O.

13.4. CASTING ANODES

The final product of fire refining is molten copper, ~0.003% S, 0.16% O, 1200 °C, ready for casting as anodes. Virtually all copper anodes are cast in open anode-shaped impressions on the top of flat copper molds. Sixteen to 32 such molds are placed on a large horizontal rotating wheel (Fig. 13.2, Table 13.2). The wheel is rotated to bring a mold under the copper stream from the anode furnace, where it rests while the anode is being poured (Fig. 13.3). When the anode impression is full, the wheel is rotated to bring a new mold into casting position and so on. Spillage of copper between the molds during rotation is avoided by placing one or two tiltable ladles between the refining furnace and casting wheel. Most casting wheels operate automatically, but with devoted human supervision.

The newly poured anodes are cooled by spraying water on the tops and bottoms of the molds while the wheel rotates (Edens & Hannemann, 2005). They are stripped from their molds (usually by an automatic raising pin and lifting machine) after a 270° rotation. The empty molds are then sprayed with a barite−water wash (30 vol.-% barite, 70% water) to prevent sticking of the next anode (Wenzl, Filzwieser, & Antrekowitsch, 2007).

Casting rates are ~50−100 tonnes of anodes/h (Ramachandran et al., 2003). The limitation is the rate at which heat can be extracted from the solidifying/cooling anodes. The flow of copper from the refining furnace is adjusted to match the casting rate by rotating the taphole up or down (rotary furnace) or by blocking or opening

FIGURE 13.2 Anode casting wheel. The molten copper is poured at 10 o'clock on the wheel. The solidified anodes are lifted out of the molds at 7 o'clock on the wheel. The mass of copper in the ladles is sensed by load cells. The sensors automatically control the mass of each copper pour without interrupting copper flow from the anode furnace. The anode molds are copper, usually cast at the smelter, occasionally machined into a cast copper block. *Photograph courtesy of Miguel Palacios, Atlantic Copper, Huelva, Spain.*

FIGURE 13.3 Molten copper being cast into an anode mold. *(Courtesy of Freeport McMoRan Copper & Gold)*

a tapping-notch (hearth furnace). New anode casting installations cast their anodes in pairs to speed up casting rates (Alarcon, 2005; Outotec, 2009).

13.4.1. Anode Molds

The molds into which anodes are cast are always copper. They are usually made by (a) casting molten anode copper (occasionally cathode copper; Wenzl et al., 2007) into large 1.2 m × 1.2 m open molds, then (b) pressing an anode-shaped die into the surface of the molten copper. The mold and die are both steel. The die is water-cooled.

A recent modification is to cast the molten copper into the steel mold without the surface impression, and then machine the anode shape into the solidified block of copper. Machining gives a smooth, crack-free surface to the anode shape, considerably increasing the number of anodes that can be cast before the mold surface deteriorates. The use of phosphorus-deoxidized copper further improves mold performance (Edens & Hannemann, 2005). Machining will probably be limited to smelters in highly industrialized locations, such as Germany.

In all cases the anode molds are perfectly leveled on the casting wheel, to ensure that the anode faces are exactly parallel.

13.4.2. Anode Uniformity

The most important aspect of anode casting, besides flat surfaces, is uniformity of thickness. This uniformity ensures that all the anodes in an electrorefining cell reach the end of their useful life at the same time. Automatic control of the mass of each pour of copper (i.e. the mass and thickness of each anode) is now used in most plants (Moats, Davenport, Demetrio, Robinson, & Karcas, 2007). The usual practice is to sense the mass of metal poured from a tiltable ladle, using load cells in the ladle supports as sensors. Anode mass is normally 360–410 kg (Moats et al., 2007). Anode-to-anode mass variation in a smelter or refinery is ± 2 to 4 kg with automatic weight control.

Recent anode designs have incorporated (a) knife-edged lugs, which make the anode hang vertically in the electrolytic cell and (b) thin tops where the anode is not submerged (i.e. where it remains undissolved during refining). The latter feature decreases the amount of un-dissolved anode scrap that must be recycled when the anode is electorefined.

13.4.3. Anode Preparation

Anode flatness, uniformity, and verticality are critical in obtaining good electrorefinery performance. For this reason, most refineries treat their anodes in an automated anode preparation machine to improve flatness and verticality (Outotec, 2007). The machine:

(a) Weighs the anodes and directs underweight and overweight anodes to re-melting
(b) Straightens the lugs and machines a knife edge on each lug (for vertical hanging)
(c) Presses the anodes flat with a *full body press*
(d) Loads the anodes in a spaced rack for dropping into an electrorefining cell.

Inclusion of these anode preparation steps has resulted in increased refining rates and current efficiencies, improved cathode purities, and decreased electrorefining energy consumption.

13.5. CONTINUOUS ANODE CASTING

Continuous casting of anodes in a Hazelett twin-belt type caster (Fig. 13.5) is being used by several smelters/refineries (Hazelett, 2011). The advantages of the Hazelett system over mold-on-wheel casting are uniformity of anode product and a high degree of mechanization/automation.

In Hazelett casting, the copper is poured at a controlled rate (30—100 tonnes/h) from a ladle into the gap between two moving water-cooled low-carbon steel belts. The product is an anode-thickness continuous strip of copper (Fig. 13.5, Table 13.3) moving at 4—6 m/min. The thickness of the strip is controlled by adjusting the gap between the belts. The width of the strip is determined by adjusting the distance between bronze or stainless steel edge blocks, which move at the same speed as the steel belts.

Recent Hazelett Contilanod casting machines have machined edge blocks into which copper flows to form anode support lugs (Fig. 13.6). The lug shape is machined to half-anode thickness in the top of these specialized blocks. The blocks are machined at a 5° angle to give a knife-edge support lug. Identical positioning of the lug blocks on opposite sides of the strip is obtained by heating or cooling the dam blocks between the specialized *lug blocks*.

The caster produces a copper strip with regularly spaced anode lugs. Individual anodes are produced from this strip by a traveling hydraulic shear (Fig. 13.4). Details of the operation are given by Hazelett (2011). Adoption of the process seems to have stopped, probably because copper fire refining (hence anode casting) is not completely continuous.

FIGURE 13.4 Copper anodes cast from a casting wheel. *(Courtesy of Palabora Mining Company, South Africa. ©Rio Tinto.)*

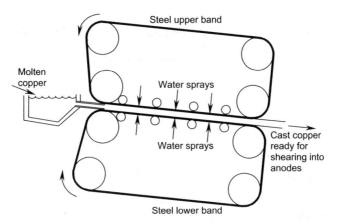

FIGURE 13.5 Hazelett twin-belt casting machine for continuously casting copper anode strip. The molten copper is contained in the casting machine by means of moving 'dam blocks' on both sides of the machine. The anode strip is ~0.047 m thick and ~1 m wide. The most recent method of cutting the strip into anodes is shown in Fig. 13.6. (*Reprinted by permission of TMS, Warrendale, PA.*)

13.6. NEW ANODES FROM REJECTS AND ANODE SCRAP

Smelters and refineries reject 2—3% of their new anodes because of physical defects or incorrect masses. They also produce 12—20% un-dissolved anode scrap after a completed electrorefining cycle (Moats et al., 2007). These two materials are re-melted and cast into fresh anodes for feeding back to the electrorefinery. The post-refining scrap is thoroughly washed before re-melting.

The reject and scrap anodes are often melted in the smelter Peirce—Smith converters. There is, however, an increasing tendency to melt them in Asarco-type shaft furnaces (Chapter 20) in the electrorefinery itself. The Asarco shaft furnace is fast and energy efficient for this purpose. Sulfur and oxygen concentrations in the product copper are kept at normal anode levels by using low sulfur fuel and by adjusting the O_2/fuel ratio in the Asarco furnace burners.

13.7. REMOVAL OF IMPURITIES DURING FIRE REFINING

Chapters 5—10 indicate that significant fractions of the impurities entering a smelter end up in the metallic copper. The fire refining procedures described above do not remove these impurities to a significant extent. The impurities report mostly to the anodes.

As long as impurity levels in the anodes are not excessive, electrorefining and electrolyte purification keep the impurities in the cathode copper product at low levels. With excessively impure blister copper, however, it can be advantageous to eliminate a portion of the impurities during fire refining (Carrasco, Figueroa, Lopez, & Alvear, 2003; Larouche, Harris, & Wraith, 2003). The process entails adding appropriate fluxes during the oxidation stage of fire refining. The flux may be blown into the copper through the refining furnace tuyeres or it may be added prior to charging the copper into the furnace.

Antimony and arsenic are removed from molten copper in the anode furnace by fluxing with alkaline fluxes such Na_2CO_3, $CaCO_3$, and CaO (Carrasco et al., 2003; Larouche et al., 2003). Removals up to 90% are obtained. Lead, on the other hand, is removed from molten anode furnace copper by adding SiO_2 flux. This was practiced

TABLE 13.3 Details of Hazelett Continuous Anode Casting Plant at Gresik, Indonesia (Sato, Adji, Prayoga, & Bouman, 2010)

Smelter	PT Smelting Co. Gresik, Indonesia
Startup data	1998
Anode production tonnes/year	300 000
Casting machine size, m	
Length between molten copper entrance and solid copper exit	3.81
Band width (total)	1.65
Width of cast copper strip (between edge dams)	0.93
Length of lug	0.18
Thickness of cast strip	0.047
Thickness of lug	0.027
Band details	
Material	ASTM A607 Grade 45 steel
Life, tonnes of cast copper	1200
Lubrication	Silicone oil
Edge block details	
Material	Hardened bronze
Life, years	~3 years (~0.5 years for anode lug blocks)
Method of controlling copper level at caster entrance	Electromagnetic level indicator
Temperatures, °C	
Molten copper	1120–1150
Cast anode (leaving caster)	880–930
Casting details	
Casting rate, tonnes/h	100
Caster use, h/day	9
Method of cutting anodes from strip	Hydraulic shear
Anode details	
Mass, kg	400
Acceptable deviation	±7 kg
% Acceptable anodes	97

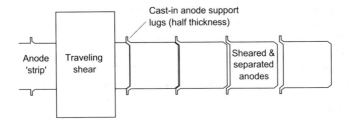

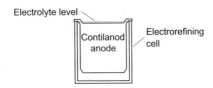

FIGURE 13.6 Sketch of system for shearing anodes from Hazelett-cast copper strip (Hazelett, 2011). Suspension of an anode in an electrolytic cell is also shown.

extensively by the Timmins smelter, now shut (Newman, MacFarlane, Molnar, & Storey, 1991). About 75% removal was obtained.

13.8. SUMMARY

The final step in Cu pyrometallurgical processing is casting of flat, thin anodes for electrorefining. The anodes must be strong and smooth-surfaced for efficient electrorefining — bubbles or blisters of SO_2 are unacceptable.

Blister formation is prevented by removing sulfur and oxygen from molten copper by air oxidation then hydrocarbon reduction. The air and hydrocarbons are usually injected into the molten copper via one or two submerged tuyeres in a rotary anode furnace.

Anodes are virtually always cast in open molds on a large horizontal rotating wheel. Uniformity of anode mass is critical for efficient electrorefining so most smelters automatically weigh the amount of copper that is poured into each anode mold. The molds are also carefully leveled on the wheel to give parallel anode faces.

The cast anodes are often straightened and flattened in automated anode preparation machines. Their lugs may also be machined to a knife-edge. Uniform, straight, flat, vertically hung anodes give pure cathodes, high productivity, and efficient energy use in the electrorefinery.

REFERENCES

Alarcon, V. R. (2005). Anode casting operation, process improvements and anode quality at the Ilo refinery. In A. Ross, T. Warner & K. Scholey (Eds.), *Converter and fire refining practice* (pp. 133–137). Warrendale, PA: TMS.

Brandes, E. A., & Brook, G. B. (1998). *Smithells metals reference book* (7th ed.). Oxford: Butterworth-Heinmann.

Carrasco, C., Figueroa, F., Lopez, J., & Alvear, G. (2003). Optimization of future of copper pyrometallurgical refining at Codelco's El Teniente division. In C. Diaz, J. Kapusta & C. Newman (Eds.), *Copper 2003—Cobre 2003, Vol. IV, Pyrometallurgy of copper, Book 1: The Hermann Schwarze symposium on copper pyrometallurgy* (pp. 405–414). Montreal: CIM.

Edens, T., & Hannemann, D. (2005). New casting moulds for anode copper. In A. Ross, T. Warner & K. Scholey (Eds.), *Converter and fire refining practice* (pp. 159−165). Warrendale, PA: TMS.

Engh, T. A. (1992). *Principles of metal refining*. Oxford: Oxford University Press.

Fagerlund, K. (2010). *Steam gas reduction system. Flash news*. http://www.outotec.com/39633.epibrw.

Filzwieser, A., Wallner, S., Caulfield, K., & Rigby, A. J. (2003). The COP KIN system part II: performance and benefits − a world wide overview. In C. Diaz, J. Kapusta & C. Newman (Eds.), *Copper 2003−Cobre 2003, Vol. IV: Pyrometallurgy of copper, Book 1: The Hermann Schwarze symposium on copper pyrometallurgy* (pp. 415−428). Montreal: CIM.

Hanusch, B. (2010). New highly efficient rotary furnace for environmentally friendly refining process. In *Copper 2010, Vol. 2. Pyrometallurgy I*. Clausthal-Zellerfeld, Germany: GDMB. 721−730.

Hazelett. (2011). *The Contilanod process and Contilanod plants*. http://www.hazelett.com/casting_machines/copper_anode_casting_machines/the_contilanod_process/the_contilanod_process.php. http://www.hazelett.com/casting_machines/copper_anode_casting_machines/contilanod_plants/contilanod_plants.php.

Heraeus Electro-Nite. (2011). *Sensors for molten metals, copper, oxygen activity control*. http://heraeus-electro-nite.com/en/sensorsformoltenmetals/copper/oxygenactivitycontrol/oxygenactivitycontrol_1.aspx.

Larouche, P., Harris, R., & Wraith, A. (2003). Removal technologies for minor elements in copper smelting. In C. Diaz, J. Kapusta & C. Newman (Eds.), *Copper 2003−Cobre 2003, Vol. IV: Pyrometallurgy of copper, Book 1: The Hermann Schwarze symposium on copper pyrometallurgy* (pp. 385−404). Montreal: CIM.

Lee, S.-S., Kim, B.-S., & Choi, S.-R. (2003). Application of the porous plug system in the anode furnace at Onsan smelter. In F. Kongoli, K. Itagaki, C. Yamaguchi & H. Y. Sohn (Eds.), *Yazawa international symposium, Vol. II: High-temperature metals production* (pp. 447−457). Warrendale, PA: TMS.

McKerrow, G. C., & Pannell, D. G. (1972). Gaseous deoxidation of anode copper at the Noranda smelter. *Canadian Metallurgical Quarterly, 11*, 629−633.

Moats, M., Davenport, W., Demetrio, S., Robinson, T., & Karcas, G. (2007). Electrolytic copper refining − 2007 world tankhouse operating data. In G. E. Houlachi, J. D. Edwards & T. G. Robinson (Eds.), *Copper 2007, Vol. V: Copper electrorefining and electrowinning* (pp. 195−241). Montreal: CIM.

Newman, C. J., MacFarlane, G., Molnar, K., & Storey, A. G. (1991). The Kidd Creek copper smelter − an update on plant performance. In C. Diaz, C. Landolt, A. Luraschi & C. J. Newman (Eds.), *Copper 91−Cobre 91, Vol. IV: Pyrometallurgy of copper* (pp. 65−80). New York: Pergamon Press.

Outotec. (2007). *Copper electrorefining refined technology for process and material handling*. http://www.outotec.com/36288.epibrw.

Outotec. (2009). *Anode casting technology*. http://www.outotec.com/38059.epibrw.

Ramachandran, V., Diaz, C., Eltringham, T., Jiang, C. Y., Lehner, T., Mackey, P. J., et al. (2003). Primary copper production − a survey of operating world copper smelters. In C. Díaz, J. Kapusta & C. Newman (Eds.), *Copper 2003−Cobre 2003, Vol. IV: Pyrometallurgy of copper, Book 1: The Hermann Schwarze symposium on copper pyrometallurgy* (pp. 3−119). Montreal: CIM.

Rinnhofer, H., & Zulehner, U. (2005). Gas-fired furnaces for copper and copper alloys. In A. von Starck, A. Mühlbauer & C. Kramer (Eds.), *Handbook of thermoprocessing technologies: Fundamentals, processes, components, safety* (pp. 358−364). Essen: Vulkan−Verlag.

Sato, H., Adji, D. S., Prayoga, A., & Bouman, T. S. (2010). PT. smelting, Gresik smelter and refinery. In *Cu 2010, Vol. 3. Pyrometallurgy II* (pp. 1143−1153), Clausthal-Zellerfeld, Germany.

Wenzl, C., Filzwieser, A., & Antrekowitsch, H. (2007). Review of anode casting − part II: physical anode quality. *Erzmetall, 60*(2), 83−88.

SUGGESTED READING

Alarcon, V. R. (2005). Anode casting operation, process improvements and anode quality at the Ilo refinery. In A. Ross, T. Warner & K. Scholey (Eds.), *Converter and fire refining practice* (pp. 133−137). Warrendale, PA: TMS.

Antrekowitsch, H., Wenzl, C., Filzwieser, I., & Offenthaler, D. (2005). Pyrometallurgical refining of copper in an anode furnace. In A. Ross, T. Warner & K. Scholey (Eds.), *Converter and fire refining practice* (pp. 191−202). Warrendale, PA: TMS.

Gamweger, K. (2010). Introduction of a slide gate system for copper anode furnaces. In *Copper 2010, Vol. 2. Pyrometallurgy I* (pp. 713–720). Clausthal-Zellerfeld, Germany: GDMB.

Pagador, R. U., Malazarte, M. C., & Gonzales, T. W. (2005). Current practices at the converter and anode furnace operations of PASAR. In A. Ross, T. Warner & K. Scholey (Eds.), *Converter and fire refining practice* (pp. 89–96). Warrendale, PA: TMS.

Potesser, M., Holleis, B., Demuth, M., Spoljaric, D., & Zauner, J. (2010). Customized burner concepts for the copper industry. In *Cu 2010, Vol. 2. Pyrometallurgy I* (pp. 1051–1062). Clausthal-Zellerfeld, Germany: GDMB.

Rigby, A. J. (2005). Controlling the processing parameters affecting the refractory requirements for Peirce–Smith converters and anode refining vessels. In A. Ross, T. Warner & K. Scholey (Eds.), *Converter and fire refining practice* (pp. 213–222). Warrendale, PA: TMS.

Rigby, A. J., & Wiessler, M. P. (2007). Interactive 3D modeling of the refractory linings of vessels for copper production. In A. E. M. Warner, C. J. Newman, A. Vahed, D. B. George, P. J. Mackey & A. Warczok (Eds.), *Copper 2007, Vol. III, Book 2: The Carlos Díaz symposium on pyrometallurgy* (pp. 661–666). Montreal: CIM.

Ross, A. G., Warner, T., & Scholey, K. (2005). *Converter and fire refining practices.* Warrendale, PA: TMS.

Wenzl, C., Filzwieser, A., & Antrekowitsch, H. (2007a). Review of anode casting – part I: chemical anode quality. *Erzmetall, 60*(2), 77–83.

Wenzl, C., Filzwieser, A., & Antrekowitsch, H. (2007b). Review of anode casting – part II: physical anode quality. *Erzmetall, 60*(2), 83–88.

Wenzl, C., Ruhs, S., Filzwieser, A., & Koncik, L. (2010). 3D-refractory engineering using the example of a Maerz tilting furnace. In *Copper 2010, Vol. 2. Pyrometallurgy I* (pp. 1233–1246). Clausthal-Zellerfeld, Germany: GDMB.

Chapter 14

Electrolytic Refining

Almost all copper is treated by an electrolytic process during its production from ore. It is either electrorefined from impure copper anodes or electrowon from leach/solvent extraction solutions. Considerable copper scrap is also electrorefined. This chapter describes electrorefining. Electrowinning is discussed in Chapter 17.

Electrorefining entails (a) electrochemically dissolving copper from impure copper anodes into an electrolyte containing $CuSO_4$ and H_2SO_4, and (b) selectively electroplating pure copper from this electrolyte *without the anode impurities*. It produces copper essentially free of impurities, and separates valuable impurities such as gold and silver from copper for recovery as byproducts.

Copper anodes with a typical purity of 98.5–99.5% Cu are electrorefined to produce cathodes with a purity of >99.997% Cu. Electrorefined copper, melted and cast, contains less than 20 parts per million (ppm) impurities, plus oxygen which is controlled at 0.018–0.025%.

Table 14.1 presents industrial ranges of copper anode and cathode compositions. Figures 14.1 and 14.2 show a typical flow sheet and modern industrial electrorefining plant, respectively.

14.1. THE ELECTROREFINING PROCESS

An electrical potential is applied between a copper anode and a metal cathode in an electrolyte containing $CuSO_4$ and H_2SO_4. The following processes occur:

(a) Copper is electrochemically dissolved from the anode into the electrolyte, producing copper cations plus electrons:

$$Cu^o_{anode} \rightarrow Cu^{2+} + 2e^- \quad E^o = -0.34\ V \quad (14.1)$$

(b) The electrons produced by Reaction (14.1) are conducted toward the cathode through the external circuit and power supply.

(c) The Cu^{2+} cations in the electrolyte migrate to the cathode by convection and diffusion.

(d) The electrons and Cu^{2+} ions recombine at the cathode surface to form copper metal (without the anode impurities):

$$Cu^{2+} + 2e^- \rightarrow Cu^o_{cathode} \quad E^o = +0.34\ V \quad (14.2)$$

Overall, copper electrorefining is the sum of Reactions (14.1) and (14.2):

$$Cu^o_{impure} \rightarrow Cu^o_{pure} \quad E^o = 0.0\ V \quad (14.3)$$

Extractive Metallurgy of Copper. DOI: 10.1016/B978-0-08-096789-9.10014-9
Copyright © 2011 Elsevier Ltd. All rights reserved.

TABLE 14.1 Industrial Range of Copper Anode and Cathode Compositions (Moats et al., 2007)

Element	Anodes, %	Cathodes, %
Cu	98.2–99.8	Up to 99.998
Ag	0.01–0.75	0.0005–0.0025
As	Up to 0.25	<0.0005
Au	Up to 0.03	Trace
Bi	Up to 0.06	<0.0003
Fe	0.001–0.030	<0.0003
O	0.035–0.35	Not determined
Ni	0.003–0.6	<0.00001–0.002
Pb	0.001–0.9	<0.00001–0.001
S	0.001–0.018	0.0002–0.013000
Sb	Up to 0.13	<0.00050
Se	0.002–0.12	<0.0002
Te	0.001–0.065	<0.0002

In practice, resistance to current flow must be overcome by applying a potential between the anode and cathode. Small overvoltages must also be applied to plate copper on the cathode (~0.05 V) and dissolve copper from the anode (~0.1 V). There are also Ohmic drops across the electrolyte and the electrical contacts. Applied industrial anode–cathode potentials are ~0.3 V (Moats, Davenport, Robinson, & Karcas, 2007).

The impurities in the anodes either do not dissolve or are not plated at the cathodes, so the purity of the copper plated onto the cathodes is much higher than that of the anode starting material.

14.2. CHEMISTRY OF ELECTROREFINING AND BEHAVIOR OF ANODE IMPURITIES

The principal impurities in copper anodes are Ag, As, Au, Bi, Co, Fe, Ni, O, Pb, S, Sb, Se, and Te (Table 14.1; Noguchi, Iida, Nakamura, & Ueda, 1992). They must be prevented from entering the cathode copper. Their behavior is governed by their position in the electrochemical series (Table 14.2).

At the anode, elements with less positive reduction potentials than Cu dissolve under the applied potential, while elements with more positive reduction potentials remain in solid form. At the cathode, elements with more positive reduction potentials deposit preferentially, while elements with more negative potentials remain in solution.

Chapter | 14 Electrolytic Refining

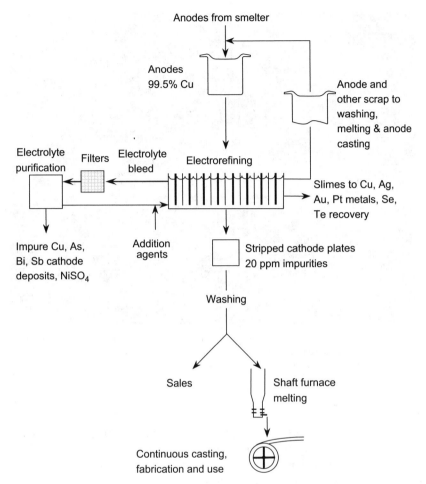

FIGURE 14.1 Copper electrorefinery flow sheet. Impure copper anodes are dissolved electrolytically in a $CuSO_4$–H_2SO_4 electrolyte and pure copper is plated from solution as cathodes. The electrolyte purification circuit treats a small fraction of the electrolyte (Section 14.5.4). The remainder is recirculated directly to refining (after reagent additions and heating).

14.2.1. Au and Platinum-group Metals

Gold and the platinum-group metals (PGMs: Pt, Pd, Rh, Ir, Ru, Os) do not dissolve in sulfate electrolyte. They form solid *slimes*, which adhere to the anode surface or fall to the bottom of the electrolytic cell. These slimes are collected periodically and sent to a Cu and byproduct metals recovery plant (Chen & Dutrizac, 2004; see Chapter 21).

14.2.2. Se and Te

Selenium and tellurium are present in anodes mainly as compounds with copper and silver. They also enter the slimes in these bound forms, which include $Cu_2(Se,Te)$, $Ag_2(Se,Te)$, and $CuAg(Se,Te)$ (Campin, 2000; Chen & Dutrizac, 2003; Hiskey & Moats, 2010; Ramírez, Ríos, & Martín, 2007).

FIGURE 14.2 A modern copper electrorefinery in China. Note the overhead crane used for moving anodes and cathodes to and from the cells. In the background, the plastic sheets covering some of the cells minimize evaporation and heat losses, thereby minimizing energy consumption. *(Courtesy of Outotec.)*

14.2.3. Pb and Sn

Lead is present in the anode both in solid solution with copper and as a secondary complex oxide phase along grain boundaries. Most lead dissolves from the anode and forms $PbSO_4$ by precipitation. Tin forms SnO_2 on the anode. Both of these compounds are sparingly soluble and report to the slimes.

14.2.4. As, Bi, Co, Fe, Ni, S, and Sb

As indicated by Table 14.2, these elements usually dissolve extensively in the electrolyte under conditions where copper is electrochemically dissolved (Atenas & Muñoz, 2007a; Beauchemin, Chen, & Dutrizac, 2008; Chen & Dutrizac, 2000; Larouche, 2001; Möller, Friedrich, & Bayanmunkh, 2010; Rodrigues Bravo, 2007; Santos Moraes, 2007). To avoid excessive buildup of these elements in the electrolyte and contamination of the cathodes, they are continuously removed from a bleed of the electrolyte stream (Fig. 14.1).

As, Sb, and Bi are particularly important. These elements affect anode passivation and cathode quality (see Section 14.6.3.1). Sb and Bi are precipitated into the slimes layer by maintaining high levels of As in the electrolyte. Arsenic in the anode has been shown to inhibit both passivation and the formation of floating slimes. Most refineries typically maintain a molar ratio of As/(Sb + Bi) > 2 in the anode.

Ni can form nickel oxide particles within the anode at concentrations greater than 2000 ppm Ni in the anode. Ni can also form a phase called *kupferglimmer* at high concentrations and in the presence of Sb and/or Sn. NiO and kupferglimmer do not dissolve rapidly and report to the slimes.

TABLE 14.2 Standard Electrochemical Potentials of Elements in Copper Electrorefining (25 °C, unit Thermodynamic Activity)

Electrochemical reaction	Standard reduction potential (25 °C), V
$Au^{3+} + 3e^- \rightarrow Au^0$	1.50
$O_2 + 2H^+ + 2e^- \rightarrow H_2O$	1.229
$Pt^{2+} + 2e^- \rightarrow Pt^0$	1.2
$Pd^{2+} + 2e^- \rightarrow Pd^0$	0.83
$Ag^+ + e^- \rightarrow Ag^0$	0.800
$Fe^{3+} + e^- \rightarrow Fe^{2+}$	0.771
$H_2SeO_3 + 4H^+ + 4e \rightarrow Se^0 + 3H_2O$	0.74
$TeO_4^- + 8H^+ + 7e \rightarrow Te^0 + 4H_2O$	0.472
$Cu^{2+} + 2e^- \rightarrow Cu^0$	**0.337**
$BiO^+ + 2H^+ + 3e^- \rightarrow Bi^0 + H_2O$	0.32
$HAsO_2 + 3H^+ + 3e^- \rightarrow As^0 + 2H_2O$	0.25
$SbO^+ + 2H^+ + 3e^- \rightarrow Sb^0 + H_2O$	0.21
$2H^+ + 2e^- \rightarrow H_2$	0.000[a]
$Pb^{2+} + 2e^- \rightarrow Pb^0$	−0.126
$Sn^{2+} + 2e^- \rightarrow Sn^0$	−0.136
$Ni^{2+} + 2e^- \rightarrow Ni^0$	−0.250
$Co^{2+} + 2e^- \rightarrow Co^0$	−0.277
$Fe^{2+} + 2e^- \rightarrow Fe^0$	−0.440
$Zn^{2+} + 2e^- \rightarrow Zn^0$	−0.763

[a] ($pH = 0$; $pH_2 = 1$ atmosphere).

14.2.5. Ag

Silver dissolves to a small extent in the electrolyte and is more noble than Cu (Table 14.2), so any Ag^+ present in solution will be reduced and codeposit at the cathode. Cathode copper typically contains 8−10 ppm (g/t) silver (Barrios, Alonso, & Meyer, 1999). Fortunately, silver is a rather benign impurity in copper.

14.2.6. O

Oxygen is found in the anode in various oxide compounds, with Cu_2O being the most prevalent. Cu_2O dissolves chemically (not electrochemically) due to the high acidity of the electrolyte. The dissolution rate of an anode is therefore greater than would be predicted from Faraday's Law. This results in the buildup of copper in the electrolyte and requires a bleed of electrolyte for de-copperization in liberator cells (Section 14.5.4).

14.2.7. Summary of Impurity Behavior

The relative deportments of the various impurity elements present in the anode are shown in Table 14.3. Au, PGMs, Se, Te, Pb, and Sn do not dissolve in $CuSO_4-H_2SO_4-H_2O$ electrolyte, so they cannot plate at the cathode. They form slimes (Chen & Dutrizac, 2007; Hait, Jana, & Sanyai, 2009).

The quantity, morphology, pore structure, and adhesion properties of the slimes are crucial factors which influence mass transfer processes at the anode—electrolyte interface (Hiskey & Moats, 2010). The presence of these elements in cathode copper is due to accidental entrapment of slime particles in the depositing copper as the deposit grows.

As, Bi, Co, Fe, Ni, S, and Sb dissolve in the electrolyte. Fortunately, Cu plates at a lower applied potential than these elements (Table 14.2) so they remain in the electrolyte while Cu is plating. Their presence in cathode copper is due to entrapment of electrolyte or slime particles (Chen & Dutrizac, 2000).

TABLE 14.3 Deportment of Elements Present in Anode to the Slimes and the Electrolyte (from Larouche, 2001)

Element	Deportment to slimes, %	Deportment to electrolyte, %
Cu	<0.2	>99.8
Au	100	0
Ag	>99	<1
Se	98	2
Te	98	2
Pb	98	2
Bi	60[a]	40
Sb	60[b]	40
As	25[c]	75
S	1	99
Ni	1	99
Co	1	99
Fe	0	100
Zn	0	100

[a]With 0.1% Pb in anode.
[b]With 0.1% As, Bi, Pb, and Sb (each) in anode.
[c]With 0.1% As in anode.

The contamination of cathode copper by these impurities is minimized by:

(a) Electrodepositing a smooth, dense copper plate on the cathode (see Section 14.5.1);
(b) Thoroughly washing the cathode product;
(c) Controlling the impurity levels in the electrolyte by bleeding electrolyte from the refinery and removing the impurities.

Small amounts of Ag dissolve in the electrolyte and are codeposited with Cu in the cathode.

14.3. EQUIPMENT

Industrial electrorefining uses large (~1 m × 1 m), thin (40–50 mm) copper anodes and thin (1–10 mm) cathodes interleaved approximately 50 mm apart in a cell filled with electrolyte (Fig. 14.3). The anodes in the cell are all at the same potential: the cathodes are all at another, lower potential. DC Power is provided by a rectifier. Anodes and cathodes are evenly spaced along the cell to ensure an even distribution of current to all electrodes.

The process is continuous except when electrodes are loaded or unloaded from the cells. Electrolyte containing $CuSO_4$ and H_2SO_4 continuously enters at the bottom end of each cell. It leaves the cells (slightly less pure) by continuously overflowing the other end of the cell into an electrolyte collection system. Anodes continuously dissolve and pure copper continuously plates on the cathodes.

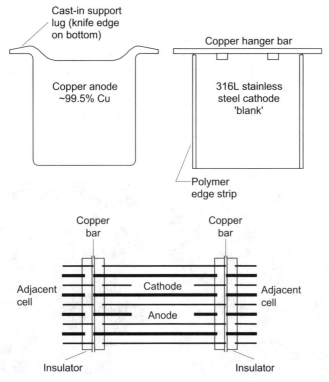

FIGURE 14.3 (a) Copper anode and stainless steel cathode. The cathode is about 1 m^2 in area. The anode is slightly smaller. (b) Schematic of electrorefining circuitry. Current flow between anodes and cathodes is through the electrolyte.

14.3.1. Anodes

The impure copper anodes are cast at a smelter or the refinery itself, as described in Chapter 13. They are typically 4−5 cm thick, with a mass of 300−400 kg. The starting masses of the anodes should be very similar so that they all dissolve at the same rate. The anodes slowly become thinner as the copper dissolves into the electrolyte. Once they reach 15−20% of their original mass (after a typical electrorefining time of 21 days), they are removed from the cell before they break up and fall into the cell. This anode scrap is washed free of slimes, dried, remelted, and cast into new anodes.

Anode casting and preparation are critical to ensuring that they dissolve evenly and that cathodes of optimum purity are produced. The following procedures should be followed (Wenzl, Antrewkowitsch, Filzweiser, & Pesl, 2007):

(a) Casting flat anodes of very similar mass;
(b) Pressing the anodes flat;
(c) Machining the anode support lugs so the anodes hang vertically;
(d) Spacing the anodes and cathodes precisely in racks before loading them into the cells.

Anode preparation machines are now widely used in the industry (Chapter 13). These allow cast anodes to be weighed, straightened, and the lugs machined automatically (Robinson, Siegmund, Davenport, Moats, & Karcas, 2010).

14.3.2. Cathodes

Modern refineries use stainless steel *blanks* as the starting cathode. This is a sheet of stainless steel welded to copper support bars (Fig. 14.4a). Copper is electrodeposited onto these cathodes for 7−10 days. The copper-plated cathodes are then removed from the cell and replaced with fresh stainless steel blanks.

The copper-plated cathodes are washed with hot-water sprays and the copper deposits (50−80 kg on each side of the blank) are machine-stripped from the stainless steel. These are packed, strapped, and sent to market or to melting and casting (Fig. 14.5). The

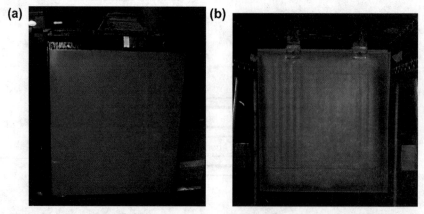

FIGURE 14.4 Electrorefined copper cathodes plated onto (a) stainless steel blanks and (b) copper starter sheets. Note the plastic edge strips on the permanent cathode and the copper loops and hangers on the starter sheet. *(Courtesy of T. Robinson.)*

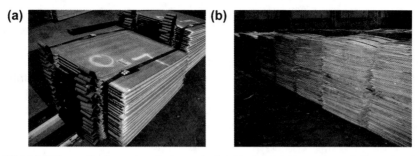

FIGURE 14.5 Electrorefined copper cathodes ready for shipping. (a) Hanger loops indicate that electrorefining has been carried out using copper starter sheets, compared to (b) stainless steel blanks. *(Courtesy of T. Robinson and White Pine Refinery, MI.)*

stripped stainless steel blanks are carefully washed and returned to the refining cells. It may be necessary to polish the stainless steel surface to maintain its smoothness and avoid plated copper from sticking onto rough patches.

The stainless steel blanks are usually flat, cold- and bright-rolled 316L stainless steel, ~3 mm thick. Electrodeposited copper attaches quite firmly to this surface so that it does not accidentally detach during refining. The vertical edges of the blanks are covered with long, tight-fitting polymer *edge strips*. These strips prevent copper from depositing completely around the cathode and allow removal of the electrorefined copper plates from the stainless steel. The bottoms of the stainless steel blanks are given a sharp-edged V groove to allow easy detachment of the plated copper from this region of the blank.

Older refineries use thin copper *starter sheet* cathodes, hung by starter sheet loops on copper support bars. These are thin sheets of copper that have been plated onto titanium or stainless steel, stripped, flattened, and then hanger bars are attached (Fig. 14.4b). These cathodes have higher mechanical stresses than stainless steel blanks and have a tendency to warp. They are often embossed to improve rigidity (Nicol, 2006). They often need to be removed after about two days of plating to be straightened in presses and then returned to the cells. Use of these cathodes therefore makes refinery management more difficult. They also short-circuit more frequently, which reduces cathode quality.

Many refineries (especially in Europe, North America, and Asia where labor costs are high) have switched from this older technology to stainless steel blanks (Eastwood & Whebell, 2007; Hashimoto, Narita, & Shimokawa, 2010; Nagai, Hashikawa, & Yamaguchi, 2010). In 2010, approximately half of the plants in the world used *permanent cathode technology*: Xstrata Technology (previously known as the Isa or Kidd processes) and Outotec supply this refinery tankhouse technology. An average labor requirement of 0.9 manhours per tonne of cathode is reported for tankhouses using permanent cathode technology, compared to 2.4 manhours/t in tankhouse that use starter sheets (Moats et al., 2007). This technology also allows use of higher current densities (>300 A/m^2, compared to about 250 A/m^2 for starter sheets), better cathode quality, higher productivity, and lower operational costs (Matsuda, Goda, Takehayashi, & Maeda, 2007).

14.3.3. Cells

Industrial refining cells are typically 3—6 m long. They are wide and deep enough (~1.1 m × 1.3 m) to accommodate the anodes and cathodes with 0.1—0.2 m underneath. Modern cells are designed with adequate height between the bottom of the electrodes and the cell floor to minimize contamination of the cathode from slimes that fall to the base of the cell. Each cell typically contains 30—60 anode—cathode pairs connected in parallel.

Modern cells are made of pre-cast polymer concrete (Moats et al., 2007). Polymer concrete cells are usually cast with built-in structural supports, electrolyte distributors, drains, *etc*. These are advantageous for fitting them into the tankhouse infrastructure.

Older cells are made of concrete, with a flexible polyvinyl chloride or lead lining. These older cells are gradually being replaced with unlined polymer concrete cells.

14.3.4. Electrical Components

The cells are connected electrically in series to form sections of 20—40 cells. Each section can be electrically isolated for inserting and removing anodes and cathodes and for cleaning and maintenance. The number of cells in each section is chosen to maximize the efficiency of these operations.

The electrical connection between cells is made by connecting the cathodes of one cell to the anodes of the adjacent cell and so on. The connection is made by seating the cathodes of one cell and the anodes of the next cell on a common copper *distributor bar* or *busbar* (Fig. 14.3). Considerable attention is paid to making good contacts between the anodes, cathodes, and distributor bar. Good contacts minimize energy loss and ensure uniform current distribution to all anodes and cathodes.

Electrorefining requires direct voltage and current. These are obtained by converting commercial alternating current (AC) to direct current (DC) at the refinery. Silicon-controlled rectifiers are used.

14.4. TYPICAL REFINING CYCLE

Production electrorefining begins by inserting a group of anodes and cathodes into the empty cells of a freshly cleaned section of the refinery. They are precisely spaced in a rack and brought to each cell by crane. This process is often completely automated. The cells are then filled with electrolyte and connected to the power supply. Electrolyte flows continuously in and out of the cells. The anodes begin to dissolve and pure copper begins to plate on the cathodes.

Each anode is electrorefined until it is 80—85% dissolved (Moats et al., 2007). Electrolyte is then drained from the cell (through an elevated standpipe), the anodes and cell walls are hosed down with water, and the slimes are drained from the bottom of the cell. The drained electrolyte is sent to filtration and storage. The slimes are sent to a Cu and byproduct metal recovery plant (Chapter 21).

Two or three cathode plating cycles are produced from a single anode. Each copper cathode typically weighs 50—100 kg. This multi-cathode plating cycle ensures that cathodes do not grow too close to the slime-covered anodes and become contaminated. Copper-loaded cathodes are removed from the cells after 7—10 days of plating and a new rack of stainless steel blanks is inserted.

The copper-loaded cathodes are washed to remove electrolyte and slimes. The plated copper is then machine-stripped from the stainless steel blanks, sampled, and stacked for shipping, typically in batches of 2.5 tonnes. Cathodes produced from starter sheets are handled similarly but do not require stripping. A robotic stripping machine in a modern refinery is shown in Fig. 14.6.

FIGURE 14.6 Xstrata Technology robotic cathode stripping machine, manufactured by Ionic Engineering Limited, installed at an Asian copper refinery. *(Courtesy of Xstrata Technologies.)*

The cells are inspected regularly during refining to locate short-circuited anode-cathode pairs. The inspection is done by infrared scanners (which locate 'hot' electrodes), gauss meters, and cell voltage monitoring systems (Mipac, 2004).

Short circuits are caused by non-vertical electrodes, bent cathodes, or nodular cathode growths between anodes and cathodes. They waste electrical current and lead to impure copper, due to settling of slimes on nodules and non-vertical cathode surfaces. Shorts are eliminated by straightening the electrodes and removing the nodules.

14.5. ELECTROLYTE

Copper refining electrolytes typically contain 40–50 g/L Cu, 170–200 g/L H_2SO_4, 10–20 g/L Ni, up to 20 g/L As, and various impurities (Table 14.4). The nature and concentration of the impurities can vary quite widely, depending on the anode composition, which, in turn, depends on the feed to the smelter (Table 14.1).

Electrical conductivity of the electrolyte increases with increasing H_2SO_4 concentration but decreases with increasing Cu and Ni concentrations (Table 14.5). High conductivity will improve energy efficiency, but higher acid concentrations also increase corrosion in the tankhouse. Too high a Cu concentration may lead to passivation of the anodes (see Section 14.6.3.1) and increase the viscosity of the electrolyte. This, in turn, can increase the solids in suspension and consequently reduce cathode purity due to incorporation of impurity particles into the cathode as it plates.

Chloride (0.02–0.05 g/L) and organic leveling and grain-refining agents (1–10 mg/L) are added to the electrolyte to improve the morphology and purity of the cathode deposit (Section 14.5.1).

The incoming electrolyte is steam-heated to 60–65 °C to improve conductivity and mass transfer. It leaves the cell about 2 °C cooler (Moats et al., 2007). Electrolyte

TABLE 14.4 Compositions of Copper Refining Electrolytes (Moats et al., 2007)

Component	Concentration
Cu	35–60 (typical 45–50) g/L
H_2SO_4	120–210 (typical 150–200) g/L
Cl	0.01–0.06 g/L
As	2–30 g/L
Bi	0.01–0.7 g/L
Fe	0.1–3 g/L
Ni	0.3–25 g/L
Sb	0.002–3 g/L
Protein colloids (glue)	35–350 g/t Cu cathode
Thiourea	30–140 g/t Cu cathode
Avitone	0–60 g/t Cu cathode

is circulated through each cell at ~1.2 m³/h. At this flowrate, the electrolyte in each cell is completely replaced within a few hours. A steady electrolyte circulation is essential to:

(a) Bring warm, purified electrolyte into the cell;
(b) Ensure uniform concentrations of Cu^{2+} and leveling and grain-refining agents across all cathode surfaces;
(c) Remove dissolving impurities from the cell.

Modern developments in electrolyte flow arrangements, such as the METTOP-BRX-Technology (Filzweiser, Filweizer, & Stibich, 2008; Wenzl, Filzwieser, Filzwieser, & Anzinger, 2010), ensure that the above requirements are met, even when employing high current densities, by minimizing the diffusion boundary layer at the electrode surfaces. Operation above 400 A/m² without loss of cathode quality or current efficiency is now routinely achieved.

14.5.1. Addition Agents

The global trend is copper electrorefining is to operate at increasingly higher current densities (Section 14.8). This reduces plant size for a new installation and increases production rate for an existing plant. However, as current density is increased, the cathode deposits become rougher and uneven surfaces are created. Impurities from the electrolyte and slimes produced at the anode adsorb onto the deposit and become encapsulated in the deposit as it grows. Under these conditions, passivation of the anodes is also promoted (see Section 14.6.3.1).

To counteract these effects, deposition of smooth, dense, pure copper is promoted by adding leveling and grain-refining agents to the electrolyte (De Maere & Winand, 1995;

TABLE 14.5 Specific Electrical Conductivity of Copper Electrorefining Electrolytes at 55 °C. (Adapted from Nicol, 2006.)

Cu, g/L	Ni, g/L	H_2SO_4, g/L	Conductivity, 1/Ω-m
40	0	175	64.6
50	0	175	61.2
60	0	175	58.1
40	10	175	60.6
50	10	175	57.9
60	10	175	54.6
40	20	175	56.7
50	20	175	53.9
60	20	175	51.2
40	0	200	70.9
50	0	200	67.2
60	0	200	53.3
40	10	200	66.4
50	10	200	63.1
60	10	200	57.9
40	20	200	62.3
50	20	200	58.9
60	20	200	55.7

Knuutila, Forsen, & Pehkonen, 1987; Nakano, Oue, Fukushima, & Kobayashi, 2010; Wang & O'Keefe, 1984). Without these, the cathode deposits would be dendritic and soft and have lower purity.

14.5.1.1. Leveling Agents

The principal *leveling agents* are protein colloid bone glues (Fig. 14.7a). All copper refineries use these glues, added at a concentration of 50–120 g per tonne of cathode copper (Moats et al., 2007). The glues consist of large protein molecules (MW 10 000–30 000), which form large cations in the electrolyte (Saban, Scott, & Cassidy, 1992). Their leveling efficacy varies so they must be thoroughly tested before being adopted by a refinery.

The glue exerts its leveling action by electrodepositing large protein molecules at the negatively charged tips of rapidly growing copper grains (Fig. 14.7b; Hu, Roser, & Rizzo, 1973). This creates an electrically resistant barrier at the tips of the protruding crystals

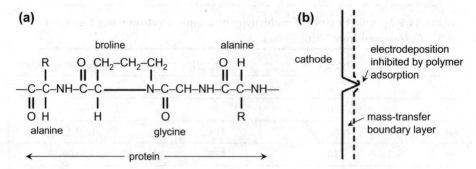

FIGURE 14.7 (a) Structure of glue molecules, comprising a series of amino acids linked together into a large protein structure, and (b) mechanism of action of glue as a leveling agent.

and encourages further copper deposition sideways and in the 'valleys' between the protrusions. In this way, the net result is dense and level growth of the copper crystals on the cathode surface.

In practice, glue in the electrolyte decomposes within 1.5—2 h at high temperatures and high acidity, and its degradation is catalyzed by H_2SO_4. It is therefore essential to constantly monitor the glue concentration in the electrolyte and ensure that it is maintained at optimum levels. Excess glue concentration leads to rough, striated, and brittle deposits; too low a glue concentration causes nodulation (Fig. 14.8).

14.5.1.2. Grain-refining Agents

The principal *grain-refining agents* are thiourea (added to the electrolyte at 30—150 g/t cathode copper) and chloride (20—50 mg/L in electrolyte, added as HCl or NaCl). Avitone, a sulfonated petroleum liquid, is also sometimes used with thiourea as a grain refiner at concentrations of ~0.5 mg/L.

Thiourea (($NH_2)_2C{=}S$) acts as a grain refiner by promoting the formation of new copper nuclei and inhibiting the growth of existing crystals. Thiourea molecules adsorb on the cathode surface and prevent crystal growth. The mechanism of action of thiourea is believed to occur via the reduction of Cu^{2+} to Cu^+ at the cathode surface. Thiourea forms a cationic complex with Cu^+ on the cathode surface. This promotes cathodes with fine-grained crystal morphology (Fig. 14.9).

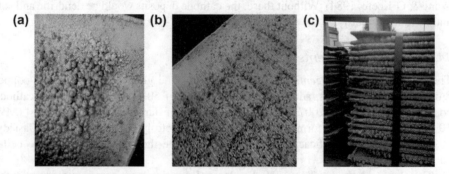

FIGURE 14.8 Poor quality of electrorefined cathode showing (a) nodules on cathode surface, (b) entrainment of anode slimes in the deposit and *roping* or striations on the cathode, and (c) dendrite growth on the edges of the cathode. *(Courtesy of T.A. Muhlare, University of Pretoria.)*

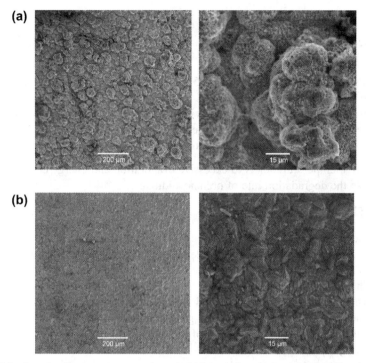

FIGURE 14.9 Scanning electron microscopy images of (a) dendritic cathode morphology, showing the rough surface created at high current density without the use of grain refiners and (b) the desired, compact cathode morphology with a smooth, dense, homogenous, and fine-grained surface. *(Courtesy of T.A. Muhlare, University of Pretoria.)*

If too high a concentration of thiourea is used, then its adsorption on the cathode surface can lead to sulfur contamination of the cathode.

Avitone is a proprietary hydrocarbon-based sodium sulfonate surfactant. It is used in combination with glue to give a dense, smooth deposit on the cathode. It also acts as a detergent, wetting oily patches on the anodes and cathodes. Too little Avitone causes slimes to adhere to the surface of the cathode, leading to entrapment of slimes and contamination of the cathode (Deni, 1994).

Chloride improves and brightens cathode deposits when combined with organic additives. It is believed to co-adsorb with thiourea. The chloride ion also acts as a grain-refining agent. Too high a Cl concentration (>50 mg/L) promotes pitting corrosion of the stainless steel cathode blanks.

14.5.1.3. Control of Addition Agents

The addition agents are dissolved in water and added to electrolyte storage tanks just before the electrolyte is sent to the refining cells (Fig. 14.1). Many refineries automatically control their reagent addition rates based on measured glue and thiourea concentrations in the stream exiting the refining cells. The electrolyte exiting a cell should contain enough addition agents to still give an excellent copper deposit to ensure a high-purity deposit on all cathodes, irrespective of their position in a cell. The CollaMat system is widely used for measuring glue concentrations (Stantke, 2002), while the Reatrol system is used for thiourea (Conard, Rogers, Brisebois, & Smith, 1990).

14.5.2. Electrolyte Temperature

Electrolyte is steam-heated to ~65 °C (using titanium or Teflon coils). This heating is expensive but its beneficial effects include:

(a) Increasing the solubility of $CuSO_4 \cdot 5H_2O$, preventing it from crystallizing on the anode (see Section 14.6.3.1);
(b) Lowering electrolyte density and viscosity (Price & Davenport, 1981), which reduces the migration of slimes in the cell;
(c) Increasing the rate of the electrochemical reactions (Eqs. (14.1) and (14.2)).

Too high a temperature leads to excessive evaporation and energy consumption, and can increase the degradation rate of organic additives.

14.5.3. Electrolyte Filtration

Insoluble solid particles present in the electrolyte can cause problems in electrorefining. Solids can originate from breakdown of the anodes, from anode slimes, or from the mold coating used when casting the anodes (see Chapter 13).

At high viscosity, insoluble particles are suspended in the electrolyte. When solids come into contact with the growing crystal surface they serve as substrates for nucleation. This enhances the local growth rate and grain roughening starts. This can initiate the growth of dendrites. Adequate filtration of the recirculating electrolyte is critical.

14.5.4. Removal of Impurities from the Electrolyte

Soluble anode impurities dissolve continuously into the electrolyte (Section 14.2). To prevent these from accumulating in the electrolyte, they are continuously removed from a bleed stream. As, Bi, Co, Fe, Ni, and Sb are the main impurities removed this way. Also, 1—2% of the Cu that dissolves from the anodes is not plated on the cathodes (see Section 14.2.6). This extra Cu is also removed from the electrolyte bleed stream. About 0.1—0.2 m^3 of electrolyte is bled and purified per tonne of product copper.

The impurities and Cu are removed in three main sequential steps (Bravo, 1995; Stantke & Leuprecht, 2010; Wang, 2004):

(a) Copper is electrowon using Pb—Sn—Ca anodes and stainless steel or copper starter sheets in *liberator cells* (Chapter 17).
(b) Water is evaporated from the Cu-depleted electrolyte and $NiSO_4$ crystals are precipitated from the concentrated solution. In some operations, $CuSO_4$ may also be recovered by evaporation and crystallization of hydrated $CuSO_4$. Vacuum and centrifuge technologies may be employed.
(c) As, Bi, and Sb are electrowon from Cu-depleted electrolyte into an impure Cu—As—Bi—Sb cathode deposit or *sludge* which can then be sent for toll refining.

The remaining concentrated acid (~1000 g/L H_2SO_4) is returned to electrolyte storage to maintain the acid balance in the refinery (Rafieipour et al., 2007). A small portion is neutralized or sold to prevent a gradual buildup of Ca, K, Mg, and Na ions in the refinery.

More sophisticated and specific technologies are increasing being used for individual and bulk recovery of impurities (Agrawal, Bagchi, Kumari, & Pandey, 2009). As, Bi, Co, Sb, and H_2SO_4 may be recovered by solvent extraction (Agrawal et al., 2008; Rondas, Scoyer, & Geenen, 1995; Stevens & Gottliebsen, 2000). Ion exchange with chelating resins is used for removal of As, Sb, and Bi (Maruyama, Furuta, Oida, Shimokawa, & Narita, 2007; Matsuda et al., 2007; Roman, Salas, Guzman, & Muto, 1999; Sheedy, Pajunen, & Wesstrom, 2007). Ion exchange is also used to remove acid prior to the recovery of nickel by precipitation with carbonate. Molecular recognition technology (MRT) is used for Bi removal (Izatt, Izatt, Bruening, & Dale, 2009).

Refineries are increasingly investigating ways to produce salable and valuable byproducts from the treatment of their bleed streams (Agrawal, Bagchi, Kumari, Kumar, & Pandey, 2007; Agrawal, Kumari, Bagchi, Kumar, & Pandey, 2007; Kumari, Agrawal, Bagchi, Kumar, & Pandey, 2006).

14.6. MAXIMIZING COPPER CATHODE PURITY

The principal technical objective of the refinery is to produce high-purity cathode copper. Other important objectives are to produce this pure copper rapidly and with a minimum consumption of energy and manpower. The rest of this chapter discusses these goals and how they are attained.

The main factors influencing the purity of cathode copper are:

(a) The physical arrangement of the anodes and cathodes in the electrolytic cells;
(b) Chemical conditions, particularly electrolyte composition, clarity, leveling and grain-refining agent concentrations, temperature, and circulation rate;
(c) Electrical conditions, particularly current density;
(d) Thorough washing of cathodes after electrorefining and their removal from the cells.

14.6.1. Physical Factors Affecting Cathode Purity

Slimes particles, with their high concentrations of impurities, are kept away from the cathodes by keeping the flow of electrolyte smooth enough so that slimes are not transported from the anodes and cell bottoms to the cathodes. Turbulence in the electrolyte should be avoided. Contamination is also avoided by filtering the electrolyte (especially that from cell cleaning) before it is recycled to electrorefining (see Fig. 14.1).

14.6.2. Chemical Factors Affecting Cathode Purity

The chemical conditions, which lead to highest-purity cathode copper are:

(a) Constant availability of high Cu^{2+} electrolyte;
(b) Constant availability of appropriate concentrations of leveling and grain-refining agents;
(c) Uniform 65 °C electrolyte temperature;
(d) Absence of slimes particles in the electrolyte at the cathode faces;
(e) Controlled concentrations of dissolved impurities in the electrolyte;
(f) Enough As in the electrolyte to prevent the formation of floating slimes and promote precipitation of Sb and Bi in the slimes layer.

A high Cu^{2+} concentration (40–50 g/L) in the electrolyte and steady recirculation of the electrolyte through the cells ensure constant availability of Cu^{2+} ions over the cathode faces. This allows a constant rate of plating, which gives uniformity of crystal grain size in the cathode. This is helped by maintaining consistent concentrations of leveling and grain-refining agents in the electrolyte (Section 14.5.1).

14.6.3. Electrical Factors Affecting Cathode Purity

The main electrical factor affecting cathode purity is cathode current density, i.e., the rate at which electricity is passed through the cathodes, measured in amperes per square meter of cathode area (A/m^2). High current densities give rapid copper plating but can also cause growth of nodules and protrusions. These can entrap slimes in the cathodes, lowering cathode purity, and can also lead to short-circuiting.

14.6.3.1. Upper Limit of Current Density

High current densities give rapid copper plating. Excessive current densities may, however, cause anodes to *passivate* by producing Cu^{2+} ions at the anode surface faster than they can convect away. The net result is a high concentration of Cu^{2+} at the anode surface and precipitation of a coherent $CuSO_4 \cdot 5H_2O$ layer on the anode (Chen & Dutrizac, 1991; Moats & Hiskey, 2010).

This $CuSO_4 \cdot 5H_2O$ layer isolates and electrically insulates the anode from the electrolyte and blocks further oxidation of Cu to Cu^{2+}, a process known as passivation (Moats & Hiskey, 2010). The likelihood of passivation increases with increasing Cu^{2+}, acid, and sulfate concentrations in the electrolyte (Moats & Hiskey, 2007). The presence of high levels of nickel in the anode can also cause passivation through the formation of a similar $NiSO_4 \cdot mH_2O$ layer (Doucet & Stafiej, 2007; Jarjoura & Kipouros, 2005). The problem is exacerbated if the impurities in the anode form a coherent slimes layer (Moats & Hiskey, 2006). Anodes with high levels of As, Ni, Sb, Sn, and Pb are also readily passivated (Atenas & Muñoz, 2007a; Moats & Hiskey, 2006; Mubarok, Antrekowitsch, & Mori, 2007; Mubarok, Antrekowitsch, Mori, Lossin, & Leuprecht, 2007). Additives in the electrolyte can also play some role in passivation (Moats & Hiskey, 2000; Nakano et al., 2010; Tantavichet, Damronglerd, & Chailapakul, 2009). Arsenic is the only known element to inhibit passivation (Krusmark, Young, & Faro, 1995; Moats & Hiskey, 2006).

Passivation can usually be avoided by operating with current densities below 300 A/m^2, depending on the impurities in the anode and maintaining a minimum arsenic anode concentration above 300 mg/L (Krusmark et al., 1995). Increasing the chloride content of the electrolyte could also be helpful (Moats & Hiskey, 2000). Warm electrolyte (with its higher $CuSO_4 \cdot 5H_2O$ solubility) also helps. Refineries in cold climates try to avoid cold regions in their tankhouse. Automatic temperature control in each cell is now widely practiced (Rantala, You, & George, 2006).

Each refinery must balance these competing economic factors.

14.6.3.2. Maximizing Current Efficiency

Cathode current efficiencies in modern copper electrorefineries are ~93 to >98%. The unused current is wasted as:

(a) Anode-to-cathode short circuits ~3%
(b) Stray current to ground ~1%
(c) Reoxidation of cathode copper by O_2 and Fe^{3+} ~1%

Short-circuiting is caused by cathodes touching anodes. It is avoided by precise, vertical electrode placement, and controlled additions of leveling and grain-refining agents to the electrolyte. Its effect is minimized by locating and immediately breaking cathode−anode contacts whenever they occur.

Stray current loss is largely due to current flow to ground via spilled electrolyte. It is minimized by good house-keeping around the refinery.

Copper plated on the cathode can be reoxidized back into the electrolyte by the presence of oxidizing agents, such as oxygen or Fe(III). The reactions are (see Table 14.2):

$$Cu^\circ + 0.5O_2 + 2H^+ + 2e^- \rightarrow Cu^{2+} + H_2O \quad (14.4)$$

$$Cu^\circ + 2Fe^{3+} \rightarrow Cu^{2+} + 2Fe^{2+} \quad (14.5)$$

This can be avoided by minimizing oxygen absorption in the electrolyte, by keeping electrolyte flow as smooth and quiet as possible while still maintaining adequate hydrodynamics for mass transfer. The presence of Fe in the electrolyte should also be avoided where possible.

Other house-keeping factors that can have a significant impact on current efficiency include:

(a) Ensuring even distribution of current to all electrodes;
(b) Ensuring that electrical contacts are clean;
(c) Ensuring that electrodes are equally spaced and hanging vertically.

14.7. MINIMIZING ENERGY CONSUMPTION

The total electrical energy consumption of an electrorefinery is 300−400 kWh per tonne of copper produced. It is minimized by maximizing current efficiency and by maintaining good electrical connections throughout the refinery.

Hydrocarbon fuel is also used in the electrorefinery, mainly for heating electrolyte and melting anode scrap. Electrolyte heating energy is minimized by insulating cells, tanks, and pipes, and by covering the electrolytic cells with canvas or plastic sheeting (Fig. 14.2).

Anode scrap melting energy is minimized by minimizing scrap production, i.e., by casting thick, equal mass anodes, and by equalizing current between all anodes and cathodes. It is also minimized by melting the scrap in an energy-efficient Asarco-type shaft furnace (see Chapter 20).

14.8. INDUSTRIAL ELECTROREFINING

Table 14.6 describes the operating conditions of seven modern electrorefineries (Robinson et al., 2010). These refineries have been selected to show the diversity of operating conditions, from one of the largest in the world (Chuquicamata, 666 000 t/a) to one of the oldest refineries in the world that is still operating (the Nippon refinery at Saganoseki, commissioned in 1916). A relatively new refinery is Gresik in Indonesia (commissioned in 1999), while the Boliden refinery in Sweden received a major upgrade in 2007. More detailed information about these and other major refineries (see Chapter 2) is provided by Moats et al. (2007).

TABLE 14.6 Selected Industrial Copper Electrorefining Data (from Moats et al., 2007)

Refinery	Aurubis Hamburg, Germany	Boliden Skelleftehamn, Sweden	Copper Refineries, Townsville, Australia	Chuquicamata Codelco Norte, Chile	Gresik, Indonesia	Sterlite Silvassa, India	Nippon Mining Saganoseki, Japan
Startup date	1989/92	1995 + expansions 2000, 2007	1959, upgrade 1998	—	1999	1996	1916
Capacity cathode Cu, t/a	380 000	225 000	300 000	666 658	200 000	199 000	235 000
Electrolytic cells							
Number (total)	1080	—	1146	2670	798	615	1133
Material	Concrete, polymer concrete	Polymer concrete	Polymer concrete	Polymer concrete	Polymer concrete	Polymer concrete	Pre-cast concrete
Lining material	Antimonial lead	—	—	—	—	—	PVC
Length × width × depth, m	5604 × 1160 × 1435	4490 × 1102 × 1130	4450 × 1160 × 1143	—	6260 × 1208 × 1380	5700 × 1200 × 1200	5190 × 1090 × 1375
Anodes/cathodes per cell	58/57	44/43	45/44	50/49	59/58	56/55	50/49
Anodes							
Cu, %	98.5–99.6	99	99.73	98.57–99.76	99.5	98.56	99.39
Ag, ppm	200–2000	2200	121	155–431	266	240	666

As, ppm	400–1500	2000	369	365–1587	932	425	820
Au, ppm	10–100	69	8	1–16	105	35	74
Bi, ppm	20–200	71	29	6–55	210	150	120
Fe, ppm	—	—	24	6–31	17	50	—
Ni, ppm	1000–4000	3800	175	13–107	364	400	1150
O, ppm	1000–2000	2400	1560	1069–1624	710	1500	1660
Pb, ppm	200–2000	100	309	18–139	3075	200	150
S, ppm	10–70	—	16	10–21	52.5	25	—
Sb, ppm	10–100	560	17	60–235	28	35	10–100
Se, ppm	100–600	560	20	86–277	610	350	250
Te, ppm	20–200	180	38	10–40	64	100	160
Length × width × thickness, mm	950 × 905 × 55	960 × 890 × 41	950 × 912 × 55	1333 × 950 × 46	979 × 930 × 47	975 × 920 × 45	1010 × 914 × 44
Initial mass of anode, kg	408	335	401	400–420	394	375	348
Center-line spacing, mm	95	100	95	114	104	100	100
Life, d	21	18	21	21	19	19.3	21
Scrap after refining, %	10–12	18	14.6	10–12	11.7–14.6	—	17.1
Anode slimes, kg/t anode	5–7	8	1.6	2	5.1	4.5–6	2.9

(Continued)

TABLE 14.6 Selected Industrial Copper Electrorefining Data (from Moats et al., 2007)—cont'd

Refinery	Aurubis Hamburg, Germany	Boliden Skellefteham, Sweden	Copper Refineries, Townsville, Australia	Chuquicamata Codelco Norte, Chile	Gresik, Indonesia	Sterlite Silvassa, India	Nippon Mining Saganoseki, Japan
Cathodes							
Type	St. steel	St. steel	St. steel	Starter	st. steel	St. Steel	Starter
Plating time (d)	6–8	6	7	12	7–12	5–6	10–11
Mass Cu plated (total) (kg)	80–152	80–100	112	176	100	90–110	140–160
Cu, %	>99.99	—	99.995	99.99	—	—	99.99
Ag, ppm	7–15	121	10.7	3.4–12	8.3	8–9	11.3
As, ppm	<1	0.5	0.8	0.1–1.7	0.2	0.3–0.6	<0.5
Bi, ppm	<0.3	0.13	0.1	0.1–0.4	0.18	0.1–0.11	<0.5
Fe, ppm	—	1.9	<0.1	0.1–4.0	<1	0.5–0.7	<0.5
Ni, ppm	<2	0.3	0.4	0.1–3.3	0.18	0.7	<0.5
Pb, ppm	<1	0.24	0.3	0.1–2.6	0.58	0.4	<0.5
S, ppm	<10	4.5	8.1	2–9	3.7	5.0–5.5	3.1
Sb, ppm	<1	0.24	0.1	0.1–1.6	0.1	0.4	<0.5
Se, ppm	<0.5	0.17	<0.1	—	0.1	0.2–0.3	<0.5
Te, ppm	<0.5	0.26	<0.1	0.1–0.9	0.1	0.2	<0.5

Electrolyte

Cu, g/L	45–49	48	48	40–44	51.1	42–45	45–50
H$_2$SO$_4$, g/L	170–200	175	160	200	176	155–170	175–185
Inlet temperature, °C	63	65	63	60	63	64	62
Outlet temperature, °C	63	68	62	52–63	63	66	62
Bone glue, g/t Cu cathode	—	75	72	35	65	100–110	82
Thiourea, g/t Cu cathode	—	65–78	88	60	90	90–100	72
Avitone, g/t Cu cathode	—	0	0	10	0	0	42

Power and energy

Cathode current density, A/m^2	347	310	331	242–252	330	312	263
Cathode current efficiency, %	94–97	97–98	93.3	—	98.4	96	93
Cell voltage, mV	340	300–400	210–410	280–320	250–350	335	320
Power, kWh/t Cu (plating only)	—	428	410	288	319	435	337

14.9. RECENT DEVELOPMENTS AND EMERGING TRENDS IN COPPER ELECTROREFINING

Most of the current developments and trends in electrorefining relate to maximizing production, safety, and product purity by upgrading of existing infrastructure and equipment, while minimizing energy consumption and ensuring enterprise sustainability. Some of the most notable developments include (Moats et al., 2007; Robinson et al., 2010):

(a) Continued adoption of polymer concrete cells;
(b) Continued adoption of permanent cathode systems (especially in countries where labor costs are high);
(c) Increased mechanization, automation, and the use of robotics systems in the tankhouse, including crane handling systems, anode preparation machines, and cathode handling machines (Aslin, Eriksson, Heferen, & Sue Yek, 2010; Djurov, 2010);
(d) Increased adoption of advanced process control technologies;
(e) New electrode contact systems to improve current distribution and current efficiency by minimizing stray currents;
(f) Improved electrical equipment and components to allow operation at increasingly higher currents (Wachendörfer, 2009);
(g) Use of on-line measurement systems, including temperature and voltage monitoring (Mipac, 2004), cell inspection devices, monitoring of glue (CollaMat system) and thiourea (Reatrol system) in the electrolyte, short-circuit detection systems, and electrolyte circulation and flowrate control (Wenzl et al., 2010);
(h) Automated tracking of individual anodes and cathodes, allowing the performance of individual electrodes and batches of cathodes to be tracked through their full service history (Phan, Whebell, & Oellermann, 2007);
(i) Use of sophisticated software and databases to optimize asset management, improve productivity, and facilitate development and implementation of operational strategies (Larinkari & Rantala, 2010; Nikus, Korpi, & Rantala, 2010; Rantala, Larinkari, & Menese, 2009).

There is a trend toward reducing the anode—cathode spacing to increase copper production in an old tankhouse at minimal cost and effort (Nagai et al., 2010). This may be associated with increasing the thickness or changing the shape of the cast anodes (Baranek, Chmielarz, & Śmieszek, 2007; Coffin & Leggett, 2007; Moats et al., 2007). This requires improved vigilance and tankhouse management to avoid increasing short circuits. Specific energy consumption can also be reduced because of the lower voltage drop across the electrolyte.

A related trend in pushing for increased production from an existing infrastructure is toward the use of steadily increasing current densities. Many tankhouses now routinely operate at 350 A/m^2 (Atenas & Muñoz, 2007b; Moats et al., 2007; Robinson et al., 2010; Stelter & Bombach, 2010). Sophisticated electrolyte flow technologies allow operation at current density as high as 420 A/m^2 (the highest in the industry in 2010) without compromising current efficiency or cathode quality (Wenzl et al., 2010).

As ore bodies become more complex, there is increasing need to treat anodes with higher levels and greater complexity of impurities. Increasingly sophisticated chemistry and methodologies continue to evolve to ensure continued high quality cathode (Fernandez & Begazo, 2010; Ríos, Delgado, Ramírez, & Martín, 2006, Ríos, Ramírez, & Arbizu, 2010; Santos Moraes, 2007).

Emerging technologies to deal with impurity recovery from electrolyte and treatment of anode slimes for recovery and creation of salable valuable byproducts are also receiving more attention (Agrawal et al., 2009; Kim & Wang, 2010; Komori, Ito, Okada, & Iwahori, 2010; Stantke & Leuprecht, 2010).

14.10. SUMMARY

This chapter has shown that electrolytic refining is the principal method of mass-producing high-purity copper. The other is electrowinning (Chapter 17). Copper from electrorefining, after melting and casting, contains less than 20 ppm impurities, plus oxygen which is controlled at 0.018–0.025%.

Electrorefining entails electrochemically dissolving copper from impure copper anodes into an electrolyte containing $CuSO_4$ and H_2SO_4 and then electrochemically depositing pure copper from the electrolyte onto stainless steel or copper cathodes. The process is continuous.

Insoluble impurities in the anode adhere to the anode or fall to the bottom of the refining cell. They are removed and sent to a Cu and byproduct metal recovery plant. Soluble impurities leave the cell in continuously flowing electrolyte. They are removed from an electrolyte bleed stream.

The critical objective of electrorefining is to produce high-purity cathode copper. It is attained using:

(a) Precisely spaced, flat, vertical anodes, and cathodes;
(b) A constant, gently flowing supply of warm, high Cu^{2+} electrolyte across all cathode faces;
(c) Provision of a constant, controlled supply of leveling and grain-refining agents.

Important modern trends are to treat anodes with more complex impurity suites at higher current density, while maximizing cathode purity, productivity, and operator safety and minimizing power consumption, waste production, and environmental damage. The adoption of pre-cast polymer concrete cells and stainless steel cathodes in both new and retrofitted tankhouses continues. Automation, robotics, and advanced process control of all aspects of electrorefining are becoming standard in the industry.

REFERENCES

Agrawal, A., Bagchi, D., Kumari, S., Kumar, V., & Pandey, B. D. (2007). Recovery of nickel powder from copper bleed electrolyte of an Indian copper smelter by electrolysis. *Powder Technology, 177*, 133–139.

Agrawal, A., Bagchi, D., Kumari, S., & Pandey, B. D. (2009). An overview of process options and behavioral aspects of the copper values recovered from copper bleed stream of a copper smelter developed at the National Metallurgical Laboratory. *Mineral Processing and Extractive Metallurgy Review, 30*, 136–162.

Agrawal, A., Kumari, S., Bagchi, D., Kumar, V., & Pandey, B. D. (2007). Recovery of copper powder from copper bleed electrolyte of an Indian copper smelter by electrolysis. *Minerals Engineering, 20*, 95–97.

Agrawal, A., Manoj, M. K., Kumari, S., Bagchi, D., Kumar, V., & Pandey, B. D. (2008). Extractive separation of copper and nickel from copper bleed stream by solvent extraction route. *Minerals Engineering, 21*, 1126–1130.

Aslin, N. J., Eriksson, O., Heferen, G. J., & Sue Yek, G. (2010). Developments in cathode stripping machines — an integrated approach for improved efficiency. In *Copper 2010, Vol. 4: Electrowinning and -refining* (pp. 1253–1268). Clausthal-Zellerfeld, Germany: GDMB.

Atenas, A. C., & Muñoz, P. A. (2007a). Electrorefining high level arsenic cast anode. In G. E. Houlachi, J. D. Edwards & T. G. Robinson (Eds.), *Copper 2007, Vol. V: Copper electrorefining and electrowinning* (pp. 3–12). Montreal: CIM.

Atenas, A. C., & Muñoz, P. A. (2007b). Copper refining electrolysis at high current densities with conventional technology. In G. E. Houlachi, J. D. Edwards & T. G. Robinson (Eds.), *Copper 2007, Vol. V: Copper electrorefining and electrowinning* (pp. 327–334). Montreal: CIM.

Baranek, W., Chmielarz, A., & Śmieszek, Z. (2007). Development of electrorefining processes in the Polish copper industry. In G. E. Houlachi, J. D. Edwards & T. G. Robinson (Eds.), *Copper 2007, Vol. V: Copper electrorefining and electrowinning* (pp. 335–344). Montreal: CIM.

Barrios, P., Alonso, A., & Meyer, U. (1999). Reduction of silver losses during the refining of copper cathodes. In J. E. Dutrizac, J. Ji & V. Ramachandran (Eds.), *Copper 99–Cobre 99, Vol. III: Electrorefining and electrowinning of copper* (pp. 237–247). Warrendale, PA, USA: TMS.

Beauchemin, S., Chen, T. T., & Dutrizac, J. E. (2008). Behaviour of antimony and bismuth in copper electrorefining circuits. *Canadian Metallurgical Quarterly, 47*(1), 9–26.

Bravo, J. L. R. (1995). Studies for changes in the electrolyte purification plant at Caraiba Metais, Brazil. In W. C. Cooper, D. B. Dreisinger, J. E. Dutrizac, H. Hein & G. Ugarte (Eds.), *Copper 95–Cobre 95, Vol. III: Electrorefining and hydrometallurgy of copper* (pp. 315–324). Montreal: CIM.

Campin, S. C. (2000). *Characterization, analysis and diagnostic dissolution studies of slimes produced during copper electrorefining*. Tucson, AZ, USA: M.S. thesis, University of Arizona.

Chen, T. T., & Dutrizac, J. E. (1991). A mineralogical study of anode passivation in copper electrorefining. In W. C. Cooper, D. J. Kemp, G. E. Lagos & K. G. Tan (Eds.), *Copper 91–Cobre 91, Vol. III: Hydrometallurgy and electrometallurgy* (pp. 369–389). New York: Pergamon Press.

Chen, T. T., & Dutrizac, J. E. (2000). A mineralogical overview of the behaviour of nickel during copper electrorefining. *Metallurgical Transactions B, 21B*, 229–238.

Chen, T. T., & Dutrizac, J. E. (2003). The behaviour of tellurium during the decopperizing of copper refinery anode slimes. In J. E. Dutrizac & C. G. Clements (Eds.), *Copper 2003, Vol. V: Copper electrorefining and electrowinning* (pp. 287–307). Montreal: CIM.

Chen, T. T., & Dutrizac, J. E. (2004). Gold in the electrorefining of copper and the decopperizing of copper anode slimes. *JOM, 56*(8), 48–52.

Chen, T. T., & Dutrizac, J. E. (2007). Mineralogical characterization of a conventional copper refinery anode slimes treatment circuit. In G. E. Houlachi, J. D. Edwards & T. G. Robinson (Eds.), *Copper 2007, Vol. V: Copper electrorefining and electrowinning* (pp. 159–172). Montreal: CIM.

Coffin, M. R., & Leggett, A. R. (2007). Increasing anode thickness to optimise copper production at the Kidd Metallurgical site. In G. E. Houlachi, J. D. Edwards & T. G. Robinson (Eds.), *Copper 2007, Vol. V: Copper electrorefining and electrowinning* (pp. 159–172). Montreal: CIM.

Conard, B. R., Rogers, B., Brisebois, R., & Smith, C. (1990). Inco copper refinery addition agent monitoring using cyclic voltammetry. In P. L. Claessens & G. B. Harris (Eds.), *Electrometallurgical plant practice* (pp. 195–209). Warrendale, PA, USA: TMS.

De Maere, C., & Winand, R. (1995). Study of the influence of additives in copper electrorefining, simulating industrial conditions. In W. C. Cooper, D. B. Dreisinger, J. E. Dutrizac, H. Hein & G. Ugarte (Eds.), *Copper 95–Cobre 95, Vol. III: Electrorefining and hydrometallurgy of copper* (pp. 267–286). Montreal: CIM.

Deni, R. M. (1994). *The effect of addition agents on cathodic overpotential and cathode quality in copper electrorefining*. Canada: M.S. thesis, Laurentian University.

Djurov, I. (2010). The new ISA 2000 refinery of Aurubis in Pirdop. In *Proceedings copper 2010, Vol. 4, Electrowinning and -refining* (pp. 1307–1325). Clausthal-Zellerfeld, Germany: GDMB.

Doucet, M., & Stafiej, J. (2007). Processing high nickel slimes at CCR refinery. In G. E. Houlachi, J. D. Edwards & T. G. Robinson (Eds.), *Copper 2007, Vol. V: Copper electrorefining and electrowinning* (pp. 173–182). Montreal: CIM.

Eastwood, K. L., & Whebell, G. W. (2007). Developments in permanent stainless steel cathodes within the copper industry. In G. E. Houlachi, J. D. Edwards & T. G. Robinson (Eds.), *Copper 2007, Vol. V: Copper electrorefining and electrowinning* (pp. 35–46). Montreal: CIM.

Fernandez, F., & Begazo, F. (2010). The effect of SPCC smelter modernization on anode quality and electrorefining process. In *Copper 2010, Vol. 4: Electrowinning and -refining* (pp. 1327−1343). Clausthal-Zellerfeld, Germany: GDMB.

Filzweiser, A., Filweizer, I., & Stibich, R. (2008). Tankhouse optimisation by METTOP GmbH-METTOP-BRX-Technology. *Erzmetall, 61*, 99−103.

Hait, J., Jana, R. K., & Sanyai, S. K. (2009). Processing of copper electrorefining anode slimes: a review. *Mineral Processing and Extractive Metallurgy (IMM Transactions Section C), C118*, 240−252.

Hashimoto, M., Narita, M., & Shimokawa, K. (2010). Recent improvements at Tamano refinery. In *Copper 2010, Vol. 4: Electrowinning and -refining* (pp. 1345−1354). Clausthal-Zellerfeld, Germany: GDMB.

Hiskey, J. B., & Moats, M. S. (2010). Periodic oscillations during electrolytic dissolution of copper anodes. In *Copper 2010, Vol. 4: Electrowinning and -refining* (pp. 1367−1377). Clausthal-Zellerfeld, Germany: GDMB.

Hu, E. W., Roser, W. R., & Rizzo, F. E. (1973). The role of proteins in electrocrystallization during commercial electrorefining. In D. J. I. Evans & R. S. Shoemaker (Eds.), *International symposium on hydrometallurgy* (pp. 155−170). New York, USA: AIME.

Izatt, S. R., Izatt, N. E., Bruening, R. L., & Dale, J. B. (2009). Bismuth removal in copper electrolytic refining operations using molecular recognition technology (MRT). In *ALTA copper 2009*. Melbourne, Australia: ALTA Metallurgical Services.

Jarjoura, G., & Kipouros, G. J. (2005). Cyclic voltammetry studies of the effect of nickel on copper anode passivation in a copper sulfate solution. *Canadian Metallurgical Quarterly, 44*, 469−482.

Kim, D., & Wang, S. (2010). Sustainable developments in copper anode slimes−wet chlorination processing. In *Copper 2010, Vol. 4: Electrowinning and -refining* (pp. 1393−1402). Clausthal-Zellerfeld, Germany: GDMB.

Knuutila, K., Forsen, O., & Pehkonen, A. (1987). The effect of organic additives on the electrocrystallization of copper. In J. E. Hoffmann, R. G. Bautista, V. A. Ettel, V. Kudryk & R. J. Wesely (Eds.), *The electrorefining and winning of copper* (pp. 129−143). Warrendale, PA, USA: TMS.

Komori, K., Ito, S., Okada, S., & Iwahori, S. (2010). Hydrometallurgical process of precious metals in Naoshima smelter & refinery. In *Copper 2010, Vol. 4: Electrowinning and -refining* (pp. 1403−1411). Clausthal-Zellerfeld, Germany: GDMB.

Krusmark, T. F., Young, S. K., & Faro, J. L. (1995). Impact of anode chemistry on high current density operation at Magma Copper's electrolytic refinery. In W. C. Cooper, D. B. Dreisinger, J. E. Dutrizac, H. Hein & G. C. I. M. Ugarte (Eds.), *Copper 95−Cobre 95, Vol. III: Electrorefining and hydrometallurgy of copper* (pp. 189−206), Montreal.

Kumari, S., Agrawal, A., Bagchi, D., Kumar, V., & Pandey, B. D. (2006). Synthesis of copper metal/salts from copper bleed solution of a copper plant. *Mineral Processing and Extractive Metallurgy Review, 27*, 159−175.

Larinkari, M., & Rantala, A. (2010). Improved efficiency, maintenance and safety with integrated tankhouse information management. In *Copper 2010, Vol. 4: Electrowinning and -refining* (pp. 1423−1435). Clausthal-Zellerfeld, Germany: GDMB.

Larouche, P. (2001). *Minor elements in copper smelting and refining*. Montreal: M. Eng. thesis, McGill University. http://digitol.librry.mcgill.ca:8881/R/?func=dbin-jump-full&object_id=33978.

Maruyama, T., Furuta, M., Oida, M., Shimokawa, K., & Narita, M. (2007). Expansion projects at Tamano refinery. In G. E. Houlachi, J. D. Edwards & T. G. Robinson (Eds.), *Copper 2007, Vol. V: Copper electrorefining and electrowinning* (pp. 291−300). Montreal: CIM.

Matsuda, M., Goda, T., Takehayashi, K., & Maeda, Y. (2007). Recent improvements at Hitachi refinery. In G. E. Houlachi, J. D. Edwards & T. G. Robinson (Eds.), *Copper 2007, Vol. V: Copper electrorefining and electrowinning* (pp. 281−290). Montreal: CIM.

Mipac. (2004). *Tankhouse cell voltage monitoring system: Typical system description*. Mipac DRN: 011−A70 Rev 1. www.mipac.com.au/_./%5Bwp%5D_Wireless_Cell_Voltage_Monitoring_-_english.

Moats, M., Davenport, W., Robinson, T., & Karcas, G. (2007). Electrolytic copper refining 2007 world tankhouse survey. In G. E. Houlachi, J. D. Edwards & T. G. Robinson (Eds.), *Copper 2007, Vol. V: Copper electrorefining and electrowinning* (pp. 195−242). Montreal: CIM.

Moats, M. S., & Hiskey, J. B. (2000). The role of electrolyte additives on passivation behaviour during copper electrorefining. *Canadian Metallurgical Quarterly, 39*, 297−306.

Moats, M. S., & Hiskey, J. B. (2006). The effect of anode composition on passivation of commercial copper electrorefining anodes. In K. Kongoli & R. G. Reddy (Eds.), *Sohn international symposium, Vol. 6: New, improved and existing technologies: Aqueous processing and electrochemical technologies* (pp. 507–518). Warrendale, PA, USA: TMS.

Moats, M. S., & Hiskey, J. B. (2007). The effect of electrolyte composition on passivation of commercial electrorefining anodes. In G. E. Houlachi, J. D. Edwards & T. G. Robinson (Eds.), *Copper 2007, Vol. V: Copper electrorefining and electrowinning* (pp. 47–58). Montreal: CIM.

Moats, M. S., & Hiskey, J. B. (2010). How anodes passivate in copper electrorefining. In *Copper 2010, Vol. 4: Electrowinning and -refining* (pp. 1463–1482). Clausthal-Zellerfeld, Germany: GDMB.

Möller, C. A., Friedrich, B., & Bayanmunkh, M. (2010). Effect of As, Sb, Bi and oxygen in copper anodes during electrorefining. In *Copper 2010, Vol. 4: Electrowinning and -refining* (pp. 1495–1506). Clausthal-Zellerfeld, Germany: GDMB.

Mubarok, A., Antrekowitsch, H., & Mori, G. (2007). Passivation behaviour of copper anodes with various chemical compositions. In P. Aryalebechi (Ed.), *Materials processing fundamentals* (pp. 225–234). Warrendale, PA, USA: TMS.

Mubarok, A., Antrekowitsch, H., Mori, G., Lossin, A., & Leuprecht, G. (2007). Problems in the electrolysis of copper anodes with high contents of nickel, antimony, tin and lead. In G. E. Houlachi, J. D. Edwards & T. G. Robinson (Eds.), *Copper 2007, Vol. V: Copper electrorefining and electrowinning* (pp. 59–76). Montreal: CIM.

Nagai, K., Hashikawa, K., & Yamaguchi, Y. (2010). Recent improvements and expansion at the Toyo Copper Refinery. In *Copper 2010, Vol. 4: Electrowinning and -refining* (pp. 1521–1529). Clausthal-Zellerfeld, Germany: GDMB.

Nakano, H., Oue, S., Fukushima, H., & Kobayashi, S. (2010). Synergistic effects of thiourea, polymer additives and chloride ions on copper electrorefining. In *Copper 2010, Vol. 4: Electrowinning and -refining* (pp. 1531–1544). Clausthal-Zellerfeld, Germany: GDMB.

Nicol, M. J. (2006). *Electrowinning and electrorefining of metals.* Short Course. Perth, Australia: Murdoch University.

Nikus, M., Korpi, M., & Rantala, J. (2010). Optimization of a tankhouse harvesting plan. In *Copper 2010, Vol. 4: Electrowinning and -refining* (pp. 1569–1584). Clausthal-Zellerfeld, Germany: GDMB.

Noguchi, F., Iida, N., Nakamura, T., & Ueda, Y. (1992). Behavior of anode impurities in copper electrorefining. *Metallurgical Review of the Mining and Metallurgical Institute of Japan, 8*, 83–98.

Phan, C., Whebell, G. W., & Oellermann, M. (2007). Operations control in strata Technology tank houses. In G. E. Houlachi, J. D. Edwards & T. G. Robinson (Eds.), *Copper 2007, Vol. V: Copper electrorefining and electrowinning* (pp. 267–280). Montreal: CIM.

Price, D. C., & Davenport, W. G. (1981). Physico-chemical properties of copper electrorefining and electrowinning electrolytes. *Metallurgical Transactions B, 12B*, 639–643.

Rafieipour, H., Keshavarz Alamdari, E., Haghshenas Fatmehsari, D., Darvishi, D., Atashdehgan, R., Daneshpajouh, Sh, et al. (2007). Recovery of sulfuric acid from Sarcheshmeh copper electrorefining tank house bleed off. In *ALTA Copper 2007*. Melbourne, Australia: ALTA Metallurgical Services.

Ramírez, R., Ríos, G., & Martín, J. (2007). Anode slime leaching and tellurium removal at Atlantic Copper refinery. In G. E. Houlachi, J. D. Edwards & T. G. Robinson (Eds.), *Copper 2007, Vol. V: Copper electrorefining and electrowinning* (pp. 125–138). Montreal: CIM.

Rantala, A., Larinkari, M., & Menese, T. (2009). Advanced wireless monitoring and integrated information management for improving electrorefinery performance. In *Hydrocopper 2009* (pp. 102–113). Santiago, Chile: Gecamin.

Rantala, A., You, E., & George, D. B. (2006). Wireless temperature and voltage monitoring of electrolytic cells. In *ALTA Copper 2006*. Melbourne, Australia: ALTA Metallurgical Services.

Ríos, G., Delgado, E., Ramírez, R., & Martín, A. (2006). Waste management and impurities control at Atlantic Copper smelter and refinery. In F. Kongoli & R. G. Reddy (Eds.), *Sohn international symposium, Vol. 9: Legal, management and environmental issues* (pp. 359–371). Warrendale, PA, USA: TMS.

Ríos, G., Ramírez, R., & Arbizu, C. (2010). Management of electrolyte impurities at the Atlantic copper refinery. In *Copper 2010, Vol. 4: Electrowinning and -refining* (pp. 1599–1616). Clausthal-Zellerfeld, Germany: GDMB.

Robinson, T., Siegmund, A., Davenport, W., Moats, M., & Karcas, G. (2010). Electrolytic copper refining 2010 world tankhouse survey. *Presented at Copper 2010*. Germany: Hamburg.

Rodrigues Bravo, J. L. (2007). Studies on bismuth content variation in Caraiba Metais' electrolytic copper. In G. E. Houlachi, J. D. Edwards & T. G. Robinson (Eds.), *Copper 2007, Vol. V: Copper electrorefining and electrowinning* (pp. 149–158). Montreal: CIM.

Roman, E. A., Salas, J. C., Guzman, J. E., & Muto, S. (1999). Antimony removal by ion exchange in a Chilean tankhouse at the pilot plant scale. In J. E. Dutrizac, J. Ji & V. Ramachandran (Eds.), *Copper 99–Cobre 99, Vol. III: Electrorefining and electrowinning of copper* (pp. 225–236). Warrendale, PA, USA: TMS.

Rondas, F., Scoyer, J., & Geenen, C. (1995). Solvent extraction of arsenic with TBP – the influence of high iron concentration on the extraction behaviour of arsenic. In W. C. Cooper, D. B. Dreisinger, J. E. Dutrizac, H. Hein & G. Ugarte (Eds.), *Copper 95–Cobre 95, Vol. III: Electrorefining and hydrometallurgy of copper* (pp. 325–335). Montreal: CIM.

Saban, M. B., Scott, J. D., & Cassidy, R. M. (1992). Collagen proteins in electrorefining: rate constants for glue hydrolysis and effects of molar mass on glue activity. *Metallurgical Transactions B, 23B,* 125–133.

Santos Moraes, I. M. (2007). Upsets on antimony contents in electrolytic copper produced at Caraiba Metais. In G. E. Houlachi, J. D. Edwards & T. G. Robinson (Eds.), *Copper 2007, Vol. V: Copper electrorefining and electrowinning* (pp. 139–148). Montreal: CIM.

Sheedy, M., Pajunen, P., & Wesstrom, B. (2007). Control of copper electrolyte impurities – overview of the short bed ion exchange technique and Phelps Dodge El Paso case study. In G. E. Houlachi, J. D. Edwards & T. G. Robinson (Eds.), *Copper 2007, Vol. V: Copper electrorefining and electrowinning* (pp. 345–358). Montreal: CIM.

Stantke, P. (2002). Using CollaMat to measure glue in copper electrolyte. *JOM, 54*(4), 19–22.

Stantke, P., & Leuprecht, G. (2010). Process change for the treatment of spent electrolyte at Aurubis Hamburg. In *Copper 2010, Vol. 4: Electrowinning and -refining* (pp. 1663–1674). Clausthal-Zellerfeld, Germany: GDMB.

Stelter, M., & Bombach, H. (2010). Copper electrorefining at high current densities. In *Copper 2010, Vol. 4: Electrowinning and -refining* (pp. 1675–1686). Clausthal-Zellerfeld, Germany: GDMB.

Stevens, G. W., & Gottliebsen, K. (2000). The recovery of sulfuric acid from copper tankhouse electrolyte bleeds – a case study on the development of solvent extraction processes. In *Minprex 2000* (pp. 57–64). Victoria, Australia: Australasian Institute of Mining and Metallurgy.

Tantavichet, N., Damronglerd, S., & Chailapakul, O. (2009). Influence of the interaction between chloride and thiourea on copper electrodeposition. *Electrochimica Acta, 55,* 240–249.

Wachendörfer, F. (2009). High-current DC-switchgear devices for copper electrolysis. In *HydroCopper 2009* (pp. 410–416). Santiago, Chile: Gecamin.

Wang, C.-T., & O'Keefe, T. J. (1984). The influence of additives and their interactions on copper electrorefining. In P. E. Richardson, S. Srinivasan & R. Woods (Eds.), *International symposium on electrochemistry in mineral and metal processing, Vol. 84-10* (pp. 655–670). Pennington, NJ, USA: The Electrochemical Society.

Wang, S. (2004). Impurity control and removal in copper tankhouse operations. *JOM, 56*(7), 34–37.

Wenzl, C., Antrewkowitsch, H., Filzweiser, I., & Pesl, J. (2007). Anode casting – chemical anode quality. In G. E. Houlachi, J. D. Edwards & T. G. Robinson (Eds.), *Copper 2007, Vol. V: Copper electrorefining and electrowinning* (pp. 91–110). Montreal: CIM.

Wenzl, C., Filzwieser, I., Filzwieser, A., & Anzinger, A. (2010). Newest developments using the METTOP-BRX technology. In *Copper 2010, Vol. 4: Electrowinning and -refining* (pp. 1713–1722). Clausthal-Zellerfeld, Germany: GDMB.

SUGGESTED READING

Copper (2010). *Vol. 4: Electrowinning and -refining.* Clausthal-Zellerfeld, Germany: GDMB.

Maio, T., Bird, S., Kim, D., Rudloff, F., Penumuri, J., & Wang, S. (2007). Process improvements at Kennecott Utah copper refinery. In G. E. Houlachi, J. D. Edwards & T. G. Robinson (Eds.), *Copper 2007, Vol. V: Copper electrorefining and electrowinning* (pp. 359–371). Montreal: CIM.

Moats, M., Davenport, W., Robinson, T., & Karcas, G. (2007). Electrolytic copper refining 2007 world tankhouse survey. In G. E. Houlachi, J. D. Edwards & T. G. Robinson (Eds.), *Copper 2007. Vol. V: Copper electrorefining and electrowinning* (pp. 195−242). Montreal: CIM.

Moats, M. S., & Hiskey, J. B. (2010). How anodes passivate in copper electrorefining. In *Copper 2010, Vol. 4: Electrowinning and -refining* (pp. 1463−1482). Clausthal-Zellerfeld, Germany: GDMB.

Orzecki, S., & Olewsinski, L. (2007). 35 years of the KGHM copper refinery plant: Głogów I. In G. E. Houlachi, J. D. Edwards & T. G. Robinson (Eds.), *Copper 2007, Vol. V: Copper electrorefining and electrowinning* (pp. 253−266). Montreal: CIM.

Pienimäki, K., & Virtanen, H. (2007). Modernisation of tankhouse technology in Boliden Harjavalta Oy. In G. E. Houlachi, J. D. Edwards & T. G. Robinson (Eds.), *Copper 2007, Vol. V: Copper electrorefining and electrowinning* (pp. 301−314). Montreal: CIM.

Standen, J. T. (2007). Trends in copper refining technology. In G. E. Houlachi, J. D. Edwards & T. G. Robinson (Eds.), *Copper 2007, Vol. V: Copper electrorefining and electrowinning* (pp. 425−438). Montreal: CIM.

Chapter 15

Hydrometallurgical Copper Extraction: Introduction and Leaching

Previous chapters describe the processes of concentration, pyrometallurgy, and electro-refining that convert copper sulfide ores into high purity electrorefined copper. These processes account for about 80% of primary copper production.

The remaining 20% of primary copper production comes from *hydrometallurgical* processing of mainly copper oxide and chalcocite ores. This copper is recovered by leaching (this chapter), followed by solvent extraction (Chapter 16) and electrowinning (Chapter 17). The final product is electrowon cathode copper, which is equal to or often greater in purity than electrorefined copper.

In 2010, about 4.5 million tonnes per year of metallic copper were hydrometallurgically produced. This production is steadily increasing as more mines begin to leach all or some of their ore and treat lower-grade materials, some of which were previously considered to be waste.

15.1. COPPER RECOVERY BY HYDROMETALLURGICAL FLOWSHEETS

Modern hydrometallurgical flowsheets all comprise three essential steps for the recovery of copper: leaching, followed by solvent extraction and electrowinning.

Leaching involves dissolving Cu^{2+} (or Cu^+) from copper-containing minerals into an aqueous H_2SO_4 solution, known as the *lixiviant*, to produce a *pregnant leach solution* (PLS). In addition to copper, the PLS will also contain other impurity species, such as Fe, Al, Co, Mn, Zn, Mg, Ca, *etc.*, that may be present in the ore and are leached with the copper. The leach *residue* (solids remaining after leaching) contains *gangue* or waste minerals, such as alumina, silica, and insoluble iron oxides/hydroxides/sulfates. The gangue is disposed of in tailings dams or dumps. The PLS is fed to the solvent-extraction circuit.

Solvent extraction (SX) treats the impure PLS to purify and upgrade the solution to produce an electrolyte suitable for electrowinning of copper. In the *extraction* step, copper is selectively loaded into an organic solvent which contains an extractant that reacts selectively with copper over other metal cations present in the PLS. The *barren* aqueous raffinate leaving the extraction circuit is higher in acid and is returned to the leach circuit as the lixiviant. In the *strip* step, copper is stripped from the loaded organic solvent into the *advance electrolyte* from which copper is electrowon. The stripped organic solvent is recycled to the extraction step.

In *electrowinning* (EW), Cu^{2+} in the purified advance electrolyte from SX is reduced to copper metal at the cathode by the application of a DC electrical current. Sulfuric acid,

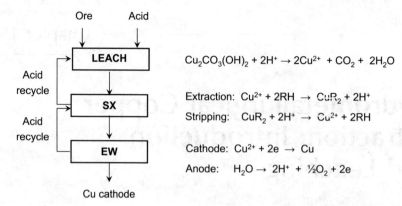

FIGURE 15.1 Simplified hydrometallurgical flowsheet, showing the acid balance between leaching, solvent extraction, and electrowinning (Sole, 2008). The leaching reaction is shown for malachite, but similar reactions can be written for other minerals. RH is the organic extractant used in SX. Solvent extraction and electro-winning are described in Chapters 16 and 17, respectively.

produced at the anode of the electrowinning cell, is returned to the SX circuit in the copper-depleted *spent electrolyte* to strip more copper from the loaded organic solvent.

The integration of the three hydrometallurgical processes is illustrated in Fig. 15.1 (Sole, 2008). Figure 15.2 shows the solution compositions for a typical heap-leach flowsheet.

15.2. CHEMISTRY OF THE LEACHING OF COPPER MINERALS

Leaching involves solubilizing the valuable components of an ore in an aqueous solution. In the case of copper minerals, sulfuric acid is almost always used as the leaching medium. Oxide copper minerals dissolve readily in acid and some sulfides, such as chalcocite and covellite, dissolve under atmospheric conditions. More refractory sulfides (chalcopyrite and bornite) require high temperatures and pressures to break down the crystal lattices to release copper from the minerals into solution.

15.2.1. Leaching of Copper Oxide Minerals

Figure 15.3 shows the Pourbaix (potential-pH) diagram for the Cu−O−S system. Copper can be dissolved as Cu^{2+} under mildly acidic conditions (pH < 5). Examples of some copper minerals that leach in this manner are:

Tenorite: $\quad CuO + H_2SO_4 \rightarrow Cu^{2+} + SO_4^{2-} + H_2O \quad$ (15.1)

Azurite: $\quad 2CuCO_3 \cdot Cu(OH)_2 + 3H_2SO_4 \rightarrow 3Cu^{2+} + 3SO_4^{2-} + 2CO_2 + 4H_2O \quad$ (15.2)

Malachite: $\quad CuCO_3 \cdot Cu(OH)_2 + 2H_2SO_4 \rightarrow 2Cu^{2+} + 2SO_4^{2-} + CO_2 + 3H_2O \quad$ (15.3)

Cuprite: $\quad Cu_2O + H_2SO_4 \rightarrow Cu^{2+} + SO_4^{2-} + Cu + H_2O \quad$ (15.4)

Chrysocolla: $\quad CuO \cdot SiO_2 + H_2SO_4 \rightarrow Cu^{2+} + SO_4^{2-} + SiO_2 \cdot nH_2O + (3-n)H_2O \quad$ (15.5)

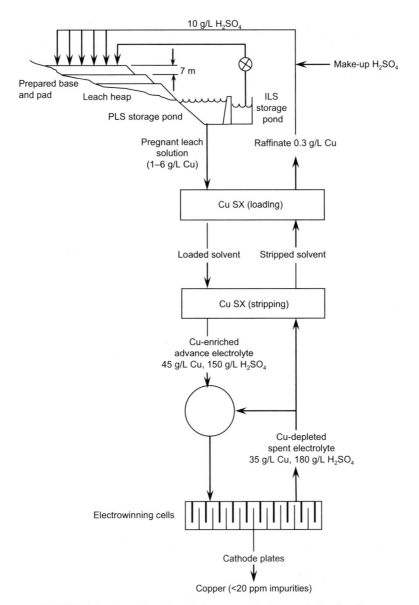

FIGURE 15.2 Copper heap leach/solvent extraction/electrowinning flowsheet.

15.2.2. Leaching of Copper Sulfide Minerals

Copper sulfide minerals require the use of acid and an oxidizing agent to break the mineral lattice and release Cu^{2+} into solution (Sherrit, Pavlides, & Weekes, 2005).

Elemental copper found in nature can be leached by either oxygen or ferric ion:

Metallic copper: $\quad Cu + 0.5O_2 + H_2SO_4 \rightarrow Cu^{2+} + SO_4^{2-} + H_2O \quad$ (15.6a)

$$Cu + 2Fe^{3+} \rightarrow Cu^{2+} + 2Fe^{2+} \quad (15.6b)$$

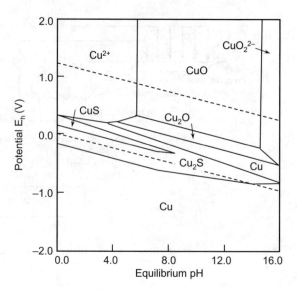

FIGURE 15.3 Pourbaix diagram of the Cu–S–O–H$_2$O system at 25 °C. [Cu] = [S] = 10^{-4} M (House, 1987). Minerals such as CuO can be leached by lowering the pH (adding acid). Sulfides and native copper require an oxidizing agent and acid.

All copper sulfides require the presence of Fe^{3+} and O$_2$ as oxidizing agents for leaching to occur. The copper sulfide is oxidized by Fe^{3+}. The resulting Fe^{2+} is reoxidized to Fe^{3+} by O$_2$. The Fe(II)/Fe(III) redox couple acts in a catalytic manner in these reactions (shown for elevated-temperature reaction conditions where the sulfur-containing product is sulfate, rather than the elemental sulfur that forms under ambient conditions):

Chalcocite: (i) $Cu_2S + Fe_2(SO_4)_3 \rightarrow Cu^{2+} + SO_4^{2-} + CuS + 2FeSO_4$ (15.7a)

(ii) $4FeSO_4 + O_2 + 2H_2SO_4 \rightarrow 2Fe_2(SO_4)_3 + 2H_2O$ (15.7b)

(overall) $Cu_2S + 0.5O_2 + H_2SO_4 \rightarrow Cu^{2+} + SO_4^{2-} + CuS + H_2O$ (15.7c)

Covellite: (i) $CuS + Fe_2(SO_4)_3 \rightarrow Cu^{2+} + SO_4^{2-} + 2FeSO_4 + S$ (15.8a)

(ii) $4FeSO_4 + O_2 + 2H_2SO_4 \rightarrow 2Fe_2(SO_4)_3 + 2H_2O$ (15.8b)

(overall) $CuS + 0.5O_2 + H_2SO_4 \rightarrow Cu^{2+} + SO_4^{2-} + S + H_2O$ (15.8c)

Under some conditions (especially biological leaching), elemental sulfur that forms as a reaction product can be converted to sulfuric acid:

Sulfur: $2S + 3O_2 + 2H_2O \rightarrow 2H_2SO_4$ (15.9)

A Pourbaix diagram for the Cu–Fe–O–S system is shown in Fig. 15.4. Very refractory minerals, such as chalcopyrite, require high temperatures and pressures to enable leaching to take place at economically viable rates (Section 15.7). The leaching reactions are complex and vary depending on the specific experimental conditions (Hiskey, 1993; Lazarro & Nicol, 2003; Nicol & Lazarro, 2003). One typical series of reactions that occur under oxidizing conditions and at high temperatures is shown below:

Chalcopyrite: (i) $2CuFeS_2 + 4Fe_2(SO_4)_3 \rightarrow 2Cu^{2+} + 2SO_4^{2-} + 10FeSO_4 + 4S$

(15.10a)

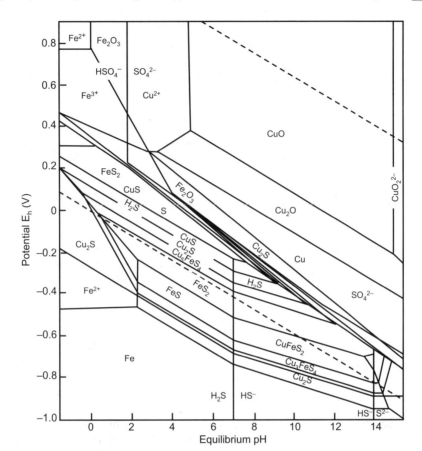

FIGURE 15.4 Pourbaix diagram of the Cu–Fe–S–O–H$_2$O system at 25 °C. [Cu] = 0.01 M; [Fe] = [S] = 0.1 M. (Peters, 1976). Most minerals require the application of both an acidic environment and an oxidizing agent to leach copper as Cu^{2+}.

(ii) $\quad 4FeSO_4 + O_2 + 2H_2SO_4 \rightarrow 2Fe_2(SO_4)_3 + 2H_2O \quad$ (15.10b)

(iii) $\quad 2S + 3O_2 + 2H_2O \rightarrow 2H_2SO_4 \quad$ (15.10c)

(overall) $\quad 4CuFeS_2 + 17O_2 + 4H_2O \rightarrow 4Cu^{2+} + 4SO_4^{2-} + 2Fe_2O_3 + 4H_2SO_4$

(15.10d)

15.3. LEACHING METHODS

There are six main methods employed for the leaching of copper minerals:

- *In-situ* leaching;
- Dump leaching;
- Heap leaching;
- Vat leaching;
- Agitation leaching;
- Pressure oxidation leaching.

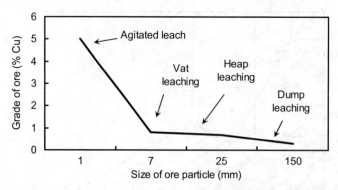

FIGURE 15.5 Relationship between copper grade and particle size of ore for different methods of atmospheric leaching (*adapted from Scheffel, 2002*).

Heap leaching accounts for the majority of copper produced hydrometallurgically. The choice of leaching method depends on the mineralogy of the copper-bearing ore, the copper grade of the ore, and the particle size (Fig. 15.5; Scheffel, 2002). The topography of the mine site, geographical location of the ore body, climatic conditions, and prevailing economic conditions may also influence this decision.

Dump leaching, *heap leaching*, and *vat leaching* all fall into the category of *percolation leaching*. In dump and heap leaching (Section 15.4), the leach solution trickles by gravity through a permeable ore mass (Murr, 1980). In vat leaching, the leach solution is pumped up through the ore and is operated in flooded fashion. This keeps fine particles in suspension and prevents them from compacting and plugging at the bottom of the vat (Section 15.5).

Permeability is achieved by blasting and crushing the ore. Fine material may be agglomerated to improve permeability. A sulfuric acid-containing solution is used for irrigation — usually raffinate or a recycled leach solution, usually called *intermediate leach solution* (ILS), followed by raffinate that is returned from the downstream SX circuit (see Chapter 16). The solution is distributed evenly over the surface of the ore mass and collected at the bottom of the heap (or the top of the vat). Copper dissolves into solution. Oxygen is present in the leach solution, but may be depleted by the leaching reactions. The PLS exiting at the base of the heap or the top of the vat is collected into large PLS storage ponds or tanks before being treated for copper recovery.

Agitation leaching is used for minerals that leach easily, such as oxides and carbonates. Leaching is carried out in stirred tanks (Section 15.6). It is a capital-intensive technique, but copper recovery can approach 100%.

Pressure oxidation (POx) leaching is a relatively new commercial technology, which uses high temperatures and pressures to promote the leaching of refractory minerals, such as chalcopyrite (Section 15.7). When the chalcopyrite copper grade is too low or too variable for a pyrometallurgical process, pressure leaching can be used to treat the concentrate product from flotation (in contrast to the other leaching methods which process whole ore).

Each of these leaching technologies is discussed in more detail in the following sections.

A less-common technology is *in-situ leaching*. This occurs when leach solution is introduced to a rock mass that is still 'in place' as it occurs in nature and has not been mined out. Because of the acidity of the leach liquors, there are environmental concerns

TABLE 15.1 Copper Minerals Normally Found in Leach Heaps

Type	Common minerals	Mineral formula
Secondary minerals		
Carbonates	Azurite	$2CuCO_3 \cdot Cu(OH)_2$
	Malachite	$CuCO_3 \cdot Cu(OH)_2$
Hydroxy-chlorides	Atacamite	$Cu_2Cl(OH)_3$
Hydroxy-silicates	Chrysocolla	$CuO \cdot SiO_2 \cdot 2H_2O$
Native copper	Metal	Cu^0
Oxides	Cuprite	Cu_2O
	Tenorite	CuO
Sulfates	Antlerite	$CuSO_4 \cdot 2Cu(OH)_2$
	Brochantite	$CuSO_4 \cdot 3Cu(OH)_2$
Supergene sulfides	Chalcocite	Cu_2S
	Covellite	CuS
	Bornite	Cu_5FeS_4
Primary sulfide minerals		
(Hypogene sulfides)	Chalcopyrite	$CuFeS_2$
	Enargite	Cu_3AsS_4
	Pyrite (source of Fe^{2+}, Fe^{3+}, and H_2SO_4)	FeS_2

with potential contamination of groundwater and aquifers, so great care must be taken to ensure appropriate geological conditions for this type of leaching. Although used extensively in the uranium industry, *in-situ* leaching is not yet widely practiced for the recovery of copper (except for recovery of residual metal valuables from mining remnants). This technology may become more widespread in the future.

15.4. HEAP AND DUMP LEACHING

Heap leaching and *dump leaching* involve trickling the H_2SO_4 lixiviant through large heaps or dumps of ore under normal atmospheric conditions. The oxide ores in Table 15.1 and chalcocite are readily leached. Bornite, covellite, and native copper are also leached under biological oxidizing conditions. Chalcopyrite is not leached to any significant extent under the mild conditions of heap leaching (Section 15.7).

In addition to the treatment of low-grade ores, heap leaching is also used to recover additional copper from tailings of heap, vat, or agitation leaching.

The chemistry of heap and dump leaching is very similar. The main differences between these two methods of leaching are the copper grade and the particle size of the ore.

Dump leaching is typically used for treating low-grade run-of-mine (ROM) ore, typically <0.5% Cu, with rock sizes ranging up to ~500 mm. It is a relatively inexpensive method of treating material that would otherwise be considered waste (see Chapter 2). The ore is usually placed on an impermeable pad or bedrock and irrigated with sulfuric acid solution (Fig. 15.6).

Heap leaching is used to treat oxides and lower-grade secondary sulfide ores, containing up to ~2% Cu. The ore is crushed to a uniform particle size (typically 12–50 mm), often agglomerated, and then stacked on heaps in a controlled manner. The heap has an impermeable pad at the base. A series of irrigation pipes distribute the solution and drainage lines collect the copper-containing solutions at the base.

15.4.1. Chemistry of Heap and Dump Leaching

As discussed in Section 15.2, non-sulfide copper minerals are leached directly by H_2SO_4 according to reactions such as:

$$CuO + H_2SO_4 \rightarrow Cu^{2+} + SO_4^{2-} + H_2O \quad (15.1)$$

Leaching of sulfide minerals, on the other hand, requires an oxidizing agent as well as H_2SO_4. The oxidizing agent is usually dissolved O_2 from air, such as:

$$Cu_2S + 1/2O_2 + H_2SO_4 \rightarrow CuS + CuSO_4 + H_2O \quad (15.7c)$$

FIGURE 15.6 Dump leaching at a mountainous site in Chile. (a) The natural topography has been used to contain the dump leach material. The leach solution flows by gravity through the dumps and is collected in PLS ponds at the base of the valley. *(Courtesy of Anglo American.)* (b) Close-up view showing the dripper pipes laid on the surface of the dump before irrigation has started.

Heap leaching of sulfides is assisted by micro-organisms. Bacteria, which occur naturally in the ore, act as a catalyst for the leaching reaction. The bacterial action increases the reaction rate so that the time frame required for leaching of these minerals becomes economic (Section 15.4.1.2).

15.4.1.1. Oxidation by Fe^{3+}

Reaction (15.7c) represents the overall sulfide leaching reaction. However, it has been shown that the Fe^{3+-} is necessary for rapid leaching. Iron-containing minerals in the ore such as pyrite (FeS_2) are oxidized by Fe^{3+} ions in the presence of *sulfur-oxidizing bacteria*, releasing Fe^{2+} ions. The Fe^{2+} ions are rapidly reoxidized to Fe^{3+} by oxygen and catalyzed by *ferrous-oxidizing bacteria* close to the surface of the pyrite to maintain a high potential. Pyrite is not only a source of Fe^{3+} but is also an important source of acid (Eq. (15.11a)). Copper sulfides (and native copper) are leached by Fe^{3+}, also assisted by direct bacterial action:

$$2FeS_2 + 7O_2 + 2H_2O \rightarrow 2Fe^{2+} + 2SO_4^{2-} + 2H_2SO_4 \quad (15.11a)$$

$$O_2 + 4Fe^{2+} + 4SO_4^{2-} + 2H_2SO_4 \rightarrow 4Fe^{3+} + 6SO_4^{2-} + 2H_2O \quad (15.11b)$$

$$Cu_2S + 10Fe^{3+} + 15SO_4^{2-} + 4H_2O \rightarrow 2Cu^{2+} + 10Fe^{2+} + 12SO_4^{2-} + 4H_2SO_4 \quad (15.11c)$$

The Fe^{2+} ions produced by Reaction (15.11c) are then reoxidized by Reaction (15.11b) and the process becomes cyclic (Brierley & Brierley, 1999). Some direct oxidation (Eq. (15.7c)) may also occur. PLS from heap leaching typically contains 1–5 g/L Fe.

15.4.1.2. Bacterial Action

The above reactions can proceed without bacterial action, but they are accelerated by orders of magnitude by the enzyme catalysts produced by bacteria. The catalytic actions are most commonly attributed to *Acidothiobacillus ferrooxidans*, *Leptospirillum ferriphilumooxidans*, and *Acidothiobacillus thiooxidans*. At moderate temperatures (40–45 °C), *Acidothiobacillus caldus* prevails, while at elevated temperature (>60 °C), *Sulfolobus metallicus* and *Metallosphaera spp* are believed to be the dominant bioleaching strains (Watling, 2006). The bacteria are rod-shaped, 0.5 × 2 μm long or round, ~1 μm diameter (Fig. 15.7).

The bacteria are indigenous to both oxide and sulfide ores and their surrounding aqueous environment (Demergasso, Galleguillos, Soto, Seron, & Iturriaga, 2010;

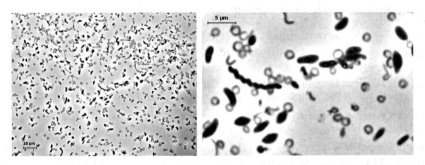

FIGURE 15.7 Naturally occurring mixed bacterial culture found in copper sulfide heap leaching. Note the presence of both rod-shaped and round bacteria. (*Courtesy of F. Perrot, CSIRO Australian Minerals Resource Centre.*)

Groudeva, Vasilev, Karavaiko, & Groudev, 2010). They are present in leach heaps in the order of 10^{12} bacteria per tonne of ore (Brierley & Brierley, 1999). Like all bacteria, they adapt readily to changes in their environment (Mutch, Watling, & Watkin, 2010).

The following conditions promote optimum activity of the bacteria:

(a) Lixiviant pH between 1.5 and 6 (optimum ~pH 2);
(b) Temperature between 5 and 45 °C (optimum ~30 °C, often generated in leach heaps and dumps by exothermic sulfide oxidation reactions);
(c) An adequate O_2 supply, often obtained by gently blowing air through perforated pipes beneath sulfide ore leach heaps (Salomon-de-Friedberg, 1998, 2000);
(d) No organics in lixiviant or heap, so adequate removal of organic entrainment from the SX raffinate is essential (Chapter 16).

It has been suggested that the bacteria might also need small amounts of minor nutrients, such as NH_4^+ and PO_4^{3-}, for cell growth (Brierley & Brierley, 1999). Once leaching has begun, the bacteria adapt so that the nutrients they need are provided by minerals in the ore: nitrogen is generally sourced from blasting chemicals used in the mine and phosphorus from the leaching of phosphates in the ore. If the ore is phosphorus-deficient, phosphate may be added to the leach solution.

Bacteria are sensitive to the presence of chloride, nitrate, and total dissolved solids in the leach solution. Where possible, it is preferable to use high quality water for make-up of the lixiviant, as this enhances the rate of leaching. Bacteria are, however, very adaptable: some Chilean operations have chloride contents up to 10 g/L, some Australian operations have up to 10,000 ppm total dissolved solids in the make-up water, and it is not unusual to have 200 g/L total sulfate content in heap leach solutions.

15.4.1.3. Rates of Leaching of Copper Minerals

Chalcocite and oxide minerals leach rapidly under heap leach conditions. Bornite and covellite leach much more slowly. Depending on the mineralogy of a heap, leach times may vary from ~90 days to three years. Chalcopyrite hardly reacts at all in heap leaching.

Heap and dump leaching are diffusion controlled — never chemical reaction-rate controlled. The long leach times are due to the changes in the diffusion rate caused by the geo-technical characteristics of the ore — especially the presence of clays, micas, and other platey and asicular-shaped particles — and not the intrinsic rate of leaching (Miller, 2002, 2008).

15.4.2. Industrial Heap Leaching

Heap leaching is the most important method of hydrometallurgical copper extraction. Copper-containing ore material (which may be agglomerated) is built into flat-topped *heaps*, typically ~7 m high and having a surface area of 0.01–1 km². The H_2SO_4-containing lixiviant is applied to the surface of the heap. The solution trickles through the heap, dissolving copper minerals by Reactions (15.1)–(15.13). The Cu^{2+}-rich PLS is collected on a sloped impermeable surface beneath the heap and directed to a PLS pond. Copper is recovered from the PLS by SX (Chapter 16) and electrowinning (Chapter 17), producing metallic copper. The acid-rich raffinate from SX is returned to the heap for further leaching (Fig. 15.2). For management of the grade of the PLS that is fed forward to SX, a recirculating intermediate leach solution (ILS) system is employed.

Table 15.2 provides some details of five heap leach operations in different localities.

TABLE 15.2 Selected Industrial Heap Leach Data (2010)

Operation	Cerro Verde, Peru	El Abra, Chile	Los Bronces, Chile	Mantoverde, Chile	Morenci, USA	Spence, Chile
Startup date	1977	1996	1998	1995	1987	2006
Cathode capacity, t/a	90 000	194 000	50 000	65 000	366 000	200 000
H_2SO_4 consumption, t/t Cu	1.5	3–4	1	3.5		3.3
Ore feed to leach						
Predominant Cu minerals	Chalcocite, covellite, chalcopyrite	Chrysocolla	Chalcocite, chalcopyrite	Brochantite, chrysocolla	Chalcocite, chrysocolla	Atacamite, chalcocite
Average Cu grade, %	0.64	0.63	0.5	0.76	0.26	1.18, 1.13
Average leachable Cu, %	0.54	0.55	0.3	0.67	0.23	0.88
Leachable Cu recovered to PLS, %	85	78	50	85	53	70–75
Ore preparation						
Crush, Y/N	Y	Y	N	Y	Y	Y
Agglomeration in rotating drum, Y/N	Y	Y	N	Y	Y	Y

(Continued)

TABLE 15.2 Selected Industrial Heap Leach Data (2010)—cont'd

Operation	Cerro Verde, Peru	El Abra, Chile	Los Bronces, Chile	Mantoverde, Chile	Morenci, USA	Spence, Chile
Ore size on heap, mm	10–15% +9.5	80% <15	ROM	95% <15	80% <12 mm crushed, 300 mm ROM	70–75% <12.5, 95%–19
Heap						
Permanent or on–off?	Permanent	On–off	Permanent	On–off	Permanent	On–off
Building machines	Conveyer stacking	Conveyer stacking	Trucks	Conveyer stacking	Conveyer stacking	Conveyer stacking
Area under leach, m²	1 000 000	30 000	200 000	—	1 300 000	500 000 and 1 800 000
No of cells under leach	45	80	20–22	—	44	16
Typical lift height, m	7	8	45	6.5	4.5	10
Aeration pipes beneath heap, Y/N	Y	N	N	N	Y	Y
Liner material	HDPE	HDPE	N	HDPE	Compacted soil liner	HDPE
Liner thickness, mm	1.5	1.5	—	1.5		1.5

Acid cure (Y/N)						
On heap or during agglomeration?	Y - rotating drum	Y - rotating drum	N	Y - rotating drum	Y - rotating drum	Y - rotating drum
H_2SO_4 in cure solution, g/L	Conc. H_2SO_4	Conc. H_2SO_4		Conc. H_2SO_4	Conc. H_2SO_4	Conc. H_2SO_4
H_2SO_4 addition, kg/t ore	3.5	15		20–26	2–3	19 and 7
Cure time before leaching, d	1	3–4		1	3	1
Lixiviant						
Source of lixiviant	Raffinate or recirculated PLS	Raffinate + H_2SO_4	ILS + H_2SO_4	Raffinate + ILS + H_2SO_4	ILS + H_2SO_4	Raffinate + ILS
H_2SO_4 concentration, g/L	4–8	7–10	4–5	10–12 and 12–14	5–8	13 and 8
Cu concentration, g/L	0.3	0.2–0.3	0.3–1.0	2–3 and 0.4–0.6	0.3	0.3–0.5
Application rate, m^3/h per m^2 area	0.012–0.065	0.08–0.12	0.012	0.12 and 0.01	0.006	0.014
Drop emitters or sprinklers?	Drippers, wobblers	Emitters	Drippers	Drippers, wobblers	Mainly emitters	Drippers
Distance apart along header pipe, m	0.2	0.46	0.5	0.55 and 8	0.45	0.6
Distance apart between header pipes, m	0.4	0.46	0.2	0.6 and 8	0.45	0.6

(Continued)

TABLE 15.2 Selected Industrial Heap Leach Data (2010)—cont'd

Operation	Cerro Verde, Peru	El Abra, Chile	Los Bronces, Chile	Mantoverde, Chile	Morenci, USA	Spence, Chile
Leach cycle sequence	Acid cure in rotating drum, build cured ore into heap, leach 90 days continuous, thereafter 15 days on, 15 days off	Acid cure in rotating drum, build cured ore into heap, rest 3–4 days, leach 120 days	Continuous leach for 60 days and thereafter 30 days off and 60 days on	Acid in cure in rotating drum, build cured ore in heap and leach 150–170 days	Mine for leach: crushed material fines agglomerated with strong acid in a rotating drum then stacked in 7 m lifts, leached for 120 days, rested 15 days, then leached for 15 days, rested for 15 days and leached for 15 days	Acid cure in rotating drum, build cured ore into heaps, rest 20 days, leach 262 days (oxide ore) and 536 days (sulfide ore)

PLS

H_2SO_4 concentration, g/L	3–4	3–4	0.5–1	2–3	2–3	2–4
Cu concentration, g/L	3	3	2–3	5–7	2–4	5–6 and 3–4
Temperature, °C	19–25	15–20	10–12	18–20	32	18–20
Flowrate from leach, m³/h	2500	10 000	1600–2600	1100	20 000	1600 and 1200

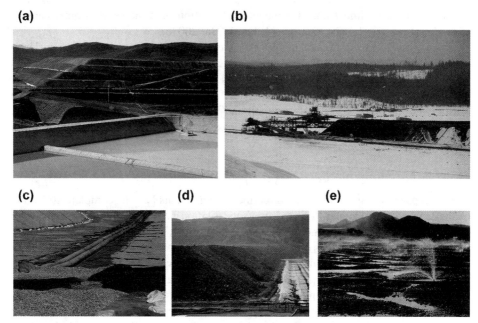

FIGURE 15.8 Heap leaching methods. (a) Multi-lift heap leach operation in Chile, showing the PLS and ILS collection ponds in the foreground *(courtesy of Miller Metallurgical Services)*. (b) Building a heap under sub-arctic conditions in Finland *(courtesy of R. Leighton, Anglo American)*. (c) Impermeable plastic liner and drainage pipes being laid down in preparation for the construction of a heap *(courtesy of J.W. Mann, Anglo American)*. (d) Plastic-lined trenches for the collection of PLS and ILS at the base of a heap. (e) Irrigation of the surface of a heap using wobblers. Some pooling of lixiviant on the surface of the heap is evident.

15.4.2.1. Construction of a Heap

Leach heaps can be either *multi-lift permanent* heaps (Fig. 15.8a) or *on/off* (*dynamic*) heaps (Fig. 15.8b). *Multi-lift heaps* consist of an initial lift built on an impermeable surface. After this lift has been leached, subsequent lifts are built on top of the first lift. Each lift is 5–8 m high and the total heap height can be up to 200 m (Connelly & West, 2009). *Dynamic heaps* consist of a single lift (7–8 m high) built on an impermeable surface. The leached material is removed after leaching and replaced by a new lift of fresh ore. Dynamic pads are the most economic installation for large-tonnage operations (5–7 million t/a). When the ore contains clay and will not support equipment to stack over the leached lift, a dynamic pad is required.

Multi-lift heaps have the advantages that (a) the ore needs to be moved only once (onto the heap), and (b) lixiviant can flow through all the lifts until leaching is moved to another area, permitting additional recovery of copper from slower-leaching minerals in the lower lifts. (More often, however, an intermediate liner is placed on top of the leached ore to limit the acid consumed by the gangue in the lower lift.) The disadvantages of a multi-lift heap are that construction of a heap, which may ultimately become 200 m high, requires a strong impermeable base and versatile heap-building equipment (Breitenbach, 1999; Iasillo & Schlitt, 1999; Scheffel, 2002). The initial base area must be large enough to hold all of the ore that will be mined and leached during the life of the operation. There can also be significant copper inventory locked up in the heap for extended periods of time. Control of the internal solution head and geo-technical stability of the heap can also be challenging (Miller, 2003a, 2003b).

Dynamic heaps have the advantages that they are simple to construct, it is easier to control the leaching at optimum conditions (Section 15.4.2.7), their base need not be as strong as those needed for multi-lift heaps, and the aeration and PLS pipe-work can be maintained when ore is emptied from the pad. Their main disadvantages are that the ore must be moved twice (on and off the pad) and the associated costs of residue disposal. Dynamic heaps are today usually the preferred choice for ore grades above 1% Cu (Scheffel, 2002).

15.4.2.2. Impermeable Base

Leach heaps are usually built on an impermeable base. This permits complete collection of the leached Cu^{2+} in the PLS and prevents solution penetration into the underlying environment and potential contamination of groundwater.

The base typically consists of a synthetic plastic liner or geomembrane sheet (such as high- or low-density polyethylene (HDPE, LDPE), polyvinyl chloride (PVC), or chlorinated PVC), 1–2 mm thick, with a layer of rolled clay or earth (0.1–0.5 m thick) beneath and a 0.5 m layer of finely crushed and screened rock (<2 cm diameter) above it. Perforated PLS collection pipes and aeration pipes are placed on top of this layer (Fig. 15.8c and d). The base is sloped to direct the PLS to a collection pond. The slope should be less than 5% (5 m drop in 100 m horizontal) to avoid slippage of the heap on the plastic liner (Breitenbach, 1999).

Considerable care is taken to avoid puncturing the plastic liner during construction of the base as this will lead to losses of copper-containing PLS and to environmental contamination. The plastic layer should be covered as soon as it is laid down to avoid the destructive effects of sunlight.

15.4.2.3. Pretreatment of the Ore

The simplest method of leaching involves placement of run-of-mine (ROM) ore directly onto the leach pad. This is the cheapest method, but it gives the slowest and least efficient recovery of copper to the PLS.

More often, ROM ore is placed on the heap and then pretreated by trickling strong H_2SO_4 solution through the heap. This *acid cure* rapidly dissolves Cu^{2+} from readily soluble oxide minerals and acidifies the heap, thereby preventing precipitation of ferric sulfate during subsequent leaching. Typically, 10–20 kg H_2SO_4 per tonne of ore is applied to the heap over a period of ~10 days (Iasillo & Schlitt, 1999). Most dump leach operations find that a preliminary acid cure economically enhances the rate of copper leaching.

The rate of copper extraction and overall copper recovery decrease with decreasing particle size of the ore because the local saturation increases (Miller, 2008). Many heap leach operations crush their ROM ore to 1–5 cm pieces; crushing to less than 1 cm does not further improve extraction, while crushing to below 0.5 cm can adversely decrease heap permeability.

In almost all modern heap leach operations, *agglomeration* is used as an intermediate step between crushing and stacking (Bouffard, 2008; Velarde, 2005). The crushed ore (10–50 mm particle size) is typically agglomerated with concentrated H_2SO_4 and SX raffinate in revolving drums (typically 3 m diameter, 9 m long, sloped at ~6°) (Fig. 15.9).

FIGURE 15.9 Rotating drum for the agglomeration of ore, usually with the addition of acid, before stacking into heaps for leaching.

Agglomeration allows uniform application of the acid to all surfaces of the mineral particles to cure the ore and improve the leach kinetics. The fines created during crushing are also aggregated into larger particles so that they do not reduce the permeability of the heap.

An appropriate bacterial culture and its nutrients may also be added to the agglomeration process (Cognis, 2010). When secondary sulfides are present in the mineralogy, a *ferric cure* may be used in conjunction with the acid cure. An acid-containing solution with >3 g/L Fe^{3+} may be added either directly to the heap or during the agglomeration process. The agglomerated material is then placed on the leach heaps.

Optimum agglomeration conditions are very ore-specific. A particular low acid-consuming disseminated chalcocite ore, for example, has optimum conditions of (Salomon-de-Friedberg, 2000): ~1 cm crush size; 60–90 s agglomeration time; ~10 rpm drum rotation speed; ~9% moisture in agglomerate; <5 kg H_2SO_4 per tonne of ore. Close attention is paid to avoiding too much clay in the ore: more than 20% clay severely decreases heap permeability and leaching kinetics.

The rapid and efficient copper extraction obtained by appropriate crushing, agglomerating, and acid curing, followed by leaching, has lead to its widespread use throughout the industry today.

15.4.2.4. Stacking of Ore on Heap

Heaps of ore to be leached are stacked on the impermeable base by dumping from trucks (the least satisfactory method) or by stacking with a wheel loader or mobile conveyor. Trucks have the advantage of simplicity but tend to compact the heap as they run over it. This reduces the permeability of the heap, decreases the leach kinetics, and limits oxygen

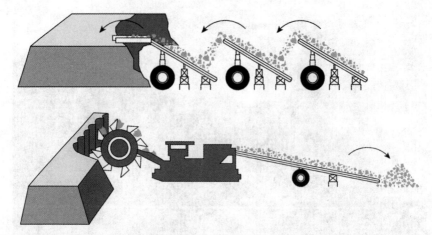

FIGURE 15.10 Diagrammatic representation of the loading of a heap using mobile conveyer belts. Unloading of a heap on completion of leaching is undertaken using bucket-wheel reclaimer.

circulation in the heap (necessary as an oxidizing agent — Section 15.2). Mobile conveyors avoid this problem and are now widely used. (Figs. 15.10 and 15.11).

The heap height is a compromise to maximize copper recovery at the lowest cost (Connelly & West, 2009). A lower heap height can require greater capital due to the greater area under leach, increased pumping requirements from more irrigation, extra drainage piping, and a lower PLS grade (due to greater solution requirements). On the other hand, a lower heap height can produce better overall copper recovery (Díaz et al., 2010), may not require forced aeration, and is unlikely to suffer permeability problems from compaction at the base of the heap.

15.4.2.5. Aeration of Heap

As discussed in Section 15.2, the leaching of sulfides requires the presence of oxygen (Reactions (15.7)–(15.10)). Oxygen also enhances bacterial activity (Watling, 2006). Oxygen is introduced to a heap by placing perforated plastic pipes about 1 m above the heap base and blowing air gently and uniformly upwards through the heap. Alternatively, low-pressure blowers are used. Many smaller size heaps and very large dumps, rely on natural air convection for oxygen supply.

The additional cost of providing oxygen to the heap is more than offset by the increase in copper recovery and enhanced kinetics that result (Salomon-de-Friedberg, 1998; Schlitt, 2006). It is, however, seldom used for leachable copper grades less than 1.0%, but always used for grades above this. Forced air supply is also used when pyrite oxidation is the main source of solution heating in colder climates.

15.4.2.6. Irrigation of Heap

In industrial heap leaching, the lixiviant used is the copper-depleted raffinate returning from SX (Fig. 15.2). This has a typical composition of 0.4 g/L Cu and ~5–10 g/L H_2SO_4. Further sulfuric acid may be added to the raffinate before it is recycled to the heap to increase the acid strength. Additional acid (above the stoichiometric requirement) is required due to the leaching of gangue minerals, such as those containing Fe, Co, Mn, Al,

FIGURE 15.11 Modern mechanized equipment for the loading and unloading of on-off heap-leach pads. (a) The TNT Super Portable mobile conveyer ~70 m long at Arizona heap leach site. (b, c, d) The world's largest fixed-pad, multiple-lift heap stacking system at El Abra in Chile (Terra Nova Technologies, 2010), which processes 8550 t/h ore using Super Portable. (e) Bucket-wheel reclaimer recovering an on-off heap after leaching. *(Courtesy of Terra Nova Technologies and Miller Metallurgical Services.)*

Ca, Mg, Na, K, P, and carbonates (Salgado, Pérez, & Alvayai, 2009; Seyedbagheri, van Staden, & McLaren, 2009). Water may also be added to replace evaporative and ore hold-up losses.

The lixiviant is added via an equi-spaced network of plastic pipes and drip emitters, wobblers, or sprinklers laid on top of the heap (Fig. 15.8e). The lixiviant addition rate is about 10 L/h per m^2 of heap surface. This low rate prevents accumulation of lixiviant on the heap surface and allows free movement of air in the heap (Connelly & West, 2009). The irrigation rate can be determined from air and water conductivity tests that are available for use in heap design (Guzman, Scheffel, & Flaherty, 2005; Guzman, Scheffel, Ahlborn, Ramos, & Flaherty, 2007, Guzman, Scheffel, & Flaherty, 2008).

The lixiviant is generally at ambient temperature. In cold regions it may be heated to enhance the kinetics of Cu extraction (Riekkola-Vanhanen, 2007; Salomon-de-Friedberg, 2000). The leaching reactions are exothermic, so the heap will increase marginally in temperature as the leach liquor percolates through the heap. This further accelerates the rate of leaching; however, significant heat is lost via evaporation from wetted surfaces and in the air flow though the heap. The acid concentration of the lixiviant, the flowrate of lixiviant, and the rate of aeration of the heap all need to be balanced to ensure that the heap does not get too hot (which will cause the bacteria to die) or too cold (which slows down the rate of leaching) (Dixon, 2000; Norton & Crundwell, 2004; Watling, 2006).

Sprinklers, wobblers, and *drip emitters* are used to irrigate the heap or dump. Sprinklers and wobblers have the advantage that they distribute solution evenly over large areas. Drip emitters require more maintenance to avoid blockages, but they avoid excessive evaporation and cooling losses.

15.4.2.7. Optimum Leach Conditions

The following conditions promote optimum leaching of oxide and sulfide material in heaps:

(a) Uniform heaps of optimally sized, cured, and agglomerated ore, which maintain their permeability throughout their life;
(b) Leach conditions, which maximize bacterial activity (30 °C, pH 1.5–2.0, 5–10 g/L H_2SO_4, limited organics);
(c) Uniform lixiviant application (~10 L/h/m^2) onto the heap surface without pooling;
(d) Well-designed impervious heap base, sloping less than 5%, with an efficient solution-collection system;
(e) Adequate heap temperature, provided in cold regions by heating raffinate, insulating pipes, and covering heaps with plastic mesh or sheet (Norton & Crundwell, 2004; Riekkola-Vanhanen, 2007; Salomon-de-Friedberg, 2000);
(f) A controlled, uniform air supply (~10 L/min/m^2 of heap surface), if used, blown in from perforated pipes beneath the heap (Lupo, 2010; Schlitt, 2006);
(g) Heap height that is appropriate to the geo-mechanical characteristics and mineralogy of the ore, to ensure unsaturated conditions and that sufficient oxygen and acid are available throughout the active leaching region (Guzman et al., 2005, 2007, 2008; Miller, 2008; Scheffel, 2002).

15.4.2.8. Collection of PLS

The PLS exiting the base of a heap typically contains 1–6 g/L Cu^{2+}. The solution flows by gravity down plastic drain pipes (~10 cm diameter) on the sloping heap base to a collection trench (Fig. 17.8d). The pipes have slits 2 mm wide, 20 mm long through which the PLS enters. The pipes are spaced 2–4 m apart, at about 45° across the slope. The PLS flows in the collection trench to a PLS pond. It is then sent to the SX circuit by gravity or pumping.

ILS is recycled for PLS grade management. The high-grade PLS is sent to the PLS drain and the low-grade ILS is re-acidified and pumped back to the heap for further copper leaching (Fig. 15.2; Lancaster & Walsh, 1997).

15.4.3. Industrial Dump Leaching

Dump leaching is used for low-grade (0.1–0.4% Cu) oxide or sulfide materials, primarily in Chile and North America. The chemistry is the same as that for heap leaching, however the material is not crushed or agglomerated. ROM ore is piled into dumps that may be as high as 200 m. The leaching, using acidified raffinate as the lixiviant, can take several years. The eventual copper recovery can vary from 30 to 75%, depending on the mineralogy and hydraulic and geo-mechanical characteristics of the ore.

15.5. VAT LEACHING

Vat leaching is carried out in large reaction vessels (*vats*). It is employed for low-grade (0.5–1% Cu) oxide and carbonate minerals, where the leach kinetics are fast and a relatively short contact time with the leach liquor is required (Mendoza, 2000; Schlitt, 1995). Table 15.3 gives the operating conditions of the only commercial vat leaching process producing copper in 2010.

TABLE 15.3 Industrial Data for the Only Copper Vat Leach Operating in 2010

Operation	Mantos Blancos, Chile
Startup date	1961
Cathode capacity, t/a	40 000
H_2SO_4 consumption, kg/t ore	12–20
Ore to leach	
Cu mineralogy	Atacamite, chrysocolla, cuprite
Cu in ore, %	0.5–1
Total Cu in ore recovered to PLS, %	60–70
Vats	
No of vats	12
Length and width of vat, m	20
Height of vat, m	7.5
Mass of dry solids in vat, t	4400
Residence time of ore in vat, h	80
Moisture content of tailings	11
Ore preparation	
Crush	Yes
Agglomeration in rotating drum	Yes
Particle size to leach, mm	30% + 6.25 mm

(Continued)

TABLE 15.3 Industrial Data for the Only Copper Vat Leach Operating in 2010—cont'd

Operation	Mantos Blancos, Chile
Acid cure	
In vat or during agglomeration	During agglomeration
H_2SO_4 addition, kg/t ore	10–12
Cure time before leaching, h	8 maximum
Leach cycle sequence	4 vats on leaching PLS, 4 vats on leaching ILS, 1 being loaded, 1 being unloaded, 1 being drained, 1 being repaired
Lixiviant	
Source of lixiviant	ILS followed by raffinate
H_2SO_4 concentration, g/L	ILS 6–8; raffinate 9–10
Cu concentration, g/L	ILS 1–3; raffinate 0.2–0.5
PLS	
H_2SO_4 concentration, g/L	4–7
Cu concentration, g/L	4–6
Total flow from leach, m³/h	600

The ore is crushed to 5–10 mm. After acid curing, 4000–9000 t ore is loaded into large square vats: 20–30 m sides, 5–8 m deep (Fig. 15.12). Vats are filled with ore from conveyer trippers that run on rails along the length of the train of vats. The vats are constructed of lined, reinforced concrete with a raised, slatted wooden floor, supported by concrete foundations, through which lixiviant is introduced.

Raffinate from SX is used as the lixiviant. This is fed up-flow into each vat until solution starts to overflow from the top of the vat. The leach liquor moves from one vat to the next in a counter-current manner so that fresh lixiviant (which contains the highest acid concentration) is contacted with ore that has been leaching the longest (only the most refractory minerals remain unleached). After an extended period of time (4–20 days) of recirculating the lixiviant through the ore in the vat, the copper-containing PLS is drained from the vats and sent for copper recovery (Zárate, Saldívar, Rojes, & Nuñez, 2004).

To avoid soluble copper losses, the wet leached solids are water-washed several times. The copper-containing spent wash water is combined with the PLS. The solids are removed from the vat using a bucket-wheel or clam-shell excavator that moves on a gantry adjacent to the vat train, loaded onto trucks, and sent to the tailings disposal area.

Because of the high ratio of ore to lixiviant volume in the vats, it is possible to achieve very high concentrations of copper in the PLS (~50 g/L Cu). The PLS from vat leaching will usually be blended with PLS from other leaching processes to provide a more dilute

FIGURE 15.12 Vat leaching at Mantos Blancos in Chile. Approximately 40 000 t/a Cu are produced from leaching in 12 vats. It is the only site in the world that still uses this leaching technology, although there is currently renewed interest in its application (Mackie & Trask, 2009; Schlitt & Johnston, 2010). *(Courtesy of Anglo American.)*

feed to SX. Alternatively, the high-grade solution can be fed to a direct electrowinning circuit (see Section 17.8).

Heap leaching typically uses a lixiviant flowrate of 10 L/h-m^2, while traditional vat leaching uses a flowrate of about 375 L/h-m^2. Lack of aeration in vats precludes the efficient leaching of sulfides.

Modern innovations in vat leaching now allow the process to be continuous, rather than the batch process described above (Mackie & Trask, 2009). Flowrates up to 30 000 L/h-m^2 enable good bed fluidization and rapid kinetics of leaching. For readily leached ores, the residence time on a heap leach is 120—500 days, while this can be reduced to as low as 24—72 h in a continuous vat operation.

15.6. AGITATION LEACHING

15.6.1. Oxide Minerals

Agitation leaching may be used for oxide minerals that have rapid leach kinetics, such as carbonates, and which have reasonable grade (0.8—5% Cu). It can be applied to ores or to milled tailings from other leach circuits. This method is widely used in the African Copperbelt and cobalt is often present in the ores (Baxter, Nisbett, Urbani, & Marte, 2008; Kordosky, Feather, & Chisakuta, 2008; Reolon, Gazis, & Amos, 2009; Roux, van Rooyen, Minnaar, Robles, & Cronje, 2010). It is also employed at Tintaya in Peru.

Following crushing and milling, the material is leached in stirred tanks. Finely ground ore is kept in suspension in the lixiviant by agitating the slurry mechanically or pneumatically. The fine particle size exposures more of the mineral to the leach solution, leading to faster kinetics and often higher copper recoveries. The slurry then reports to a solid—liquid separation, usually a counter-current decantation (CCD) circuit, in which some further leaching can take place. The eventual recovery of acid-soluble copper is usually above 98% of the readily acid-soluble minerals, generating a PLS that contains 2–12 g/L Cu (Cognis, 2010). Ore high in chrysocolla will need extended leach times as this mineral has slow leach rates because the copper is removed from within the residual silica matrix.

The single largest copper loss in many of these plants is due to low recovery of PLS from the leach residue due to poor washing. Efficient operation of the CCD circuits is essential to maximize copper recovery and maintain the water balance in the circuit. These circuits can also have high operating costs due to the need for fine milling and increased acid requirements for the fine ore, followed by the subsequent need to neutralize the tailings for disposal (Nisbett, Feather, & Miller, 2007). Various flowsheet configurations have been proposed to minimize these disadvantages (Baxter et al., 2008; Reolon et al., 2009).

Agitation leaching is a capital-intensive technique for these ores, but copper recovery can approach 100% in a time frame of hours, rather than the months or years of the percolation leaching techniques.

15.6.2. Sulfide Minerals

Secondary sulfides of reasonable grade (>6% Cu) can also be treated using agitation leaching. The method is similar to that described for oxide leaching, but elevated temperature and an oxidizing agent are required (see Section 15.2.2). Most operations use air (oxygen), but recently commissioned plants in Australia, Laos, and Spain use Fe^{3+} as the oxidizing agent (Baxter, Pavlides, & Dixon, 2004; Fleury, Delgado, & Collao, 2010).

The PLS generated from sulfide leaching is usually more concentrated than that of oxide leaching, 12–25 g/L Cu (Cognis, 2010). Copper recovery varies from 95–98% with leach times of less than 24 h to several days. Table 15.4 gives typical operating data for several plants in which agitation leaching is practised.

15.7. PRESSURE OXIDATION LEACHING

15.7.1. Economic and Process Drivers for a Hydrometallurgical Process for Chalcopyrite

Chalcopyrite is essentially unleached under the mild oxidizing conditions of heap leaching (very slow leach kinetics). It can, however, be leached under oxidizing conditions at elevated temperatures or by microbial leaching under low-potential conditions (Gericke, Govender, & Pinches, 2010). This has led to extensive study into leaching of chalcopyrite *concentrates* as an alternative to smelting. Some economic drivers for treating chalcopyrite concentrates hydrometallurgically, rather than by smelting, include (King & Dreisinger, 1995; Marsden & Brewer, 2003; Marsden, Wilmot, & Hazen, 2007a; Marsden, Wilmot, & Hazen, 2007b; Marsden, Wilmot, & Mathern, 2007; Marsden, Wilmot, & Smith, 2007; Robles, 2009):

(a) The relatively high and cyclically variable cost of external smelting and refining charges;
(b) Limited availability of smelting capacity in some locations;
(c) High capital cost of installing new smelting capacity;

TABLE 15.4 Industrial Data for Selected Copper Agitation Leach Plants (2010)

Operation	Kansanshi, Zambia	Las Cruces, Spain	Sepon, Laos
Startup date	2004	2009	2005
Cathode Cu, t/a	87 000	72 000 (design)	80 000
H_2SO_4 consumption, t/t Cu	3.45	20	0.4–0.7 additional acid not produced by POx
Ore feed to leach			
Predominant Cu minerals	Major malachite, minor azurite, tenorite, cuprite, chrysocolla, chalcopyrite and secondary sulfides (mainly chalcocite and covellite)	65–75% pyrite, 6% digenite and chalcocite, minor covellite, chalcopyrite, galena and blende	Chalcocite, covellite, some zones of carbonate
Average Cu grade, %	Cu 2.26 (total), Cu 1.48 (acid-soluble), Cu 0.78 (acid-insoluble)	6.22	4.6
Average S grade, %	—	41	8–12
Agglomeration, Y/N	N	N	N
Acid curing, Y/N	N	N	N
Ore size to leach, μm	80%–150	100%–150	80% <106
Ore fed to leach, t/h	700–1200		230

(Continued)

TABLE 15.4 Industrial Data for Selected Copper Agitation Leach Plants (2010)—cont'd

Operation	Kansanshi, Zambia	Las Cruces, Spain	Sepon, Laos
Agitation tank			
No of tanks	5	8	5
Tank height, m	12.25		12.5
Tank diameter, m	11		12
Impeller type	Mixtec		4-blade axial
Impeller power, kW	132		185
Solids' density, %	35		15–19
Residence time of solids in leach, h	2.8		6–8
Lixiviant			
Source	HG raffinate, HPOX, conc. H$_2$SO$_4$	Ferric sulfate, raffinate	CCD wash liquor + POx
H$_2$SO$_4$ concentration, g/L	19–20, 36–40, conc. H$_2$SO$_4$	~60	10–15
Cu concentration, g/L	0.5–0.7, 9–11, –	~10	8–12

Chapter 15 Hydrometallurgical Copper Extraction: Introduction and Leaching

Fe concentration, g/L	3–5, 14–16; –	25 Fe^{2+}, 25 Fe^{3+} 30–35
Temperature, °C	25–35, 65–70, ambient	90 80
Feed rate to leach, m³/h	1000, 270, 84	900–1000
PLS		
H_2SO_4 concentration, g/L	8–10	6–8
Cu concentration, g/L	6–11	43 15–17
Fe concentration, g/L	3–5	50 30–35
Temperature, °C	45	40
Flowrate from leach, m³/h	2000	313 600–6700
Leach residue		
Predominant mineralogy	Major chalcopyrite, minor covellite, chrysocolla	Covellite
Cu in residue, %	0.3–0.4	0.4–0.6
Leach efficiency		
Cu extraction (leach only), %	82	92.5 (target) —
Cu extraction (Leach + CCDs), %	90–95	— 90–95

(d) Increasingly stringent environmental legislation, particularly avoiding the generation of gaseous effluents such as SO_2;
(e) The ability to treat high-impurity concentrates, particularly those that contain elements that are deleterious in electrorefining, such as As, Sb, F, and others (Chapter 14);
(f) The ability to treat lower-grade concentrates on site, increasing overall recovery from the ore body
(g) The construction of small leach plants at mine sites rather than shipping concentrate to large, distant smelters, thereby eliminating freight and transport costs;
(h) The need to cost-effectively generate sulfuric acid at mine sites for use in heap and stockpile leaching operations;
(i) Overall lower cost production of copper (if credit can be obtained for the acid production).

15.7.2. Elevated Temperature and Pressure Leaching

Temperature affects both the thermodynamics and the kinetics of leaching. An increase in temperature and pressure promotes leaching because it permits much greater concentration of oxygen in solution and allows faster kinetics for reactions that are slow or thermodynamically unfavorable at atmospheric pressures.

Elevated temperature and pressure leaching is carried out in an autoclave. The pressure inside the leaching vessel can be increased by pressurizing the autoclave, usually using oxygen; the temperature is typically increased by steam injection on start-up. The autoclave is usually horizontal, with a series of compartments through which the lixiviant and solids move in a co-current manner (Fig. 15.13). The material to be leached is ground to a fine particle size (typically ~10 μm) for maximum exposure of the lixiviant to the mineral surfaces.

Various approaches have been taken to finding an economically viable process for the hydrometallurgical treatment of refractory copper sulfides (Dreisinger, 2006; McElroy & Young, 1999). These encompass sulfate-based processes (including bacterially assisted processes and low-, medium-, and high-temperature pressure leaching),

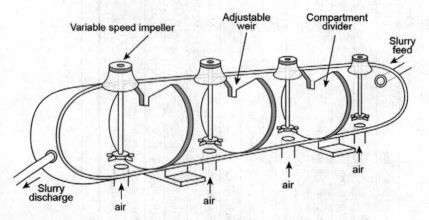

FIGURE 15.13 Schematic representation of a four-compartment horizontal autoclave (baffles inside each compartment not shown).

chloride-assisted processes, and ammonia- and nitrate-based processes. The status of some of the more well-known processes is given in Table 15.5. Medium-temperature, medium-pressure and high-temperature, high-pressure processes have recently been developed to commercial scale (Keokhounsy, Mooreand, & Liu, 2006; McClelland, 2007; Marsden et al., 2007; Sherrit et al., 2005).

15.7.2.1. High-temperature, High-pressure Oxidation Leaching

In high-temperature and -pressure leaching (200–230 °C, ~3000 kPa), leaching is rapid and >99% of the copper is typically dissolved. However, all of the sulfide sulfur in the concentrate is converted to H_2SO_4, which means higher oxygen consumption and the need to neutralize this acid ahead of downstream processing, unless the acid can be made available for heap leaching in other parts of the flowsheet. Iron in the concentrate reports to the leach residue in the form of hematite (Fe_2O_3), which is environmentally stable and easily filterable. In the presence of sufficient dissolved oxygen, the leach reaction can be represented as (McDonald & Muir, 2007a; Marsden & Brewer, 2003):

$$4CuFeS_2 + 17O_2 + 4H_2O \rightarrow 4Cu^{2+} + 4SO_4^{2-} + 2Fe_2O_3 + 4H_2SO_4 \quad (15.12)$$

High-temperature and -pressure oxidation of copper concentrates has been practiced at a commercial scale at Bagdad, Arizona (16 000 t/a Cu cathode, now closed), at Kansanshi, Zambia (30 000 t/a Cu), and for the treatment of a copper-containing pyrite concentrate at Sepon, Laos (80 000 t/a Cu) (Baxter et al., 2004; Bell, Hoey, & Lui, 2010). Details of two of these plants are given in Table 15.6.

15.7.2.2. Medium-temperature, Medium-pressure Oxidation Leaching

Medium-temperature and -pressure oxidation processes (140–150 °C, 1000–1200 kPa) require fine grinding of the feed material (80%-10 μm or finer) and often rely on the addition of chloride to attain reasonable kinetics of leaching (for example, see Defreyne et al., 2008; McDonald & Muir, 2007b; Nicol, Miki, & Velasquez-Yevenes, 2010; Senanayake, 2009; Velasquez-Yevenes, Miki, & Nicol, 2010; Velasquez-Yevenes, Nicol, & Miki, 2010; Yoo et al., 2010). Sulfides are converted to a mixture of H_2SO_4 and elemental sulfur:

$$4CuFeS_2 + 5O_2 + 4H_2SO_4 \rightarrow 4Cu^{2+} + 4SO_4^{2-} + 2Fe_2O_3 + 8S + 4H_2O \quad (15.13)$$

Oxidation can also occur by the reaction:

$$4CuFeS_2 + 17O_2 + 2H_2SO_4 \rightarrow 4Cu^{2+} + 10SO_4^{2-} + 4Fe^{3+} + 2H_2O \quad (15.14)$$

The melting point of sulfur is 119 °C, so it is critical to avoid passivation of the surface of the chalcopyrite particles by molten sulfur and to avoid agglomeration of the finely ground solids. Surfactants and other additives are used (Hackl, Dreisinger, & King, 1995; Hackl, Dreisinger, Peters, & King, 1995; Steyl, 2004).

Depending on the temperature and free acid present in solution, iron will be rejected as basic ferric sulfate ($Fe(OH)SO_4$) or jarosite ($Fe_3(OH)_6(SO_4)_2 \cdot 2H_2O$), although hematite can form at higher temperatures. It is desirable to produce an insoluble iron product because residual Fe^{3+} in solution that is transferred through SX can affect current efficiency in electrowinning (Chapter 17).

Medium-temperature and -pressure oxidation processes have been proven at a demonstration scale at Bagdad, Arizona (14 000 t/a Cu) (Marsden, Wilmot, &

TABLE 15.5 Description and Status of Selected Chalcopyrite and Chalcopyrite Concentrate Leach Processes

Generic process	Name of process	Description	Status	References
Sulfate atmospheric, bacterially assisted	BioCOP™	Thermophilic bacteria used to oxidize and leach copper at 65–80 °C	20 kt/a Cu bioleach plant built in 2002 in Chile (Alliance Copper)	Batty and Rorke (2006); Dew and Batty (2003)
	Bactech/Mintek	Low-temperature (35–50 °C) tank leach process with very fine grind to overcome passivation	Demonstrated at 500 kg/d Cu scale in Mexico (Met-Mex Peñoles). Also heap leaching in China and Sar-chesmah, Iran	Gericke et al. (2009, 2010)
Low-temperature, medium pressure	Activox	5–15 μm grind, 100 °C, 1000 kPa	Several piloting trials on various concentrates containing Cu, Ni, Co, Au	Corrans et al. (1995); Evans and Johnson (1999)
Medium-temperature, medium pressure	Anglo American Corporation/University of British Columbia (AAC/UBC)	Fine grind (10 μm), use of surfactants to avoid S ball formation	Demonstrated at pilot scale	Dempsey and Dreisinger (2003); Dreisinger et al. (2003); Steyl (2004)
	CESL	P_{95} 45 μm grind, chloride catalyzed. Basic copper sulfate salt produced that requires releach to produce PLS	Demonstrated in two-year trial at UHC Sossego, Brazil	Defreyne et al. (2008); Hayton et al. (2009); Jones et al. (2006); Omena et al. (2007)
	Galvanox	Atmospheric 80 °C leach galvanically assisted by the reaction between chalcopyrite and pyrite to avoid passivation	Pilot plant trials	Dixon (2007); Dixon et al. (2007); Miller et al. (2008)
	Phelps Dodge (now Freeport McMoRan)	13–15 μm grind, surfactants to disperse molten S, 140–180 °C to maximize Cu extraction and minimize acid consumption. Fe_2O_3 is major residue product	7-month trial at Bagdad, AZ, producing 16 000 t/a Cu	Marsden et al. (2007a, 2007b); Marsden et al. (2007); Marsden et al. (2007)

Process	Company/Plant	Description	Status/Notes	References
High temperature, high pressure	Phelps Dodge (now Freeport McMoRan)	Fine grind, 220 °C, 3000 kPa	16 000 t/a Cu plant run at Bagdad, AZ, for several years (now closed)	Marsden and Brewer (2003)
	Kansanshi (First Quantum)	Similar to above but does not use fine grind	14 000 t/a production, capacity 30 000 t/a in 2011	Anon (2006)
Chloride	HydroCopper	Chlorine leaching of concentrate, NaOH precipitation of Cu_2O, which is then reduced with H_2 to Cu. NaOH, Cl_2, and H_2 are regenerated in chloralkali cell		Hyvärinen and Hämäläinen (2005); Outotec (2007)
	Intec	Cu sulfides are leached in a mixed halide medium. LME A Cu cathode is electrowon from NaCl-NaBr medium. The $BrCl_2^-$ species generated at the anode is recycled as lixiviant	350 t/a demonstration plant	Moyes et al. (2002); Intec (2010)
Nitrate	NSC	Nitrogen-species catalyzed oxidation using NO^+ (derived from $NaNO_2$)	Limited commercial application to the leaching of chalcopyrite concentrates	Anderson (2000); Anderson et al. (2010)
Ammonia	Arbiter	Ammonia-oxygen agitation pressure leach followed by SXEW	100 t/d Cu plant started in 1974 but closed due to technical difficulties and high costs	Arbiter and McNulty (1999)
	Escondida	Ammonia-air agitation leach of Cu_2S, followed by SXEW	80 000 t/a Cu plant started in 1994 but closed due to slow rate of copper production	Duyvesteyn and Sabacky (1993, 1995)

TABLE 15.6 Industrial Data for Two Copper Pressure Oxidation Leach Plants (2010). Kansanshi Leaches Chalcopyrite Directly, while Laos Uses Pressure Leaching Primarily to Produce Ferric ion, Which Then is Used to Leach Chalcocite

Operation	Kansanshi, Zambia	Sepon, Laos
Startup date	2007	2005
Cathode capacity, t/a	30,202 design, 14,257 actual 2010	80 000
H_2SO_4 consumption, kg/t ore	1.9	—
Ore/concentrate feed to autoclave		
Source of solids' feed to autoclave	Chalcopyrite concentrate	Pyrite concentrate
Cu mineralogy	70–80% chalcopyrite, 3–10% chalcocite	76% pyrite, refractory sulfides including covellite, enargite, and chalcopyrite
Cu in ore/concentrate, %	23–27	2–3
S in ore/concentrate, %	28–35	40–45
Particle size feed to leach, μm	P_{80} 75 μm	P_{80} 125 μm
Solids' feed rate to leach, t/h	7.65 t/h per clave on two claves running, 10.6 t/h on single clave running	10
Autoclave		
No of autoclaves	2	1
No of compartments per autoclave	5	4
Length, m	26.8	14
Diameter, m	3.35	4.2
Shell material	Carbon steel	Mild steel

Lining material	Lead layer, 2 layers of refractory bricks	Glass-fiber-reinforced furan membrane, 3 layers of refractory bricks
Impeller type	8-blade Rushton turbine (compartment 1), 4-blade Rushton turbine (compartments 2–5)	8-blade Rushton turbine (compartment 1 & 2), 4-blade Rushton turbine (compartments 3 & 4)
Impeller power, kW	34.2	110
Impeller specific power, kW/m^3	1.44	3–4
Temperature of operation, °C	220	220
Pressure of operation, kPa	3000	2800–3000
Oxygen overpressure, kPa	800	600–800
Oxygen grade, %	>99	98
Solids' density, %	45–50 (slurry feed to autoclave & excludes raffinate for cooling)	45
Residence time of solids in leach, min	98	60
Temperature control mechanism	Raffinate injection for cooling (exothermic reaction)	Raffinate injection for cooling (exothermic reaction)
Lixiviant (solution feed to autoclave)		
Source of lixiviant	High-grade raffinate + H$_2$SO$_4$ make-up	Raffinate
H$_2$SO$_4$ concentration, g/L	64–66	30
Cu concentration, g/L	<1.3	1.0–1.5
Fe concentration, g/L	<3	30–35

(*Continued*)

TABLE 15.6 Industrial Data for Two Copper Pressure Oxidation Leach Plants (2010). Kansanshi Leaches Chalcopyrite Directly, while Laos Uses Pressure Leaching Primarily to Produce Ferric ion, Which Then is Used to Leach Chalcocite—cont'd

Operation	Kansanshi, Zambia	Sepon, Laos
Temperature, °C	<40	Ambient
Flowrate to leach, m³/h	75–90	110
PLS (discharge from autoclave)		
H_2SO_4 concentration, g/L	56–62	60
Cu concentration, g/L	20–25	3
Fe concentration, g/L	7–12	40
Temperature, °C	95	100–102 after flash
Flowrate from leach, m³/h	80–110 per clave	120–140
Total Cu in solids feed recovered to PLS, %	99	97
Residue		
Predominant mineralogy of residue	Basic ferric sulfate	Basic ferric sulfate
Cu concentration (g/L) Cu in residue, %	<0.3	0.05

Mathern, 2007) and at UHC Carajas, Brazil (10 000 t/a Cu) (Defreyne et al., 2008; Hayton, Defreyne, & Murray, 2009; Jones, Xavier, Gonçalves, Costa, & Torres, 2006; Omena, Geraldo, Lima, & Cabral, 2007). A commercial operation was commissioned at Morenci, Arizona (67 000 t/a Cu) in 2007 (Freeport McMoRan, 2008; Marsden et al., 2007).

15.8. FUTURE DEVELOPMENTS

Phenomenal growth in technological advances and commercialization of heap leaching of ores has taken place during the last three decades, particularly for the leaching of acid-soluble oxide minerals and secondary sulfides, such as chalcocite. The contribution to copper production by heap leaching is expected to continue, as this technology is relatively simple, robust, flexible, and low cost.

Improvements in heap leaching efficiency continue through optimization of crushing, acid curing, agglomeration, understanding liquid and air conductivities, heap construction, aeration, lixiviant composition, lixiviant application rate, bacterial activity, and temperature. Modeling of heap leaching is focused on maximizing the performance of heaps by optimizing operational and environmental parameters to maximize both the rate of leaching and overall copper recovery (Dixon, 2003; Guzman et al., 2007, 2008; Miller, 2003). The use of forced aeration to supply adequate oxygen requirements throughout a heap is also increasing (Schlitt, 2006).

Significant advances in sophisticated materials-handling equipment that can assist in rapidly and uniformly stacking and reclaiming on–off heaps and minimize labor requirements continue to be brought to the industry (see, for example, Terra Nova Technologies, 2010).

Development of commercially viable technologies for the hydrometallurgical processing of chalcopyrite remains the holy grail of copper metallurgy. Enormous research efforts continue to seek better understanding of the chemistry and mechanisms of leaching and the behavior of reaction products and byproducts (Berezowsky & Trytten, 2002; Cordoba, Muñoz, Blázquez, González, & Ballester, 2008a; Cordoba, Muñoz, Blázquez, González, & Ballester, 2008b; Cordoba, Muñoz, Blázquez, González, & Ballester, 2008c; Cordoba, Muñoz, Blázquez, González, & Ballester, 2008d; Hiroyoshi, Miki, Hirajima, & Tsunekawa, 2001; Hiroyoshi, Kuroiwa, Miki, Tsunekawa, & Hirajima, 2004; Hiroyoshi, Kitagawa, & Tsunekawa, 2008; Klauber, 2008; Padilla, Pavez, & Ruiz, 2008; Wang, 2005).

Bioleaching of chalcopyrite ores in heaps and dumps remains a topic of continuing research interest (Dreisinger, 2006; Pradhan, Nathsarma, Srinivasa Rao, Sukla, & Mishra, 2008). Although no commercial operation exists in 2010, several successful piloting and demonstration plants have been reported and understanding of the critical factors to make this an economic proposition is improving (Gericke, Neale, & van Staden, 2009, 2010; Konishi et al., 2010; López & Encínas, 2004; Miller, 2001; Mutch et al., 2010; Petersen & Dixon, 2006; Watling, 2006).

Pressure oxidation leaching has not yet reached full commercial acceptance and has been commercially applied only in niche applications to date. It is likely that this technology will become better understood and more widely used during the next decade.

In-situ leaching may also increase in use, as it may be a relatively low-cost option for the recovery of copper under appropriate geo-technical and mineralogical conditions.

15.9. SUMMARY

Hydrometallurgical extraction accounts for about 4.5 million tonnes of cathode copper per year (about 20% of total primary copper production). Most of this is produced by heap leaching.

Heap leaching consists of trickling H_2SO_4-containing lixiviant uniformly through flat-surfaced heaps of crushed ore, agglomerate, or run-of-mine ore. Oxide ores are leached by H_2SO_4 without the need for oxidation. Chalcocite (and, to a much lesser extent, bornite and covellite) needs to be oxidized and leached by H_2SO_4-containing solutions in the presence of oxygen and/or Fe^{3+}.

Leaching of sulfide ores at economic rates is made possible by indigenous bacteria, which increase the mineral leaching kinetics by several orders of magnitude. The rate-determining kinetics is that of diffusion in the saturated zones of the ore. The bacterial activity is maximized at a pH of 1.5–2, a temperature of ~30 °C, and an adequate O_2 supply.

The product of heap leaching is PLS containing 1–6 g/L Cu^{2+}. The PLS is collected on a sloping impervious HDPE liner beneath the leach heaps and sent to solvent extraction and electrowinning for copper production (Chapters 16 and 17). The Cu^{2+}-depleted raffinate from SX is recycled to leaching, with the acid strength usually increased by the addition of concentrated H_2SO_4.

The rate of leaching and the overall copper recovery are maximized by optimizing the conditions for diffusion control: crush size, acid curing, agglomeration, heap permeability, lixiviant composition, aeration, and bacterial activity.

TABLE 15.7 Comparison of Different Leaching Methods for Hydrometallurgical Recovery of Copper

Parameter	Dump	Heap	Vat	Agitation	Agitation	Pressure
Material treated	ROM, oxide, or secondary sulfide	Oxides, secondary sulfides, tailings, usually milled, agglomerated, and acid cured, bacteria-assisted	Oxides, secondary sulfides, secondary leach of tailings, agglomerated and acid cured	Oxides and tailings	Sulfides	Primary and secondary sulfides, refractory flotation concentrates
Cu grade, %	0.1–0.4	0.3–2.3	>0.8	0.8–5.0	>6	Up to 25
Particle size, mm	Up to 1000	12–50	0.5–2	0.1–0.2	0.1–0.2	0.01–0.75
Leach time	Years	Months to years	Weeks	Days	Days	Hours
Cu in PLS, g/L	0.5–3	1.5–8	6–40	2–30	12–25	25–80
Cu recovery	35–75	Up to 90%	>90	85–100	95–98	>96

Low-grade ores are treated by dump leaching, while vat leaching can be used to treat higher grade ore. Only one vat leaching plant is currently operating (in Chile). Agitation leaching in tanks is applied to both oxide and sulfide materials: the PLS concentrations can be up to 25 g/L Cu. Agitation leaching is a more expensive technology but can give very high copper recoveries. It is applied mainly on the African Copperbelt.

Most copper minerals are amenable to leaching under atmospheric conditions. A critical exception is chalcopyrite ($CuFeS_2$), which does not dissolve under percolation leach conditions. It can be leached under oxidizing conditions at elevated temperatures, either using the assistance of thermophilic microbes or high temperatures and pressures. Several process routes for the hydrometallurgical treatment of chalcopyrite have been proposed and are the topics of ongoing research, but few have yet been successfully commercialized. Total pressure oxidation (220 °C, 3000 kPa) circuits to treat chalcopyrite-containing concentrates have recently been commissioned in Laos and Zambia.

Table 15.7 summarizes the main differences between the different leaching methodologies.

REFERENCES

Anderson, C. G. (2000). The treatment of chalcopyrite concentrates with nitrogen species catalyzed oxidative pressure leaching. In P. R. Taylor (Ed.), *EPD congress 2000* (pp. 489–501). Warrendale, PA: TMS.

Anderson, C. G., Fayran, T. S., & Twidwell, L. G. (2010). Industrial NSC pressure oxidation of combined copper and molybdenum concentrates. In *Copper 2010, Vol. 7: Plenary, mineral processing, recycling, posters* (pp. 2589–2620). Clausthal-Zellerfeld, Germany: GDMB.

Anon. (2006). Kansanshi gets high pressure leach facility. *Modern Mining*, 30–35. http://www.first-quantum.com/i/pdf/ModernMining-Kansanshi12-06.pdf (Nov).

Arbiter, N., & McNulty, T. (1999). Ammonia leaching of copper sulfide concentrates. In S. K. Young, D. B. Dreisinger, R. P. Hackl & D. G. Dixon (Eds.), *Copper 99–Cobre 99, Vol. IV, Hydrometallurgy of copper* (pp. 197–212). Warrendale, PA: TMS.

Batty, J. D., & Rorke, G. V. (2006). Development and commercial demonstration of the BioCOP™ thermophile process. *Hydrometallurgy, 83*, 83–89.

Baxter, K., Nisbett, A., Urbani, M., & Marte, K. (2008). Flowsheet alternatives for high acid consuming copper ores. In *ALTA copper 2008*. Melbourne, Australia: ALTA Metallurgical Services.

Baxter, K. G., Pavlides, A. G., & Dixon, D. G. (2004). Testing and modelling a novel iron control concept in a two-stage ferric leach/pressure oxidation process for the Sepon Copper Project. In M. J. Collins & V. G. Papangelakis (Eds.), *Pressure hydrometallurgy* (pp. 57–76). Montreal: CIM.

Bell, M., Hoey, M., & Lui, M. (2010). Design, construction and commissioning of the Sepon POX II circuit. In *ALTA nickel cobalt copper 2010*. Melbourne, Australia: ALTA Metallurgical Services.

Berezowsky, R., & Trytten, L. (2002). Commercialization of the acid pressure leaching of chalcopyrite. In *ALTA 2002, copper-7 forum*. Melbourne, Australia: ALTA Metallurgical Services.

Bouffard, S. C. (2008). Agglomeration for heap leaching: equipment design, agglomerate quality control, and impact on the heap leach process. *Minerals Engineering, 21*, 1115–1125.

Breitenbach, A. J. (1999). The good, the bad, and the ugly lessons learned in the design and construction of heap leach pads. In G. V. Jergensen, II (Ed.), *Copper leaching, solvent extraction and electrowinning technology* (pp. 139–147). Littleton, CO: SME.

Brierley, C. L., & Brierley, J. A. (1999). Bioheap processes – operational requirements and techniques. In G. V. Jergensen, II (Ed.), *Copper leaching, solvent extraction, and electrowinning technology* (pp. 17–27). Littleton, CO: SME.

Cognis. (2010). *Copper leaching: Training course material*. Tucson, AZ, USA: Cognis Mining Chemicals Technology.

Connelly, D., & West, J. (2009). Trips and traps for copper heap leaching. In *ALTA copper 2009*. Melbourne: ALTA Metallurgical Services.

Cordoba, E. M., Muñoz, J. A., Blázquez, M. L., González, F., & Ballester, A. (2008a). Leaching of chalcopyrite with ferric ion. Part I: general aspects. *Hydrometallurgy, 93*, 81–87.

Cordoba, E. M., Muñoz, J. A., Blázquez, M. L., González, F., & Ballester, A. (2008b). Leaching of chalcopyrite with ferric ion. Part II: effect of redox potential. *Hydrometallurgy, 93*, 88–96.

Cordoba, E. M., Muñoz, J. A., Blázquez, M. L., González, F., & Ballester, A. (2008c). Leaching of chalcopyrite with ferric ion. Part III: effect of redox potential on the silver-catalyzed process. *Hydrometallurgy, 93*, 97–105.

Cordoba, E. M., Muñoz, J. A., Blázquez, M. L., González, F., & Ballester, A. (2008d). Leaching of chalcopyrite with ferric ion. Part IV: the role of redox potential in the presence of mesophilic and thermophilic bacteria. *Hydrometallurgy, 93*, 106–115.

Corrans, I. J., Angove, J. E., & Johnson, G. D. (1995). The treatment of refractory copper-gold ores using Activox® processing. In *Randol gold forum Perth '95 - Gold metallurgy & environmental management* (pp. 221–224). Golden, CO: Randol International.

Defreyne, J., Brace, T., Miller, C., Omena, A., Matos, M., & Cobral, T. (2008). Commissioning UHC: a Vale copper refinery based on CESL technology. In C. A. Young, P. R. Taylor, C. G. Anderson & Y. Choi (Eds.), *Hydrometallurgy 2008* (pp. 357–366). Littleton, CO: SME.

Demergasso, C., Galleguillos, F., Soto, P., Seron, M., & Iturriaga, V. (2010). Microbial succession during a heap bioleaching cycle of low grade copper sulfides: does this knowledge mean a real input for industrial process design and control? *Hydrometallurgy, 104*, 382–390.

Dempsey, P., & Dreisinger, D.B. (2003). Process for the extraction of copper. US Patent 6,503,293.

Dew, D., & Batty, J. (2003). *Biotechnology in mining. Development of the BIOCOP process. Hydrometallurgy 2003 Short Course Notes*. Montreal, Canada: CIM.

Díaz, M., Salgado, C., Pérez, C., Alvayai, C., Herrera, L., & Zárate, G. (2010). Mine to heap in Mantoverde Anglo American Division. In *Copper 2010, Vol. 5: Hydrometallurgy* (pp. 2693–2701). Clausthal-Zellerfeld, Germany: GDMB.

Dixon, D. D., Baxter, K., & Sylwestrzak, L. (2007). Galvanox™ treatment of copper concentrates. In *ALTA copper 2007*. Melbourne, Australia: ALTA Metallurgical Services.

Dixon, D. G. (2000). Analysis of heat conservation during copper sulfide heap leaching. *Hydrometallurgy, 58*, 27–41.

Dixon, D. G. (2003). Heap leach modelling: the current state of the art. In C. A. Young, A. M. Alfantazi, C. G. Anderson, D. B. Dreisinger, B. Harris & A. James (Eds.), *Hydrometallurgy 2003, Vol. 1., Leaching and solution purification* (pp. 289–314). Warrendale, PA: TMS.

Dixon, D. G. (2007). *Galvanox, a novel process for the treatment of copper concentrates. Copper 2007 short course notes*. Montreal, Canada: CIM. www.uilo.ubc.ca/galvanox/reference_materials.html.

Dreisinger, D. (2006). Copper leaching from primary sulfides: Options for biological and chemical extraction of copper. *Hydrometallurgy, 83*, 10–20.

Dreisinger, D. B., Steyl, J. D. T., Sole, K. C., Gnoinski, J., & Dempsey, P. (2003). The Anglo American Corporation/University of British Columbia (AAC/UBC) chalcopyrite process: integrated pilot-plant evaluation. In P. A. Rivieros, D. Dixon, D. B. Dreisinger & J. Menacho (Eds.), *Copper 2003–Cobre 2003, Vol. VI, Book 2: Hydrometallurgy of copper* (pp. 223–238). Montreal: CIM.

Duyvesteyn, W. P. C., & Sabacky, B. J. (1993). The Escondida process for copper concentrates. In R. G. Reddy & R. N. Weizenbach (Eds.), *Extractive metallurgy of copper, nickel and cobalt (the Paul E. Queneau international symposium), Vol. I: Fundamental aspects* (pp. 881–910). Warrendale, PA: TMS.

Duyvesteyn, W. P. C., & Sabacky, B. J. (1995). Ammonia leach process for Escondida concentrates. *Transactions of the Institution of Mining and Metallurgy, 104*, C125–C140.

Evans, H. A., & Johnson, G. (1999). Activox technology for treatment of copper-gold sulfide concentrates. In *Oretest Colloquium '99*. Victoria, Australia: Australian Institute of Mining and Metallurgy.

Fleury, F., Delgado, E., & Collao, N. (2010). A new technology for processing hydrometallurgical copper ore: Cobre Las Cruces Project. *Proceedings copper 2010, Vol. 5* (pp. 1871–1898). Clausthal-Zellerfeld, Germany: GDMB.

Freeport McMoRan. (2008). Construction and start-up of the Freeport-McMoRan concentrate leach plant at Morenci, AZ. In *ALTA copper 2008*. Melbourne, Australia: ALTA Metallurgical Services.

Gericke, M., Govender, Y., & Pinches, A. (2010). Tank bioleaching of low-grade chalcopyrite concentrates using redox control. *Hydrometallurgy, 104*, 414−419.

Gericke, M., Neale, J. W., & van Staden, P. J. (2009). A Mintek perspective of the past 25 years in minerals bioleaching. *Journal of the Southern African Institute of Mining and Metallurgy, 109*, 567−585.

Groudeva, V. I., Vasilev, D. V., Karavaiko, G. I., & Groudev, S. N. (2010). Changes in an ore dump during a 40-year period of commercial scale and spontaneous natural bioleaching. *Hydrometallurgy, 104*, 420−423.

Guzman, A., Scheffel, R., Ahlborn, G., Ramos, S., & Flaherty, S. (2007). Geochemical profiling of a sulfide leaching operation: the rest of the story. In G. E. Houlachi, J. D. Edwards & T. G. Robinson (Eds.), *Copper 2007, Vol. VI: Hydrometallurgy*. Montreal, Canada: CIM.

Guzman, A., Scheffel, R., & Flaherty, S. (2005). *Geochemical profiling of a sulfide leaching operation: A case study*. http://www.infomine.com/publications/docs/Guzman2005.pdf.

Guzman, A., Scheffel, R., & Flaherty, S. (2008). The fundamentals of physical characterisation of ore for leach. In C. A. Young (Ed.), *Hydrometallurgy 2008* (pp. 937−966). Littleton, CO: SME.

Hackl, R. P., Dreisinger, D. B., & King, J. A. (1995). Effect of sulfur-dispersing surfactants on the oxygen pressure leaching of chalcopyrite. In W. C. Cooper, D. B. Dreisinger, J. E. Dutrizac, J. Hein & G. Ugarte (Eds.), *Cobre 95−Copper 95, Vol. III: Electrorefining and hydrometallurgy of copper* (pp. 559−578). Montreal: CIM.

Hackl, R. P., Dreisinger, D. B., Peters, E., & King, J. A. (1995). Passivation of chalcopyrite during oxidative leaching in sulfate media. *Hydrometallurgy, 39*, 25−48.

Hayton, N., Defreyne, J., & Murray, K. (2009). UHC copper refinery: an update on the Vale project based on CESL technology. In E. Domic & J. Casas (Eds.), *HydroCopper 2009* (pp. 114−123). Santiago, Chile: Gecamin.

Hiroyoshi, N., Kitagawa, H., & Tsunekawa, M. (2008). Effect of solution composition on the optimum redox potential for chalcopyrite leaching in sulfuric acid solutions. *Hydrometallurgy, 91*, 144−149.

Hiroyoshi, N., Kuroiwa, S., Miki, H., Tsunekawa, M., & Hirajima, T. (2004). Synergistic effect of cupric and ferrous ions on active-passive behaviour in anodic dissolution of chalcopyrite in sulfuric acid solutions. *Hydrometallurgy, 74*, 103−116.

Hiroyoshi, N., Miki, H., Hirajima, T., & Tsunekawa, M. (2001). Enhancement of chalcopyrite leaching by ferrous ions in acidic ferric sulfate solutions. *Hydrometallurgy, 60*, 185−197.

Hiskey, J. B. (1993). Chalcopyrite semiconductor electrochemistry and dissolution. In R. G. Reddy & R. N. Weizenbach (Eds.), *Extractive metallurgy of copper, nickel and cobalt (the Paul E. Queneau international symposium), Vol. I: Fundamental aspects* (pp. 949−969). Warrendale, PA: TMS.

House, C. I. (1987). Potential-pH diagrams and their application to hydrometallurgical systems. In G. A. Davies (Ed.), *Separation processes in hydrometallurgy* (pp. 3−19). Chichester: Ellis Horwood.

Hyvärinen, O., & Hämäläinen, M. (2005). Hydrocopper − a new technology producing copper directly from concentrate. *Hydrometallurgy, 77*, 61−65.

Iasillo, E., & Schlitt, W. J. (1999). Practical aspects associated with evaluation of a copper heap leach project. In G. V. Jergensen, II (Ed.), *Copper leaching, solvent extraction, and electrowinning technology* (pp. 123−138). Littleton, CO: SME.

Intec. (2010). *The Intec copper process*. http://www.intec.com.au/public_panel/copper_process.php.

Jones, D., Xavier, F., Gonçalves, L. R., Costa, R., & Torres, V. (2006). CESL process - semi-industrial operation at CVRD Sossego plant. In *ALTA copper 2006*. Melbourne, Australia: ALTA Metallurgical Services.

Keokhounsy, S., Moore, T., & Liu, M. (2006). Hydromet at Sepon. In *ALTA copper 2006*. Melbourne, Australia: ALTA Metallurgical Services.

King, J. A., & Dreisinger, D. B. (1995). Autoclaving of copper concentrates. In J. E. Dutrizac, H. Hein & G. Ugarte (Eds.), *Copper 95−Cobre 95, Vol. III: Electrorefining and electrowinning of copper* (pp. 511−533). Montreal: CIM.

Klauber, C. (2008). A critical review of the surface chemistry of acidic ferric sulfate dissolution of chalcopyrite with regards to hindered dissolution. *International Journal of Mineral Processing, 86*, 1−17.

Konishi, Y., Saitoh, N., Shuto, M., Ogi, T., Kawakita, K., & Kamiya, T. (2010). Bioleaching of crude chalcopyrite ores by the thermophilic archean *Acidianus brieleyi* in a batch reactor. In *Copper 2010, Vol. 5: Hydrometallurgy* (pp. 2781−2792). Clausthal-Zellerfeld, Germany: GDMB.

Kordosky, G., Feather, A., & Chisakuta, G. (2008). The Copperbelt of Africa — new plants and new technology in copper hydrometallurgy. In *ALTA copper 2008*. Melbourne, Australia: ALTA Metallurgical Services.

Lancaster, T., & Walsh, D. (1997). The development of the aeration of copper sulfide ore at Giralambone. In *BioMine '97*. Glenside, Australia: Australian Mineral Foundation. M5.4.1—10.

Lazarro, I., & Nicol, M. (2003). The mechanism of the dissolution and passivation of chalcopyrite: an electrochemical study. In C. A. Young, A. M. Alfantazi, C. G. Anderson, D. B. Dreisinger, B. Harris & A. James (Eds.), *Hydrometallurgy 2003, Vol. 1: Leaching and solution purification* (pp. 405—417). Warrendale, PA: TMS.

López, J. L., & Encínas, M. (2004). Mexicana de Cananea: Producing SX Cu from chalocite and chalcopyrite. *Cytec Solutions, 10*, 2—3.

Lupo, J. (2010). Assessment of natural and forced air in ore heaps. In J. Casas, P. Ibáñez & M. Jo (Eds.), *Hydroprocess 2010* (pp. 64—65). Santiago, Chile: Gecamin.

McClelland, D. (2007). Pressure oxidation at Sepon: opportunities abound. In *Copper 2007 short course copper hydrometallurgy, paper 2*. Montreal: CIM.

McDonald, R. G., & Muir, D. M. (2007a). Pressure oxidation leaching of chalcopyrite. Part I. Comparison of high and low temperature reaction kinetics and products. *Hydrometallurgy, 86*, 191—205.

McDonald, R. G., & Muir, D. M. (2007b). Pressure oxidation leaching of chalcopyrite: Part II: comparison of medium temperature kinetics and products and effect of chloride ion. *Hydrometallurgy, 86*, 206—220.

McElroy, R., & Young, W. (1999). Pressure oxidation of complex copper ores and concentrates. In G. V. Jergensen, II (Ed.), *Copper leaching, solvent extraction and electrowinning technology* (pp. 29—40). Littleton, CO: SME.

Mackie, D., & Trask, F. (2009). Continuous vat leaching — first copper pilot trials. In *ALTA copper 2009*. Melbourne, Australia: ALTA Metallurgical Services.

Marsden, J. O., & Brewer, R. E. (2003). Hydrometallurgical processing of copper concentrates by Phelps Dodge at the Bagdad mine in Arizona. In *ALTA copper 2003*. Melbourne, Australia: ALTA Metallurgical Services.

Marsden, J. O., Wilmot, J. C., & Hazen, N. (2007a). Medium-temperature pressure leaching of copper concentrates — Part I: chemistry and initial process development. *Minerals and Metallurgical Processing, 24*, 193—204.

Marsden, J. O., Wilmot, J. C., & Hazen, N. (2007b). Medium-temperature pressure leaching of copper concentrates — Part II: development of direct electrowinning and an acid-autogenous process. *Minerals and Metallurgical Processing, 24*, 205—217.

Marsden, J. O., Wilmot, J. C., & Mathern, D. R. (2007). Medium-temperature pressure leaching of copper concentrates — Part III: commercial demonstration at Bagdad, Arizona. *Minerals and Metallurgical Processing, 24*, 218—225.

Marsden, J. O., Wilmot, J. C., & Smith, R. J. (2007). Medium-temperature pressure leaching of copper concentrates — Part IV: application at Morenci, Arizona. *Minerals and Metallurgical Processing, 24*, 226—236.

Mendoza, V. R. (2000). Lixiviación en Mantos Blancos. In *Seminario innovación tecnológica en minería*. Santiago, Chile: Instituto de Ingenieros de Minas de Chile.

Miller, G. M. (2002). Recent advances in copper heap leach analysis and interpretation. In *Proceedings ALTA copper 2002*. Melbourne, Australia: ALTA Metallurgical Services.

Miller, G. (2003a). Ore geo-technical effects on copper heap leach kinetics. In C. A. Young, A. M. Alfantazi, C. G. Anderson, D. B. Dreisinger, B. Harris & A. James (Eds.), *Hydrometallurgy 2003, Vol. 1: Leaching and solution purification* (pp. 329—342). Warrendale, PA: TMS.

Miller, G. (2003b). Ore geo-mechanical effects on copper heap leach kinetics. In *ALTA copper 2003*. Melbourne, Australia: ALTA Metallurgical Services.

Miller, G. M. (2003). Analysis of commercial heap leaching data. In P. A. Riveros, D. Dixon, D. B. Dreisinger & J. Menacho (Eds.), *Copper 2003—Cobre 2003, Vol. VI, Book 1: Hydrometallurgy of copper* (pp. 531—545). Montreal: CIM.

Miller, G. (2008). The effective particle size in heap leaching. In *ALTA copper 2008*. Melbourne, Australia: ALTA Metallurgical Services.

Miller, K. J., Sylwestrzak, L. A., & Baxter, K. G. (2008). Treatment of copper sulfide deposits − evaluation of a Galnvanox™ versus Sepon circuit configuration. In *ALTA copper 2008*. Melbourne, Australia: ALTA Metallurgical Services.

Miller, P. (2001). Bioleaching of chalcopyrite − will it happen? *Cytec Solutions, 4*, 2−5.

Moyes, J., Sammut, D., & Houllis, F. (2002). The Intec copper process: superior and sustainable metals production. In *ALTA 2002 copper-7 forum*. Melbourne, Australia: ALTA Metallurgical Services.

Murr, L. E. (1980). Theory and practice of copper sulfide leaching in dumps and in-situ. *Minerals Science and Engineering, 12*, 121−189.

Mutch, L. A., Watling, H. R., & Watkin, E. L. J. (2010). Microbial population dynamics of inoculated low-grade chalcopyrite bioleaching columns. *Hydrometallurgy, 104*, 391−398.

Nicol, M., & Lazarro, I. (2003). The role of non-oxidative processes in the leaching of chalcopyrite. In P. A. Riveros, D. Dixon, D. B. Dreisinger & J. Menacho (Eds.), *Copper 2003−Cobre 2003, Vol. VI, Book 1: Hydrometallurgy of copper* (pp. 367−381). Montreal: CIM.

Nicol, M., Miki, H., & Velasquez-Yevenes, L. (2010). The dissolution of chalcopyrite in chloride solutions: Part 3. Mechanisms. *Hydrometallurgy, 103*, 86−95.

Nisbett, A., Feather, A., & Miller, G. (2007). Implementation of the Split Circuit™ concept into copper agitation leach plants. In J. M. Menacho & J. M. Casas (Eds.), *HydroCopper 2007* (pp. 185−190). Santiago, Chile: Gecamin.

Norton, A. E., & Crundwell, F. K. (2004). The Hotheap™ process for the lhepa leaching of chalcopyrite. In *Colloquium: Innovations in leaching technologies*. Johannesburg: Southern African Institute of Mining and Metallurgy.

Omena, A., Geraldo, J., Lima, L., & Cabral, T. (2007). Technical feasbility study to produce cathodes in the UHC Carajas plant by the hydrometallurgical CESL route. In J. M. Menacho & J. M. Casas (Eds.), *HydroCopper 2007* (pp. 67−73). Santiago, Chile: Gecamin.

Outotec. (2007). *HydroCopper®*. www.outotec.com/36286.epibrw.

Padilla, R., Pavez, P., & Ruiz, M. C. (2008). Kinetics of copper dissolution from sulfidized chalcopyrite at high temperatures in $H_2SO_4-O_2$. *Hydrometallurgy, 91*, 113−120.

Peters, E. (1976). Direct leaching of sulfides: chemistry and applications. *Metallurgical Transactions B, 7*, 505−517.

Petersen, J., & Dixon, D. G. (2006). Competitive bioleaching of pyrite and chalcopyrite. *Hydrometallurgy, 83*, 40−49.

Pradhan, N., Nathsarma, K. C., Srinivasa Rao, K., Sukla, L. B., & Mishra, B. K. (2008). Heap bioleaching of chalcopyrite: a review. *Minerals Engineering, 21*, 355−365.

Reolon, M., Gazis, T., & Amos, S. (2009). Optimizing acid utilisation and metal recovery in African Cu/Co flowsheets. In *Hydrometallurgy Conference 2009* (pp. 1−20). Johannesburg, South Africa: Southern African Institute of Mining and Metallurgy. http://www.saimm.co.za/Conferences/Hydro2009/365-384_Reolon.pdf.

Riekkola-Vanhanen, M. (2007). Talvivaara project − bioheap leaching of a complex sulfide ore in boreal conditions. In *ALTA copper 2007*. Melbourne, Australia: ALTA Metallurgical Services.

Robles, E. (2009). Economical considerations in the analysis of alternative processes for the treatment of copper sulfides. In E. Domic & J. Casas (Eds.), *HydroCopper 2009* (pp. 217−226). Santiago, Chile: Gecamin.

Roux, L., van Rooyen, M., Minnaar, E., Robles, E., & Cronje, I. (2010). Copper oxide agitated leaching on the African Copperbelt and possible heap leach application. In *Copper 2010, Vol. 5: Hydrometallurgy* (pp. 1999−2016). Clausthal-Zellerfeld, Germany: GDMB.

Salgado, C., Pérez, C., & Alvayai, C. (2009). Acid management in heap leaching. Are we doing it right? In E. Domic & J. Casas (Eds.), *HydroCopper 2009* (pp. 68−76) Santiago, Chile: Gecamin.

Salomon-de-Friedberg, H. (1998). Design aspects of aeration in heap leaching. In *Randol copper hydromet roundtable '98* (pp. 243−247). Golden, CO: Randol International.

Salomon-de-Friedberg, H. (2000). Quebrada Blanca: lessons learned in high altitude leaching. Paper presented to Instituto de Ingenieros de Minas de Chile, Expomin 2000, Santiago, Chile, May 2000.

Scheffel, R. E. (2002). Copper heap leach design and practice. In A. L. Mular, D. N. Halbe & D. J. Barratt (Eds.), *Mineral processing plant design, practice, and control, Vol. 2* (pp. 1571−1605). Littleton, CO: SME.

Schlitt, J. (1995). Copper leaching and recovery: 1000 years and counting. In *ALTA copper hydrometallurgy forum*. Melbourne, Australia: ALTA Metallurgical Services.

Schlitt, J., & Johnston, A. (2010). The Marcobre vat leach system: a new look at an old process. In *Proceedings copper 2010, Vol. 5: Hydrometallurgy* (pp. 2039–2057). Clausthal-Zellerfeld, Germany: GDMB.

Schlitt, W. J. (2006). The history of forced aeration in copper sulfide leaching. In *Proceedings SME annual meeting*. Littleton, CO: SME. Preprint 06–019.

Senanayake, G. (2009). A review of chloride-assisted copper sulfide leaching by oxygenated sulfuric acid and mechanistic considerations. *Hydrometallurgy*, 98, 21–32.

Seyedbagheri, A., van Staden, P., & McLaren, C. (2009). A study of acid-gangue reactions in heap leach operations. In E. Domic & J. Casas (Eds.), *HydroCopper 2009* (pp. 68–76). Santiago, Chile: Gecamin.

Sherrit, R., Pavlides, A. G., & Weekes, B. L. (2005). Design and commissioning of the Sepon copper pressure oxidation circuit. In *1st extractive metallurgy operators' conference* (pp. 21–27). Victoria: Australasian Institute of Mining and Metallurgy.

Sole, K. C. (2008). Solvent extraction in the hydrometallurgical processing and purification of metals: process design and selected applications. In M. Aguilar & J. L. Cortina (Eds.), *Solvent extraction and liquid membranes: fundamentals and applications in new materials* (pp. 141–200). New York: Taylor and Francis.

Steyl, J. D. T. (2004). The effect of surfactants on the behaviour of sulfur in the oxidation of chalcopyrite at medium temperature. In M. J. Collins & V. G. Papangelakis (Eds.), *Pressure hydrometallurgy 2004* (pp. 101–117). Montreal: CIM.

Terra Nova Technologies. (2010). *Featured project: El Abra.* www.tntinc.com.

Velarde, G. (2005). Agglomeration control for heap leaching processes. *Mineral Processing and Extractive Metallurgy Review*, 26, 219–231.

Velasquez-Yevenes, L., Miki, H., & Nicol, M. (2010). The dissolution of chalcopyrite in chloride solutions: part 2: effect of various parameters on the rate. *Hydrometallurgy*, 103, 80–85.

Velasquez-Yevenes, L., Nicol, M., & Miki, H. (2010). The dissolution of chalcopyrite in chloride solutions: part 1. the effect of solution potential. *Hydrometallurgy*, 103, 108–113.

Wang, S. (2005). Copper leaching from chalcopyrite concentrates. *JOM*, 57(7), 48–51.

Watling, H. R. (2006). The bioleaching of sulfide minerals with emphasis on copper sulfides – a review. *Hydrometallurgy*, 84, 81–108.

Yoo, K., Kim, S.-K., Lee, J.-C., Ito, M., Tsunekawa, M., & Hiroyoshi, N. (2010). Effect of chloride ions on leaching rate of chalcopyrite. *Minerals Engineering*, 23, 471–477.

Zárate, G., Saldívar, E., Rojes, C., & Nuñez, J. (2004). Vat leach cycle optimisation at Anglo American Chile Mantos Blancos division. *Cytec Solutions*, 10, 4–5.

SUGGESTED READING

Habashi, F. (2009). Kinetics and mechanisms of leaching copper minerals. In E. Domic & J. Casas (Eds.), *HydroCopper 2009* (pp. 176–193). Santiago, Chile: Gecamin.

Klauber, C. (2008). Why chalcopyrite is so difficult to dissolve. In *ALTA copper 2008*. Melbourne, Australia: ALTA Metallurgical Services.

Ojumu, T. V., Petersen, J., Searby, G. E., & Hansford, G. S. (2006). A review of rate equations proposed for microbial ferrous-iron oxidation with a view to application to heap bioleaching. *Hydrometallurgy*, 83, 21–28.

Scheffel, R. E. (2002). Copper heap leach design and practice. In A. L. Mular, D. N. Halbe & D. J. Barratt (Eds.), *Mineral processing plant design, practice, and control, Vol. 2* (pp. 1571–1605). Littleton, CO: SME.

Schlitt, J. (1995). Copper leaching and recovery: 1000 years and counting. In *ALTA copper hydrometallurgy forum*. Melbourne, Australia: ALTA Metallurgical Services.

Schlitt, J. (2005). Things learned in 35 years of leaching. In *ALTA copper 2005*. Melbourne, Australia: ALTA Metallurgical Services.

Chapter 16

Solvent Extraction

Solvent extraction (SX) purifies and upgrades the pregnant leach solution (PLS) produced by the leaching operation (Chapter 15) to generate an electrolyte from which high quality copper cathode can be electrowon (Chapter 17).

Most leach liquors, irrespective of the leaching process by which they are produced, are too dilute in copper (1–10 g/L Cu), too impure (containing 1–20 g/L Fe, as well as other deleterious species), and not sufficiently conductive for direct electrodeposition of high purity cathode copper. Electrolysis of these solutions would produce soft, powdery, and impure copper deposits.

Industrial electrowinning (EW) requires pure copper-rich electrolytes (45–55 g/L Cu) with high conductivity. This high concentration of copper ensures that (a) Cu^{2+} ions are always available for plating at the cathode surface at current densities that allow economical rates of plating, being readily renewed by mass transfer, and (b) gives smooth, dense, high purity, readily marketable cathode copper. The high conductivity of the electrolyte is provided by a high acid concentration, typically 175–190 g/L H_2SO_4.

SX provides the means for producing pure, concentrated Cu electrolytes from dilute, impure leach liquors. It is a crucial step in the production of ~4.5 million tonnes of copper cathode per year (ICSG, 2010). The combined technology of SX-EW continues to grow in importance as more copper is produced by leaching and lower grade materials are treated for recovery of copper (Kordosky, 2002).

16.1. THE SOLVENT-EXTRACTION PROCESS

SX flowsheets used in purifying of copper leach liquors comprise two essential continuous processes: *extraction* and *stripping*, with the following steps (Fig. 16.1):

(a) In the *extraction* process, the PLS (feed to the SX process; 1–10 g/L Cu^{2+}, 0.5–5 g/L H_2SO_4) is contacted with an organic phase containing a Cu-specific organic *extractant*. The extractant complexes with Cu^{2+}, resulting in the transfer of the metal ion from the aqueous phase into the organic phase, leaving all other impurity species present in the PLS in the aqueous phase.
(b) The organic phase, now *loaded* with Cu, and the depleted aqueous phase (*raffinate*), are separated by gravity.
(c) The raffinate is recycled back to the leach circuit where the acid generated in the extraction process can be used.
(d) In the *stripping* process, the Cu-loaded organic phase is contacted with a strong acid *spent* (or *lean, barren, weak*) *electrolyte* (175–190 g/L H_2SO_4) from the EW circuit, which strips Cu from the organic into the electrolyte.

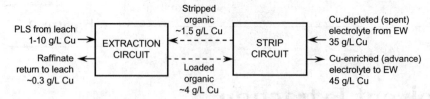

FIGURE 16.1 Schematic copper SX process. The inputs are pregnant leach solution (PLS) and Cu-depleted (spent) electrolyte. The products are Cu-enriched (advance) electrolyte and low-Cu raffinate. Aqueous streams are shown as solid lines, while organic flows are indicated by dashed lines. The extraction and strip circuits may each contain several stages, through which the aqueous and organic streams flow in counter-current directions (see Section 16.6).

(e) The Cu-depleted organic phase (known as the *stripped* or *barren organic*) and the Cu-enriched aqueous phase (known as the *advance*, *rich*, *loaded*, or *pregnant electrolyte*) are separated by gravity.

(f) The stripped organic phase is recycled back to the extraction circuit for renewed contact with fresh PLS.

(g) The advance electrolyte is sent to EW where Cu^{2+} is reduced and electrodeposited on the cathode as pure metallic copper.

16.2. CHEMISTRY OF COPPER SOLVENT EXTRACTION

The organic extractant removes Cu^{2+} from the PLS by the extraction reaction:

$$2RH(org) + Cu^{2+}(aq) + SO_4^{2-}(aq) \rightarrow R_2Cu(org) + 2H^+(aq) + SO_4^{2-}(aq) \quad (16.1)$$

stripped organic   PLS (1–10 g/L Cu)   loaded organic   raffinate (0.05–0.5 g/L Cu)

where RH represents the aldoxime or ketoxime extractant (see Section 16.3).

From the Law of Mass Action (Le Chatelier's Principle) loading of the extractant with Cu is favored by a low acid concentration (H^+) in the aqueous phase. The dilute H_2SO_4 concentration of the PLS (pH 1.5–2.5) permits extraction of Cu from the aqueous phase into the organic phase. Two protons (H^+) are generated for each Cu^{2+} that is extracted, so the extraction reaction produces acid.

The organic and aqueous phases are then separated by gravity. The Cu^{2+}-depleted, H^+-enriched raffinate is returned to the leach circuit to dissolve more copper. The copper-loaded organic phase is sent forward to the strip circuit where copper is stripped back into an aqueous phase using the high acid content of Cu^{2+}-depleted spent electrolyte. The strip reaction is the reverse of Eq. (16.1):

$$2H^+(aq) + SO_4^{2-}(aq) + R_2Cu(org) \rightarrow 2RH(org) + Cu^{2+}(aq) + SO_4^{2-}(aq) \quad (16.2)$$

high acid, Cu–depleted   loaded organic   stripped organic   Cu–replenished electrolyte
electrolyte (~180 g/L H_2SO_4,   extractant   extractant   (~165 g/L H_2SO_4,
~35 g/L Cu)   ~45 g/L Cu)

The equilibrium of the strip reaction is pushed to the right by the high acid concentration of the spent electrolyte. Cu is stripped from the organic extractant, regenerating the extractant to be recycled to the extraction circuit, and enriching the electrolyte to the required high Cu^{2+} concentration necessary for EW.

Chapter | 16 Solvent Extraction

In summary, the extractant-containing organic phase is:

(a) Loaded with Cu^{2+} from low acid-containing PLS;
(b) Separated from the PLS by gravity;
(c) Contacted with high acid-containing electrolyte and stripped of copper.

SX is an equilibrium-controlled process.

16.3. COMPOSITION OF THE ORGANIC PHASE

The organic phase comprises two essential components:

(a) The *extractant* is the active molecule that complexes with Cu^{2+}, enabling it to be transferred from the aqueous to the organic phase;
(b) The *diluent* is an inert hydrocarbon carrier for the extractant, which reduces the viscosity of the organic phase (so that it can be easily mixed with the aqueous phases to allow mass transfer to occur) and reduces the density of the organic phase (to allow for gravity separation of the two phases).

The organic phase should be immiscible with $CuSO_4$–H_2SO_4 solutions and fluid enough (viscosity = 2–10 cP) for pumping, continuous mixing, and gravity separation.

16.3.1. Extractants

Modern copper extractants are based on the *oxime* functionality. These molecules complex with the Cu^{2+} ion in a process known as *chelation* (Fig. 16.2). It is the nature of this chelating reaction that allows these extractants to react very selectively with Cu^{2+}, rejecting other species in the PLS to the raffinate.

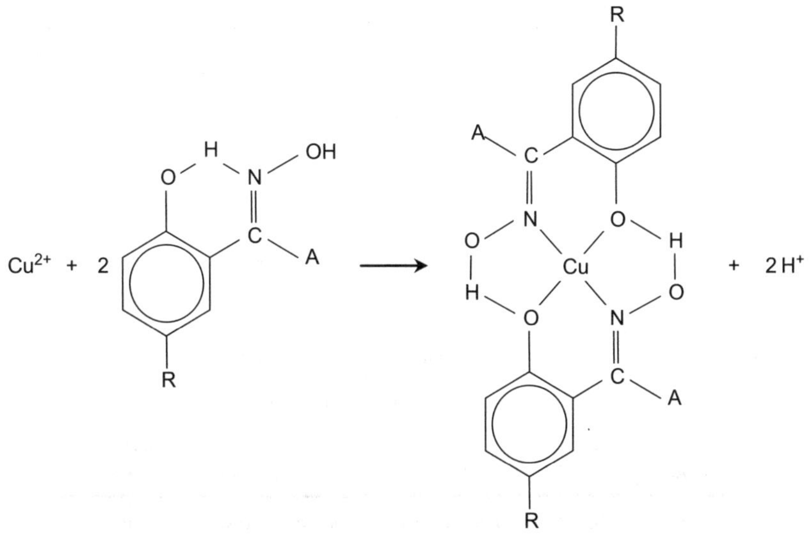

FIGURE 16.2 Structure of oxime extractant and Cu-oxime complex. The copper complex is formed from reaction of a Cu^{2+} cation with two oxime molecules, releasing two protons (Eq. (16.1)). For aldoximes: R = C_9H_{19} or $C_{12}H_{25}$, A = H. For ketoximes: R = C_9H_{19}, A = CH_3 (Kordosky, 2008; Sole, 2008).

An extractant should fulfill the following process requirements (Kordosky, Sudderth, & Virnig, 1999):

(a) Efficiently extract Cu from the PLS at typical pH values of leach liquors;
(b) Efficiently strip Cu into the EW electrolyte under the acid concentrations required for typical electrolytes;
(c) Have economically rapid extraction and strip kinetics;
(d) Allow rapid and complete disengagement of the aqueous and organic phases in both the extraction and strip circuits, i.e., not form a stable emulsion;
(e) Be insoluble in the aqueous phases;
(f) Be stable under extraction and strip conditions so that it can be recycled many times;
(g) Not absorb sulfuric acid;
(h) Extract Cu preferentially over other metals in the PLS, particularly Fe and Mn (which are detrimental to the subsequent EW step — see Chapter 17);
(i) Not transfer deleterious species from PLS to electrolyte, particularly Cl^- (which causes pitting of copper cathodes in EW);
(j) Be soluble in an inexpensive petroleum distillate diluent;
(k) Be non-flammable, non-toxic, and non-carcinogenic.

TABLE 16.1 Properties of Modern Copper Extractants (Adapted from Sole, 2008)

Property	Modified aldoximes[a]	Aldoxime–ketoxime mixtures[b]	Modified aldoxime–ketoxime mixtures[b]	Ketoximes
Extractive strength	Strong/customized	Strong/customized	Customized	Weak
Stripping ability	Customized	Customized	Very good	Excellent
Cu/Fe selectivity	Excellent	Moderate	Good	Low
Kinetics of extraction	Very fast	Fast	Fast	Slower
Rate of phase separation	Very fast	Very fast	Very fast	Very fast
Chemical stability	Very good	Very good	Very good	Excellent
Crud generation	Dependent on PLS and operation			
Entrainment losses	Dependent on operation and diluent			
Examples	Acorga M5640	LIX 984N	Acorga OPT5510	LIX 84-I

[a] Esters are most commonly employed as the modifier, although tridecanol-modified aldoxime extractants are occasionally employed. The strength of the extractant depends on the relative aldoxime:modifier ratio.
[b] Aldoxime–ketoxime extractants are customized by their relative quantities: the greater the proportion of aldoxime, the stronger the extractant, i.e., the more Cu it will extract at lower pH values.

There are four main classes of modern copper extractants (Table 16.1): *ketoximes, modified aldoximes, modified aldoxime−ketoxime mixtures,* and *aldoxime−ketoxime mixtures*. The selection of which extractant is most appropriate for a given application will depend on the properties of the PLS, the characteristics of the advance electrolyte required, the number of extraction and strip stages available, and, often, on the prevailing economic conditions. Economic conditions may influence whether a particular plant is trying to achieve maximum copper recovery from the PLS or maximum copper throughput (Kordosky, 2008).

Ketoximes are relatively weak, low selectivity extractants, but have the best chemical stability. They are therefore favored for niche applications, such as the treatment of leach solutions that contain species like nitrates (common at some plants in the Atacama Desert of Chile) that are strong oxidizing agents and would otherwise rapidly degrade the extractant (Hurtado-Guzmán & Menacho, 2003; Virnig, Eyzaguirre, Jo, & Calderon, 2003).

Aldoximes are strong extractants. However they can only be stripped by very high acid concentrations (+225 g/L H_2SO_4). This level of acid is too corrosive for industrial EW and also tends to degrade the extractant. For these reasons, aldoximes are usually mixed with ketoximes and/or modifiers, such as highly branched alcohols or esters (Kordosky & Virnig, 2003; Maes, Tinkler, Moore, & Mejías, 2003). Modified aldoximes do not require very high acid strengths for stripping and can be tailored to the desired extractant strength. They exhibit better selectivity for Cu over Fe than aldoxime−ketoxime mixtures and are often favored when the PLS has a high iron content. Aldoxime−ketoxime mixtures are stripped at lower acid concentrations, and are appropriate when PLS Cu concentrations are lower and there is no significant concentration of Fe in the PLS. Mixtures of modified aldoximes and ketoximes have properties intermediate between modified aldoximes and aldoxime−ketoxime mixtures (Soderstrom, 2006).

16.3.2. Diluents

Oximes are thick viscous liquids that are unsuitable for pumping, mixing, and phase separation. The extractants are therefore formulated to make them fluid and then added to a diluent at concentrations of 5−35 vol.-% for use. The extractant concentration chosen depends on the copper concentration of the PLS (see Section 16.7).

The selection of the diluent will depend on local availability, the concentration of extractant employed, and safety, health, and environmental (SHE) considerations. Diluents are moderately refined, high flash point petroleum distillation byproducts (purified kerosene) from the petrochemical industry. They range in aromaticity from aliphatic (fully saturated, hydrocarbon chains) to fully aromatic (unsaturated ring structures). Some of the advantages and disadvantages of choosing an aliphatic over partially aromatic diluent for Cu SX are given in Table 16.2.

Modern copper plants are concerned with minimizing their environmental impact and carbon footprint, as well as ensuring the safest possible working conditions for their operators (Hopkins, 2005; Miller, 2005; Smith, 2009). For these reasons, a diluent with the lowest possible aromaticity that still gives favorable chemical and physical behavior will usually be selected. Diluents with 0−20% aromatic content are typically preferred, as these provide the best compromise of properties: good solvation of the copper-extractant complex in the organic phase, good phase separation from the aqueous phases, and acceptable SHE characteristics.

TABLE 16.2 Properties of Aliphatic Diluents Compared to Partially Aromatic Diluents

Advantages of aliphatic diluents

- Higher flash point assists in minimizing fire risk
- Lower volatility reduces evaporation losses and improves operator environment
- Lower rate of oxidative degradation
- Lower carbon footprint

Disadvantages of aliphatic diluents

- Increased viscosity, especially at lower temperatures
- Can reduce solubilization of complexes, especially at high extractant concentrations
- Reduced solubility of impurities, which may increase crud formation
- Less selective for Fe transfer by diluent
- Slower rate of phase separation
- Higher organic-in-aqueous entrainment, leading to higher extractant and diluent consumptions
- May have poorer crud management

16.4. MINIMIZING IMPURITY TRANSFER AND MAXIMIZING ELECTROLYTE PURITY

To produce high purity copper cathode in EW (Chapter 17), the advance electrolyte produced by the SX step must minimize contamination from impurities. An efficient extractant must therefore transfer Cu from the PLS to the electrolyte while *not* transferring impurities, particularly Fe, Mn, and Cl.

There are two mechanisms by which impurities can be carried forward to the electrolyte:

(a) *Chemical transfer*, in which cationic impurity species are co-extracted by the extractant along with Cu^{2+};
(b) *Physical transfer*, in which finely divided droplets of aqueous phase are not completely separated from the organic phase after mixing and are transferred to the strip circuit in the loaded organic phase. Both cationic and anionic impurities can be transferred in this manner, known as *aqueous-in-organic entrainment*.

Fortunately, modern oxime extractants, particularly modified aldoximes, are very selective for Cu over other cations, especially Fe, so chemical transfer of impurities can be minimized by appropriate choice of extractant for a particular PLS (Tinkler, Cronje, Soto, Delvallee, & Hangoma, 2009).

Entrainment can be minimized by:

(a) Appropriate choice of diluent to promote acceptable phase disengagement and to minimize transfer of neutral species by solvation by the diluent;
(b) Choice of a low density, low viscosity extractant, particularly for circuits in which a high extractant concentration is required;

(c) Avoiding over-mixing and air entrainment into the mixing boxes (see Section 16.5.1);
(d) Installing picket fences, coalescence media, or other features in the settler to promote phase disengagement (see Section 16.5.2);
(e) Ensuring adequate residence time for phase disengagement in the settlers;
(f) Contacting the loaded organic phase with a hydrophobic media (such as polymer scrap) that will assist coalescence of the aqueous droplets;
(g) *Washing out* droplets of PLS from the loaded organic in a *wash* stage (see Section 16.6.3).

16.5. EQUIPMENT

Copper SX is carried out in mixer—settlers (Fig. 16.3). While details of the equipment design vary (see, for example, Anderson, Giralico, Post, Robinson, & Tinkler, 2002; Gigas & Giralico, 2002; Giralico, Gigas, & Preston, 2003; Miller, 2006; Vancas, 2003), the essential functions always consist of:

(a) Pumping the aqueous and organic phases into a mixer at predetermined rates;
(b) Mixing the two phases with an impeller in two or three mixing stages;
(c) Overflowing the aqueous—organic mixture from the mixer through flow distributors into a flat settler where the phases separate by gravity based on the difference in their specific gravities (organic and aqueous phases have specific gravities of approximately 0.85 and 1.1, respectively);
(d) Overflowing the organic phase and underflowing the aqueous phase at the far end of the settler once complete phase disengagement has taken place.

Flowrates of aqueous and organic phases through mixer—settlers can range from 2—4000 m^3/h.

16.5.1. Mixer Designs

The *mixer* is designed to create a well-mixed aqueous—organic dispersion to ensure good mass transfer for the extraction and stripping reactions. The drop-size distribution should

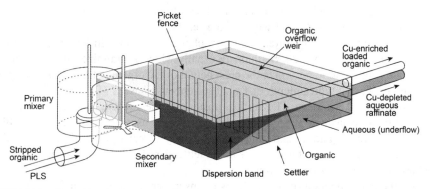

FIGURE 16.3 Conventional mixer—settler used for copper SX. Note the two mixing compartments, the large settler, and the organic overflow/aqueous underflow system. Flow is distributed evenly across the settler by passing the emulsion from the mixer first through a chevron (not shown) and then through one or more picket fences.

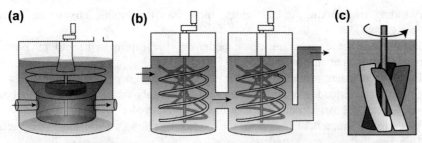

FIGURE 16.4 Novel mixer designs. The pumping and mixing functions are separated in Outotec technology, using first (a) a dispersion overflow pump (DOP) mixer, able to pump large flow volumes into the mixer which achieves high mass transfer efficiency while controlling the size of droplets formed, followed by (b) Spirok mixers which allow gentle mixing of the emulsion and minimize crud formation (Pekkala, Laitala, Nyman, Ekma, & Weatherseed, 2007). In Bateman's turbulent settling design (c) homogeneous mixing with a narrow drop-size distribution enhances the subsequent settling steps and minimizes entrainment and crud losses (Kokotov et al., 2010).

be as narrow as possible, with minimum air entrainment (which can stabilize crud; see Section 16.9.2) and without forming a stable emulsion (which inhibits subsequent phase separation).

Modern mixers consist of two or three mixing chambers. The smaller first chamber will contain a pumping impeller (~0.02 kWh/m^3), which is required to *pump* the two phases into the mixer from an adjacent mixer—settler or a holding tank. Subsequent mixers are larger and operate more gently (~0.01 kWh/m^3), serving to maximize mass transfer. The extraction and stripping kinetics of copper SX systems are relatively slow: contact times of 2—3 min are required for the chemistry to approach equilibrium. Entrainment of very fine droplets is avoided by using low tip-speed (<350 m/min) impellers. Modern mixer designs, illustrated in Fig. 16.4, enable good mass transfer while minimizing the formation of fine drops (Kokotov et al., 2010; Outotec, 2007).

16.5.2. Settler Designs

The *settler* is designed to separate the dispersion into separate aqueous and organic layers. When the dispersion overflows from the mixer, it passes through one or more flow distributors to give smooth, uniform forward flow, which allows the aqueous and organic phases to separate across the full width of the settler.

Large modern settlers are often square in plan and can vary in length up to 30 m, depending on the PLS flowrate. Settlers are usually wide to minimize linear velocity at low organic depths and to minimize the crest height of the organic phase overflowing the weirs. Settlers have a typical depth of 1 m. The aqueous phase is ~0.5 m deep, while the organic phase is ~0.3 m deep. The vertical position of the aqueous—organic interface is controlled by an adjustable weir. Copper SX settlers are usually designed for a settling rate of 4—5 m^3/m^2 h and a maximum linear velocity of each phase of 0.06 m/s. This means that residence times of each phase in the settlers are 10—20 min, which is sufficient to achieve complete phase separation.

Since the main function of the settler is to ensure smooth horizontal flow and clean separation (no entrainment) of the two phases, considerable effort has gone into various designs to promote coalescence and phase disengagement. The most common of these are flow distributors, known as *chevrons* or *picket fences*, which can be installed at various

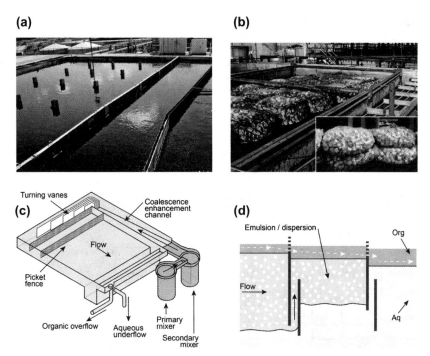

FIGURE 16.5 Modern settler designs include features that enhance droplet coalescence and phase separation from the emulsion overflowing from the mixer. (a) Chevron and picket fence across an open mixer–settler. (b) Coalescence packs installed in the settler can act as a coalescence media (Courtesy of Miller Metallurgical Services). (c) In reverse-flow settlers, the emulsion is directed along a launder on the side of the settler, then is further slowed by turning vanes before settling uniformly across the settler in a reverse direction (Vancas, 2003). (d) Deep dispersion gate (DDG) system for slowing flow and assisting coalescence (Outotec, 2007).

positions along the length of the settler (Fig. 16.5a). They assist in slowing down the velocity of the emulsion, which backs up behind the installation and the emulsion itself acts as a coalescence medium. In some cases, a hydrophobic or hydrophilic coalescence material can be inserted into a settler (coalescence packs, Fig. 16.5b), which can assist with coalescence of the dispersed phase, but are also prone to mechanical blockage if solids enter the settler or precipitation occurs.

Engineering innovations include the use of *reverse-flow settlers* (Fig. 16.5c) in which the mixer overflow emulsion is directed into a launder running along the side of a settler, before passing through an angled chevron and then settling in a direction back toward the mixer. The *deep dispersion gate* (DDG®) is another patented design that promotes depth coalescence, allowing additional time and assistance for the coalescence process (Fig. 16.5d).

16.6. CIRCUIT CONFIGURATIONS

16.6.1. Series Circuit

The simplest Cu SX circuit typically has two extraction stages in *series*, with countercurrent flow of the aqueous and organic phases, and a single strip stage (Fig. 16.6a). The two extraction stages typically transfer ~90% of Cu present in the PLS onto the extractant. However, the remaining Cu in the raffinate is not lost: it merely circulates around the leach circuit. The single strip stage strips 50–65% of the Cu from the

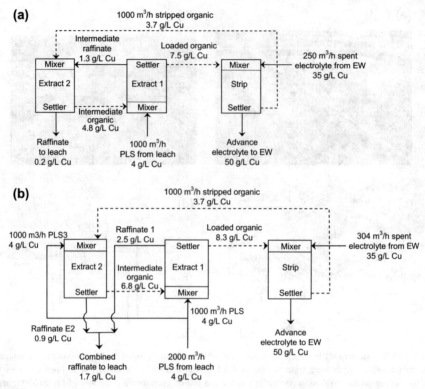

FIGURE 16.6 Plan view of (a) series and (b) parallel SX circuits with two extraction mixer–settlers and one strip mixer–settler. Industrial mixer–settlers are built close together to minimize plant area and flow distances. This is also the reason that most plants are arranged in a 'head-to-tail' arrangement, as shown. Note the increased copper production that can result from a series-parallel arrangement, albeit with a higher raffinate concentration (lower overall copper recovery from PLS). Details of the flowrates and Cu concentrations are given in Tables 16.3 and 16.4.

loaded organic into the advance electrolyte (depending on the extractant and operating conditions). The remainder circulates around the SX circuit. This *2E-1S* configuration tends to predominate in the industry, although 2E-2S circuits are very common in Chile.

The efficiency of Cu transfer can be increased by adding extraction and strip stages to the circuit, increasing the extractant concentration, or changing the relative flowrates of the organic and aqueous flowrates (the *O/A ratio*). Depending on the characteristics of the PLS (pH, Cu concentration, impurities), up to three extraction stages and two strip stages may be employed.

In plants which have very large PLS flowrates, the PLS may be split into several *trains*. Each train is processed in an identical SX circuit, capable of treating 500–4000 m^3/h. Each train may transfer 20–380 t Cu from PLS to electrolyte per day, depending on the Cu content and flowrate of the PLS. Modern engineering practice often chooses this flow arrangement. It allows for minimizing production losses that may occur should one circuit become non-functional in case of fire or other damage, because additional flow can then be routed through the remaining trains (Hopkins, 2005).

Chapter | 16 Solvent Extraction

16.6.2. Parallel and Series-parallel Circuits

To increase copper production, a *parallel* or *series-parallel* flow arrangement may be used instead of series flow in the extraction circuit (Fig. 16.6b). In this example, PLS is fed in parallel into the mixers of both extraction stages, while the flow of the organic phase remains the same as in the series circuit.

This arrangement allows the same circuit to process double the PLS flowrate, provided that the extractant concentration can be increased to transfer the additional copper. The overall rate of copper transfer from the PLS to the EW circuit can therefore be increased. Introducing a parallel flow in a circuit is a low-cost way of increasing the productivity of an existing series plant. A plant will often be converted from a series to parallel or series-parallel flow configuration toward the end of the plant life, when the ore grades are lower. This conversion allows the leach volume to remain the same even though the Cu concentration in the leach liquor is decreasing.

The disadvantage of this circuit is that overall copper recoveries are reduced (i.e., raffinate Cu concentrations increase), due to lower efficiency of extracting Cu from PLS into organic than that of a series circuit. This is because (a) the parallel portion of the PLS goes through only one extraction mixer—settler (rather than two) before returning to leach and (b) the series portion of the PLS is contacted with organic phase that is already partially loaded with Cu in the second extraction stage, so extraction efficiency in the first extraction stage will be reduced.

The patented Optimum Series-Parallel circuit is also being increasingly adopted, particularly in Chilean operations (Nisbett, Peabody, & Kordosky, 2008, Nisbett, Peabody, & Crane, 2009; Rojas, Araya, Picardo, & Hein, 2005).

16.6.3. Inclusion of a Wash Stage

In some circuits, a *wash* stage is included between the extraction and strip circuits. The purpose of this circuit is to *wash out* impurities that are physically entrained in the loaded organic phase to avoid these being transferred to the electrolyte. The wash liquor is typically dilute H_2SO_4 (pH 2). The spent wash liquor is returned to the extraction circuit to ensure that no copper is lost.

Examples of impurities that may be carried across are chlorides and nitrates (anionic species that are not chemically complexed by the extractant), as well as Fe, Mn, and Co that may be present in the PLS. Many plants in the Atacama Desert of Chile, where the PLS contains significant quantities of chlorides and nitrates, use a wash stage (Hein, 2005).

16.7. QUANTITATIVE DESIGN OF A SERIES CIRCUIT

This section describes how a preliminary design of a series SX circuit is carried out. It is based on the design criteria in Table 16.3 and the circuit illustrated in Fig. 16.6a.

16.7.1. Determination of Extractant Concentration Required

As a starting point, one rule of thumb used in the industry is that for each 1 g/L Cu extraction required, the concentration of extractant required will be approximately 4 vol.-%. Therefore, for a PLS containing 4 g/L Cu, an extractant concentration of 16% is required. The SX plant requires 1000 m^3/h of 16% extractant diluted in an appropriate diluent (see Section 16.3.2) to be pumped into the extraction circuit.

TABLE 16.3 Preliminary Specifications for the Design of a Cu SX Plant

No of extraction stages	2
No of strip stages	1
PLS composition	4 g/L Cu, pH 2
PLS flowrate to be processed	1000 m^3/h
Specified O/A ratio in extraction circuit	1/1 (m^3/h organic flow/m^3/h aqueous flow)[a]
Composition of spent electrolyte	35 g/L Cu, 180 g/L H$_2$SO$_4$
Composition of advance electrolyte required	50 g/L Cu

[a]O/A ratios of ~1 permit easy switching between aqueous-continuous and organic-continuous operation of the mixer (see Section 16.9.3).

This calculation is the first step in selecting the composition of the organic phase for a proposed SX circuit. The chosen organic must then be tested with actual leach and EW solutions to ensure suitability for the proposed operation.

16.7.2. Determination of Extraction and Stripping Isotherms

Experimental laboratory testwork is carried out to determine the equilibrium loading and stripping characteristics of the selected organic phase with the PLS and electrolyte compositions. An extraction isotherm is measured by contacting the PLS and stripped organic phase batch-wise at different O/A ratios, allowing each system to reach equilibrium. The Cu concentrations in each phase are analyzed. By plotting the Cu in the aqueous phase against the Cu in the organic phase, the maximum loading of Cu into that particular organic phase can be determined (Fig. 16.7a). Stripping isotherms can be similarly determined using loaded organic and spent electrolyte as the input streams.

16.7.3. Determination of Extraction Efficiency

Using the experimental extraction isotherm and the specified extraction O/A, a *McCabe–Thiele diagram* can be employed to determine the raffinate concentration that will be produced from two stages of counter-current contact of PLS and stripped organic. The graphical output is seen in Fig. 16.7a. This shows that, for the operating conditions specified in Table 16.3, the raffinate will contain 0.2 g/L Cu, corresponding to an extraction efficiency of 94.8%. This calculation also shows that the maximum loading of the extractant is 7.5 g/L Cu under these conditions.

16.7.4. Determination of Equilibrium Stripped Organic Cu Concentration

The high-H$_2$SO$_4$ concentration in spent electrolyte (180 g/L) strips Cu from the loaded organic into the electrolyte (Eq. 16.2). From the strip isotherm (Fig. 16.7a), it is determined that the 50 g/L Cu electrolyte specified for return to the EW tankhouse will be at equilibrium with ~3.8 g/L Cu in the stripped organic phase.

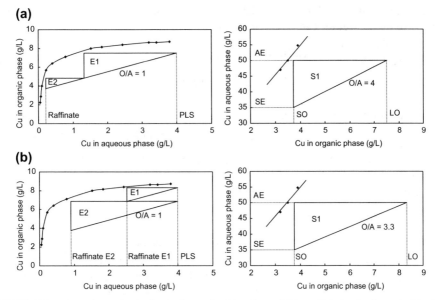

FIGURE 16.7 Extraction and stripping isotherms for (a) series and (b) parallel circuits shown in Fig. 16.6. McCabe—Thiele constructions are used to determine the Cu concentrations of the raffinate and stripped organic (SO). AE: advance electrolyte; SE: spent electrolyte; LO: loaded organic. Note that a *stage efficiency* of 90% has been assumed for both extraction and stripping. This means that mixing is only 90% efficient, as is common in practice. For quantitative mixing, the left-hand sides of the McCabe—Thiele horizontal construction lines would coincide with the equilibrium isotherm lines.

16.7.5. Transfer of Cu Extraction into Organic Phase

The overall mass flowrate at which Cu is extracted into the organic phase is given by the mass balance:

$$\text{Cu extraction (kg/h Cu)} = \text{PLS flowrate}(m^3/h) \times (\text{Cu in PLS} - \text{Cu in raffinate})(g/L) \quad (16.3)$$

(One g/L is equal to one kg/m^3.) In this case (and converting to appropriate units):

$$\begin{aligned}\text{Cu extraction (kg/h Cu)} &= 1000 \, m^3/h \, \text{PLS} \\ &\times (4.0 \, g/L \, \text{Cu in PLS} - 0.2 \, g/L \, \text{Cu in raffinate}) \\ &= 3800 \, kg/h \, \text{Cu}\end{aligned}$$

This is also the overall mass flowrate at which metallic copper will have to be plated in the EW tankhouse. It allows the design engineer to calculate the cathode area and current density required for the proposed EW plant (see Chapter 17).

16.7.6. Determination of Electrolyte Flowrate Required to Strip Cu Transferred

Section 16.7.5 indicates that the EW plant must plate 3800 kg/h metallic copper. For the strip circuit, this means that 3800 kg/h Cu must be transferred from the loaded organic phase to the electrolyte. By mass balance and using the specified compositions of spent and advance electrolyte (Table 16.3), the rate at which electrolyte must flow into and out of the strip mixer—settler can be calculated (Fig. 16.1):

$$\genfrac{}{}{}{}{\text{Rate of Cu plating}}{\text{in EW(kg/h Cu)}} = \text{electrolyte flowrate}(m^3/h)$$

$$\times (\text{Cu in advance electrolyte} - \text{Cu in spent electrolyte})$$

The electrolyte flowrate required is therefore:

electrolyte flowrate = (3800 kg/h Cu) ÷ (50 − 35) kg/m^3 Cu = 253 m^3/h.

In other words, the O/A in the strip circuit will be 1000 m^3/h (organic flowrate) ÷ 253 m^3/h (aqueous flowrate) = 4.

Fig. 16.6a summarizes the flows and Cu concentrations in the newly designed plant.

16.7.7. Alternative Approach

In the previous example, the relative flowrates of the aqueous and organic phases were specified (O/A ratio) and the extraction efficiency that could be achieved using these operating parameters was calculated from the McCabe—Thiele diagram. An alternative approach to the circuit design is to specify the extraction efficiency required, say 97%, and then calculate what O/A is required to achieve this value. It may be necessary to add a third extraction stage to achieve the specified extraction efficiency. Alternatively, by using two strip stages (instead of one), the stripping will be more efficient, so the Cu concentration on the loaded organic phase will be lower, and greater transfer of Cu will be possible.

The major extractant vendors have sophisticated software that can model such scenarios very quickly and accurately. This allows a circuit to be designed and tested for robustness on paper, which minimizes the experimental work required to optimize a circuit design for a particular project.

16.8. QUANTITATIVE COMPARISON OF SERIES AND SERIES-PARALLEL CIRCUITS

The influence of the circuit flow configuration on SX performance is illustrated by comparing the series circuit discussed in Section 16.7 with an equivalent series-parallel circuit. All parameters remain the same as indicated in Table 16.3. The only difference is that PLS is fed to both extraction stages (Fig. 16.6b).

The extraction and stripping isotherms are shown in Fig. 16.7b and the performance results are shown in Table 16.4. The series-parallel circuit allows a higher rate of Cu throughput (i.e., increased overall Cu production), at the expense of low overall Cu recovery from the PLS. A series circuit will usually be converted to a series-parallel circuit when the Cu concentration of the PLS drops significantly (see Section 16.6.2).

16.9. OPERATIONAL CONSIDERATIONS

16.9.1. Stability of Operation

Industrial SX circuits are easily controlled and forgiving, allowing a consistent electrolyte composition to be transferred to EW.

TABLE 16.4 Comparison of the Performance of Series and Parallel Flow Configurations Using Two Extraction Stages and One Strip Stage. The Operating Conditions are given in Table 16.3, the Circuits are Illustrated in Fig. 16.6, and the Isotherms in Fig. 16.7

	Series	Parallel
Cu in PLS, g/L	4	4
Extractant concentration, vol.-%	16	16
Cu in raffinate E2, g/L	0.2	0.9
Cu in raffinate E1, g/L	–	2.5
Cu in total (combined) raffinate, g/L	0.2	1.7
Cu on loaded organic, g/L	7.5	8.3
Volume of PLS processed, m^3/h	1000	2000
Cu recovery from PLS, %	94.8	57.1
Cu transferred to strip, KG/H	3800	4580

Consider, for example, how the circuit examined in Section 16.7 responds to an increase in Cu concentration in PLS (which would happen if more easily leached ore is encountered in the mine). Suppose that the PLS concentration increases from the 4 g/L Cu in Fig. 16.6a to 4.4 g/L Cu. Using the extraction isotherm of Fig. 16.7a with the new PLS composition, it can be shown that (Fig. 16.8a):

(a) The raffinate will now contain 0.3 g/L Cu rather than 0.2 g/L Cu;
(b) The overall extraction efficiency decreases from 95 to 93%;
(c) But 4090 kg/h Cu will be transferred to the loaded organic phase instead of 3800 kg Cu.

The resulting flows and Cu concentrations are shown in Fig. 16.8b. Of course, the rate at which copper is plated in EW will also have to increase to 4090 kg/h Cu. This can be done by increasing current density and/or by bringing unused cells into operation.

16.9.2. Crud

Almost all SX circuits experience a phenomenon known as *crud* formation. Crud is a solid-stabilized emulsion that is formed from various contaminants in the SX circuit. These can include solids in the PLS (colloidal silica from dust or from silica present in the PLS), vegetation, mold, organic degradation precipitates, minor organic constituents, and air entrainment (Ritcey, 1980; Sole, 2008; Virnig, Olafson, Kordosky, & Wolfe, 1999; Wang, 2005).

Crud usually forms a layer at the aqueous–organic interface of the settler, although it can also float on the surface of the organic phase. Small amounts of interfacial crud can benefit phase separation, by acting as a coalescence medium.

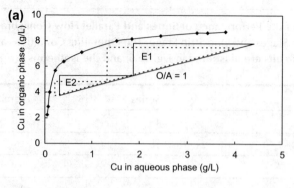

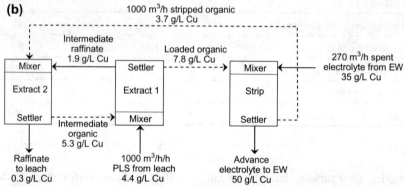

FIGURE 16.8 The SX circuit of Fig. 16.6a has been perturbed by receiving a PLS containing 4.4 g/L Cu instead of 4 g/L Cu. It is assumed that Cu EW rate has been increased (by increasing current density) to match the rate at which Cu is being transferred from PLS to electrolyte. (a) Modified McCabe–Thiele construction for the extraction isotherm shown in Fig. 16.8a for higher Cu concentration in PLS (4.4 g/L Cu). The construction for a PLS containing 4 g/L Cu is shown in dotted lines. (b) Modified concentrations and solution flows for the new PLS. Note that the only operating variable that has to be changed is the electrolyte recycle flowrate.

Large amounts can occupy a significant volume of the settler and reduce the rate of phase separation. Floating crud is more harmful, as this can move from one stage to another with the organic phase, thereby transferring adsorbed and entrained impurities through the circuit to the electrolyte.

Crud formation is minimized by minimizing the amount of solids in the PLS. This is done by allowing the PLS to settle in ponds and by directing rain runoff away from PLS collection areas. Vat, agitation, and pressure leaching solutions can be processed through clarification equipment, such as pinned-bed clarifiers, ahead of the SX circuit. Minimizing air entrainment to the mixers and dust egress to the circuit (by installed roofs over the SX mixer–settlers) and careful selection of extractant and diluent for a particular application can also be helpful in mitigating the detrimental effects of crud (Cytec, 2006; Sole, 2008).

Crud is removed from the settler by regular vacuum pumping or by periodically emptying the settler and pumping the crud from the settler floor. It is usually separated from its host organic by filtering on a plate-and-frame or vacuum-drum filter. A centrifuge can also be used (Hartmann, Horbach, & Kramer, 2010; Mukutuma, Mbao, Schwartz, Chisakuta, & Feather, 2007). The liquid portion can be washed out with diluent, organic, or water in a reslurry tank.

The crud solids are usually sent to the leach heaps. The host organic is regenerated by mixing it with acid-activated bentonite clay or zeolite then filtering (Dudley & Crane, 2006). This process also removes degradation products (formed by oxidative degradation of the extractant and diluent) from the organic phase, which can also be harmful to phase separation and extraction kinetics. The organic is then returned to the SX circuit.

16.9.3. Phase Continuity

When two phases are mixed, the emulsion that forms will be either *aqueous-continuous*, containing organic droplets (the *dispersed* phase) in an aqueous matrix (the *continuous* phase), or it may be *organic-continuous*, with aqueous droplets dispersed in an organic phase (Sole, 2008; Vancas, 2005). The operating rule of thumb is to run *in the continuity of the phase that is more critical to control*. For example, to minimize organic contamination of the raffinate and advance electrolyte streams, it would be appropriate to operate the two stages where these streams leave the SX circuit in organic continuity. On the other hand, E1 and wash stages may be run aqueous-continuous to avoid transferring small aqueous droplets (containing impurities) to the strip circuit in the loaded organic phase.

Copper SX plants usually operate with organic-continuous mixing. The rate of phase separation is usually faster from an organic-continuous emulsion. Organic continuity also ensures compact crud formation and minimizes crud movement through the circuit. In certain instances, however, it may be more appropriate to operate specific stages in aqueous continuity.

16.9.4. Organic Losses and Recovery

One of the main operating costs of an SX plant is the replacement of extractant and diluent. The main cause of diluent loss is evaporation. Both extractant and diluent are lost to crud and by entrainment in the raffinate and advance electrolyte.

Appropriate choice of diluent and a roof over settlers can minimize evaporative losses (Bishop, Gray, Greene, Bauer, Young, May, Evans, & Amerson–Treat, 1999). Losses to crud can be prevented by minimizing crud formation in the circuit and recovering the contained organic phase (Section 16.9.3). Organic entrainment losses can be minimized or recovered by passing the aqueous stream through an after-settler, or through a coalescer, flotation cell, activated carbon column, or dual media or CoMatrix filter. These types of equipment all allow additional settling time and promote coalescence, so that the organic phase can be recovered and returned to the SX circuit (Greene, 2002; Greene & Rubio, 2010; Minson, Jiménez, & Gilmour, 2005).

16.10. INDUSTRIAL SOLVENT-EXTRACTION PLANTS

SX plants are designed to match the production rate at which Cu is leached in the preceding leach operation. They vary in capacity from 5 to 400 kt Cu per year (Table 2.7). Table 16.5 gives operational details of seven SX plants. These include plants in which the leach liquors have been generated by dump leaching, heap leaching, and pressure leaching. The range of operating conditions is evident, demonstrating the versatility of this technology. Further details are given by Robinson et al. (2008).

TABLE 16.5 Selected Industrial Copper Solvent-extraction Plant Data (2008). Additional Details of Other Major SX Plant Operations (see Table 2.7) are Given by Robinson et al. (2008)

Operation	Bwana Mkubwa, Zambia	El Abra, Chile	Escondida Oxide, Chile	Lomas Bayas, Chile	Mantos Blancos, Chile	Morenci Stargo, USA	Sepon, Laos
Startup date	2005	—	—	—	1995	2008	2005
Cathode production, t/a	51 000	225 000	150 000	64 300	60 000	125 400	60 000
Total PLS flowrate, m³/h	700–750	1200–1400	1123	2300	1140	6041	600
SX plant							
Number of SX trains	2	4	3	1	1	1	1
Extraction stages per train	3	3	2	3	2	3	3
Extraction configuration	Series-parallel	Series-parallel	Series	Optimum series-parallel	Series	Series-parallel	Series
Strip stages per train	1	1	1	2	2	1	2
Wash stages per train	1	1	1	1	1	1	0
Mixers							
Type	Mixtec	1 pump, 2 aux.	2-stage	Conventional	Lightnin, 2 + 1	Spirok + DOP	1 DOP, 2 VSF
Impeller diameter, m	1.4	1.9	—	1.6	—	2.9	1.8
Power (kW)	69	55	—	66	21	40	50

Chapter | 16 Solvent Extraction

Material	Stainless steel	Polymer concrete	Stainless steel	Steel	Vinyl ester-lined concrete	Stainless steel	Stainless steel
Extraction residence time, s	2.5–3	2.5	3	2	3.2	2.5	3
Strip residence time, s	3	3	4.8	1.65	1.6	0.13	3
Settlers							
Length × width × depth, m	10.7 × 17.25 × 1.2	28 × 29 × 1.1	24.7 × 26.4 × 1.06	26 × 18 × 1.1	25 × 25 × 1	31.7 × 28.7 × 1.25	20.35 × 17.7 × 2.0
Picket fences per settler	2	2	1	1	3	2	4
Settling velocity, m³/m² h	4.9	3.7	3.8	4.8	3.6	3.3–3.6	4.7
Material	Stainless steel	HDPE-lined concrete	Stainless steel	HDPE-lined concrete	HDPE-lined concrete	Stainless steel	Stainless steel
Organic depth, mm	—	270–320	300	300	250	355	350
Aqueous depth, mm	200–380	730–680	600	600	640	610	1650
Organic phase							
Extractant	LIX 984NC	LIX 84 IC+ LIX 612 NLV	Acorga M5640	LIX 84-IC	Acorga M5640 + LIX 8186	LIX 616	LIX 974N LVS
Diluent	Shellsol 2325	Shellsol 2046AR	Shellsol 2046AR	Shellsol 2046AR	Solbrax M	Penrico 170ES	Shellsol 2046TH
[Extractant] (vol.-%)	25–28	16.2–16.6	17	21	15.8	28	32–26

(Continued)

TABLE 16.5 Selected Industrial Copper Solvent-extraction Plant Data (2008). Additional Details of Other Major SX Plant Operations (see Table 2.7) are Given by Robinson et al. (2008)—cont'd

Operation	Bwana Mkubwa, Zambia	El Abra, Chile	Escondida Oxide, Chile	Lomas Bayas, Chile	Mantos Blancos, Chile	Morenci Stargo, USA	Sepon, Laos
Flowrate per train, m³/h	550/650	1500	1250	1100	1120	2043	1050
PLS							
Flowrate per train, m³/h	700–750	1300	1123	1150	1140	6041	600
Cu, g/L	8 + 13	2.5	7	4.1	5.2	2.7	14–17
Fe, g/L	–	3.0	2.5	1.5	1.7	3.1	30–35
Mn, g/L	0.5	0.65	0.45	3.4	–	0.5	2.0
Cl, g/L	–	1100	0.25	10.6	42	–	–
pH	2.0	2–2.1	1.6–1.8	2.1	1.5	1.9	1.8–2.5
Raffinate							
Cu, g/L	0.6 + 1.5	0.18	0.5	0.65	0.64	0.2	0.6–1.5
H₂SO₄, g/L	11 + 17	7	12	6	11.7	7.2	30–35

Entrained organic, mg/L	—	20	40	30	43	40	20–30

Spent electrolyte							
Cu, g/L	38	41	38	43	34	32	35
H_2SO_4, g/L	187	187	180	182	198	195	185
Co, g/L	180	180	160	195	150	155	200
Flowrate per train, m^3/h	300/180	520	450	720	426	1250	570

Advance electrolyte							
Cu, g/L	49	54	50	53	45	45	51
H_2SO_4, g/L	178	170	170	165	180	172	165
Entrained organic, mg/L	—	4	<10	13	8	10	<15

Performance							
Extraction efficiency, %	—	93	93–96	85	90	94	93
Stripping efficiency, %	65–68	68–70	70	84	67	63	—
Cu transfer, g/L-vol.-%	0.23	0.27	0.24	0.36	0.30	0.30	0.24

16.11. SUMMARY

Impure leach solutions containing relatively low concentrations of copper are processed to high copper concentration, pure EW electrolytes by SX. It is a crucial step in producing high purity electrowon copper from leached ores.

The SX process consists of extracting Cu^{2+} from aqueous PLS into a copper-specific organic extractant, separating the aqueous and organic phases by gravity, and then stripping Cu^{2+} from the organic extractant into high-H_2SO_4 electrolyte returned from the EW circuit. The process is continuous and is carried out in large mixer—settlers.

The SX process is equilibrium-controlled. Loading of Cu into the organic extractant is favored by the low acid concentration (0.5—5 g/L H_2SO_4) of the PLS. Stripping of Cu from the organic extractant is favored by the high acid concentration (180—190 g/L H_2SO_4) of the electrolyte.

The copper-specific organic extractants used in the copper industry are modified aldoximes, ketoximes, and modified aldoxime—ketoxime and aldoxime—ketoxime mixtures, the relative strength of which can be tailored for the particular application. The extractant is dissolved in a kerosene diluent to a concentration appropriate to the Cu concentration of the PLS (typically 5—35 vol.%).

About 4.5 million tonnes of Cu per year are treated by SX followed by EW. Copper production by SX-EW continues to grow as increasing volumes of ore and concentrates are processed by leaching technologies.

REFERENCES

Anderson, C. G., Giralico, M. A., Post, T. A., Robinson, T. G., & Tinkler, O. S. (2002). Selection and sizing of copper solvent extraction and electrowinning equipment and circuits. In A. L. Mular, D. N. Halbe & D. J. Barratt (Eds.), *Mineral processing plant design, practice, and control, Vol. 2* (pp. 1709—1744). Littleton, CO: SME.

Bishop, M. D., Gray, L. A., Greene, M. G., Bauer, K., Young, T. L., May, J., Evans, K. E., & Amerson–Treat, I. (1999). Investigation of evaporative losses in solvent extraction circuits. In S. K. Young, D. B. Dreisinger, R. P. Hackl & D. G. Dixon (Eds.), *Copper 99, Vol. IV: Hydrometallurgy of copper* (pp. 277—289). Warrendale, PA: TMS.

Cytec. (2006). *Crud: How it forms and techniques for controlling it.* Cytec Brochure MCT-1102. www.cytec.com/specialty-chemicals/PDFs/SolventExtraction/CrudFormationandControlinC%20SX.pdf.

Gigas, B., & Giralico, M. (2002). Advanced methods for designing today's optimum solvent extraction mixer settler unit. In K. C. Sole, P. M. Cole, J. S. Preston & D. J. Robinson (Eds.), *International solvent extraction conference ISEC 2002, Vol. 2* (pp. 1388—1395). Johannesburg: Southern African Institute of Mining and Metallurgy.

Giralico, M., Gigas, B., & Preston, M. (2003). Optimised mixer settler designs for tomorrow's large flow production requirements. In P. A. Rivieros, D. G. Dixon, D. B. Dreisinger & J. H. Menacho (Eds.), *Copper—Cobre 2003, Vol. VI, Book 2. Hydrometallurgy of copper: Modelling, impurity control and solvent extraction* (pp. 775—794). Montreal: CIM.

Greene, W. A. (2002). Organic recovery with CoMatrix filtration. In K. C. Sole, P. M. Cole, J. S. Preston & D. J. Robinson (Eds.), *International solvent extraction conference ISEC 2002, Vol. 2* (pp. 1428—1433). Johannesburg, South Africa: Southern African Institute of Mining and Metallurgy.

Greene, W., & Rubio, D. (2010). Aqueous coalescer for electrolyte and raffinate entrained organic removal. In *Hydroprocess 2010* (pp. 90—92). Santiago, Chile: Gecamin.

Hartmann, T., Horbach, U., & Kramer, J. (2010). Copper crud treatment, concentration-dependent pond depth adjustment for decanter centrifuges, DControl®. In *Hydroprocess 2010* (pp. 82—83). Santiago, Chile: Gecamin.

Hein, H. (2005). The importance of a wash stage in copper solvent extraction. In J. M. Menacho & J. Casas de Prada (Eds.), *HydroCopper 2005* (pp. 425–436). Santiago, Chile: Universidad de Chile.

Hopkins, W. (2005). Fire hazards in SX plant design – an update. In *ALTA SX/IX world summit on SX fire protection*. Melbourne, Australia: ALTA Metallurgical Services.

Hurtado-Guzmán, C., & Menacho, J. M. (2003). Oxime degradation chemistry in copper solvent extraction plants. In P. A. Rivieros, D. G. Dixon, D. B. Dreisinger & J. H. Menacho (Eds.), *Copper–Cobre 2003, Vol. VI, Book 2: Hydrometallurgy of copper: Modelling, impurity control and solvent extraction* (pp. 719–734). Montreal: CIM.

ICSG. (2010). *Directory of copper mines and plants 2008 to 2013*. Lisbon, Portugal: International Copper Study Group. http://www.icsg.org.

Kokotov, Y., Braginsky, L., Shteinman, D., Slonim, E., Barfield, V., & Grinbaum, B. (2010). Turbulent settling (TS) technology for solvent extraction. In *ALTA copper 2010*. Melbourne, Australia: ALTA Metallurgical Services.

Kordosky, G. A. (2002). Copper recovery using leach/solvent extraction/electrowinning technology: forty years of innovations, 2.2 million tonnes of copper annually. In K. C. Sole, P. M. Cole, J. S. Preston & D. J. Robinson (Eds.), *International solvent extraction conference ISEC 2002, Vol. 2* (pp. 853–862). Johannesburg: Southern African Institute of Mining and Metallurgy.

Kordosky, G. A. (2008). Development of solvent-extraction processes for metal recovery – finding the best fit between the metallurgy and the reagent. In B. A. Moyer (Ed.), *Solvent extraction: fundamentals to industrial applications. International solvent extraction conference ISEC 2008, Vol. 1* (pp. 3–16). Montreal: CIM.

Kordosky, G., & Virnig, M. (2003). Equilibrium modifiers in copper solvent extraction reagents – friend of foe? In C. A. Young, A. M. Alfantasi, C. G. Anderson, D. B. Dreisinger, G. B. Harris & A. James (Eds.), *Hydrometallurgy 2003, Vol. 1: Leaching and solution purification* (pp. 905–916) Warrendale, PA: TMS.

Kordosky, G. A., Sudderth, R. B., & Virnig, M. J. (1999). Evolutionary development of solvent extraction reagents: real life experiences. In G. V. Jergensen, II (Ed.), *Copper leaching, solvent extraction, and electrowinning technology* (pp. 259–271). Littleton, CO: SME.

Maes, C., Tinkler, O., Moore, T., & Mejías, J. (2003). The evolution of modified aldoxime copper extractants. In P. A. Rivieros, D. G. Dixon, D. B. Dreisinger & J. H. Menacho (Eds.), *Copper–Cobre 2003, Vol. VI, Book 2. Hydrometallurgy of copper: Modelling, impurity control and solvent extraction* (pp. 753–760). Montreal: CIM.

Miller, G. (2005). Engineering design for lowering fire risk. In *ALTA SX/IX world summit on SX fire protection*. Melbourne, Australia: ALTA Metallurgical Services.

Miller, G. (2006). Design of mixer-settlers to maximise performance. In *ALTA copper 2006*. Melbourne, Australia: ALTA Metallurgical Services.

Minson, D., Jiménez, C., & Gilmour, J. (2005). Filtros Co-Matrix – resultados de una plata pilota y sus applicaciones en las plantas de extracción por solventes. In J. M. Menacho & J. Casas de Prada (Eds.), *HydroCopper 2005* (pp. 437–448). Santiago, Chile: Universidad de Chile.

Mukutuma, A., Mbao, B., Schwartz, N., Chisakuta, G., & Feather, A. (2007). A case study on the operation of a Flottweg Tricanter® centrifuge for solvent-extraction crud treatment at Bwana Mkubwa, Ndola, Zambia. In *Africa's base metals resurgence* (pp. 393–405). Johannesburg, South Africa: Southern African Institute of Mining and Metallurgy.

Nisbett, A., Peabody, S., & Crane, P. (2009). Concepts for developing high performing copper solvent extraction circuits. In *ALTA Copper 2009*. Melbourne, Australia: ALTA Metallurgical Services.

Nisbett, A., Peabody, S., & Kordosky, G. A. (2008). Developing high-performing copper solvent-extraction circuits: strategies, concepts, and implementation. In B. A. Moyer (Ed.), *Solvent extraction: Fundamentals to industrial applications. International solvent extraction conference ISEC 2008, Vol. 1* (pp. 93–100). Montreal: CIM.

Outotec. (2007). *Outotec copper SX-EW technology*. www.outotec.com/38582.epibrw.

Pekkala, P., Laitala, H., Nyman, B., Ekma, E., & Weatherseed, M. (2007). Performance highlights of a modern VSFTM SX plant. In *ALTA copper 2007*. Melbourne, Australia: ALTA Metallurgical Services.

Ritcey, G. M. (1980). Crud in solvent extraction processing – a review of causes and treatment. *Hydrometallurgy, 5*, 97–107.

Robinson, T., Moats, M., Davenport, W., Karcas, G., Demetrio, S., & Domic, E. (2008). Copper solvent extraction—2007 world operating data. In B. A. Moyer (Ed.), *Solvent extraction: Fundamentals to industrial applications. International solvent extraction conference ISEC 2008, Vol. 1* (pp. 435–440). Montreal: CIM.

Rojas, E., Araya, G., Picardo, J., & Hein, H. (2005). Serie paralelo óptimo: un neuvo concepto de confiuracíon en extraccíon por solventes. In M. Menacho & J. Casas de Prada (Eds.), *HydroCopper 2005* (pp. 449–458). Santiago, Chile: Universidad de Chile.

Smith, B. (2009). Fires in SX plants — recent trends in fire prevention and fire protection. In *ALTA copper 2009*. Melbourne, Australia: ALTA Metallurgical Services.

Soderstrom, M. (2006). New developments in copper SX reagents. In *ALTA copper 2006*. Melbourne, Australia: ALTA Metallurgical Services.

Sole, K. C. (2008). Solvent extraction in the hydrometallurgical processing and purification of metals: process design and selected applications. In M. Aguilar & J. L. Cortina (Eds.), *Solvent extraction and liquid membranes: Fundamentals and applications in new materials* (pp. 141–200). New York: Taylor and Francis.

Tinkler, O. S., Cronje, I., Soto, A., Delvallee, F., & Hangoma, M. (2009). The ACORGA OPT-series: industrial performance vs. aldoxime:ketoxime reagents. In *ALTA copper 2009*. Melbourne, Australia: ALTA Metallurgical Services.

Vancas, M. F. (2003). Solvent extraction settlers — a comparison of various designs. In P. A. Rivieros, D. G. Dixon, D. B. Dreisinger & J. H. Menacho (Eds.), *Proceedings Copper–Cobre 2003. Vol. VI, Book 2: Hydrometallurgy of Copper: Modelling, impurity control and solvent extraction* (pp. 707–718). Montreal: CIM.

Vancas, M. F. (2005). A practical guide to SX plant operation. In *Solvent extraction for sustainable development, international solvent extraction conference ISEC 2005*. Beijing, China: China Academic Journal Electronic Publishing House.

Virnig, M., Eyzaguirre, D., Jo, M., & Calderon, J. (2003). Effects of nitrates on copper SX circuits: a case study. In P. A. Rivieros, D. G. Dixon, D. B. Dreisinger & J. H. Menacho (Eds.), *Copper–Cobre 2003, Vol. VI, Book 2: Hydrometallurgy of copper: Modelling, impurity control and solvent extraction* (pp. 795–810). Montreal: CIM.

Virnig, M. J., Olafson, S. M., Kordosky, G. A., & Wolfe, G. A. (1999). Crud formation: field studies and fundamental studies. In S. K. Young, D. B. Dreisinger, R. P. Hackl & D. G. Dixon (Eds.), *Copper 99–Cobre 99, Vol. IV: Hydrometallurgy of copper* (pp. 291–304). Warrendale, PA: TMS.

Wang, C. Y. (2005). Crud formation and its control in solvent extraction. In *Solvent extraction for sustainable development. International solvent extraction conference ISEC 2005*. Beijing, China: China Academic Journal Electronic Publishing House.

SUGGESTED READING

Blass, E. (2004). Engineering design and calculation of extractors for liquid-liquid systems. In J. Rydberg, M. Cox, C. Musikas & G. R. Choppin (Eds.) *Solvent extraction principles and practice* (2nd ed.). (pp. 367–414) New York: Marcel Dekker.

Cognis. (2007). *MCT redbook: Solvent extraction reagents and applications*. Tucson, AZ, USA: Cognis Corporation. www.cognis.com/NR/rdonlyres/7A687186-A305-4969-B6EA-0CF869930714/0/MCT_Redbook_English.pdf.

Cox, M. (2004). Solvent extraction in hydrometallurgy. In J. Rydberg, M. Cox, C. Musikas & G. R. Choppin (Eds.), *Solvent extraction principles and practice* (2nd ed.). (pp. 455–506) New York: Marcel Dekker.

Cytec. (2010). *Mining chemicals handbook*. Woodland Park, NJ: Cytec Industries. 321–390.

Giralico, M., Gigas, B., & Preston, M. (2007). Optimized solvent extraction mixer settler design for tomorrow's operations. In *ALTA copper 2007*. Melbourne, Australia: ALTA Metallurgical Services.

Godfrey, J. C., & Slater, M. J. (1994). *Liquid–liquid extraction equipment*. Chichester, UK: John Wiley & Sons.

Komulainen, T., Doyle, F. J., III, Rantala, A., & Jamsa-Jounela, S.-L. (2009). Control of an industrial copper solvent extraction process. *Journal of Process Control, 19*, 1350–1358.

Kordosky, G. A. (2003). Copper SX circuit design and operation: current advances and future possibilities. In *ALTA copper 2003*. Melbourne, Australia: ALTA Metallurgical Services.

Laitala, H., Hakkarainen, J., & Rodríguez, C. (2009). Plant design and operational aspects in solvent extraction impurity control. In *HydroCopper 2009* (pp. 311–317). Santiago, Chile: Gecamin.

Miller, G. (2007). SX contactors for the future. In *ALTA copper 2007*. Melbourne, Australia: ALTA Metallurgical Services.

Moreno, C. M., Perez-Correa, J. R., & Otero, A. (2009). Dynamic modelling of copper solvent extraction mixer-settler units. *Minerals Engineering, 22*(15), 2–15.

Ritcey, G. M. (2006). In *Solvent extraction: Principles and applications to process metallurgy* (2nd ed.). Ottawa, Canada: Gordon M. Ritcey and Associates.

Robles, E., Poulter, S., Haywood, R., Cronje, I., & León, M. (2009). The future of SX plant design for modern hydrometallurgical refineries. In *HydroCopper 2009* (pp. 392–399). Santiago, Chile: Gecamin.

Taylor, A. (2007). Review of mixer-settler types and other possible contactors for copper SX. In *ALTA copper 2007*. Melbourne, Australia: ALTA Metallurgical Services.

Chapter 17

Electrowinning

Chapters 15 and 16 show how leaching and solvent extraction (SX) produce a high-purity $CuSO_4$–H_2SO_4 electrolyte containing ~45 g/L Cu^{2+}. This chapter explains how the Cu^{2+} ions in this electrolyte are reduced and electrowon as pure metallic copper in the form of cathodes.

The purity specifications of copper cathode are dictated by the various international metal exchanges. Table 20.1 shows the requirements for copper cathode sold to the London Metal Exchange (LME) and the New York Mercantile Exchange (Comex) (LME, 2010).

17.1. THE ELECTROWINNING PROCESS

The electrowinning (EW) process entails:

(a) Immersing metal *cathodes* and inert (but conductive) *anodes* into a purified electrolyte containing $CuSO_4$ and H_2SO_4;
(b) Applying a direct electrical current from an external source such as a rectifier, which causes current to flow through the electrolyte between the cathodes and anodes;
(c) Plating pure metallic copper from the electrolyte onto the cathodes using the energy provided by the electrical current to drive the reduction of the Cu^{2+} ions to $Cu°$ metal.

The cathodes are usually stainless steel blanks. The anodes are usually rolled Pb-alloy sheets. Copper is electroplated onto the cathodes for 6–7 days, after which the plated copper is machine-stripped from the stainless steel cathode blanks, washed, and sold.

About 4.5 million tonnes of copper are electrowon per year (ICSG, 2010). Production of copper using this technology continues to increase due to the growth of leaching as a process technology for copper.

17.2. CHEMISTRY OF COPPER ELECTROWINNING

The EW cathode reaction is the same as for electrorefining:

$$Cu^{2+} + 2e^- \rightarrow Cu° \quad E° = +0.34 \text{ V} \quad (17.1)$$

The anode reaction is, however, completely different. Water is decomposed at the inert anode to form oxygen gas and release protons:

$$H_2O \rightarrow H^+ + OH^- \rightarrow 0.5 O_2 + 2H^+ + 2e^- \quad E° = -1.23 \text{ V} \quad (17.2)$$

The overall EW reaction is the sum of Reactions (17.1) and (17.2) in the presence of sulfate ions:

$$Cu^{2+} + SO_4^{2-} + H_2O \rightarrow Cu^{\circ} + 0.5\,O_2 + 2H^+ + SO_4^{2-} \quad E^{\circ} = -0.89\text{ V} \quad (17.3)$$

The EW products are:

(a) Pure copper metal at the cathode;
(b) Oxygen gas at the anode;
(c) Regenerated sulfuric acid in the electrolyte.

The copper is stripped from the cathode blank, washed, packed, and strapped into bundles of 1–3 t, and sent to market. The oxygen bubbles burst at the surface of the electrolyte and are released to the atmosphere. The acid is returned to SX in Cu-depleted spent electrolyte where the free H^+ are used to strip Cu^{2+} from the loaded organic phase into the advance electrolyte (Eq. (16.2)).

17.3. ELECTRICAL REQUIREMENTS

The electrical potential or *cell voltage* needed for EW is ~2.0 V, as compared to ~0.3 V for electrorefining (Chapter 14). As illustrated in Fig. 17.1 (Nicol, 2006), it is made up of:

Theoretical voltage for Reaction (17.3), E° ~ 0.9 V.
Overvoltage for copper deposition at cathode, η_c ~ 0.05–1.0 V.
Overvoltage for oxygen evolution at anode, η_a ~ 0.5 V.
Ohmic potential drop across electrolyte, V_s ~ 0.25–0.3 V
Ohmic potential drops across cell hardware and rectifier, $V_h + V_r$ ~ 0.3 V

The current requirement for EW includes:

Stoichiometric current requirement for Eq. (17.1), nF
Current inefficiencies due to side reactions, CI_e
Current inefficiencies due to shorts, CI_s
Stray currents in the tankhouse, SC

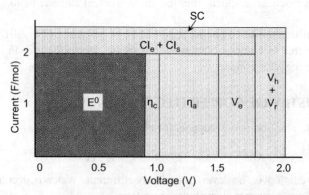

FIGURE 17.1 Current and voltage contributions to power consumption in a Cu EW cell (Nicol, 2006). Although the theoretical power requirement is ~750 kWh/t Cu (dark-shaded area), the actual power requirements are approximately 2000 kWh/t (dark- + light-shaded area) — an energy efficiency of about 30%. The importance of minimizing the effects of side reactions, shorts, and stray currents is evident.

The power requirement is the product of current and voltage (Eq. (17.4)), and is equal to the total area under the graph in Fig. 17.1:

$$P = V \times I \quad (17.4)$$

The rate of plating of Cu from solution is given by Faraday's Law:

$$m = MIt\xi/nF \quad (17.5)$$

where m is the mass of Cu plated (g); M is the molar mass of Cu (63.55 g/mol); I is the current passed (A); t is the time for which the current is passed (s); ξ is the current efficiency (i.e., the fraction of the total current used in producing Cu; see Section 17.6); n is the number of electrons involved in the plating of Cu (equal to 2 from Eq. (17.1)); and F is the Faraday constant (96 485 C/mol of charge = 96 485 A/mol).

The theoretical energy consumption for plating one tonne of copper is therefore

$$P = (1.23 - 0.34) \text{ V} \times 2/\text{mol} \times 96{,}485 \text{ A}/(63.54 \text{ g/mol}) = 2703 \text{ W/g} = 754 \text{ kWh/t Cu}$$

However, in practice the energy requirement for electrowinning is ~2000 kWh/t. This is considerably greater than that for copper electrorefining (200–300 kWh/t Cu). This difference is due to the large voltage requirements for electrowinning, as well as current losses. The energy efficiency of Cu EW is therefore only about 30%.

17.4. EQUIPMENT AND OPERATIONAL PRACTICE

17.4.1. Cathodes

Most copper EW tankhouses today employ reusable 316L stainless steel blank cathodes, virtually identical to electrorefining cathodes. This is known as *permanent cathode technology* (Fig. 17.2). Some older EW tankhouses use copper starter sheet cathodes. The starter sheets are obtained from an electrorefinery or are made in a separate section of the EW tankhouse itself, using stainless steel or titanium as starter cathode blanks. As with refining, many of these older plants are switching to stainless steel when the benefits of modernization can justify the capital investment (Robinson, Davenport, Moats, Karcas, & Demetrio, 2007).

Plastic edge strips are used on the sides of the cathode blanks to avoid plating around the sides of the cathode. The bottom of the blanks are either wax-dipped or have a V groove to allow for easy removal of the plated copper.

The size of the cathodes is determined by industry standards (to fit efficiently into shipping containers) and by the openings in the furnaces in which the copper will be remelted (see Chapter 20).

17.4.2. Anodes

Modern copper EW anodes are cold-rolled lead (Pb) alloyed with 1.35% Sn and 0.07–0.08% Ca.

Sn provides strength, corrosion resistance, and corrosion layer conductivity. Ca improves the mechanical properties and decreases the anode potential. Cold rolling adds strength (Prengaman & Siegmund, 1999). A few plants still use Pb–Sn or Pb–Sb alloys, but these are being gradually replaced with Pb–Sn–Ca anodes.

The Pb–Sn–Ca plates are soldered onto slotted copper hanger bars for support in the EW cells (Fig. 17.3). Lead is then electrodeposited around the joints to protect them from

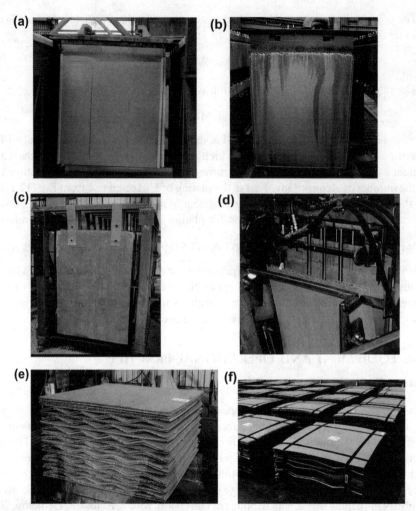

FIGURE 17.2 Stainless steel cathode blank (a) before and (b) after plating and (c) copper starter sheet. Note the hanger tabs on the starter sheet and the edge strips on the permanent cathodes. (d) Copper cathode sheets are stripped from the blank by an automatic stripping machine, (e) corrugated and packed, and (f) strapped for shipping. The corrugations are introduced so that the cathode stack melts more easily when the cathodes are melted and cast (see Chapter 20).

corrosion. The Pb–Sn–Ca alloy forms an adherent corrosion layer. This minimizes Pb contamination of the cathode copper and extends anode life. The anode size is slightly smaller than the cathode size (about 30 mm on the width and height) to ensure that Cu plating does not occur around the sides of the cathode.

Anodes are often fitted with several polymer spacer insulators, buttons, or straps that prevent contact with the cathode. This avoids short-circuiting and minimizes lead corrosion product carryover from anodes to electrodepositing copper.

The life expectancy of an anode is typically 6–9 years (Beukes & Badenhorst, 2009). The anodes slowly corrode in the sulfuric acid electrolyte, forming an inert PbO_2 layer on their surface (Nikoloski, Nicol, & Stuart, 2010). The components of the anode either dissolve

FIGURE 17.3 Pb-alloy anode. Note the insulator strips hanging on each side of the anode and the insulator buttons on the side. These prevent short circuiting between the anode and cathode and assist in placing the anodes vertically and straight in the cells.

in the electrolyte or fall to the bottom of the EW cell as insoluble anode sludge. This is periodically removed from the cells and sent to a recycling facility to recover the lead.

17.4.3. Cell Design

The electrolytic cells, anodes, and cathodes are approximately the same size as for electrorefining:

Cells: polymer concrete, 6.5–8 m long × 1.2 m wide × 1.5 m deep
Anodes: rolled Pb–Sn–Ca alloy, 1.1 m long × 0.9 m wide × 0.006 m thick
Cathodes: stainless steel, 1.2 m long × 1.0 m wide × 0.003 m thick.

Most modern cells are constructed of polymer concrete. Some designs use reinforced concrete with a chemically resistant lining, such as welded or spray-coated polyvinyl chloride (PVC). Older tankhouses have cell linings of rubber, lead, or epoxy resin coating. The length of the cell depends on the spacing between the electrodes. Modern designs tend to reduce the anode–cathode spacing to accommodate more plating capacity for a given floor area, with 60–84 cathodes per cell. Typical anode–cathode spacing is ~50 mm.

The anodes and cathodes are interleaved as in electrorefining (Fig. 17.4). The anodes in a cell are all at one potential. The cathodes are all at another, lower potential. The electrodes in each cell are connected in parallel but the cells of a tankhouse are connected in series. Current is carried between electrodes and between cells using copper busbars. To minimize heat losses through the busbars, the current is limited to 1–1.2 A/m^2 of cathode surface. Where parallel busbars are run, there is an air gap between them to allow for air cooling.

Electrolyte flows continuously from an active storage tank into each cell through a manifold around the bottom of the cell (Fig. 17.5). It overflows the cell into a collection

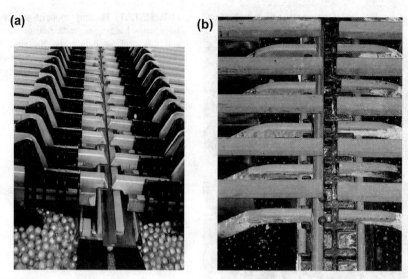

FIGURE 17.4 Anode—cathode arrangement showing different contact systems that minimize the occurrence of poor contacts of electrodes with the busbar. Note the plastic beads on the surface of the electrolyte, which minimize the effects of acid mist (see Section 17.4.5). *(Courtesy of Outotec and M.S. Moats, University of Utah.)*

system and eventually recycles to SX for Cu^{2+} replenishment. Electrolyte is added at ~15 m^3/h into each cell. This gives constant availability of warm, high Cu^{2+} electrolyte over the surfaces of all the cathodes. The specific flowrate of electrolyte past the face of the cathode (the *face velocity*) is typically designed at 0.08 m^3/h per m^2 of cathode area (Beukes & Badenhorst, 2009).

Efficient circulation of the electrolyte to all cathodes in a cell is critical to good cathode quality. As electrode spacing distances decrease and current densities increase, this becomes even more important. Modern cell designs may introduce the electrolyte at

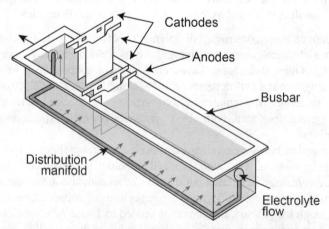

FIGURE 17.5 Schematic view of an electrowinning cell. Anodes and cathodes are arranged alternately. Electrical contact is via busbars. An electrolyte input manifold for uniformly distributing electrolyte over all the cathode faces is shown. Single, multi-holed horizontal pipes on the cell floor are also used.

several places into the cell, including at multiple points from a manifold at the base of the cell, or from feed points at the side or the end of a cell.

17.4.4. Current Density

Cathode current density in modern plants varies from 200 to above 450 A/m^2 (Robinson et al., 2007). This gives a plating rate of 0.4—1.0 kg/h of copper on each cathode.

Cathodes are plated to a thickness of 5—8 mm (typically 50—55 kg Cu on each side of the stainless steel blank), which means that the cathodes are harvested every 4—7 days, depending on the applied current density. In a modern tankhouse using an overhead crane, every third cathode is harvested from a particular cell at a time, leaving the remaining two-thirds of the cathodes available to continue plating (Fig. 17.6). This procedure maintains an adherent corrosion product layer on the anodes and minimizes lead contamination of the electrodepositing copper. The longer plating is continued, the greater the likelihood of shorts occurring due to the growth of nodules or dendrites. If starter sheets are used, these are often pulled out after two days of plating, the cathodes straightened in a press, and then reinstalled to continue the plating cycle. This minimizes growth of nodules or dendrites due to bent cathodes and avoids short circuiting.

The rate of copper plating increases with increasing current density. However, excessive current density can promote rough, nodular cathode deposits, and decreased copper purity if not properly managed. The use of air sparging of the electrolyte helps to minimize inhomogeneities in the electrolyte composition throughout each cell and permits high quality copper to be plated at increasingly higher current densities.

A new technology, known as SELE, is in operation in Chile (New Tech Copper, 2010). This comprises a plastic rack that is inserted into each cell with grooves into which each cathode slides (Fig. 17.7). This avoids the need for edge strips and allows the cathodes to be kept very straight, vertical, and evenly spaced, which means that current distribution is even throughout the tankhouse and higher current densities can be employed.

FIGURE 17.6 Harvesting of plated cathodes by crane. Note that only every third cathode is removed from the cell each time. This allows the cell to continue plating without having to stop the current supply to that cell.

FIGURE 17.7 SELE cathode holder maintains accurate positioning and electrode alignment in EW cells for high current density applications.

17.4.5. Acid Mist Suppression

The oxygen bubbles that are produced at the anodes (Eq. (17.2)) rise to the top of the electrolyte where they burst into the atmosphere, carrying tiny droplets of sulfuric acid with them. This can create a corrosive, unpleasant, and unhealthy environment in the tankhouse, and may cause health problems for operators. For these reasons, combinations of various measures are employed to minimize the effects of this *acid mist*.

Good ventilation is essential. In some plants, the mist is cleared from the workplace by large blowers or extraction fans on the sides of the tankhouse building or, where the climate permits, the sides of the building may be open. Modern tankhouse designs often include hoods over the cells. The acid mist is captured and sent to a scrubber, while evaporation of the electrolyte is also reduced (Fig. 17.8; Vainio & Weatherseed, 2009).

Hollow polyethylene balls (~2 cm diameter) are usually floated, 5−10 cm deep, on the electrolyte (Fig. 17.4). Polypropylene beads (3 mm diameter) are sometimes used. These systems are used in combination with chemical mist suppression agents, such as a foaming agent or fluorocarbon surfactant, that modify the characteristics of the bursting bubbles, reducing their energy on their release at the surface of the electrolyte (Bender, 2010; San Martin, Otero, & Cruz, 2005).

17.4.6. Electrolyte

The *advance electrolyte* from SX typically contains 45 g/L Cu^{2+} and 170 g/L H_2SO_4. This is combined with a bleed of *spent electrolyte* leaving the EW cells to generate the *recirculating electrolyte* that is the feed to the EW cells. The spent electrolyte contains ~5 g/L Cu^{2+} less than the recirculating electrolyte, because a single pass through an EW cell takes a *bite* of 5 g/L Cu^{2+} from the electrolyte.

Depleted spent electrolyte is replenished with Cu^{2+} by sending it back to SX where it strips Cu^{2+} from loaded organic extractant (Chapter 16). About 25−50% of the recirculating electrolyte is sent to SX. The remainder is sent to an active storage tank where it is mixed with Cu-enriched advance electrolyte returning from SX.

FIGURE 17.8 Tankhouse equipment to avoid and mitigate the effects of acid mist. (a) Older tankhouses use extraction fans and ventilation across the cells, while (b) modern tankhouses use hoods to cover the cells. These are lifted for harvesting of cathodes (c), requiring very accurate positioning, so computer control of the crane movements is essential. (d) The acid mist is captured and scrubbed to avoid releasing corrosive gases to the environment. *(Photos (b) and (d) courtesy of Outotec.)*

The advance electrolyte from SX is filtered and treated in carbon columns and flotation cells before it reenters the EW tankhouse. This removes most of the organics that may have been carried over from SX. The presence of organics is detrimental to Cu EW as the cathodes may become discolored at the electrolyte−air interface (so-called *organic burn*) and it is a fire risk.

To minimize power consumption (see Section 17.3), the conductivity of the electrolyte is an important consideration (Tozawa, Umetsu, Su, & Li, 2003; Tozawa, Su, & Umetsu, 2003). A temperature of 45−55 °C is optimum for conductivity, although the spent electrolyte returning to SX needs to be kept below 40 °C to avoid degradation of the organic phase. The incoming advance electrolyte is often heated to 45−50 °C by heat exchange using outgoing spent electrolyte that has been resistively heated in the EW cell.

Higher acidities can also improve electrolyte conductivity, but too high an acidity can cause hydrolytic degradation of the organic phase in SX. For this reason, the spent electrolyte acid concentration is generally limited to 190−200 g/L. Higher acidities also increase the corrosivity of acid mist and decrease anode lifetime. A higher Cu concentration in the electrolyte will improve cathode quality and have lower organic entrainment from SX, but can also lead to $CuSO_4$ crystallization. The effects of Cu, H_2SO_4, and temperature variation on electrolyte conductivity are shown in Table 17.1.

TABLE 17.1 Specific Electrical Conductivity of Copper Electrowinning Electrolytes (from Nicol, 2006)

Cu (g/L)	H_2SO_4 (g/L)	Temperature (°C)	Resistivity (Ω-m)	Conductivity (1/Ω-m)
30	150	45	0.0192	52.0
30	160	45	0.0187	53.6
30	170	45	0.0181	55.2
30	180	45	0.0175	57.0
35	150	45	0.0196	51.1
35	160	45	0.0190	52.6
35	170	45	0.0185	54.2
35	180	45	0.0179	55.9
40	150	45	0.0200	50.1
40	160	45	0.0194	51.6
40	170	45	0.0188	53.1
40	180	45	0.0183	54.7
35	150	40	0.0203	49.2
35	160	40	0.0198	50.6
35	170	40	0.0192	52.1
35	180	40	0.0186	53.7
35	150	50	0.0189	53.0
35	160	50	0.0183	54.7
35	170	50	0.0177	56.4
35	180	50	0.0172	58.2

Some iron may be present in electrolytes when the upstream SX step has not adequately rejected iron (see Chapter 16). The reduction of Fe^{3+} at the cathode:

$$Fe^{3+} + e^- \rightarrow Fe^{2+} \qquad (17.6)$$

competes with the Cu EW reaction (Eq. (17.1)). This lowers the current efficiency for plating of copper by ~2.5% for each 1 g/L Fe. For example, 5 g/L Fe^{3+} in the electrolyte can drop the current efficiency to as low as 80% (Das & Gopala Krishna, 1996; Dew & Phillips, 1985; Khouraibchia & Moats, 2009, 2010).

Manganese is often present in electrolytes, particularly in South America. This contributes to the formation of MnO_2 on the anode which can redissolve when power is lost, contributing to entrapment of Pb and Mn in the cathode and sludge formation at the

Chapter | 17 Electrowinning

TABLE 17.2 Some Standard Reduction Potentials Relevant to Cu Electrowinning

Reaction	Standard potential (V)
$Mn^{2+} + 2e \rightarrow Mn$	−1.18
$Fe^{2+} + 2e \rightarrow Fe$	−0.44
$PbSO_4 + 2e \rightarrow Pb + SO_4^{2-}$	−0.35
$Co^{2+} + 2e \rightarrow Co$	−0.28
$Ni^{2+} + 2e \rightarrow Ni$	−0.26
$Pb^{2+} + 2e \rightarrow Pb$	−0.13
$2H^+ + 2e \rightarrow H_2$	0.00
$AgCl + e \rightarrow Ag + Cl^-$	0.22
$Cu^{2+} + 2e \rightarrow Cu$	**0.34**
$Fe^{3+} + e \rightarrow Fe^{2+}$	0.77
$Ag^+ + e \rightarrow Ag$	0.80
$O_2 + 4H^+ + 4e \rightarrow 2H_2O$	**1.23**
$MnO_2 + 4H^+ + 2e \rightarrow Mn^{2+} + 2H_2O$	1.33
$Cl_2 + 2e \rightarrow 2Cl^-$	1.36
$Mn^{3+} + e \rightarrow Mn^{2+}$	1.49
$MnO_4^- + 8H^+ + 5e \rightarrow Mn^{2+} + 4H_2O$	1.51
$PbO_2 + SO_4^{2-} + 4H^+ + 2e \rightarrow PbSO_4 + 2H_2O$	1.70

bottom of the cell (Tjandrawan & Nicol, 2010). Anodes need to be periodically cleaned with high-pressure water to remove this layer. Mn(II) is also problematic in SXEW circuits because it can be oxidized to Mn(III), Mn(IV), or Mn(VII) at the anodes (Cheng, Hughes, Barnard, & Larcombe, 2000). To avoid the formation of the permanganate ion (MnO_4^-) at the anode — a highly oxidizing species that can have very detrimental effects on the SX organic phase — a ratio of Fe: Mn in the electrolyte of approximately 10:1 is often maintained (Miller, 1995).

The behavior of other metal species present in the electrolyte can be understood with reference to their position in the electrochemical series (Table 17.2). Those species with a more positive reduction potential than Cu can plate at the cathode, while those with a more negative potential will remain in solution.

The buildup of impurities in the electrolyte is controlled by bleeding a small stream from the tankhouse and removing impurities. This is usually achieved by precipitation but ion exchange can also be used (Shaw, Dreisinger, Arnold, Weekes, & Tinsman, 2008; also see Chapter 14). Typically ~50% of the bleed goes back to SX for recovery of the copper, while the remainder goes to the leach circuit where the acid is used for leaching (Chapter 15).

17.4.7. Electrolyte Additives

All Cu EW plants dissolve smoothing agents in their electrolytes to promote the plating of dense, smooth copper deposits with minimum entrapment of electrolyte impurity. Guar gum (150–400 g per tonne of cathode) is most widely used (Fig. 17.9). It behaves much as glue in electrorefining (Chapter 14), but it is compatible with SX organics. Modified starches (polysaccharides) and polyacrylamides are also widely used. Smoothing agents originating from natural products tend to break down above current densities of about 300 A/m^2. Recent advances in the development of synthetic smoothing agents allow high quality plating at current densities well above 400 A/m^2.

Chloride ions are either naturally present in the electrolyte (particularly in Chilean operations) or are sometimes added as HCl. They promote the growth of dense, fine-grained, low impurity copper deposits on the cathode. With stainless steel cathodes, the chloride ion concentration must be kept below ~30 mg/L to avoid *pitting corrosion* at the top of the stainless steel blade at the electrolyte–air interface, which causes the depositing copper to stick and resist detachment. Leach operations with high concentrations of Cl$^-$ in their leach liquors have a wash stage in the SX circuit to prevent excessive Cl$^-$ transfer to the advance electrolyte (Section 17.6.3).

Cobalt sulfate is also added to provide 100–200 mg/L Co^{2+} in the electrolyte. Co^{2+} promotes O$_2$ evolution at the anode (Eq. (17.2)) rather than Pb oxidation (Nikolosi & Nicol, 2008a). This stabilizes the PbO$_2$ layer on the anode surface, minimizes Pb contamination of the depositing copper, and extends anode life (Prengaman & Siegmund, 1999). As the current density increases, the Co concentration required also typically increases (Robinson et al., 2007).

17.5. MAXIMIZING COPPER PURITY

The demands of a modern EW tankhouse are to produce copper of increasingly higher purity, using higher current densities, while minimizing power consumption (Marsden, 2008).

From a thermodynamic point of view, only metals that have a higher electrochemical potential will codeposit with Cu (Table 17.2). Because Cu is relatively noble, it is easily plated with high purity.

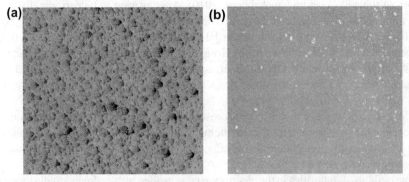

FIGURE 17.9 The effect on the morphology of copper cathode plated in the (a) absence and (b) presence of guar. Guar, added at concentrations of 150–400 g/t cathode, acts as a smoothing agent, allowing refinement of the copper crystals grains as they are plated. This gives the cathode a smooth morphology and avoids the growth of dendrites and nodules that could short circuit the current between anode and cathode. *(Courtesy of M. Nicol, Murdoch University.)*

The three main impurities in electrowon cathode copper are:

(a) Pb (1 or 2 ppm) from physical entrapment of anode corrosion product;
(b) S (4 or 5 ppm) from PbSO$_4$ anode corrosion product and electrolyte entrapment;
(c) Fe (1 or 2 ppm) from electrolyte entrapment.

Cathode purity is maximized by (Maki, 1999):

(a) Maintaining straight, vertical, equi-spaced cathodes and anodes with no anode–cathode contact;
(b) Immediate thorough washing of copper cathodes with water at 70–80 °C;
(c) Using rolled Pb–Sn–Ca anodes with adherent corrosion products;
(d) Frequent removal of anode corrosion products from the bottom of the EW cells;
(e) Addition of Co^{2+} to the electrolyte to minimize Pb corrosion;
(f) Minimizing the Mn^{2+} concentration in the electrolyte to avoid rapid Pb corrosion and flaky corrosion products (Miller, 1995);
(g) Maintaining a steady electrical current at all times to maintain an adherent anode corrosion layer;
(h) Keeping the electrolyte free of solids and SX organics;
(i) A steady, uniform flow of warm, pure, high Cu^{2+} electrolyte across all cathode faces;
(j) An electrolyte temperature of 45–50 °C electrolyte to give pure fine-grained copper deposits;
(k) Use of guar and chloride addition agents to give dense, smooth copper deposits (but with Cl^- < 30 mg/L in the electrolyte);
(l) Maintaining Fe in electrolyte to <2 g/L to minimize Fe in cathode copper and reduce losses in current efficiency.

17.6. MAXIMIZING ENERGY EFFICIENCY

Energy consumption in EW is determined by (a) the cell voltage, (b) the applied current (or current density), and (c) the current efficiency. Cell voltage depends largely on the cell design, electrode materials, and the reactions occurring at the anode and cathode. The applied current is usually determined by the production rate required. Tankhouse operators can, however, influence and manage energy consumption by understanding how to maximize current efficiency (Moats, in press).

The *theoretical* mass of Cu that can be plated for a given current is given by Faraday's Law (Eq. (17.5)). The ratio of the *actual* mass of Cu plated, compared to this theoretical value, is defined as the *current efficiency*, ξ:

$$\xi(\%) = 100 \text{ (measured mass of Cu plated/theoretical mass of Cu plated)} \quad (17.7)$$

Current efficiencies in modern Cu EW plants range from 85–95% (Robinson, Siegmund, Davenport, Moats, & Karcas, 2010). The unused current is wasted by (Moats, in press; Nicol, 2006):

(a) Anode-to-cathode short circuits;
(b) Stray current to ground;
(c) Reduction of Fe^{3+} or dissolved oxygen at the cathode;
(d) Redissolution/oxidation of the deposited metal;
(e) Deposited metal falling off the cathode and which cannot be harvested.

High current efficiency is important because it maximizes copper plating rate and minimizes electrical energy consumption.

Strict attention to good housekeeping practices is the most important criterion in maximizing cathode purity and energy efficiency (Joy, Staley, Moats, Perkins, Uhrie, & Robinson, 2010; Kumar, Zimba, Sakala, & Zambika, 2010; Pfalzgraff, 1999). Electrolyte spills should be avoided. Anode—cathode short circuits are detected by frequent infrared and gauss meter inspections and by individual voltage measurements on cells. Shorts are minimized by ensuring that the electrodes are evenly spaced, straight, and vertical (Blackett & Nicol, 2010). Bending of stainless steel blanks during copper stripping must be avoided. Automation and advanced instrumentation in modern tankhouses can assist greatly in maintaining good electrode arrangements, uniformity of current distribution, and rapid short detection.

The $Fe^{3+} \rightarrow Fe^{2+}$ *chemical short-circuit* is minimized by minimizing the iron concentration in the electrolyte (Section 17.4.6). This is done by preventing leach solution carryover to electrolyte during SX and by regularly bleeding a small portion of Fe-bearing electrolyte from the tankhouse.

Current efficiency generally increases with increasing Cu and acid concentrations in the electrolyte and with current density (j), but decreases with increasing Fe^{3+} concentration and temperature. A useful practical correlation that can be used to optimize current efficiency on a daily basis in a tankhouse has been provided by Khouraibchia and Moats (2009, 2010) for operation at 40 °C:

$$\xi(\%) = 88.19 - 4.91[Fe^{3+}] + 0.52[Cu^{2+}] + 1.81 \times 10^{-3}j - 6.83 \\ \times 10^{-3}[Cu^{2+}]^2 + 0.028[Fe^{3+}][Cu^{2+}] + 4.015 \times 10^{-3}j[Fe^{3+}] \quad (17.8)$$

where all concentrations are in g/L and current density is in A/m^2.

17.7. MODERN INDUSTRIAL ELECTROWINNING PLANTS

Electrowinning differs from electrorefining due to (a) the use of inert Pb-alloy anodes instead of soluble copper anodes; and (b) the higher applied voltage.

The use of inert anodes means that (a) the anodes remain in the cells for 5+ years, (b) there is little anode scrap, and (c) no precious metal slimes are produced (although lead sludge accumulates at the bottom of the EW cells and needs to be periodically removed).

Table 17.3 shows the operating conditions and equipment in seven EW tankhouses that process electrolyte from an SX circuit. Cerro Verde and San Francisco (using SELE — Section 17.4.4) currently operate among the highest current densities in the world, while Toquepala has one of the highest current efficiencies. Radomiro Tomić has one of the largest tankhouses in the world (Table 2.7). Gaby, Safford, and Las Cruces were commissioned in 2008 and represent current state-of-the art in tankhouse design. Further information regarding these and other EW plants is available from Robinson et al. (2007).

17.8. ELECTROWINNING FROM AGITATED LEACH SOLUTIONS

A few operations, mainly in developing African countries, treat leach solutions directly by electrowinning for Cu recovery. The liquors, obtained by agitation leaching of high-grade oxides and roasted sulfide concentrates or mattes, contain 30—40 g/L Cu. These

Chapter | 17 Electrowinning

TABLE 17.3 Selected Industrial Copper Electrowinning Plant Data (2009). Additional Details of Other Major EW Plant Operations (see Table 2.7) are given by Robinson et al. (2007, 2010)

Operation	Cerro Verde, Peru	Gaby, Chile	Cobre Las Cruces, Spain	Radomiro Tomić, Chile	Safford, USA	San Francisco, Chile	Toquepala, Peru
Startup date	1977	2008	2008	1997	2008	2002	—
Cathode capacity, t/a	90 000	150 000	72 000	300 000	109 000	37 000	56 200
Electrolytic cells							
Total number	448	504	144	1000	170	72	162
Construction material	Polymer concrete	Polymer concrete	Polymer concrete	Polymer concrete	Polymer concrete	Polymer concrete	Polymer concrete
Length × width × depth, mm	5000 × 1250 × 1400	6420 × 1300 × 1386	9150 × 1300 × 1700	—	7620 × 1180 × 1690	6590 × 1270 × 1450	6390 × 1470 × 1540
Anodes/cathodes per cell	51/50	60/61	85/84	61/60	73/72	64/63	63/62
Anode–cathode spacing, mm	50	50	47.5	50	—	50	47.5
Cell inspection system	IR scanner, Cellsense	e	Manual IR scanner	IR camera	—	Manual IR scanner	IR scanner
Acid mist suppression system							
Polypropylene balls	x	x		x	x	x	
BBs							x
Ventilation	Natural	Outotec		Desom	Natural	Hoods	MistElim
Chemical suppressant	FC 1100			Mistop	FC 1100	FC1100	

(*Continued*)

TABLE 17.3 Selected Industrial Copper Electrowinning Plant Data (2009). Additional Details of Other Major EW Plant Operations (see Table 2.7) are given by Robinson et al. (2007, 2010)—cont'd

Operation	Cerro Verde, Peru	Gaby, Chile	Cobre Las Cruces, Spain	Radomiro Tomić, Chile	Safford, USA	San Francisco, Chile	Toquepala, Peru
Hoods						SAME	
Anodes							
Pb, %	99.9	98.4–98.6	98.6	98.4	—	98.44–98.69	98.54
Sn, %	0.04	1.25–1.5	1.3	1.5	—	1.25–1.50	1.2–1.5
Ca, %	0.06	0.06–0.10	0.1	0.08	—	0.06–0.07	0.05–0.06
Al, %	0	0.005–0.02	0	0.02	0	0	0
Cast or rolled	Rolled	Rolled	Rolled	Rolled	Rolled	Rolled	Rolled
Length × width × thickness, mm	1075 × 940 × 6	1080 × 938 × 6	1105 × 940 × 6	1192 × 940 × 7.5	1319 × 975 × 6.35	1095 × 940 × 7	1171 × 950 × 6.4
Lifetime, yr	5	5	—	5	—	6	4
Cathodes							
Type	Starter	St. steel (Xstrata)	St. steel (Outotec)	St. steel (Xstrata)	St. steel (Xstrata)	St. steel	St. steel (Xstrata)
Length × width × thickness, mm	1060 × 1021 × 1.4	1132 × 1018 × 3.25	1235 × 1025 × 3	1232 × 1018 × 3.25	1300 × 1000 × 3.18	1210 × 1048 × 3.25	1245 × 1036 × 3.25
Plating time, d	6	6.5	7	6	5.5	3–4	7

Mass Cu plated (total, kg)	130	—	110–120	80–90	—	41	108–112
Fe, ppm	<3	—	—	<3	—	<1	<1
Pb, ppm	<3	—	—	<2	—	<2	<1
S, ppm	<8	—	—	<8	—	<20	<3
Cathode stripping method							
Machine manufacturer	Hand-made	—	Outotec	Xstrata	—	Xstrata	Wenmec
Rate, blanks/h	200	—	350	500	400	120	300
Water temperature, °C	80	70		70	55–65	>75	85
Washing	x	x	x	x		x	x
Stripping	x	x	x	x	x	x	x
Stacking	x	x	x	x		x	x
Sampling	x	x	x	x	x		x
Weighing		x	x	x		x	x
Strapping		x	x		x		x
Blank buffing	x						
Power and energy							
Cathode current density, A/m²	400	276–333	320	230–290	366	384	285
Cathode current efficiency, %	89	90	>93	92–94	90	92	93
Cell voltage, V	2.12	1.95–2.14	—	1.87	2.1	2.1	1.9–2.0

(Continued)

TABLE 17.3 Selected Industrial Copper Electrowinning Plant Data (2009). Additional Details of Other Major EW Plant Operations (see Table 2.7) are given by Robinson et al. (2007, 2010)—cont'd

Operation	Cerro Verde, Peru	Gaby, Chile	Cobre Las Cruces, Spain	Radomiro Tomić, Chile	Safford, USA	San Francisco, Chile	Toquepala, Peru
Power consumption, kWh/t Cu	2189	—	—	1800	—	2100	2500
Electrolyte							
Circulation into each cell, m³/min	0.26	0.29	up to 0.45	0.28	—	0.43	0.3
Cu into cells, g/L	43	50	37–38	50–55	40	54	45
H₂SO₄ into cells, g/L	160	157	180	180–190	177	148	165
Temperature into cells, °C	44–45	44	45–50	47	49	56	45
Cu out of cells, g/L	40	35	—	38	38	38	36

H_2SO_4 out of cells, g/L	170	180	—	180	180	172	180
Temperature out of cells, °C	46	46	—	47.5	39–50	57	46
Fe^{2+} in electrolyte, g/L	1.05	—	—	0.1	—	—	0.33
Fe^{3+} in electrolyte, g/L	3.35	—	—	1.0–1.2	—	—	1.87
Co^{2+} in electrolyte, mg/L	180	150	—	180–200	120	—	110–120
Cl^- in electrolyte, mg/L	27	<30	—	15–20	<25	23	<10
Guar addition, g/t Cu	500	250	—	200	300	297	380
Co addition, g/t Cu	880	365	—	1130	—	562	1766
Electrolyte bleed rate, m³/d	218	—	—	600	—	1.1	1635
Bleed to control	Fe	Fe	—	Cl	—	Fe, Mn, Cl	—

leach liquors can be purified or upgraded by SX but are also processed by direct EW to produce Cu cathode. Some 120 000 t of copper were produced in this manner in 2010 (ICSG, 2010).

Direct EW electrolytes are not as pure as those obtained using SX so (a) the purity of the copper cathode is lower and (b) current and energy efficiencies tend to be lower, mainly because of the high Fe content of the electrolytes.

Because there is no recycle of the spent electrolyte to SX in direct EW, it is necessary to electrowin Cu from the electrolyte to much lower concentrations. This is usually carried out in a two-step EW process: the first EW circuit will reduce the Cu from ~40 to ~20 g/L (a bit lower than SX-EW concentrations) and the second circuit, known as *liberator cells*, will reduce the Cu to as low as 5 g/L. The quality of the cathodes produced in liberator cells is very poor: this material is usually sent to anode furnaces for melting and electrorefining. The spent electrolyte is very high in acid and is returned directly to the leach circuit.

17.9. CURRENT AND FUTURE DEVELOPMENTS

New developments in EW are dominated by (a) improving energy efficiency and (b) producing high purity copper at higher current densities (Durand, Cerpa, & Campbell, 2003; Marsden, 2008; Nieminen, Virtanen, Ekman, Palmu, & Tuuppa, 2010). Advances are being made in new anode materials that minimize voltage requirements (Asokan & Subramanian, 2009; Moats, Hardee, & Brown, 2003; Morimitsu, Oshiumi, & Wada, 2010; Pagliero, 2010). Developments in hardware, software, automation, and robotic control promote improved tankhouse management, reduced labor requirements, and consistent production of high quality copper (Aslin, Eriksson, Heferen, & Sue Yek, 2010).

The commercial application as of 2011 of alternative anodes (non-lead anodes) in two tankhouses in the United States (Sandoval, Clayton, Dominguez, Unger, & Robinson, 2010) and a third in Chile is a significant advance. These anodes use a thin layer of precious metal oxides (usually IrO_2) coated onto titanium blanks. Advantages of these anodes include minimization of Pb contamination, elimination for Co^{2+} addition to the electrolyte, elimination of cell cleaning due to the elimination of anode sludge, and a 0.3−0.4 V decrease in oxygen overpotential (see Section 17.3), which lowers energy consumption by ~15%. Workplace hygiene is improved by elimination of exposure to lead and sludge. The disadvantages of this anode are its higher cost (due to the precious metal content) and the need for gentle handling (to avoid damaging the IrO_2/Ti surface layer). Manganese, present in electrolytes from certain locations, can also be detrimental to these anodes.

Another new development is the inclusion of Co directly into conventional Pb-alloy anodes, rather than adding Co^{2+} to the electrolyte to promote the formation of a coherent PbO_2 layer (Nikolosi & Nicol, 2008b; Prengaman & Ellis, 2010a, 2010b). The cobalt requirement is much lower and cobalt losses from the circuit in electrolyte bleed are minimized, thereby reducing operating costs.

Alternatives to 316L stainless steel for cathode blanks are also receiving attention. Other materials, such as 304 stainless and LDX duplex steels, have recently been introduced as cathode blanks. These new materials cost less and may have improved properties for EW, such as higher conductivity (lower resistivity) and better weldability to copper header bars (Eastwood & Whebell, 2007).

Crane automation facilitates the use of hoods over cells for acid mist mitigation, allows accurate positioning of electrodes, and minimizes the need for operator activities in the tankhouse. Online monitoring of cell temperatures and voltages and wireless interfaces enable rapid detection of shorts (Mipac, 2004; Rantala, You, & George, 2006). Periodically using alternating current (AC) instead of direct current (DC) can improve the smoothness of the deposit and minimize electrolyte entrapment when operating at high current density (Hecker et al., 2010). Advanced process control is also increasingly used in many tankhouses (Romero, Avila, Fester, & Yacher, 2007). All of these developments improve current efficiency, productivity, and operator health and safety.

17.10. SUMMARY

Electrowinning produces pure metallic copper from leach/SX electrolytes. In 2010, about 4.5 million tonnes of pure copper were electrowon.

Electrowinning entails applying an electrical potential between inert Pb-alloy anodes and stainless steel (occasionally copper) cathodes in $CuSO_4-H_2SO_4$ electrolyte. Pure copper electroplates on the cathodes. O_2 and H_2SO_4 are generated at the anodes.

The copper is stripped from the cathode and sold. The O_2 joins the atmosphere. The Cu^{2+}-depleted electrolyte is returned to SX as the spent electrolyte for stripping Cu^{2+} from the loaded organic phase and replenishment of Cu^{2+} in the advance electrolyte.

Electrowon copper is as pure as or purer than electrorefined copper. Its only significant impurities are S (4 or 5 ppm) and Pb and Fe (1 or 2 ppm each). Careful control and attention to detail in housekeeping and tankhouse management can decrease these impurity concentrations to the low end of these ranges.

Modern trends in copper EW include (a) increasing current densities above 450 A/m^2, (b) reducing energy consumption, (c) increasing automation and robotic control and monitoring of the process, and (d) the use of a new generation of anodes.

REFERENCES

Aslin, N. J., Eriksson, O., Heferen, G. J., & Sue Yek, G. (2010). Developments in cathode stripping machines. An integrated approach for improved efficiency. In *Copper 2010: Vol. 4. Electrowinning and -refining* (pp. 1253–1271). Clausthal-Zellerfeld, Germany: GDMB.

Asokan, K., & Subramanian, K. (2009). Copper electrowinning using noble metal oxide coated titanium-based bipolar electrodes. In N. R. Neelemeggham, R. G. Reddy, C. K. Belt, & E. E. Vidal (Eds.), *Emerging technology perspectives* (pp. 201–212). Warrendale, PA: TMS.

Bender, J. T. (2010). Evaluation of mist suppression agents for use in copper electrowinning. In *Copper 2010: Vol. 4. Electrowinning and -Refining* (pp. 1271–1279). Clausthal-Zellerfeld, Germany: GDMB.

Beukes, N. T., & Badenhorst, J. (2009). Copper electrowinning: theoretical and practical design. *Journal of the Southern African Institute of Mining and Metallurgy, 109*, 343–356.

Blackett, A., & Nicol, M. (2010). The simulation of current distribution in cells during the electrowinning of copper. In *Copper 2010: Vol. 4. Electrowinning and -refining* (pp. 1291–1292). Clausthal-Zellerfeld, Germany: GDMB.

Cheng, C. Y., Hughes, C. A., Barnard, K. R., & Larcombe, K. (2000). Manganese in copper solvent extraction and electrowinning. *Hydrometallurgy, 58*, 135–150.

Das, S. P., & Gopala Krishna, P. (1996). Effect of Fe(II) during copper electrowinning at higher current density. *International Journal of Mineral Processing, 46*, 91–105.

Dew, D. W., & Phillips, C. V. (1985). The effect of Fe(II) and Fe(III) on the efficiency of copper electrowinning from dilute acid Cu(II) sulphate solutions with the Chemelec cell. *Hydrometallurgy, 14*, 331–367.

Durand, O., Cerpa, M., & Campbell, J. (2003). Production of quality electrowon copper with high current densities. *Copper 2003–Cobre 2003: Vol. V. Copper electrorefining and electrowinning* (pp. 509–521). Montreal: CIM.

Eastwood, K. L., & Whebell, G. W. (2007). Developments in permanent stainless steel cathodes with the copper industry. In G. E. Houlachi, J. D. Edwards, & T. G. Robinson (Eds.), *Copper 2007: Vol. V. Copper electrorefining and electrowinning* (pp. 35–46). Montreal: CIM.

Hecker, C., Vera, M., Martínez, G., Bustos, J. B., Villavicencio, C., & Beas, E. (2010). Improvement in copper EW and ER processes by using a multi-frequency AC + DC current. In *Copper 2010: Vol. 4. Electrowinning and -refining* (pp. 1355–1367). Clausthal-Zellerfeld, Germany: GDMB.

ICSG. (2010). *Directory of copper mines and plants 2008 to 2013*. Lisbon, Portugal: International Copper Study Group. http://www.icsg.org.

Joy, S., Staley, A., Moats, M., Perkins, C., Uhrie, J., & Robinson, T. (2010). Understanding and improvement of electrowinning current efficiency at FMI Bagdad. In *Copper 2010: Vol. 4. Electrowinning and -refining* (pp. 1379–1392). Clausthal-Zellerfeld, Germany: GDMB.

Khouraibchia, Y., & Moats, M. (2009). Effective diffusivity of ferric ions and current efficiency in stagnant synthetic copper electrowinning solutions. *Minerals and Metallurgical Processing, 26*, 176–190.

Khouraibchia, Y., & Moats, M. (2010). Evaluation of copper electrowinning parameters on current efficiency and energy consumption using surface response methodology. In F. M. Doyle, R. Woods, & G. H. Kesall (Eds.), *Electrochemistry in Mineral and Metal Processing VIII ESC Transactions* (pp. 295–306). Pennington, NJ: The Electrochemical Society.

Kumar, S., Zimba, W., Sakala, C., & Zambika, F. (2010). Evolution of cathode quality at Konkola copper mines. In *Copper 2010: Vol. 4. Electrowinning and -refining* (pp. 1413–1422). Clausthal-Zellerfeld, Germany: GDMB.

LME. (2010). *London metal exchange: Special contract rules for copper grade A*. Available from: http://www.lme.com/downloads/metalspecs/LMEspecification_Copper_111010.pdf.

Maki, T. (1999). Evolution of cathode quality at Phelps Dodge mining company. In G. V. Jergensen, II (Ed.), *Copper leaching, solvent extraction, and electrowinning technology* (pp. 223–225). Littleton, CO: SME.

Marsden, J. O. (2008). Energy efficiency and copper hydrometallurgy. In C. Young, P. R. Taylor, C. G. Anderson, & Y. Choi (Eds.), *Hydrometallurgy 2008* (pp. 29–42). Littleton, CO: SME.

Miller, G. M. (1995). The problem of manganese and its effects on copper SX-EW operations. In J. E. Dutrizac, H. Hein, & G. Ugarte (Eds.), *Copper 95—Cobre 95: Vol. III. Electrorefining and electrowinning of copper* (pp. 649–663). Montreal: CIM.

Mipac. (2004). *Tankhouse cell voltage monitoring system: Typical system description*. Mipac DRN: 011–A70 Rev 1. www.mipac.com.au/_./%5Bwp%5D_Wireless_Cell_Voltage_Monitoring_-_english.

Moats, M., Hardee, K., & Brown, C., Jr. (2003). Mesh-on-lead anodes for copper electrowinning. *JOM, 55*(7), 46–48.

Moats, M.S. How to evaluate current efficiency in copper electrowinning. In R.-H. Yoon Symposium SME, Littleton, CO, in press.

Morimitsu, M., Oshiumi, N., & Wada, N. (2010). Smart anodes for electrochemical processing of copper production. In *Copper 2010: Vol. 4. Electrowinning and -refining* (pp. 1511–1520). Clausthal-Zellerfeld, Germany: GDMB.

New Tech Copper. (2010). *SELE technology*. http://www.ntcsa.cl.

Nicol, M. (2006). *Electrowinning and electrorefining of metals*. Short Course. Perth, Australia: Murdoch University.

Nieminen, V., Virtanen, H., Ekman, E., Palmu, L., & Tuuppa, E. (2010). Copper electrowinning with high current density. In *Copper 2010: Vol. 4. Electrowinning and -refining* (pp. 1545–1558). Clausthal-Zellerfeld, Germany: GDMB.

Nikolosi, A. N., & Nicol, M. J. (2008a). Effect of cobalt ions on the performance of lead anodes used for the electrowinning of copper — a literature review. *Mineral Processing and Extractive Metallurgy Review, 29*, 143–172.

Nikolosi, A. N., & Nicol, M. J. (2008b). Addition of cobalt to lead anodes used for oxygen evolution — a literature review. *Mineral Processing and Extractive Metallurgy Review, 31*, 30–57.

Nikoloski, A., Nicol, M., & Stuart, A. (2010). Managing the passivation layer on lead alloy anodes in copper electrowinning. In *Copper 2010: Vol. 4. Electrowinning and -refining* (pp. 1559–1568). Clausthal-Zellerfeld, Germany: GDMB.

Pagliero, A. (2010). Evolution of the behaviour of anodes for copper electrowinning. In J. Casas, P. Ibáñez, & M. Jo (Eds.), *HydroProcess 2010* (pp. 76–77). Santiago, Chile: Gecamin.

Pfalzgraff, C. L. (1999). Do's and don'ts of tankhouse design and operation. In G. V. Jergensen, II (Ed.), *Copper leaching, solvent extraction, and electrowinning technology* (pp. 217–221). Littleton, CO: SME.

Prengaman, D., & Ellis, T. (2010a). New lead anode for copper electrowinning. In J. Casas, P. Ibáñez, & M. Jo (Eds.), *HydroProcess 2010* (pp. 88–89). Santiago, Chile: Gecamin.

Prengaman, R. D., & Ellis, T. (2010b). New lead anode for copper electrowinning. In *Copper 2010: Vol. 4. Electrowinning and -refining* (pp. 1593–1598). Clausthal-Zellerfeld, Germany: GDMB.

Prengaman, R. D., & Siegmund, A. (1999). Improved copper electrowinning operations using wrought Pb–Ca–Sn anodes. In J. E. Dutrizac, J. Ji, & V. Ramachandran (Eds.), *Copper 99–Cobre 99: Vol. III. Electrorefining and electrowinning of copper* (pp. 561–573). Warrendale, PA: TMS.

Rantala, A., You, E., & George, D. B. (2006). Wireless temperature and voltage monitoring of electrolytic cells. In *ALTA copper 2006*. Melbourne, Australia: ALTA Metallurgical Services.

Robinson, T., Davenport, W., Moats, M., Karcas, G., & Demetrio, S. (2007). Electrolytic copper electrowinning – 2007 world tankhouse operating data. In G. E. Houlachi, J. D. Edwards, & T. G. Robinson (Eds.), *Copper 2007: Vol. V. Copper electrorefining and electrowinning* (pp. 375–423). Montreal: CIM.

Robinson, T., Siegmund, A., Davenport, W., Moats, M., & Karcas, G. (2010). *Copper electrowinning 2010 world tankhouse survey*. Hamburg, Germany: Presented at Copper 2010.

Romero, F., Avila, G., Fester, R., & Yacher, L. (2007). Using multivariable analysis for cathode quality improvement in C.M. Doña Ines de Collahuasi EW plant. In G. E. Houlachi, J. D. Edwards, & T. G. Robinson (Eds.), *Copper 2007: Vol. V. Copper electrorefining and electrowinning* (pp. 77–90). Montreal: CIM.

San Martin, R. M., Otero, A. F., & Cruz, A. (2005). Use of quillaja saponins (*Quillaja saponaria* Molina) to control acid mist in copper electrowinning process. Part 2: pilot plant and industrial scale evaluation. *Hydrometallurgy, 77*, 171–181.

Sandoval, S., Clayton, C., Dominguez, S., Unger, C., & Robinson, T. (2010). Development and commercialization of an alternative anode for copper electrowinning. In *Copper 2010: Vol. 4. Electrowinning and -refining* (pp. 1635–1648). Clausthal-Zellerfeld, Germany: GDMB.

Shaw, R., Dreisinger, D., Arnold, S., Weekes, B., & Tinsman, H. (2008). New developments of ion exchange in copper processing. In *ALTA copper 2008*. Melbourne, Australia: ALTA Metallurgical Services.

Tozawa, K., Su, Q., & Umetsu, Y. (2003). Surface tension of acidic copper sulfate solution simulating electrolytic solution for copper electrolysis with/without addition of gelatine. In F. Kongoli, K. Itagaki, C. Yamauchi, & H. Y. Sohn (Eds.), *Yazawa international symposium: Vol. III. Aqueous and electrochemical processing* (pp. 127–137). Warrendale, PA: TMS.

Tozawa, K., Umetsu, Y., Su, Q., & Li, Z. (2003). Constitution of electrical conductivity of acidic copper sulfate electrolytes acidified with sulfuric acid for copper electrowinning and -refining processes. In F. Kongoli, K. Itagaki, C. Yamauchi, & H. Y. Sohn (Eds.), *Yazawa international symposium: Vol. III. Aqueous and electrochemical processing* (pp. 105–119). Warrendale, PA: TMS.

Tjandrawan, V., & Nicol, M. J. (2010). The oxidation of manganese ions on lead alloys during the electrowinning of copper. In *Copper 2010: Vol. 4. Electrowinning and -refining* (pp. 1699–1712). Clausthal-Zellerfeld, Germany: GDMB.

Vainio, T., & Weatherseed, M. (2009). Acid mist capture and recycling for copper electrowinning tankhouses. In *ALTA copper 2009*. Melbourne, Australia: ALTA Metallurgical Services.

SUGGESTED READING

Beukes, N. T., & Badenhorst, J. (2009). Copper electrowinning: theoretical and practical design. *Journal of the Southern African Institute of Mining and Metallurgy, 109*, 343–356.

Hopkins, W. (2009). Electrowinning – some recent developments. In *ALTA copper 2009*. Melbourne, Australia: ALTA Metallurgical Services.

ICSG. (2010). *Directory of copper mines and plants 2008 to 2013*. Lisbon, Portugal: International Copper Study Group. http://www.icsg.org.

Lillo, A. R. (2010). Intensification of copper electrowinning. In *Copper 2010: Vol. 4. Electrowinning and -refining* (pp. 1449–1462). Clausthal-Zellerfeld, Germany: GDMB.

Minakshi, M., & Nicol, M. (2009). The mechanism of the incorporation of lead into copper cathodes. In E. Domic & J. Casas (Eds.), *HydroCopper 2009* (pp. 382–391). Santiago, Chile: Gecamin.

Moats, M., & Free, M. (2007). A bright future for copper electrowinning. *JOM, 59*(10), 34–36.

Pfalzgraff, C. L. (1999). Do's and don'ts of tankhouse design and operation. In G. V. Jergensen, II (Ed.), *Copper leaching, solvent extraction, and electrowinning technology* (pp. 217–221). Littleton, CO: SME.

Pagliero, A. (2010). Evolution of the behaviour of anodes for copper electrowinning. In J. Casas, P. Ibáñez & M. Jo (Eds.), *HydroProcess 2010* (pp. 76–77). Santiago, Chile: Gecamin.

Robinson, T., Davenport, W., Moats, M., Karcas, G., & Demetrio, S. (2007). Electrolytic copper electrowinning – 2007 world tankhouse operating data. In G. E. Houlachi, J. D. Edwards & T. G. Robinson (Eds.), *Copper 2007: Vol. V. Copper electrorefining and electrowinning* (pp. 375–423). Montreal: CIM.

Chapter 18

Collection and Processing of Recycled Copper

Previous chapters describe production of *primary* copper — extraction of copper from ore. This chapter and the next describe production of *secondary* copper — recovery of copper from scrap. About half the copper reaching the marketplace has been scrap at least once, so scrap recycle is of the utmost importance.

This chapter describes scrap recycling in general, major sources and types of scrap, and physical beneficiation techniques for isolating copper from its coatings and other contaminants. Chapter 19 describes the chemical aspects of secondary copper production and refining.

18.1. THE MATERIALS CYCLE

Figure 18.1 shows the materials cycle flowsheet. It is valid for any material not consumed during use. Its key components are:

(a) Raw materials — ores from which primary copper is produced
(b) Primary production — processes described in previous chapters of this book
(c) Engineering materials — the final products of smelting/refining, mainly cast copper, and pre-draw copper rod, ready for manufacturing
(d) Manufacturing — production of goods to be sold to consumers
(e) Obsolete products — products that have been discarded or otherwise taken out of use
(f) Discard — sending of obsolete products to a discard site, usually a landfill.

Obsolete copper products are increasingly being recycled rather than sent to landfills. This is encouraged by the value of their copper, and the increasing cost and decreasing availability of landfill sites.

18.1.1. Home Scrap

The arrow marked (1) in Fig. 18.1 shows the first category of recycled copper, known as *home* or *run-around* scrap. This is copper that primary producers cannot further process or sell. Off-specification anodes, cathodes, bar, and rod are examples of this type of scrap. Anode scrap is another example.

The arrow shows that this material is reprocessed directly by the primary producer, usually by running it through a previous step in the process. Off-specification copper is usually put back into a converter or anode furnace then electrorefined. Physically defective rod and bar is re-melted in the vertical shaft furnace and recast.

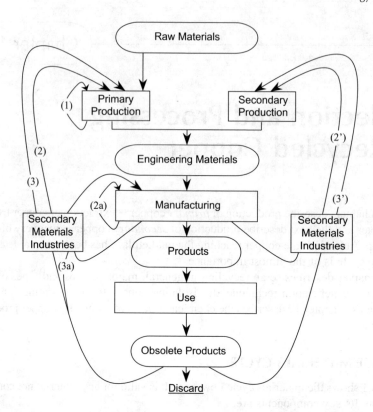

FIGURE 18.1 Flowsheet of 'materials cycle'. This is valid for any material not consumed during use. The arrow marked (1) shows *home* or *run-around* scrap. The arrows marked (2), (2a) and (2′) denote *new, prompt industrial* or *internal arising* scrap. The paths marked (3), (3a) and (3′) show *old, obsolete, post-consumer*, or *external arising* scrap.

The annual amount of home scrap production is not known because it is not reported. However, industrial producers try to minimize its production to avoid recycle expense.

18.1.2. New Scrap

The arrows marked (2), (2a) and (2′) in Fig. 18.1 denote *new, prompt industrial* or *internal arising* scrap. This is scrap that is generated during manufacturing. The primary difference between this and home scrap is that it may have been adulterated during processing by alloying or by applying coatings and coverings. Examples of *new* scrap are as numerous as the products made with copper, since no manufacturing process is 100% efficient.

The pathway taken by new scrap depends on its chemical composition and the degree to which it has become entwined with other materials. The simplest approach is to recycle it internally (2a). This is common practice with gatings and risers from castings. They are simply re-melted and cast again. Direct recycling has the advantages of (a) retaining the value of added alloying elements such as zinc or tin, which would be lost if the alloy were sent to a smelter, and (b) eliminating the cost of removing the alloying elements, which would be required if the metal were reprocessed at a smelter.

Similar re-processing is done for scrap copper tube and uncoated copper wire. Path (2a) is the most common recycling route for new scrap. As much as 90% of new U.S. copper scrap is recycled along this path (Edelstein, 1999).

If the new scrap has coatings or attachments that cannot easily be removed, or if the manufacturing facility cannot directly reuse its new scrap (e.g., a wire-drawing plant without its own melting facilities), then paths (2) and (2′) are followed.

The secondary materials industries described in Fig. 18.1 fill the role that mining and ore beneficiation facilities fill for primary copper production. In many cases they simply remove the coatings, insulation, or attachments from the scrap to make it suitable for reuse by the manufacturing facility. If purification or refining is needed, the cleaned-up scrap is sent to a primary or secondary smelter/refinery. Since these facilities produce cathode grade copper, alloying elements present in the scrap are lost (Fig. 18.2).

Specific activities of secondary materials industries are described later in this chapter.

18.1.3. Old Scrap

The final category of copper scrap (paths (3), (3a) and (3′) in Fig. 18.1) is termed *old, obsolete, post-consumer*, or *external arising* scrap. It is obtained from products that have ended their useful life. Old scrap is a huge potential source of recyclable copper. It is also difficult to process. The challenges for processing old scrap include:

(a) Low Cu grades — old copper scrap is often mixed with other materials and must be separated from this waste
(b) Unpredictability — deliveries of materials and objects vary from day to day, making processing difficult
(c) Location — old scrap is scattered about the landscape rather than being concentrated in a specific location like primary ore or new scrap.

The sources of old scrap have been divided into six classes (Bertram, Graedel, Rechberger, & Spatari, 2002; Ruhrberg, 2006):

End-of-life vehicles (ELV) include cars, trucks, and buses, but not planes, ships, or trains. The copper in ELV consists primarily of wiring, in particular electric motors and the wiring 'harness' that connects the lights and motors (Von Zitzewitz, 2004). The coverall copper content of ELV is 1.0–2.1%.

Construction and demolition (C&D) waste includes copper from the wiring of buildings, pipes and fittings from the plumbing, and brass from door handles and frames, among other items. C&D waste as a whole contains about 0.3% copper. The current high price of copper encourages copper recovery from buildings; unfortunately, this often occurs as theft from job sites or vandalism of buildings.

Waste from electrical and electronic equipment (WEEE) is one of the largest sources of old copper scrap. WEEE has three subcategories (European Copper Institute, 2007): *white goods* (domestic electrical appliances such as washing machines, refrigerators, dishwashers, etc.), *brown goods* (audiovisual appliances such as televisions and DVD players), and *gray goods* (computers and telecommunication appliances). The estimated copper content of WEEE ranges from 2 to 20%.

FIGURE 18.2 Grades of copper scrap: (a) No. 1 copper wire (ISRI code Barley/Berry), (b) No. 1 heavy copper (Candy), (c) No. 2 heavy copper (Cliff), (d) Chopped copper wire (Clove/Cobra/Cocoa), (e) Upgraded shredded electronic scrap (Dallas), (f) Light copper scrap (Dream), (g) Insulated wire scrap (Druid), (h) Shredded electric motors (Shelmo), (i) Upgraded auto shredder residue (Zebra). All photographs courtesy ISRI, International, except (e) Wallach Iron & Metal, and (i) Lynx Recyclers.

TABLE 18.1 Common Scrap Grades (ISRI, 2009)

Scrap type (ISRI name)	Description
No. 1 copper wire (Barley)	Shall consist of No. 1 bare, uncoated, unalloyed copper wire, not smaller than No. 16 B & S wire gauge. Green copper wire and hydraulically compacted material to be subject to agreement between buyer and seller.
No. 1 copper wire (Berry)	Shall consist of clean, untinned, uncoated, unalloyed copper wire and cable, not smaller than No. 16 B & S wire gauge, free of burnt wire, which is brittle. Hydraulically briquetted copper subject to agreement.
No. 2 copper wire (Birch)	Shall consist of miscellaneous, unalloyed copper wire having a nominal 96% copper content (minimum 94%) as determined by electrolytic assay. Should be free of the following: Excessively leaded, tinned, soldered copper wire; brass and bronze wire; excessive oil content, iron, and non-metallics; copper wire from burning, containing insulation; hair wire; burnt wire which is brittle; and should be reasonably free of ash. Hydraulically briquetted copper subject to agreement.
No. 1 heavy copper (Candy)	Shall consist of clean, unalloyed, uncoated copper clippings, punchings, bus bars, commutator segments, and wire not less than 1/16 of an inch thick, free of burnt wire which is brittle; but may include clean copper tubing. Hydraulically briquetted copper subject to agreement.
No. 2 copper (Cliff)	Shall consist of miscellaneous, unalloyed copper scrap having a nominal 96% copper content (minimum 94%) as determined by electrolytic assay. Should be free of the following: Excessively leaded, tinned, soldered copper scrap; brasses and bronzes; excessive oil content, iron and non-metallics; copper tubing with other than copper connections or with sediment; copper wire from burning, containing insulation; hair wire; burnt wire which is brittle; and should be reasonably free of ash. Hydraulically briquetted copper subject to agreement.
No. 1 copper wire nodules (Clove)	Shall consist of No. 1 bare, uncoated, unalloyed copper wire scrap nodules, chopped or shredded, free of tin, lead, zinc, aluminum, iron, other metallic impurities, insulation, and other foreign contamination. Minimum copper 99%. Gauge smaller than No. 16 B & S wire and hydraulically compacted material subject to agreement between buyer and seller.
No. 2 copper wire nodules (Cobra)	Shall consist of No. 2 unalloyed copper wire scrap nodules, chopped or shredded, minimum 97% copper. Maximum metal impurities not to exceed 0.50% aluminum and 1% each of other metals or insulation. Hydraulically compacted material subject to agreement between buyer and seller.
Copper wire nodules (Cocoa)	Shall consist of unalloyed copper wire scrap nodules, chopped or shredded, minimum 99% copper. Shall be free of excessive insulation and other non-metallics. Maximum metal impurities as follows: Al 0.05%, Sn 0.25%, Ni 0.05%, Sb 0.01%, Fe 0.05%. Hydraulically compacted material subject to agreement between buyer and seller.

(Continued)

TABLE 18.1 Common Scrap Grades (ISRI, 2009)—cont'd

Scrap type (ISRI name)	Description
Electronic scrap metals 3 (Dallas)	Shredded copper/precious metal bearing from an end-of-life electronic products (EOLEP) shredding operation, with the majority of iron and aluminum removed. Material may contain plastic. The size will be less than one inch and the material will be free of mercury, toner, and batteries. Typically sold on a recovery basis, subject to terms between the Buyer and Seller.
Electronic scrap metals 2 (Depot)	May include any whole or partially demanufactured EOL electronic products that are destined for a recycling processing operation. Material may contain printed circuit boards, ribbon cable, monitor yokes, and other copper and/or precious metal bearing components. Final acceptance subject to agreement between Buyer and Seller.
Light copper (Dream)	Shall consist of miscellaneous, unalloyed copper scrap having a nominal 92% copper content (minimum 88%) as determined by electrolytic assay and shall consist of sheet copper, gutters, downspouts, kettles, boilers, and similar scrap. Should be free of the following: Burnt hair wire; copper clad; plating racks; grindings; copper wire from burning, containing insulation; radiators and fire extinguishers; refrigerator units; electrotype shells; screening; excessively leaded, tinned, soldered scrap; brasses and bronzes; excessive oil, iron and non-metallics; and should be reasonably free of ash. Hydraulically briquetted copper subject to agreement. Any items excluded in this grade are also excluded in the higher grades above.
Insulated copper wire scrap (Druid)	Shall consist of copper wire scrap with various types of insulation. To be sold on a sample or recovery basis, subject to agreement between buyer and seller.
Mixed electric motors (Elmo)	Shall consist of whole electric motors and/or dismantled electric motor parts that are primarily copper wound. May contain aluminum-wound material, subject to agreement between buyer and seller. No excessive steel attachments such as gear reducers, iron bases, and pumps, or loose free iron allowed. Specification not to include sealed units or cast iron compressors.
Shredded electric motors (Shelmo)	Shall consist of mixed copper-bearing material from ferrous shredding, comprised of motors without cases. May contain aluminum-wound material and insulated copper harness wire, subject to agreement between buyer and seller. Trace percentages of other contaminants and fines may be present. No free iron or sealed units.
High-density shredded non-ferrous (Zebra)	Shall consist of high-density non-ferrous metals produced by media separation technology containing brass, copper, zinc, nonmagnetic stainless steel, and copper wire. Material to be dry and free from excess oxidation. The percentage and types of metals other than these, as well as the percentage and types of non-metallic contamination, are to be agreed upon between the buyer and seller.

Industrial electrical equipment waste (IEW) includes power cables (underground and surface), transformers, and other electrical equipment. The relative amount of this material is small, but its copper content is high (5–80%).

Industrial non-electrical equipment waste (INEW, sometimes grouped with IEW) includes large transportation equipment (planes, ships and trains), spent ammunition and ordnance, and other machinery. Although no formal estimates of its copper content exist, the percentage is likely to be small (Ruhrberg, 2006).

Municipal solid waste (MSW) is the hardest waste stream to process, and contains very little copper (0.05–0.20%). Loose coins in MSW contain much of its copper, and small appliances contain much of the rest.

Recovery of old copper scrap from waste streams depends on the type of waste, the location of the waste stream, and the price of copper (Vexler, Bertram, Kapur, Spatari, & Graedel, 2004). High-grade streams such as IEW are profitable to process and recycle, so nearly all the copper from this waste is recovered. Although few statistics are kept, the same is likely true for INEW. The presence of infrastructure for dismantling and processing ELV means that recovery of copper from this stream is higher in the developed world than in less-developed areas. Recovery of copper from MSW is always low. Recovery of copper from WEEE has also been low, but is increasing as a result of government mandates to process it.

Numerous investigations of material stocks and flows analysis (SFA) have been performed for various countries in the past decade. As a result, estimated total copper recoveries from waste materials have been reported for:

- Africa: 40% in 1994 (Van Beers, Bertram, Fuse, Spatari, & Graedel, 2003)
- China: 70% in 2005 (Yue, Lu, & Zhi, 2009)
- Europe: 73% in 1999 (Ruhrberg, 2006)
- Latin America and the Caribbean: 84% in 1994 (Vexler et al., 2004)
- USA: 49% in 1994 (Lifset, Gordon, Graedel, Spatari, & Bertram, 2002).

18.2. SECONDARY COPPER GRADES AND DEFINITIONS

The Institute of Scrap Recycling Industries (ISRI, 2009) currently recognizes 45 grades of copper-base scrap. However, most of these are for alloy scrap, which is much less available than 'pure' copper scrap. Alloy scrap is also more likely to be directly recycled than copper scrap. As a result, ISRI designations are not always important to copper recyclers. Table 18.1 describes the most significant grades of non-alloy copper scrap, including ISRI and trade names.

In addition, copper recycling often includes the treatment of *process wastes*. The definition of this word is a matter of debate in industrialized countries, because the sale and transportation of materials designated as waste is more heavily regulated than that of materials designated as scrap. In fact, material graded as copper-bearing scrap is defined in many countries as waste, despite the fact that it can be recycled profitably. Wastes generally have a low copper content, low economic value, and a high processing cost per kg of contained copper. As a result, recyclers sometimes charge a fee for processing these materials (Lehner, 1998). Examples of process wastes include the sludge generated from copper plating, refining slags from foundries, and the residue from leaching of anode slimes.

18.3. SCRAP PROCESSING AND BENEFICIATION

18.3.1. Wire and Cable Processing

Wire and cable are by far the most common forms of old scrap. It is these forms for which the most advanced re-processing technology exists. Nijkerk and Dalmijn (1998) divide scrap wire and cable into three types:

(a) *Above-ground*, mostly high-tension power cable. These cables are high-grade (mainly copper, little insulation) and fairly consistent in construction. They are easy to recycle.
(b) *On-the-ground*, with a variety of coverings and sizes. These are usually thin wires, so the cost of processing per kg of recovered copper is higher than that for cable. Wire is also more likely to be mixed with other waste, requiring additional separation. Automotive harnesses and appliance wire are examples.
(c) *Below-ground*, which feature complex construction and many coverings. These cables often contain lead sheathing, bitumen, grease, and mastic. This means that fairly complex processing schemes are required to recover their copper without creating safety and environmental hazards.

Copper recovery from scrap cable by shredding (also known as chopping or granulating) has its origins in World War II when it was developed to recover rubber coatings (Sullivan, 1985). Shredding has since become the dominant technology for scrap wire and cable processing (Nijkerk & Dalmijn, 1998).

Figure 18.3 shows a typical cable-chopping flowsheet. Before going to the first *granulator*, the scrap cable is sheared into lengths of one meter or less (Nijkerk & Dalmijn, 1998; Sullivan, 1985). This is especially important for larger cables. The first granulator, or *prechopper*, is typically a rotary knife shear with one rotating shaft. The knives on this shaft cut against a second set of stationary knives. Rotation speed is about 120 rpm, and a screen is provided to return oversize product to the feed stream. Its primary task is size reduction rather than separation of the wire from its insulation. Depending on the type of material fed to the prechopper, the length of the product pieces is 10–100 mm (Reid, 2003).

The primary granulator also liberates any pieces of steel that may be attached to the scrap cable. These are removed from the product by a magnetic separator (Reid, 2003).

The partially chopped cable is then fed to a second granulator (Sullivan, 1985). The second granulator is similar to the first, but operates at much higher speeds (400 rpm) and has more knives and smaller blade clearances (as small as 0.05 mm). It chops the cable to lengths of 6 mm or smaller, mostly liberating the copper from its insulation. Again, a screen is used to return oversize material. The recent trend is to operate the second granulator at lower speeds, which reduces metal losses to the plastics fraction (Reid, 2003).

The final unit process in scrap cable and wire processing is separation of copper from insulation. This is normally accomplished using the difference between the specific gravity of the copper (8.96) and that of the insulating plastic and rubber (1.3–1.4). Figure 18.3 shows a *specific gravity separator*, which typically produces three fractions:

(a) A 'pure' plastic fraction
(b) Copper *chops* meeting Number 1 or Number 2 scrap purity specifications (Table 18.1).

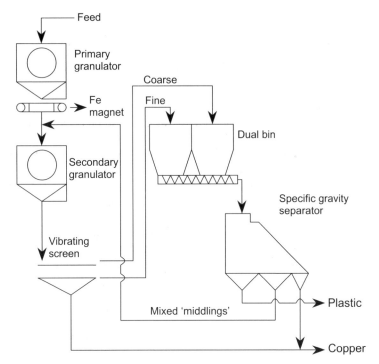

FIGURE 18.3 Recycle cable 'chopping' flowsheet for separating copper from insulation and ferrous material. The specific gravity separator is an air table, which blows insulation upwards while allowing copper to sink.

(c) A middlings fraction, which is returned to the second granulator for re-processing.

The separator is most commonly an air table (Sullivan, 1985), which ultimately recovers 80−90% of the input copper. Compact units that include the granulator and air separator in one unit are increasingly popular for smaller processors (Fowler, 2008).

In larger chopping operations, elutriation or electrostatic separation are sometimes used as 'cleaner' steps to recover the last bit of copper from the plastic fraction (Reid, 2003). A significant difference between copper chops and plastic is particle shape − the copper chops are longer than the plastic. Efforts have been made to develop separation processes based on this characteristic (Koyanaka, Ohya, Endoh, Iwata, & Ditl, 1997). Results from De Araújo, Chaves, Espinosa, & Tenorio et al. (2008) suggest that elutriation is the most effective of these techniques.

Underground cable processing is complicated by the complexity of its construction, the flammability of its coverings, and the presence of aluminum or lead in the shredding product. Nijkerk and Dalmijn (1998) describe the use of cable stripping for larger cables. This follows shearing and involves slicing open the cable and removing the copper wire by hand. Smaller underground cables (< 0.05 m diameter) can be successfully chopped without prior stripping (Reid, 2003). Attempts have been made to introduce cryogenic shredding to reduce the flammability hazard. Eddy-current separators can be used following shredding to separate lead and aluminum from copper.

18.3.2. Automotive Copper Recovery (ELV)

Figure 18.4 shows a flowsheet for recovering materials from junked automobiles (Suzuki et al., 1995). There are three potential sources of recyclable copper in this flowsheet.

The first is the radiator, which is manually removed from the car before shredding. Most automotive radiators are now made of aluminum or plastic (Von Zitzewitz, 2004); however, copper radiators from older vehicles still show up on occasion. Regardless of material, the recycling rate for radiators is nearly 100%.

The second source of copper in Fig. 18.4 is the non-ferrous metal scrap stream remaining after the car has been shredded and its iron and steel have been magnetically removed. Three metals dominate this stream: aluminum, copper, and zinc. The copper consists mostly of wire from the electrical circuits (Von Zitzewitz, 2004).

Because aluminum and zinc are more easily oxidized than copper, the unseparated non-ferrous fraction (also known as Zorba) can be sold to copper smelters without complete separation. However, this eliminates the value of the aluminum and zinc and

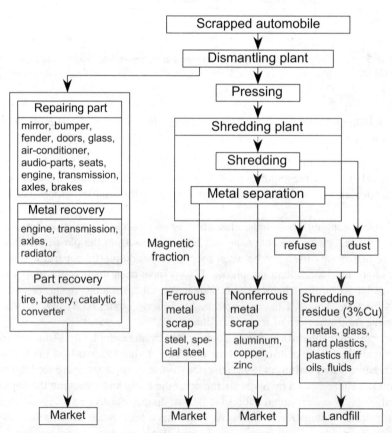

FIGURE 18.4 Flowsheet for recovering metals from scrap automobiles. Procedures for recovering copper are detailed in Section 18.3.2.

increases the cost of smelting. As a result, several techniques are used to separate the metals in this fraction (Dalmijn & De Jong, 2007; Nevada Shredder Consulting, 2009). Eddy-current separation (ECS) uses the different electrical conductivities of the metallic and non-metallic particles in the shredded material to remove most of the non-metallics. Heavy-media separation removes the lower-density aluminum and magnesium from the metallic fraction. What remains is Zebra (see Table 18.1), a mixture of mostly zinc, brass, and copper.

Two methods can be used to upgrade Zebra, both based on the difference in color of the copper, brass, zinc, and other metals in it (Pretz & Mutz, 2006). The first is hand-sorting, which is widely practiced in China and India but may become less viable as labor costs there increase over time. The second is automated sorting, which has been commercially practiced for over 10 years.

Better efforts have been made over the past ten years to recover copper from the ferrous scrap stream produced by ELV shredding. This stream contains copper in the form of *meatballs*, the intermingled remains of electric motors (Kiser, 2007). The copper is a significant contaminant in the production of steel, and can be smelted to produce secondary copper if the meatballs are collected (see next chapter). Induction sorters have been developed that can sense the differing electrical conductivity of copper-containing scrap, and separate this from the rest of the ferrous material (Fig. 18.5).

The final potential source of copper in Fig. 18.4 is the auto shredder residue (ASR), which remains after the metals have been removed. This residue consists primarily of dust and organic matter — plastic from the dashboard and steering wheel, fluff from the seat cushions, and pieces of carpet and fabric. ASR also contains up to 3% copper and has some fuel value (Forton, Harder, & Moles, 2006). Current practice is to landfill this material, but this practice is increasingly unacceptable (Kurose, Okuda, Nishijima, & Okada, 2006). As a result, extensive research efforts have been made over the past decade to find a means of processing it to recover valuables. Commercial facilities now operate

FIGURE 18.5 *Meatballs* are created when electric motors are shredded, resulting in a commingled mix of copper wire and iron motor core. *(Photo courtesy Zane Voss, Nucor, Inc.)*

specifically for processing ASR, including copper recovery (Fischer, 2006; Toyota, 2010). ASR is also directly charged to smelting furnaces (Fujisawa, 2005; Nihsiwaki & Hayashi, 2005).

18.3.3. Electronic Scrap Treatment

Although it consists of a variety of items, the overall composition of electronic scrap can be divided into three categories: (a) plastic (~30% in 1991), (b) refractory oxides (~30%), and (c) metals (~40%). About half its metal content is copper. It also contains significant amounts of gold and silver. Copper smelting/refining is already set up to recover gold and silver, so it is a logical destination for treating electronic scrap.

A potential problem with smelting electronic scrap is incomplete combustion of its plastic fraction and consequent evolution of organic compounds. However, high temperature oxygen smelting completely avoids this problem.

A more serious problem is the declining metal content of electronic scrap. The producers of circuit boards (PCB) and other assemblies have learned over time to reduce the amount of metal needed in their products (Zhang & Forssberg, 1999). This makes the scrap increasingly difficult to profitably recycle.

The result has been development of 'minerals processing' strategies for isolating the metals in electronic scrap. The approach is similar to that used for automobiles (Chen, 2009; Cui & Forssberg, 2003):

(a) Disassembly to recover large items
(b) Shredding to reduce the size of the remaining material and liberate the metal
(c) Separation of metals based on differences in density and electrical conductivity.

Shredding is typically done using low-speed, high-torque knife shredders (Anon, 2004). A primary shredding step generates relatively large particles, from which copper is recovered by air tabling, screening, or hand-picking. Secondary shredding generates a finer material with better liberation. Copper recovery from this stream is accomplished using ECS or electrostatic separation (Anon, 2005; Cui & Forssberg, 2003; Huang, Guo, & Xu, 2009).

However, shredding does not liberate metal that has been plated onto surfaces, such as printed circuit boards. In addition to copper, precious metals such as gold and silver also remain behind in this fraction, and this makes it worth recovering. As a result, the plastic residue from PCB processing is also frequently treated as a raw material. There are two general approaches for processing it, the *formal* and *non-formal*.

The non-formal approach for metals recovery from PCB surfaces is a hydrometallurgical process commonly used in developing countries, particularly China and India (Chatterjee & Kumar, 2009; Huang et al., 2009; Widmer, Oswald-Krapf, Sinha-Khetriwal, Schnellman, & Böni, 2005). This process replaces mechanical processing with burning in open air to liberate pieces of copper. Two acid treatments follow, the first to dissolve solder and the second to dissolve copper. The solution is then boiled away to recover copper sulfate. This process is simple to operate, especially on a small scale. It is also environmentally destructive, and increasingly unacceptable to local governments and international bodies.

The formal approach uses mechanical processing, and sends the residues to secondary smelters, which use it as feed in their process. The gold and silver report to the impure copper metal (along with any copper), and are recovered using the same flowsheet used for gold and silver recovery from primary concentrates (Hagelüken, 2006; Lehner, 1998). This is described in more detail in the next chapter.

18.4. SUMMARY

About half the copper reaching the market today has been scrap at least once. Scrap is generated at all stages in the life span of a copper product, including production (home scrap), manufacturing (new scrap), and post-consumer disposal (old scrap).

The purest copper scrap is simply re-melted and recast in preparation for manufacture and use. Less pure copper scrap is re-smelted and re-refined. Alloy scrap is usually recycled directly to make new alloy.

Considerable scrap must be physically treated to isolate its copper from its other components. An important example of this is recovery of copper from wire and cable. It is done by:

(a) Chopping the wire and cable into small pieces to liberate its copper
(b) Physically isolating its copper by means of a specific gravity separation (air table).

Copper recovery from used automobiles and electronic devices follows a similar pattern, i.e.:

(a) Liberation by size reduction (shredding)
(b) Isolation of copper by magnetic, specific gravity, and eddy-current separation.

The copper from these processes is then re-smelted and re-refined.

Old (obsolete) scrap is often discarded in landfills. There is, however, an increasing tendency to recycle this material due mainly to the increased cost and decreased availability of landfill sites.

REFERENCES

Anon. (2004). Equipment spotlight: Electronic shredders. *American Recycler*. www.americanrecycler.com/0104spotlight.html 26 April 2004.

Anon. (2005). Equipment spotlight: Electronics recycling separation systems. *American recycler*. www.americanrecycler.com/0305spotlight.shtml 5 August 2010.

Bertram, M., Graedel, T. E., Rechberger, H., & Spatari, S. (2002). The contemporary European cycle: Waste management subsystem. *Ecological Economics, 42*, 43—57.

Chatterjee, S., & Kumar, K. (2009). Effective electronic waste management and recycling process involving formal and non-formal sectors. *International Journal of the Physical Sciences, 4*, 893—905.

Chen, B. X. (2009). Where gadgets go to die: Facility strips, rips, and recycles. *Condé Nast Digital*. www.wired.com/gadgets/miscellaneous/news/2009/03/gallery_ewaste_recycling?currentPage=all 2 April 2009.

Cui, J., & Forssberg, E. (2003). Mechanical recycling of waste electric and electronic equipment: a review. *Journal of Hazardous Materials, B99*, 243—263.

Dalmijn, W. L., & De Jong, T. P. R. (2007). The development of vehicle recycling in Europe: sorting, shredding, and separation. *JOM, 59*(11), 52—56.

De Araújo, M. C. P. B., Chaves, A. P., Espinosa, D. C. R., & Tenorio, J. A. S. (2008). Electronic scraps — recovering of valuable materials from parallel wire cables. *Waste Management, 28*, 2177—2182.

Edelstein, D. L. (1999). Copper. In *Recycling — Metals*. Washington, DC: U.S. Geol. Survey. http://minerals.usgs.gov/minerals/pubs/commodity/recycle/870499.pdf.

European Copper Institute. (2007). *Recycling of copper waste and WEEE: A response to growing demand for raw materials*. European Copper Institute. www.eurocopper.org/files/presskit/pk_en_recycling_31.05.07_final.pdf 31 May 2007.

Fischer, T. (2006). Getting a return from residue. *Scrap, 63*(3), 57–62.

Forton, O. T., Harder, M. K., & Moles, N. R. (2006). Value from shredder waste: ongoing limitations in the UK. *Resources Conservation and Recycling, 46*, 104–113.

Fowler, J. (2008). Equipment focus: compact wire processors. *Scrap, 65*(2), 231–242.

Fujisawa, T. (2005). The non-ferrous metals industry and social sustainability in Japan. *World of Metallurgy – Erzmetall, 58*, 263–268.

Hagelüken, C. (2006). Recycling of electronic scrap at Umicore's integrated metals smelter and refinery. *World of Metallurgy – Erzmetall, 59*, 152–161.

Huang, K., Guo, J., & Xu, Z. (2009). Recycling of waste printed circuit boards: a review of current technologies and treatment status in China. *Journal of Hazardous Materials, 164*, 399–408.

ISRI. (2009). *Scrap specifications circular 2009*. Institute of Scrap Recycling Industries. http://iguana191.securesites.net/paperstock/files/2010/Scrap_Specifications_Circular.pdf 19 November 2009.

Kiser, K. (2007). Motors & meatballs: from trash to treasure. *Scrap, 64*(3), 50–60.

Koyanaka, S., Ohya, H., Endoh, S., Iwata, H., & Ditl, P. (1997). Recovering copper from electric cable wastes using a particle shape separation technique. *Advanced Powder Technology, 8*, 103–111.

Kurose, K., Okuda, T., Nishijima, W., & Okada, M. (2006). Heavy metals removal from automobile shredder residues (ASR). *Journal of Hazardous Materials, B137*, 1618–1623.

Lehner, T. (1998). Integrated recycling of non-ferrous metals at Boliden Ltd. Rönnskar Smelter. In *1998 IEEE international symposium on electronics and the environment* (pp. 42–47). Piscataway: IEEE.

Lifset, R. J., Gordon, R. B., Graedel, T. E., Spatari, S., & Bertram, M. (2002). Where has all the copper gone: the stocks and flows project, part 1. *JOM, 54*(10), 21–26.

Nevada Shredder Consulting. (2009). *Shredder technology overview*. Nevada Shredder Consulting, Inc. http://form-letters.com/shredder_technology.htm 18 December 2009.

Nihsiwaki, M., & Hayashi, S. (2005). Stabilization of recycling technology for shredder residue in copper smelting process. *Shigen-to-Sozai, 121*, 357–362.

Nijkerk, A. A., & Dalmijn, W. L. (1998). *Handbook of recycling techniques*. The Hague, Netherlands: Nijkerk Consultancy.

Pretz, T., & Mutz, S. (2006). Metal recovery in automobile recycling. *Acta Metallurgica Slovaca, 12*, 456–462.

Reid, R. L. (2003). Equipment focus: wire processing. *Scrap, 60*(5), 63–69.

Ruhrberg, M. (2006). Assessing the recycling efficiency of copper from end-of-life products in Western Europe. *Resources Conservation and Recycling, 48*, 141–165.

Sullivan, J. F. (1985). Recycling scrap wire and cable: the state of the art. *Wire Journal International, 18*(11), 36–50.

Suzuki, M., Nakajima, A., & Taya, S. (1995). Recycling scheme for scrapped automobiles in Japan. In P. B. Queneau & R. D. Peterson (Eds.), *Third International Symposium on Recycling of Metals and Engineered Materials* (pp. 729–737). Warrendale, PA: TMS.

Toyota. (2010). *Development of ASR recycling technologies*. Toyota Motor Corporation. http://www2.toyota.co.jp/en/tech/environment/recycle/asr 29 March 2010.

Van Beers, D., Bertram, M., Fuse, K., Spatari, S., & Graedel, T. E. (2003). The contemporary African copper cycle: one year stocks and flows. *Journal of the Southern African Institute of Mining and Metallurgy, 93*, 147–162.

Vexler, D., Bertram, M., Kapur, A., Spatari, S., & Graedel, T. E. (2004). The contemporary Latin American and Caribbean copper cycle: 1 year stocks and flows. *Resources Conservation and Recycling, 41*, 23–46.

Von Zitzewitz, A. (2004). Recycling of copper from cars – the wiring harness as an example. *World of Metallurgy – Erzmetall, 57*, 211–216.

Widmer, R., Oswald-Krapf, H., Sinha-Khetriwal, D., Schnellman, M., & Böni, H. (2005). Global perspectives on e-waste. *Environmental Impact Assessment Review, 25*, 436–458.

Yue, Q., Lu, S. W., & Zhi, S. K. (2009). Copper cycle in China and its entropy analysis. *Resources Conservation and Recycling, 53*, 680–687.

Zhang, S., & Forssberg, E. (1999). Intelligent liberation and classification of electronic scrap. *Powder Technology, 105*, 295–301.

SUGGESTED READING

Cui, J., & Forssberg, E. (2003). Mechanical recycling of waste electric and electronic equipment: a review. *Journal of Hazardous Materials, B99*, 243–263.

Nijkerk, A. A., & Dalmijn, W. L. (1998). *Handbook of recycling techniques*. Nijkerk Consultancy.

Ruhrberg, M. (2006). Assessing the recycling efficiency of copper from end-of-life products in Western Europe. *Resources Conservation and Recycling, 48*, 141–165.

Sullivan, J. F. (1985). Recycling scrap wire and cable: the state of the art. *Wire Journal International, 18*(11), 36–50.

Chapter 19

Chemical Metallurgy of Copper Recycling

About one-third of the copper currently produced in the world is derived from *secondary materials* — the copper scrap and waste described in the previous chapter. Secondary material is recycled in numerous ways. New scrap is often recycled directly back to the melting furnace, where it was produced in the first place. Old scrap and waste streams (and some new scrap) travel a more complex path. They can either be added to one of the furnaces used to produce primary copper — the smelters, converters, and anode furnace described in previous chapters. They can also be reprocessed by secondary smelters specifically designed to handle such material. This chapter looks at the different methods used to turn old scrap and waste back into new copper metal.

19.1. CHARACTERISTICS OF SECONDARY COPPER

Several factors determine what happens to copper-containing scrap and waste. These include:

- *The condition of the scrap.* The different types of No. 1 scrap (see Table 18.1) contain very small levels of oxidation or impurities, and need only be remelted in a non-oxidizing environment. As a result, this scrap commands a higher price, and is more likely to be charged to furnaces requiring a purer grade of input material. No. 2 scrap is more oxidized, and requires refining. Heavily oxidized waste materials like plating sludges require remelting in a reducing environment to avoid copper loss.
- *Location.* Copper scrap is often collected at locations which are very far from primary smelters. The cost of transporting it determines where it winds up.
- *Alloying elements.* Impurity elements in secondary copper often have sufficient value to be recovered in their own right. This includes the gold and silver content of electronic scrap, the lead in copper drosses, and the lead, tin, and zinc in brass and bronze scrap. Recovering these elements often dictates recycling strategy.
- *Government regulations.* Some governments place export tariffs on copper-containing scrap to encourage recycling at home. Restrictions on the exporting of some electronic scrap have also encouraged local reprocessing efforts. Restrictions on air emissions from processing plants have had an impact on processing strategy.

19.2. SCRAP PROCESSING IN PRIMARY COPPER SMELTERS

Secondary copper can be added at three locations in the primary coppermaking process. The most common is the converting furnace, but additions are also made in the smelting and anode furnaces.

19.2.1. Scrap Use in Smelting Furnaces

Because secondary copper has no sulfur in it, adding it to a furnace is a net energy consumer. This makes it difficult to add to matte smelting furnaces, which already require some fuel use. Making the size of scrap or waste particles small enough to use in flash-furnace concentrate burners is also difficult. As a result, the extensive use of waste materials in the flash furnace at Kosaka (Watanabe & Nakagawara, 2003) is rare. However, small-size shredded scrap can be added in limited quantities to flash furnaces (Maeda, Inoue, Kawamura, & Ohike, 2000). In fact, the smelting furnace is preferred to the converter for feeding electronic scrap, due to its plastic content. There are two reasons for this:

(a) The plastic has fuel value, which provides heat for smelting
(b) When burned intermittently, plastic often gives off smoke and other particulates, which might escape through the mouth of a Peirce–Smith converter, adversely affecting workplace hygiene. Burned in a sealed flash furnace, these particulates are efficiently captured by dust collection devices.

Feeding scrap and waste to other types of primary smelting furnace is easier, and both the Noranda and Mitsubishi smelting processes have been adapted to include scrap in the feed (Reid, 1999; Huitu, 2003). Figure 19.1 illustrates the flowsheet for scrap usage in the Mitsubishi process at Naoshima (Komori, Shimizu, Usami, & Kaneda, 2007; Oshima, Igarashi, Hasegawa, & Kumada, 1998). Particulate scrap is mixed with concentrate and blown into the smelting furnace through its rotating lances. Larger scrap pieces are charged to the smelting and converting furnaces through roof and wall chutes. Mitsubishi converting is particularly exothermic, allowing large amounts of scrap to be melted in the converting furnace. The Onahama smelter in Japan has for some time used reverberatory furnaces to add ASR (see Section 18.3.2) to sulfide matte. The heat generated by combustion of the organic matter in the residue helps offset the extra fuel required for this (Kikumoto, Abe, Nishiwaki, & Sato, 2000).

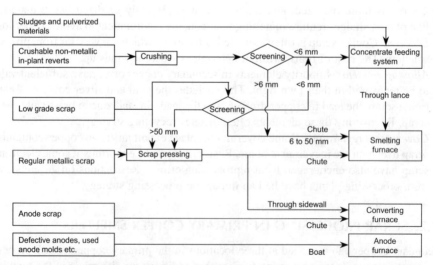

FIGURE 19.1 Scrap smelting in the Mitsubishi smelter, Naoshima (Oshima et al., 1998). Low-grade particulate scrap is fed into the smelting furnace through its rotating lances. Larger scrap is fed through roof and sidewall chutes. Large high-grade scrap is fed into the anode furnace.

19.2.2. Scrap Additions to Converters and Anode Furnaces

In contrast to smelting furnaces, scrap is usually needed in copper converters, to use the excess heat created by the converting reaction. Depending on matte grades, oxygen enrichment, and scrap quality, the fraction of scrap used in converter charges can range from 0 to 35% or higher.

A higher quality of scrap is needed for primary converters: low-alloy scrap, No. 1 and No. 2 scrap (Section 18.3.1) if available, along with compressed turnings and anode scrap. Low-grade material and plant reverts may also be fed if their plastic content is not too large (Oshima et al., 1998).

Scrap can also be added to anode furnaces, although the most common addition is directly recycled scrap anodes, rather than externally purchased scrap. Because anode furnaces were designed to handle molten blister copper, scrap melting can be slow (Potesser, Antrekowitsch, & Hollets, 2007). New furnace and burner designs can make anode furnaces better for melting scrap (Deneys & Enriquez, 2009).

19.3. THE SECONDARY COPPER SMELTER

Although each secondary copper smelter is unique, they can be divided into two groups. The first type are *metal* smelters, which treat only higher-grade metallic scrap. Many of these smelters are located in China (Risopatron, 2007). The second are *black-copper* smelters, which process low-grade scrap and waste along with higher-value scrap. Hoboken in Belgium and Brixlegg in Austria are examples of this type of smelter (Hageluken, 2006; Messner, Antrekowitsch, Pesl, & Hofer, 2005).

19.3.1. High-Grade Secondary Smelting

Rinnhofer and Zulehner (2005) described the typical process for scrap remelting in tiltable reverberatory furnaces (*reverbs*). The furnaces typically have a small capacity (5–30 tonnes) and a fairly long cycle time (up to 24 h tap-to-tap; Risopatron, 2007). They have the advantage of flexibility, and can be used for both oxidation of impurities and deoxidation afterward. Side tuyeres are used both for injection of air during oxidation and natural gas during reduction.

While reverbs are effective furnaces for refining operations, their efficiency as scrap melters is low (Potesser et al., 2007), and this has created incentive for more effective furnace design. Recent designs have added a reverb to a shaft furnace (see next chapter), which delivers molten copper to the reverb at a much quicker rate (Air Products, 2007; Properzi & Djukic, 2008). The refining capability of the reverb allows the use of more scrap in casting facilities than normal.

19.3.2. Smelting to Black Copper

Figure 19.2 is a flowsheet for pyrometallurgical processing of low-grade scrap in a secondary black-copper smelter. The 'Melting/Reducing furnace' at the top accepts the 'copper bearing scrap' described in Section 18.2. This scrap includes:

(a) Automobile shredder product from which the copper and iron cannot be separated, along with motors, switches and relays (Elmo and Shelmo from the previous chapter)
(b) Dross from decopperizing lead bullion
(c) Dusts from copper melting and alloying facilities

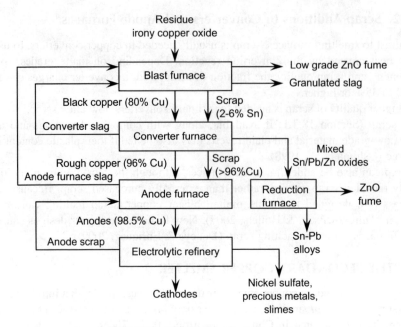

FIGURE 19.2 Flowsheet for Cu scrap treatment.

(d) Sludges from copper electroplating operations
(e) Auto shredder residue

The feed to these furnaces is low grade and highly oxidized. It requires reduction to metallic copper. Major metallic impurities are lead and tin (from bronze scrap, solder, and decopperizing dross), zinc (from scrap brass), iron (from automotive scrap), and nickel (from scrap monel and other alloys). These elements are often present as mixtures of metal and oxide.

Heat and CO reductant are supplied to this type of furnace by combusting metallurgical coke or natural gas, e.g.:

$$C(coke) + 0.5O_2 \rightarrow CO + heat \tag{19.1}$$

The carbon monoxide in turn reduces the oxides of the feed to metal or a lower oxide, i.e.:

$$CO + Cu_2O \rightarrow CO_2 + 2Cu(l) \tag{19.2}$$
$$CO + ZnO \rightarrow CO_2 + Zn(g) \tag{19.3}$$
$$CO + PbO \rightarrow CO_2 + Pb(l,g) \tag{19.4}$$
$$CO + NiO \rightarrow CO_2 + Ni(l) \tag{19.5}$$
$$CO + SnO_2 \rightarrow CO_2 + SnO(l,g) \tag{19.6}$$
$$CO + SnO \rightarrow CO_2 + Sn(l) \tag{19.7}$$

Metallic iron in the scrap also performs some reduction, especially of easily reduced oxides like Cu_2O:

$$Fe + Cu_2O \rightarrow FeO(l) + Cu(l) \tag{19.8}$$

As a result of these reactions, secondary copper reducing furnaces generate three products. They are:

(a) Molten *black copper*, 74−80% Cu, 6−8% Sn, 5−6% Pb, 1−3% Zn, 1−3% Ni and 5−8% Fe (Ayhan, 2000)
(b) Molten slag containing FeO, CaO, Al_2O_3, SiO_2 along with 0.6−1.0% Cu (as Cu_2O), 0.5−0.8% Sn (as SnO), 3.5−4.5% Zn (as ZnO), and small amounts of PbO and NiO
(c) Offgas containing CO, CO_2, H_2O, and N_2, plus metal and metal oxide vapors.

Cooling and filtering of offgas (c) recovers oxide dust containing 1−2% Cu, 1−3% Sn, 20−30% Pb, and 30−45% Zn. The dust also contains chlorine from chlorinated plastics in the feed. It is always processed to recover its metal content (Hanusch & Bussmann, 1995).

Traditionally a blast furnace was used as the vessel for initial melting and reducing secondary copper. However, this furnace has all but disappeared in recent years, due to its thermal inefficiency and the need for expensive metallurgical coke. The blast furnaces at Brixlegg in Austria and Kovohuty in Slovakia are among the few remaining examples (Kovohuty, 2010; Pesl, 2010). Two other types of furnace have become more prominent for this purpose.

The first is a Kaldo furnace (Outotec, 2009; Fig. 19.3). This is an upgraded version of the well-known top-blown rotary converter (TBRC). The Kaldo furnace is a batch processor, which burns natural gas or oil. The use of oxygen or enriched air reduces heat losses, and the oxygen/fuel ratio can be adjusted to produce a reducing or oxidizing atmosphere. The furnace rotates during operation, which improves kinetics, and the overall fuel usage is less than that of a blast furnace. Because they rotate, they are generally preferred for smaller operations. Kaldo furnaces are currently in use for secondary copper processing in Belgium, China, India, and Sweden (Lehner & Vikdahl, 1998; Pesl, 2010; Tandon, Guha, & Kamath, 2007; Zhu & Liu, 2000). However, none have been put into operation in the past decade, and this may be outmoded technology.

The second type of commonly used furnace for secondary copper melting and reduction is the top submerged lance (TSL) furnace that is increasingly prevalent for sulfide matte smelting and converting (Edwards & Alvear, 2007; Matusewicz & Reuter, 2008; also see Chapter 9). Like the Kaldo furnace, Ausmelt and Isasmelt furnaces accept a wide variety of feeds, and can process almost anything with copper in it (Ayhan, 2000). As stationary furnaces, they can burn coal or coke fines as a fuel, along with fuel oil and natural gas. They can also use the plastic in electronic scrap and ASR as a fuel (Brusselaers, Mark, & Tange, 2006). They can change from a reducing to an oxidizing environment like a Kaldo furnace, which allows them to remove the impurities from black copper without needing another furnace. Isasmelt furnaces are being used for secondary copper production in Belgium and Germany, and three Ausmelt furnaces are in service in Asia.

19.3.3. Converting Black Copper

The impurities in black copper can be divided into two groups − those that are more easily oxidized than copper (Fe, Pb, Sn, Zn) and those that are difficult or impossible to remove by oxidation (Ni, Ag, Au, platinum-group metals). These impurities are removed sequentially by a strategy similar to that for purifying primary copper.

The first step in refining black copper is oxidation. The black copper produced by blast furnaces or Kaldo units is typically oxidized in a Peirce−Smith converter (Fig. 1.6).

Black copper produced in a TSL furnace is refined as a second stage of operation in that furnace (Edwards & Alvear, 2007). In either case, air is blown into the molten black copper, oxidizing Fe, Pb, Sn, and Zn along with some Ni and Cu. Alloyed copper scrap (the light copper described in Section 18.2) is also added to the converter. Most of its impurities are also oxidized.

This oxidation generates slag containing 10–30% Cu, 5–15% Sn, 5–15% Pb, 3–6% Zn and 1–5% Ni, depending on the composition of the converter feed (Ayhan, 2000). The slag can be returned to the blast furnace to recover the copper and nickel (Messner et al., 2005; Pesl, 2010; Tandon et al., 2007), or processed first to recover the lead and tin (Hageluken, 2006). Lead, tin, and zinc will ultimately be recovered either from slag or dust generated by one of the furnaces, but several trips through the circuit may be required to route these elements to the appropriate phase.

An offgas is also generated which, when cooled and filtered, yields dust analyzing 0.5–1.5% Cu (as Cu_2O), 0.5–1.5% Sn (as SnO), 10–15% Pb (as Pb and PbO), and 45–55% Zn (as ZnO). This dust is usually reduced to recover its Pb and Sn as solder (Ayhan, 2000).

Oxidation of black copper provides little heat to the converter (unlike oxidation of matte, Chapter 9). Heat for the converting process must, therefore, be provided by burning hydrocarbon fuel.

19.3.4. Fire Refining and Electrorefining

The main product of the converter is molten *rough copper*, 95–97% Cu. It is added to an anode furnace for final, controlled oxidation before casting it as anodes. No. 1 scrap (Table 18.1) is also often added to the anode furnace for melting and casting as anodes. Plant practice is similar to that for fire refining of primary copper (Chapter 13).

The rough copper in Fig. 19.2 usually contains nickel or tin, which is never completely removed by oxidation converting. It may also contain appreciable amounts of gold, silver, and platinum-group metals from the original scrap (Reid, 1999). Recovery of these metals is important to the profitability of a recycling facility. As a result, the anode furnace product is almost always cast as anodes for electrorefining. The impurity level in secondary anodes is higher than that in most anodes from primary smelting operations. As a result, electrolyte purification facilities need to be larger. Otherwise, plant practice is similar to that described in Chapter 14.

The principal electrorefining products are high purity cathode copper, nickel sulfate from electrolyte purification, and anode slimes. Processing of anode slimes is described in Chapter 21.

19.4. SUMMARY

Copper scrap is smelted in primary (concentrate) and secondary (scrap) smelters. Primary smelters mainly smelt concentrate. Some, however, are well adapted to smelting all grades of scrap. Smelters with Ausmelt/Isasmelt, Mitsubishi, Noranda, reverberatory, and top-blown rotary converter smelting furnaces are particularly effective.

Scrap is also extensively recycled to the converters in primary smelters. The heat from the converter's exothermic Fe and S oxidation reactions is particularly useful for melting scrap, especially if considerable oxygen is used for the oxidation reactions.

Secondary scrap smelters primarily use TSL furnaces and top-blown rotary converters for smelting low Cu-grade scrap. The main smelting product is molten black copper

(80% Cu), which is converted to rough copper (96% Cu) then fire refined and cast into anodes (98.5% Cu).

These processes do not completely remove Ni and Sn from Cu, so the refining furnace product must be electrorefined. Electrorefining also recovers Ag, Au, and platinum-group metals.

Secondary copper electro refining is similar to primary copper electro refining. However, scrap may contain more impurities than concentrates, so larger electrolyte purification and slimes treatment facilities may be required.

REFERENCES

Air Products. (2007). *Melting and refining technology for copper produced from scrap charge*. http://www.airproducts.com/nr/rdonlyres/906a3970-3a98-461c-89b2-498549239ff0/0/refining_scrap_copper_33407004us.pdf.

Ayhan, M. (2000). Das neue HK—verfahren für die verarbeitung von kupfer—sekundärmaterialien. In *Intensivierung Metallurgische Prozesse, 197*—207.

Brusselaers, J., Mark, F. E., & Tange, L. (2006). *Using metal-rich WEEE plastics as feedstock/fuel substitute for an integrated metals smelter*. http://www.plasticseurope.org/Documents/Document/20100312155603-Using_metal-rich_WEEE_plastics_as_feedstock_fuel_substitute_for_an_integrated_metals_smelter.pdf.

Deneys, A., & Enriquez, A. (2009). Scrap melting in the anode furnace and the development of coherent jet technology in copper refining. In J. Kapusta & T. Warner (Eds.), *International Peirce–Smith converting centennial* (pp. 321–338). Warrendale, PA: TMS.

Edwards, J. S., & Alvear, G. R. F. (2007). *Converting using Isasmelt™ technology*. http://www.isasmelt.com/downloads/Converting%20Using%20ISASMELT%20Technology.pdf.

Hageluken, C. (2006). Recycling of electronic scrap at Umicore's integrated metals smelter and refinery. *World of Metallurgy — Erzmetall, 59*, 152–161.

Hanusch, K., & Bussmann, H. (1995). Behavior and removal of associated metals in the secondary metallurgy of copper. In P. B. Queneau & R. D. Peterson (Eds.), *Third international symposium on recycling of metals and engineered materials* (pp. 171–188). Warrendale, PA: TMS.

Huitu, K. (2003). Using of secondary raw materials in copper processing. *Helsinki University of Technology publications in materials science and metallurgy, TKK-MK-152*, 43–51.

Kikumoto, N., Abe, K., Nishiwaki, M., & Sato, T. (2000). Treatment of industrial waste material in reverberatory furnace at Onahama smelter. In P. R. Taylor (Ed.), *EPD congress 2000* (pp. 19–27). Warrendale, PA: TMS.

Komori, K., Shimizu, T., Usami, N., & Kaneda, A. (2010). Recent approach to recycling business in Naoshima smelter & refinery. In D. Rodier & W. Adams (Eds.), *Copper 2007 — Vol. VI: Sustainable development, HS&E, and recycling* (pp. 229–240). Montreal: CIM.

Kovohuty, a.s. (2010). *Production programme: Pyrometallurgy*. Kovohuty, a.s. http://www.kovohuty.sk/En/pyrometallurgy.html. 21 July 2010.

Lehner, T., & Vikdahl, A. (1998). Integrated recycling of non-ferrous metals at Boliden Ltd. Rönnskär Smelter. In J. A. Asteljöki & R. L. Stephens (Eds.), *Sulfide smelting '98* (pp. 353–362). Warrendale, PA: TMS.

Maeda, Y., Inoue, H., Kawamura, S., & Ohike, H. (2000). Metal recycling at Kosaka Smelter. In D. L. Stewart, R. Stephens & J. C. Daley (Eds.), *Fourth international symposium on recycling of metals and engineered materials* (pp. 691–700). Warrendale, PA: TMS.

Matusewicz, R., & Reuter, M. A. (2008). The role of top submerged lance (TSL) technology in recycling and closing the material loop. In B. Mishra, C. Ludwig & S. Das (Eds.), *REWAS 2008* (pp. 951–961). Warrendale, PA: TMS.

Messner, T., Antrekowitsch, H., Pesl, J., & Hofer, M. (2005). Prozessoptimierung durch stoffstromanalyse am schachtofen der Montanwerke Brixlegg AG. *Berg- und Hüttenmännische Monatshefte, 150*, 251–257.

Oshima, E., Igarashi, T., Hasegawa, N., & Kumada, H. (1998). Recent operation for treatment of secondary materials at Mitsubishi process. In J. A. Asteljöki & R. L. Stephens (Eds.), *Sulfide smelting '98* (pp. 597–606). Warrendale, PA: TMS.

Oututec. (2009). *Outotec® Kaldo furnace for copper and lead.* http://www.outotec.com/39557.epibrw.
Pesl, J. (2010). *State of slag cleaning in secondary copper smelting.* Presented at Short Course on Slag Cleaning at Copper 2010. Germany: Hamburg. 5 June 2010.
Potesser, M., Antrekowitsch, H., & Hollets, B. (2007). The elliptical anode furnace — a metallurgical comparison — part one. In A. E. M. Warner, C. J. Newman, A. Vahed, D. B. George, P. J. Mackey & A. Warczok (Eds.), *Copper 2007, Vol. III Book 1: The Carlos Díaz symposium on pyrometallurgy* (pp. 667–680). Montreal: CIM.
Properzi, G., & Djukic, V. (2008). Fire refined copper rod production in a clean environment. *Wire Journal International, 41*(8), 88–92.
Reid, R. L. (1999). High-tech low-grade recycling. *Scrap, 56*(3), 76–80.
Rinnhofer, H., & Zulehner, U. (2005). Gas-fired furnaces for copper and copper alloys. In A. von Starck, A. Mühlbauer & C. Kramer (Eds.), *Handbook of thermoprocessing technologies: Fundamentals, processes, components, safety* (pp. 358–364). Essen: Vulkan-Verlag.
Risopatron, C. (2007). *Copper scrap recycling project.* International Copper Study Group. http://www.economia-dgm.gob.mx/dgpm/doctos/cedoc/Tres.pdf. 16 November 2007.
Tandon, S., Guha, K., & Kamath, B. (2007). Secondary copper smelting. In A. E. M. Warner, C. J. Newman, A. Vahed, D. B. George, P. J. Mackey & A. Warczok (Eds.), *Copper 2007, Vol. III Book 1: The Carlos Diaz symposium on pyrometallurgy* (pp. 235–240). Montreal: CIM.
Watanabe, K., & Nakagawara, S. (2003). The behavior of impurities at Kosaka smelter. In F. Kongoli, K. Itagaki, C. Yamauchi & H. Y. Sohn (Eds.), *Yazawa international symposium, Vol. II: High-temperature metals production* (pp. 521–531). Warrendale, PA: TMS.
Zhu, Y., & Liu, W. (2000). Kaldo technology for copper scrap treatment. *Youse Jinshu, 52*, 72–74.

SUGGESTED READING

Huitu, K. (2003). Using of secondary raw materials in copper processing. *Helsinki University of Technology publications in materials science and metallurgy, TKK-MK-152,* 43–51.
Oshima, E., Igarashi, T., Hasegawa, N., & Kumada, H. (1998). Recent operation for treatment of secondary materials at Mitsubishi process. In J. A. Asteljöki & R. L. Stephens (Eds.), *Sulfide smelting '98* (pp. 597–606). Warrendale, PA: TMS.

Chapter 20

Melting and Casting

About 95% of the copper currently produced in the United States has existed as cathode copper at some time during its processing (Edelstein, 2000). The cathodes are produced by electrorefining pyrometallurgical anodes (from ore and scrap) and by electrowinning copper leached from oxide and chalcocite ores. To make it useful, this copper must be melted, alloyed as needed, cast, and fabricated.

Much of the fabrication process for copper and its alloys is beyond the scope of this book; see Joseph (1999) and Davis (2001) for more information. However, melting and casting are often the last steps in a copper smelter or refinery. A discussion of these processes is, therefore, in order.

20.1. PRODUCT GRADES AND QUALITY

The choice of melting and casting technology is defined by (a) the quality of the input copper, (b) the required chemistry of the desired product, and (c) the type of final product, e.g. wire or tube. Table 20.1 lists the *copper cathode* impurity limits specified by various national standards (ASTM, 2010; Pepperman, 2008: Qingyuan, 2008). Customers usually require purer copper than these specifications (Nairn, 2001). Fortunately, recent adoption of stainless-steel cathodes for electrorefining and electrowinning has improved cathode purity to match these customer requirements.

The tightest impurity limits in copper cathode are for selenium, tellurium, and bismuth. All three of these elements are nearly insoluble in solid copper. They form distinct grain boundary phases upon casting and solidification. Selenium and tellurium form Cu_2Se and Cu_2Te, while bismuth exists as pure Bi (Zaheer, 1995). These phases are brittle and cause rod cracking and poor drawability (Nairn, 2001). Figure 20.1 illustrates the effect of impurities on the resistivity of solid copper. Several of the impurities in Table 20.1 have a significant impact (Pops, 2008), which is why they are included in the specifications.

The Unified Numbering System currently recognizes 45 grades of wrought coppers (99.3% Cu or better) and six grades of cast coppers (Davis, 2001). Several of these coppers are alloyed with small amounts of phosphorus to combine with oxygen when they are being welded.

Unalloyed coppers can be divided into two general classes. The first is *electronic tough pitch* copper (ETP), which purposefully contains 100–650 ppm dissolved oxygen (Table 20.2; Davis, 2001; Nairn, 2001). Dissolving oxygen in molten copper accomplishes two goals. The first is removal of inadvertently adsorbed hydrogen during melting by the reaction (Cochrane & Yeoman, 2006; Pops, 2008):

$$[O] + 2[H] \rightarrow H_2O(g) \tag{20.1}$$

TABLE 20.1 Upper Impurity Limits for Copper Cathodes (ppm = Mass Parts Per Million) as Specified in the United States (ASTM), Great Britain (BS EN), and China (GB). EN 1978 is Equivalent to LME Grade A (Cu-CATH-1)

Element	ASTM B115−10 Grade 1	ASTM B115−10 Grade 2	BS EN 1978:1998	GB/T467−1997
Cu + Ag, %		99.95		
Se, ppm	2	10	2	2
Te, ppm	2	5	2	2
Bi, ppm	1	3	2	2
Sb, ppm	4	15	4	4
Pb, ppm	5	40	5	30
As, ppm	5	15	5	5
Fe, ppm	10	25	10	40
Ni, ppm	10	20		20
P, ppm				10
Sn, ppm	5	10		
S, ppm	15	25	5	40
Ag, ppm	25	70	25	
Co, ppm				
Mn, ppm				
Zn, ppm				30
Bi + Se + Te, ppm	3		3	
As + Cd + Cr + Mn + P + Sb, ppm			15	
Co + Fe + Ni + Si + Sn + Zn, ppm			20	
Total, ppm	65		65	

This reduces the amount of porosity created by water vapor formation during casting and welding. The second is reaction of the oxygen with metallic impurities, precipitating them as oxides at grain boundaries during solidification (Cochrane & Yeoman, 2006; Pops, 2008). These oxide precipitates have a smaller adverse effect on drawability than the compounds that would form if oxygen were not present. Removing the impurities from solution also reduces their impact on conductivity.

Chapter | 20 Melting and Casting

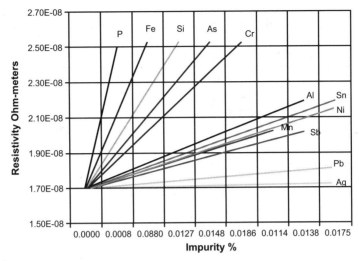

FIGURE 20.1 Effect of impurities of electrical resistivity of copper (Nairn, 2001).

Most copper is cast and fabricated as ETP, *chemically-refined tough pitch* (CRTP), or *fire-refined high conductivity* (FRHC). The chemical specifications for these grades (which are identical) are shown in Table 20.2 (ASTM, 2003; Davis, 2001; Nairn, 2001; Ningbo, 2010).

The second class of pure coppers are the *oxygen-free* (oxygen-free copper [OFC] or oxygen-free high conductivity copper [OFHC]) grades. The amount of oxygen in these grades is so low that no visible amount of Cu_2O is present in the solid copper microstructure. The maximum permissible oxygen level in OFC Grade 2 is 10 ppm. In the best grade (oxygen-free electronic, OFE Grade 1), it is only 5 ppm (ASTM, 2010; Davis, 2001).

Because no Cu_2O is generated in the grain boundaries, the electrical conductivity of OFC is higher than that of tough pitch copper. As a result, OFC is primarily used for demanding electrical applications, such as bus tube and wave guides (Joseph, 1999). The lack of Cu_2O precipitates also reduces the chance of wire breakage during fine wire drawing. As a result, magnet wire is often produced with OFC (Nairn, 2001), although ETP copper can be used if the impurity levels in the melted cathode are low enough (Cochrane & Yeoman, 2006; Southwire, 2010). The fraction of wire-rod copper produced as OFC appears to be growing; data presented by Keung (2009) suggest that it may be as high as 18%, although other sources suggest that it is much smaller.

20.2. MELTING TECHNOLOGY

20.2.1. Furnace Types

According to CRU International (Keung, 2009), about 50% of world copper production in 2006–2007 was used for wire-rod manufacture. Another 25% is also unalloyed copper. It is mostly fabricated into pipe and tube. As a result, most current melting and casting technology produces (a) copper rod for drawing into wire or (b) billets for extrusion to pipe and tube. The vast majority of this copper is tough pitch.

Most tough pitch copper is produced from cathode in shaft furnaces, shown in Fig. 20.2. The furnace operates counter-currently, with rising hot hydrocarbon combustion gas heating and melting the descending copper cathodes. High-quality scrap can also be melted.

TABLE 20.2 Upper Impurity Limit Specifications for Electronic Tough Pitch Copper (ASTM International, 1998, 2003; Davis, 2001; Nairn, 2001). (ASTM = American Society for Testing and Materials; GB = Chinese Standards; ppm = parts per Million.)

Element	C11000 (ETP) C11020 (FRHP) C11030 (CRTP) CW004A	C11040	ASTM B216−97	GB T2
Cu (min.), %	99.90	99.90	99.88	99.95
Ag, ppm		25		
As, ppm	0.5	5	120	20
Sb, ppm		4	30	20
Bi, ppm	0.2	2	30	10
Fe, ppm		10		50
Pb, ppm	5	5	40	50
Ni, ppm		10	500	
O, ppm	200−400	100−650		
Se, ppm	0.2	2		
S, ppm		15		50
Sn, ppm		5		
Te, ppm	0.2	2		
Se + Te, ppm			250	
Se + Te + Bi, ppm		3		
Total, ppm		65		

Natural gas is the usual fuel; propane, butane, naphtha, and LPG (liquefied petroleum gas) are also used (Kamath, Adiga, Sharma, & Gupta, 2003; Montgomery, 2008). Avoiding high-sulfur fuels prevents sulfur pickup in the metal (Meseha & Meseha, 2005). Oxygen enrichment is also practiced in some of these furnaces. The melting process is continuous.

An important feature of the furnace is its burners. Each burner uses a high-velocity premix flame in a burner tile. This eliminates the need for an external manifold (Amato, 2009; Ware, 2007). This design reduces accretions, shortens downtime for cleaning, and allows individual control of each burner. The burners are frequently mounted in two or more rows. Firing the upper row at a higher rate may improve fuel efficiency (Spence & Hugens, 2005).

Automatic burner control using CO or hydrogen analysis of the offgas is a common feature of these furnaces (Kamath et al., 2003; Montgomery, 2008; Ware, 2007). A flame with a moderately reducing atmosphere is produced, resulting in molten metal with about

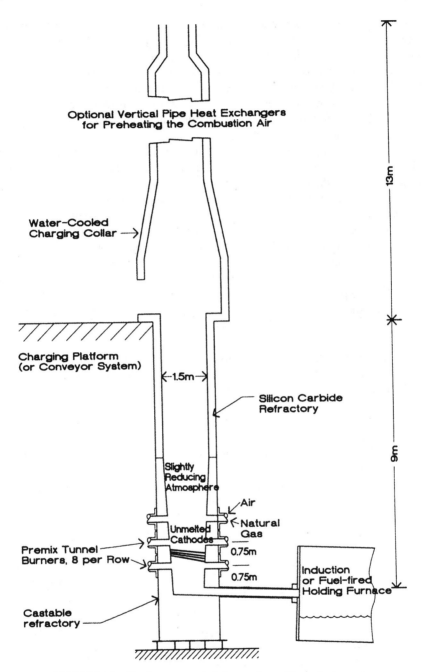

FIGURE 20.2 The Asarco-type shaft furnace for melting cathode copper.

60–90 ppm oxygen and 0.3–0.4 ppm hydrogen (Amato, 2009). Other impurity concentrations are largely unaffected. The reducing conditions allow the use of silicon carbide refractories, usually silicon-nitride bonded (Huffman, 2007). These refractories are more wear-resistant than oxide refractories, reducing the number of inclusions in the product metal.

Three types of shaft furnace are currently in use. The most common is the Asarco furnace, of which more than 100 are currently operating (Montgomery, 2008). Approximately 70 Southwire shaft furnaces are in operation (Ware, 2007). The third type is a Properzi/Luxi furnace, which is less common. The basic principle of the three furnaces is similar. Asarco furnaces have individual fuel and air mixing points for each burner, while Southwire furnaces use multiple premix manifolds. As a result, Asarco furnaces can be adjusted more easily for changing conditions, while Southwire furnaces are simpler to operate.

Efforts to improve the shaft furnace center around increasing thermal efficiency, which is currently about 50% (Spence & Hugens, 2005). Heat transfer from the hot combustion gases to the charge is a current limitation of the furnace. Recent developments include increasing the height of the furnace, which increases contact time (Kamath et al., 2003). A new cyclonic burner has been developed, which enhances circulation in the middle of the furnace (Amato, 2009). Spacing the cathodes and keeping them in a vertical position during charging also seems to help (Spence & Hugens, 2005). With sufficient improvement, thermal efficiencies of 80% might be possible.

Lower-quality scrap is less suitable for shaft furnaces, which have no refining ability. As a result, some producers have combined reverberatory furnaces with a shaft furnace (Air Products, 2007; Arderiu & Properzi, 1996). Metal charged to these furnaces can be fire refined. This allows the furnaces to melt lower grade cathode and better grades of scrap.

Another melting option is the induction furnace, either the channel or coreless type (Bebber & Phillipps, 2001; Eklin, 2005; Nairn, 2008). Induction furnaces are usually used to melt oxygen-free or phosphorus-deoxidized grades, since the absence of a combustion atmosphere prevents oxygen and hydrogen from being absorbed into the molten copper. Feed to induction furnaces, which produce oxygen-free copper is limited to high-quality cathode and scrap. Melting capacities are small, typically 10 000–20 000 tonnes per year. Electric resistance heating using graphite elements is also used (Cochrane & Yeoman, 2006; Nairn, 2001, 2008). The use of graphite crucibles rather than oxide refractory linings in resistance furnaces reduces the number of inclusions that end up in the metal.

Molten copper tapped from melting furnaces flows into a holding furnace before being directed to continuous casting. This ensures a steady supply of molten copper to the casting machines. It also creates more consistent metal temperature and composition. Holding furnaces vary considerably in size and type, but are usually induction-heated to minimize hydrogen pickup from combustion gases. The copper may also be covered with charcoal to minimize oxygen pickup (Bebber & Phillipps, 2001; Eklin, 2005; Nairn, 2008). Automation of the holding furnace to produce a steady flow of metal at constant temperature is an important part of casting operations (Meseha & Meseha, 2005; Ware, 2007).

Although slag production during melting is minimal, the desire to eliminate inclusions from the cast rod has made its removal increasingly important. As a result, a deslagging vessel is now placed between the shaft furnace and the holding furnace (Bebber & Phillipps, 2001; Ware, 2007). Ceramic filters are also increasingly used to remove inclusions caused by erosion of the furnace system refractories or precipitation of solid impurities from the molten copper (Kamath et al., 2003; Zaheer, 1995). These inclusions are a leading cause of wire breaks (Joseph, 1999; Meseha & Meseha, 2005).

Multi-chamber induction furnaces are increasingly popular (Brey, 2005). The 'storage' chambers in these furnaces eliminate the need for multiple holding furnaces.

20.2.2. Hydrogen and Oxygen Measurement/Control

As previously mentioned, control of hydrogen and oxygen in molten copper is critical. Oxygen is monitored one of two ways. The first is Leco infrared absorbance, which measures the amount of CO_2 generated when the oxygen in a heated sample of copper reacts with admixed carbon black. This method requires external sample preparation, and so does not offer an immediate turnaround.

The second approach is an oxygen sensor, which is applied directly to the molten copper (Richerson, 2005). The electrode potential of the dissolved oxygen in the copper is measured against a reference electrode in the sensor. This relative potential is converted to an equivalent oxygen content in the metal at the measurement temperature. Dion, Sastri, & Sahoo (1995) have shown that the two methods yield similar results. The amount of oxygen in the molten copper is controlled by adjusting burner flames and by injecting compressed air into the copper.

Hydrogen is more difficult to monitor and control. Analysis of solid samples is the usual practice (Strand, Breitling, & DeBord, 1994), but efforts have been made to adapt aluminum industry technology to on-line measurement of hydrogen in molten copper (Hugens, 1994). Japanese investigators have had success using α-alumina as a solid electrolyte for a hydrogen sensor (Okuyama et al., 2009).

20.3. CASTING MACHINES

Casting machines can be divided into three main types:

(a) Billet (log) casting, for extrusion and drawing to tube (Fig. 20.3)

FIGURE 20.3 Semi-continuous vertical chill caster for extrusion. *(Courtesy SMS Meer, 2008.)*

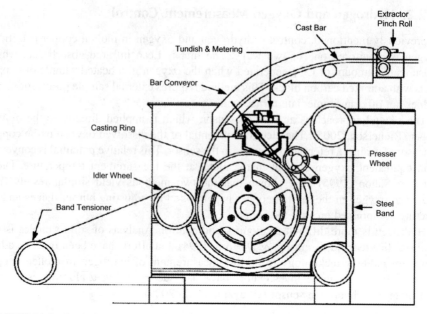

FIGURE 20.4 Southwire casting machine for continuously casting copper bar (Adams & Sinha, 1990).

(b) Bar casting, for rolling to rod and drawing to wire (Figs. 20.4 and 20.5, Table 20.3)
(c) Strip casting, for rolling to sheet and forming of welded tube.

20.3.1. Billet Casting

Billet casting is usually performed in vertical direct-chill casters, such as that shown in Fig. 20.3 (SMS Meer, 2008). Graphite-lined copper or graphite–ceramic molds are used. Diameters up to 30 cm are cast (Hugens & DeBord, 1995). Oscillation of the water-cooled molds (60–360/min) improves surface quality and prevents sticking in the mold.

Billets are often cast from phosphorus-deoxidized grades of copper (DHP), rather than tough pitch (Induga, 2007). The first reason is that tough pitch often performs poorly when extruded from billet. The second reason is the use of graphite molds. The dissolved oxygen in tough pitch copper reacts with the graphite to form carbon monoxide, which corrodes the mold.

Over the past decade, horizontal casters have begun to replace vertical billet casters, due to their lower cost (Müller & Schneider, 2005). Horizontal continuous casting of hollow billets has become an attractive alternative for smaller operations (Induga, 2007; Schwarze, 2007). These billets are rolled directly to tube, eliminating the need for extrusion and piercing.

20.3.2. Bar and Rod Casting

Copper bar is mostly cast in continuous wheel-and-band and twin-band casting machines (Table 20.4; Figs. 20.4 and 13.5). Figure 20.4 shows a Southwire wheel-and-band caster. Its key features are (a) a rotating copper–zirconium alloy rimmed wheel

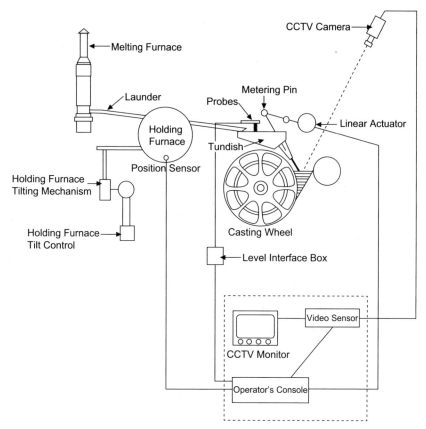

FIGURE 20.5 System for controlling molten copper level in Southwire continuous casting machine (Adams & Sinha, 1990). *(Reprinted courtesy TMS.)*

with a mold shape machined into its circumference, and (b) a cold-rolled steel band which moves in the same direction and at the same speed as the wheel circumference.

Molten copper is poured from a *pour pot* into the mold just as the steel band joins the wheel to form the fourth side of the mold (Ware, 2007). The wheel and band move together through water sprays as the copper solidifies. After 180–250° of rotation, the band moves off to an idler wheel and the solidified copper bar is drawn away (under minimum tension) to a rolling mill. Pouring to bar separation takes about 15 s. The cast bar is removed at about 0.25 m/s. The Properzi casting machine is similar, but uses individually controlled zones for spray cooling the solidifying bar (Meseha & Meseha, 2005).

The Hazelett twin-band caster is shown in Fig. 13.5 in its role as an anode-casting machine. Molten copper is fed from a pour pot into the space between two sloped moving steel bands (Von Gal, 2007). The bands are held apart by moving alloyed copper dam blocks on each side, creating a mold cavity ranging between 5–15 cm in width and 5–10 cm in thickness. Both separations are adjustable, allowing variable product size. Solidification times are similar to those of the Southwire and Properzi machines (Strand et al., 1994).

TABLE 20.3 Operating Details of Asarco Cathode Melting Shaft Furnaces

Melting plant	Nexans Canada Montreal	FMI Refinery El Paso, USA	Halcor S.A. Greece	Encore Wire USA
Inputs	Cathodes and 'runaround' scrap rod	Cathodes	Cathodes and internal scrap	Cathodes and recycled scrap
Molten copper destination	Hazelett caster & rod mill	Hazelett caster & rod mill	Hazelett caster & rod mill	Southwire caster & rod mill
Melting rate, tonnes of copper per hour	48	58	10.7	22
Feed system	Skip hoist	Elevator with automatic trip	Forklift truck & skip hoist	Forklift truck & skip hoist
Furnace details, m				
Height, taphole to charge floor	13	12.2	7.5	10
Inside diameter at charge floor	1.7	1.75	1.68	1.68
Inside diameter at taphole	1.3	1.37	1.37	1.37
Burner details				
Number of burners	23	27	10	15
Rows of burners	3	4	2	2
Fuel	Natural gas	Natural gas	Natural gas	Natural gas
Combustion rate, Nm³/hour		2400	400	
Natural gas burnt per tonne of copper melted	1.9 GJ	1.8 GJ	53 Nm³ (including holding furnace and launders)	1435 MJ/t
Refractory life, tonnes of copper				
Above burners	500 000	500 000	300 000	400 000
Below burners	250 000	300 000	200 000	400 000

TABLE 20.4 Operating Details of Hazelett and Southwire Continuous Casting Machines

Casting plant	Nexans Canada	FMI Refinery	Halcor S.A.	Encore Wire
Casting machine	Hazelett twin band	Hazelett twin band	Hazelett	Southwire wheel & band
Bar size, cm × cm	7 × 13	7 × 13.0	3.5 × 6	5.4 × 7.9
Casting rate of this bar, tonnes/hour	48	63	10.7	20
Molten copper level control in caster	Electromagnetic pool level measurement	Electromagnetic pool level measurement	Electromagn. pool level control	Infrared scanner
Casting temp., °C	1125	1130	1135–1140	1100
Bar temperature leaving caster, °C	~950	1015	930–950	930
Target O in copper, ppm	250	250	180–320	270
Measurement Technique	Electro-nite cell in launder; Tempolab in holding furnace; Leco on rod	Leco on rod	Ströhlein on finished rod	Leco on rod
Control system	Manual	Compressed air injection into molten Cu	Controlled natural vent on transfer launder	Compressed air injection into molten Cu
Wheel and band details				
Wheel diameter, m				2.01
Rotation speed, rpm				1.8
Rim materials				Cu–Cr–Zr

(Continued)

TABLE 20.4 Operating Details of Hazelett and Southwire Continuous Casting Machines—cont'd

Casting plant	Nexans Canada	FMI Refinery	Halcor S.A.	Encore Wire
Rim life, tonnes of cast				
Copper				28,000
Band material				Cold-rolled steel
Band life				2200 t
Lubrication				Acetylene soot
Twin-band details				
Caster length, m	3.7	3.7	3	
Band material	Low carbon steel	Titanium steel	Titanium steel	
Life	24 h	1300 tonnes Cu	36–48 h	
Lubrication	Oil	Union Carbide Lb-300x oil	Oil Breox	
Dam block material	Si bronze	Cu with 1.7–2% Ni & 0.5–0.9% Si	CuNiSiCr	
Dam block life	100 000 tonnes cast copper	~1000 h	20 000 t	

The three types of moving-band casting devices have several features in common. All require lubrication of the bands and mold wheel or dam blocks, using silicone oil or acetylene soot (Kamath et al., 2003; Meseha & Meseha, 2005). Leftover soot is removed from the bands after each revolution, then reapplied. This ensures an even lubricant thickness and a constant heat transfer rate. The casters all use similar input metal temperatures, 1110–1130 °C (Table 20.3). All require smooth, low-turbulence metal feed into the mold cavity, to reduce defects in the solidified cast bar. Lastly, all require steady metal levels in the pour pot and mold.

Control of mold metal level is done automatically (Fig. 20.5). Metal level in the mold cavity is measured electromagnetically (Hazelett) or with a television camera (Southwire). It is controlled with a stainless-steel or ceramic metering pin in the pour pot. Metal level in the pour pot is determined using a bubble tube or load cell. It is controlled by changing the tilt of the holding furnace, which feeds it (Ware, 2007). The temperature of the solidified copper departing the machine is controlled to 940–1015 °C by varying the casting machine cooling-water flow rate.

Copper cast in the Hazelett, Properzi, and Southwire machines is usually fed directly into a rolling machine for continuous production of copper rod. Southwire Continuous Rod and Hazelett Contirod are prominent (Bebber & Phillipps, 2001; Ware, 2007; Zaheer, 1995). Both systems produce up to 60 tonnes of 8–22 mm diameter rod per hour (Table 20.3).

20.3.3. Oxygen-free Copper Casting

The low oxygen and hydrogen content of oxygen-free copper minimizes porosity when these grades are cast. As a result, the rolling step, which is used to turn tough pitch copper bar into rod is not necessary. This has led to the development of processes for direct casting of OFC copper rod. Nearly all of these are vertical casting machines. Upcast OY® (formerly Outokumpu, now a separate company) is the best known OFC caster type; others include Rautomead and Conticast® (Fig. 20.6; Cochrane & Yeoman, 2006; Eklin, 2005; Nairn, 2008). Numerous Chinese vertical casting machines are also in use (Keung, 2009).

Upward vertical casting machines use a vacuum to draw metal into water-cooled graphite-lined dies partially submerged in the molten copper. As it freezes, the rod is mechanically drawn upward and coiled (Nairn, 2008). It is about the same size as rolled rod. First-generation vertical casting machines produce 3–12 rod strands simultaneously, and have a capacity of 3000–12 000 tonnes per year (Nairn, 2008). These are integrated with a single electrically heated furnace, which is used for both melting and casting (Cochrane & Yeoman, 2006). Newer machines have separate furnaces for melting and casting, and produce up to 32 strands. Their production capacity is as high as 40 000 tonnes per year (Eklin, 2005).

20.3.4. Strip Casting

The development of strip casting for copper and copper alloys parallels developments in the steel industry, in that continuous processes are favored. The newer the technology, the less rolling is required. One approach taken by small-volume producers is to roll strip from the bar produced by a Hazelett caster (Roller, Kalkenings, & Hausler, 1999). This can be combined with continuous tube rolling/welding to make optimum use of the casting machine for a mix of products.

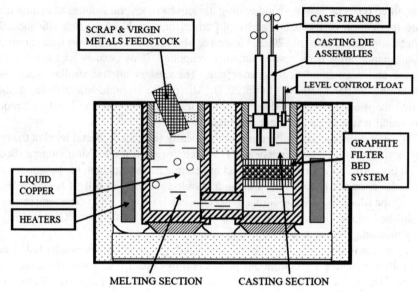

FIGURE 20.6 Conticast® vertical casting machine for OFC (Cochrane & Yeoman, 2006). *(Courtesy Hormes–Conticast Group.)*

However, direct strip casting which avoids rolling is the goal. Current horizontal casters can produce *thick strip* (15–20 mm), which requires some rolling (Roller & Reichelt, 1994). Development efforts are being made to develop *thin-strip* (5–12 mm) casting to avoid rolling completely. Hazelett (2010) recently put into service the first continuous strip caster/in-line rolling facility in Hettstedt, Germany.

20.4. SUMMARY

The last step in copper extraction is melting and casting of electrorefined and electrowon cathodes. The main products of this melting and casting are:

(a) Continuous rectangular bar for rolling to rod and drawing to wire
(b) Round billets *(logs)* for extrusion and drawing to tube
(c) Flat strip for rolling to sheet and forming into welded tube.

The copper in these products is mostly tough pitch copper, cathode copper into which 100–500 ppm oxygen has been dissolved during melting/casting. The remainder of unalloyed copper production is in the form of oxygen-free high conductivity copper with 5–10 ppm dissolved oxygen.

These pure copper products account for about 70% of copper use. The remainder is used in the form of copper alloy, mainly brass and bronze.

The principal melting tool for cathodes is the shaft furnace. It is thermally efficient and provides good oxygen-in-copper control. Its molten copper is mainly cast as rectangular bar in continuous wheel-and-band and twin-band casters, or as logs in horizontal and vertical direct-chill casters.

The quality of cathode copper is tested severely by its performance during casting, rolling, and drawing to fine wire. Copper for this use must have high electrical conductivity, good drawability and good annealability. These properties are all favored by maximum cathode purity.

REFERENCES

Adams, R., & Sinha, U. (1990). Improving the quality of continuous copper rod. *JOM, 42*(5), 31–34.

Air Products. (2007). *Melting and refining technology for copper produced from scrap charge.* http://www.airproducts.com/nr/rdonlyres/906a3970-3a98-461c-89b2-498549239ff0/0/refining_scrap_copper_33407004us.pdf.

Amato, K. (2009). *SFIP – Using cyclonic burner technology.* http://www.namfg.com/images/CS-7A.pdf.

Arderiu, G., & Properzi, G. (1996). Continuous copper rod production from 100 percent scrap. *Wire Journal International, 29*(3), 60–67.

ASTM International. (1998). Standard specification for copper rod drawing stock for electrical purposes (B49-98). In *Annual book of standards, section 2, nonferrous metal products*. Philadelphia, PA: ASTM.

ASTM International. (2003). Standard specification for tough pitch fire-refined copper – refinery shapes (B216–97). In *Annual book of standards, section 2, nonferrous metal products*. Philadelphia, PA: ASTM.

ASTM International. (2010). Standard specification for electrolytic cathode copper (B115-10). In *Annual book of standards, section 2, nonferrous metal products*. Philadelphia, PA: ASTM.

Bebber, H., & Phillipps, G. (2001). *The CONTIROD furnace concept at MKM.* http://www.induga.com/images/05_gesamtanlagen/02_herstellung_kupferdraht/Contirod_e.pdf.

Brey, M. (2005). Horizontal continuous casting of copper alloy billets. In K. Ehrke & W. Schneider (Eds.), *Continuous casting* (pp. 325–332). Weinheim, Germany: Wiley–VCH.

Cochrane, S., & Yeoman, G. (2006). *Cu-OF or Cu-ETP?: Some general comparisons of small diameter copper wire rod products.* http://www.conticast.com/documents/CuOF%20or%20CuETP%20-%20GENERAL%20COMPARISONS.pdf.

Davis, J. R. (Ed.). (2001). *ASM specialty handbook: Copper and copper alloys* (pp. 15). Materials Park, OH: ASM International.

Dion, J. L., Sastri, V. S., & Sahoo, M. (1995). Critical studies on determination of oxygen in copper anodes. *Transaction of the American Foundrymen's Society, 103*, 47–53.

Edelstein, D. E. (2000). Copper. In *1999 Minerals yearbook*. United States Geological Survey. http://minerals.usgs.gov/minerals/pubs/commodity/copper/240499.pdf.

Eklin, L. (2005). The Outokumpu UPCAST® system. In K. Ehrke & W. Schneider (Eds.), *Continuous casting* (pp. 333–340). Weinheim: Wiley–VCH.

Hazelett Strip-Casting. (2010). *Copper strip casting machines.* http://www.hazelett.com/casting_machines/strip_casting_machines/copper_strip_casting_machines/copper_strip_casting_machines.php.

Huffman, B. (2007). New concepts in refractory linings for copper shaft furnaces, 1997–2007. In J. Hugens (Ed.), *Copper 2007, Vol. I: Plenary/copper and alloy casting and fabrication/copper – Economics and markets* (pp. 427–438). Montreal: CIM.

Hugens, J. R. (1994). An apparatus for monitoring dissolved hydrogen in liquid copper. In G. W. Warren (Ed.), *EPD congress 1994* (pp. 657–667). Warrendale, PA: TMS.

Hugens, J. R., & DeBord, M. (1995). Asarco shaft melting and casting technologies '95. In W. J. Chen, C. Díaz, A. Luraschi & P. J. Mackey (Eds.), *Copper 95–Cobre 95, Vol. IV: Pyrometallurgy of copper* (pp. 133–146). Montreal: CIM.

Induga. (2007). *Induga copper tube furnaces.* http://www.furnacesinternational.com/download/IndugaCopperTubeFurnaces.pdf.

Joseph, G. (1999). In K. J. A. Kundig (Ed.), *Copper: Its trade, manufacture, use and environmental status, Vol. 141–154* (pp. 193–217). Materials Park, OH: ASM International.

Kamath, B. P., Adiga, R., Sharma, L. K., & Gupta, S. (2003). Productivity improvements at continuous casting rod plant of Sterlite Copper at Silvassa. In G. E. Lagos, M. Sahoo & J. Camus (Eds.), *Copper 2003–Cobre 2003, Vol. 1: Plenary lectures, economics and applications of copper* (pp. 720–727). Montreal: CIM.

Keung, S. (2009). *Recent trends in copper wirerod markets.* http://www.icsg.org/index.php?option=com_docman&task=doc_download&gid=201&Itemid=62.

Meseha, J., & Meseha, G. (2005). Comparison of competing continuous casting processes. *Wire Journal International, 38*(3), 185−193.

Montgomery, M. C. (2008). *Shaft melting furnace.* http://www.asarco.com/Asarco_Shaft_Melting_Furnace.html.

Müller, W., & Schneider, P. (2005). Horizontal casting technology for copper products. In H. R. Müller (Ed.), *Continuous casting* (pp. 329−335). Weinheim, Germany: Wiley−VCH.

Nairn, M. (2001). *Graphite crucible furnace technology as a basis of highest quality copper redraw rod.* http://www.rautomead.com/files/WireSing_Reprint.pdf.

Nairn, M. (2008). *Adding value and growing markets through wire rod production.* http://www.rautomead.com/files/Microsoft_Word_-_Metal_Bulletin_Sofia_2008_-_paper_with_pics.pdf.

Ningbo Jintian Copper (Group) Co., Ltd. (2010). *High-precision copper strip exhibition.* China (Mainland). http://www.hisupplier.com/sell-148954-High-precision-Copper-Strip.

Okuyama, Y., Kurita, N., Yamada, A., Takami, H., Ohshima, T., Katahira, K., et al. (2009). A new type of hydrogen sensor for molten metals usable up to 1600K. *Electrochimica Acta, 55*, 470−474.

Pepperman, H. (2008). *London metal exchange: Special contract rules for copper grade A.* http://www.LME_Specification_for_Copper_041108.pdf.

Pops, H. (2008). *The metallurgy of copper wire.* http://www.litz-wire.com/pdf%20files/Metallurgy_Copper_Wire.pdf.

Qingyuan JCCL EPI Copper Ltd. (2008). *Products.* http://www.jecopper.com/Product.htm.

Richerson, D. W. (2005). *Modern ceramic engineering: Properties, processing, and use in design* (3rd ed.). Boca Raton, FL: CRC Press.

Roller, E., Kalkenings, P., & Hausler, K. H. (1999). Continuous narrow strip production line for welded copper tubes. *Tube International, 18*, 28−31.

Roller, E., & Reichelt, W. (1994). Strip casting of copper and copper alloys. In *Proc. METEC congress 94, Vol. 1* (pp. 480−486). Düsseldorf, Germany: Verein Deutscher Eisenhüttenleute.

Schwarze, M. (2007). Innovations in the field of non-ferrous metal processing. In J. Hugens (Ed.), *Copper 2007, Vol. I: Plenary/copper and alloy casting and fabrication/copper − Economics and markets* (pp. 151−172). Montreal: CIM.

SMS Meer. (2008). *Continuous casting plants TECHNICA.* http://www.sms-meer.com/produkte/kupfer/demag-technica/produkte/ver_pr_bo_hk_e.html.

Southwire. (2010). *Southwire Copper Rod Systems.* http://www.southwire.com/scr/scr-copper-rod-systems.htm

Spence, G., & Hugens, J. (2005). Copper shaft furnace melting and charging: a new approach. *Wire Journal International, 38*(4), 110−115.

Strand, C. I., Breitling, D., & DeBord, M. (1994). Quality control system for the manufacture of copper rod. In *1994 conf. proc. wire assoc. inter.* (pp. 147−151). Guilford, CT: Wire Association International.

Von Gal, R. (2007). Innovations in twin-belt compliant mold casting technology. In J. Hugens (Ed.), *Copper 2007, Vol. I: Plenary/copper and alloy casting and fabrication/copper − Economics and markets* (pp. 189−202). Montreal: CIM.

Ware, P. W. (2007). The SCR® process produces high quality copper rod around the world. In J. Hugens (Ed.), *Copper 2007, Vol. I: Plenary/copper and alloy casting and fabrication/copper − Economics and markets* (pp. 175−188). Montreal: CIM.

Zaheer, T. (1995). Reduction of impurities in copper. *Wire Industry, 62*(742), 551−553.

SUGGESTED READING

Cochrane, S., & Yeoman, G. (2006). *Cu-OF or Cu-ETP?: Some general comparisons of small diameter copper wire rod products.* Conticast Group. http://www.conticast.com/documents/CuOF%20or%20CuETP%20-%20GENERAL%20COMPARISONS.pdf. 7 August 2006.

Joseph, G. (1999). In K. J. A. Kundig (Ed.), *Copper: Its trade, manufacture, use and environmental status* (pp. 193−217). Materials Park, OH: ASM International.

Meseha, J., & Meseha, G. (2005). Comparison of competing continuous casting processes. *Wire Journal International, 38*(3), 185−193.

Nairn, M. (2008). *Adding value and growing markets through wire rod production*. Rautomead Ltd. http://www.rautomead.com/files/Microsoft_Word_-_Metal_Bulletin_Sofia_2008_-_paper_with_pics.pdf. 24 June 2008.

Spence, G., & Hugens, J. (2005). Copper shaft furnace melting and charging: a new approach. *Wire Journal International, 38*(4), 110−115.

Ware, P. W. (2007). The SCR® process produces high quality copper rod around the world. In J. Hugens (Ed.), *Copper 2007, Vol. I: Plenary/copper and alloy casting and fabrication/copper − Economics and markets* (pp. 175−188). Montreal: CIM.

Chapter 21

Byproduct and Waste Streams

The production of copper generates a number of byproduct and waste streams. In some cases, these streams contain valuable components that are profitable to recover. In other cases, treatment of these streams is needed to prevent the release of toxic or hazardous chemicals. The processing of byproduct streams is therefore a significant activity at copper concentrators, smelters, and refineries. The treatment of several common byproducts is the subject of this chapter.

21.1. MOLYBDENITE RECOVERY AND PROCESSING

Many porphyry Cu deposits contain economic concentrations of Mo, virtually all of it as molybdenite, MoS_2. Table 21.1 shows the Cu and Mo contents of several porphyry Cu–Mo ores.

Molybdenite floats with chalcopyrite, chalcocite, and other Cu–S and Cu–Fe–S minerals. It can then be separated from the resulting Cu–Mo bulk concentrate by depressing the Cu minerals with sodium hydrosulfide (NaHS), while re-floating the MoS_2 in a Mo/Cu separator plant (Amelunxen & Amelunxen, 2009). Mo/Cu separator flotation plants are much smaller than their preceding bulk flotation plants because their feed (bulk Cu–Mo concentrate) mass is only two or three percent of the initial ore feed.

Typical Cu–Mo bulk concentrates and their equivalent Mo/Cu separator molybdenite and Cu concentrates are given in Table 21.2. A representative flowsheet is presented in Fig. 21.1. The molybdenite recovered using this process often contains rhenium as an impurity. This further enhances the economic viability of this process.

21.2. FLOTATION REAGENTS

The principal collector for molybdenite is diesel fuel or similar oil, ~0.01 kg/tonne of ore. It is added to the bulk concentrator ore slurry feed to maximize molybdenite recovery from ore. Its effect carries over to flotation of molybdenite in the Mo/Cu separator flotation plant. Some Mo/Cu separation plants add more in the Mo/Cu separator plant.

The most common Cu–Fe–S mineral depressant is NaHS (Table 21.2). It is added to the Mo/Cu separator plant bulk concentrate slurry feed, at 4–10 kg per tonne of bulk concentrate.

Molybdenite flotation frothers are conventional. Two examples are Dowfroth 250 and pine oil. The pH is raised with CaO and lowered with sulfuric acid, occasionally CO_2. The pH requirement varies from ore to ore.

21.3. OPERATION

Flow through the Mo/Cu separator plant is continuous. Depressant and frother (sometimes oily collector) are continuously added where needed. Product Mo froth and Cu

TABLE 21.1 Ore, Tailings and Concentrate Compositions for Four Cu/Mo Concentrate Producers (Data Courtesy Peter Amelunxen, Amelunxen Mineral Engineering Ltda., Santiago, Chile)

Ore		Ore concentrator tailing		Ore concentrator bulk concentrate (Mo/Cu separator flotation plant feed)		Final Cu concentrate		Final Mo concentrate	
%Cu	%Mo	%Cu	%Mo	%Cu	%Mo	%Cu	%Mo	%Cu	%Mo
0.3	0.01	0.035	0.003	34	00.9	34.6	0.0	3	48
0.4	0.025	0.05	0.006	30	1.67	30.9	0.1	4.3	48
0.6	0.015	0.07	0.004	26	0.53	26.2	0.1	2.5	50
0.8	0.025	0.09	0.009	30	0.68	30.3	0.1	5	45

TABLE 21.2 Reagents, Cells, and pH Used in Floating Molybdenite from Bulk Cu/Mo Concentrate (Amelunxen & Amelunxen, 2009)

Plant	Cu−S and Cu−Fe−S mineral depressants	NaHS addition amount, kg per tonne of bulk concentrate feed	Flotation gas	Cells enclosed?	Rougher feed (Fig. 21.1) pH
Andina, Chile	NaHS + H_2SO_4	4.0	Recycled air	Yes	7−8
Bagdad, USA	NaHS	4.5	N_2 + air	No	11.5
Cerro Verde, Peru	NaHS + CO_2	12	Recycled air	Yes	8−9.5
Chuquicamata, Chile	NaHS + H_2SO_4	6	N_2	No	7−8
Cuajone 87, Peru	NaHS	4	N_2	No	10−11
El Salvador, Chile	NaHS + As_2O_3	7.4	N_2	No	
Gibraltar 84, Canada	NaHS + COOH	5	N_2	No	7−8
Las Tortolas, Chile	NaHS + H_2SO_4	4	Recycled air	Yes	7−8
Pelambres, Chile	NaHS + H_2SO_4	4.6	N_2	No	7−8
Sierrita, U.S.A.	NaHS	5−10	N_2	No	11−12
Teniente, Chile	$Na_3PS_2O_2$ (Nokes reagent) + NaHS	4	N_2	No	9

underflow continuously depart the plant for dewatering and subsequent transport to production of metal and chemical products. Flow rates and Mo and Cu concentrations are continuously measured and applied to process optimization.

Mo/Cu separator flotation plants often contain regrind mills. Their objective is to liberate molybdenite grains from Cu−Fe−S mineral particles. However these grains have already gone through grinding in the preceding ore concentrator, so the efficacy of grinding in the Mo/Cu separation plant is questionable.

21.4. OPTIMIZATION

The objectives of a Mo/Cu bulk concentrator are to (a) maximize Cu and Mo recovery to the bulk concentrate and (b) maximize rejection of non-Cu, non-Mo minerals to the tailings. The objectives of an Mo/Cu separation flotation plant are to (a) maximize recovery of molybdenite to the Mo concentrate, and (b) minimize recovery of Cu−Fe−S minerals to the Mo concentrate. The bulk concentrator objectives are met by adequate grinding, efficient rougher scavenger flotation, and quiescent concentrate cleaning in column cells.

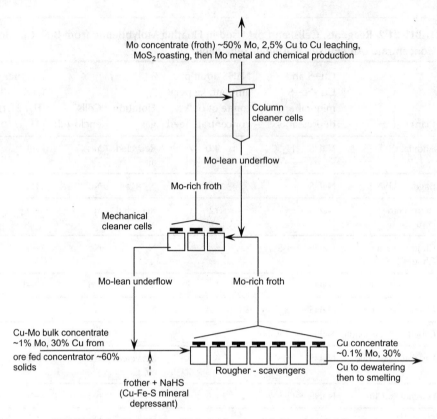

FIGURE 21.1 Simplified molybdenite flotation (Mo/Cu separator) flowsheet.

The Mo/Cu separator plant objectives are obtained by

(a) Efficient collection of molybdenite and efficient depression of Cu−Fe−S minerals in the rougher scavenger cells with adequate concentrations of oily collector and NaHS depressant
(b) Efficient cleaning of molybdenite concentrate in ~5 column flotation cells in series (Amelunxen & Amelunxen, 2009)
(c) Use of N_2 in the flotation cells.
(d) Adequate washing of cleaner froth with fresh water (Fig. 4.6).

Several companies market proprietary reagents for enhancing molybdenite recovery (Chevron Phillips, 2009; Cytec, 2011). Independent appraisals of these reagents are not available.

21.5. ANODE SLIMES

21.5.1 Anode Slime Composition

Table 21.3, taken largely from the review by Moats, Robinson, Davenport, Karcas, and Demetrio (2007), tabulates the composition of several anode slimes from electrolytic refineries around the world. These compositions reflect the composition of the raw

TABLE 21.3 Composition of Anode Slimes, Mass %

	Cu	Ag	Au	Se	Te	As	Sb	Bi	Pb	Fe	Ni	Sn	S	Zn
Primary smelters														
La Caridad (Mexico)	12.7	15.4	0.036	4.2	0.37	5.0	5.0	1.2	14.8	2.1	0			0.81
BGMK (Kazakhstan)	3.0	20	1.2	6.6	1.2	5.3	19		11.9					
Olympic Dam (Australia)	30	9.9	1.19	14.9	3.5	3.9		2.5	12.3					
Port Kembla (Australia)	29	3.1	1.0	8	0.9	1.6	0.5	2.5	22					
Townsville (Australia)	21	6.1	0.63	3	0.5	3.9	0.6	0.7	9.8	0.04	0.3		2.5	
Bahia (Brazil)	4.38	7.59	0.40	11.5	2.6	3.8	4.1	0.7	6.7		0.1			
Med (Bulgaria)	25.27	4.86	0.49	9.25	0.83	2.96	1.16	0.18	10.3	0.06	0.136		10	
Timmins (Canada)	17	23	0.26	8.4	0.7	2.3	0.9	0.8	23		0.9			
Montreal (Canada)	15	29	1.4	5	2	1.5	4.5	1.5	16		3.8			
Potrerillos (Chile)	7.80	15.42	0.47	8.65	0.66	9.22	10.45	0.41	1.16	0.19	0.02			
Las Ventanas (Chile)	24.27	14.77	5.4	7.9	0.8	6.2	5.5	0.3	8.1	0.1	0.1			
Gresik (Indonesia)	0.6	4.9	2.1	11.1	0.2	1.2	0.3	2.8	51.4	0.1	0			

(Continued)

TABLE 21.3 Composition of Anode Slimes, Mass %—cont'd

	Cu	Ag	Au	Se	Te	As	Sb	Bi	Pb	Fe	Ni	Sn	S	Zn
Saganoseki (Japan)	25.2	19	2.1	12.1	3.9	3.3	1.8	0.7	4.0	0.21	0.4		3.2	
Nishibara (Japan)	21.4	10.8	1.4	6.9	1.7	3.3	1.7	1.4	14		0.7		8.1	
Almalyk (Uzbekistan)	29			9	11	4	2	0.2	24	0.1			6	
Amarillo USA	19.6	18.6	0.12	14.1	1.45	2.91	2.86	0.44	0.15		0.74			
El Paso (USA)	1	22	0.2	20	0.4	2	4	0.7	5	0.04	0.05		12	
Knnecott (USA)	30	5	0.5	5	1	5	1	3	30	0.25	0.05			
Kayseri (Turkey)	18	1.5	0.08	1	0.4	0.3	0.4	0.04	16.9		0.2			
Secondary smelters														
Lünen (Germany)	1	8	0/16	0.7	0.7	2.5	9	0.4	30		1	9.5	8.5	
Beerse (Belgium)	19	5.5	0.06	0.5	0.5	2.5	16	1.5	4					
Boliden (Sweden)	15.69	22.95	0.578	5.17	0.94	3.21	4.15	0.777	9.69		5.09			
Montanwerke Brixlegg (Austria)	7.00	8.751	0.125	0.30	0.59		3.80	0.41	24.56		10.86	12.14		

materials input to the coppermaking process. Concentrates with high levels of gold and silver will produce anode slimes with high levels of these elements; the same is true for other impurities. Secondary smelters use little or no concentrate in their feed, and so produce slimes with little or no sulfur, selenium, or tellurium. However, anode slimes from secondary smelters often contain alloying elements from the scrap and waste streams fed to the furnace. This means higher levels of lead than contained in anode slimes from primary smelters, and significant levels of elements such as tin and nickel. The use of higher matte grades in smelters has led to increased levels of arsenic, antimony, and bismuth in anode slimes over recent decades (Hoffmann, 2008). The reason for this is the reduced amount of converter slag produced when higher matte grades are used.

The mineralogy of anode slimes varies with content, but is essentially a mixture of oxide and chalcogenide solid solutions (Chen & Dutrizac, 2005; Hait, Jana, & Sanyal, 2009). Lead is found mostly as insoluble $PbSO_4$, and silver and gold occur mostly as selenide and telluride phases. Antimony, arsenic, and bismuth are found as a series of multiple oxides. Copper occurs in many phases.

21.5.2. The Slime Treatment Flowsheet

Figure 21.2 Illustrates the slimes treatment process used by CCR in Canada (Hait et al., 2009), which is somewhat typical of current practice. The first step is *decoppering*, which was previously done by leaching under atmospheric conditions in spent electrolyte from the electrorefining cells (Hoffmann, 2008). The use of oxygen pressure leaching at 125–150 °C improves both process kinetics and overall copper recovery from the slime. However, it also dissolves much of the tellurium in the slimes. This is followed by roasting with sulfuric acid at ~300 °C, which converts most of the selenium in the slime to a vapor:

$$Se + 2H_2SO_4 \rightarrow SeO_2(g) + 2SO_2 + 2H_2O \tag{21.1}$$

This process has largely replaced the soda ash roasting previously practiced. It eliminates the generation of elemental Se vapor during roasting, which is difficult to recover.

In Fig. 21.2 the remaining slime is smelted to produce a slag containing the arsenic, antimony, and bismuth, along with a metal phase called *Doré* bullion containing the gold, silver, and remaining copper (Hait et al., 2009; Hoffmann, 2008). Some of the precious metal content winds up in the slag, which is treated to recover the silver and gold or returned to the smelter. Doré smelting has traditionally been performed in a small reverberatory furnace, but top-blown rotary converters have become popular (Becker, 2006). Doré bullion is hard to refine, and the smelting process also produces selenium vapor species, which are an environmental challenge.

As a result, a change over recent decades has been the development of hydrometallurgical processes for anode slimes. Several different processes are in use; the best known is the wet chlorination process (Kurokawa, Asano, Sakamoto, Heguri, & Hashikawa, 2007), which leaches the slime with chlorine gas bubbled into a pulp of as-received slime. This produces leach liquor containing gold, selenium, tellurium, and some of the copper, and a residue containing silver chloride and the rest of the copper. Solvent extraction recovers the gold from the leach liquor, and precipitation with SO_2 sequentially recovers the Se and Te. The slime is sometimes decopperized before leaching, and a mixture of hydrogen peroxide and hydrochloric acid is a less expensive

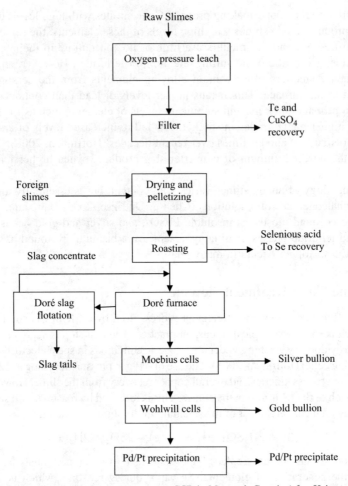

FIGURE 21.2 The slimes treatment process at CCR in Montreal, Canada (after Hait et al., 2009).

alternative to chlorine gas (Hait et al., 2009; Kim, Wang, & Brees, 2010; Thuy et al., 2004).

21.6. DUST TREATMENT

Chapter 8 briefly described the formation of dust during matte converting operations. Dusts are generated during all stages of pyrometallurgical processing of copper concentrates, and the mass of dust can be as much as 10% of the mass of input concentrate. These dusts are collected by heat recovery boilers, electrostatic precipitators (ESP), or baghouses (see Chapter 12). Common practice has been to recycle these dusts to the smelter to recover the copper content (Montenegro, Sano, & Fujisawa, 2011). However, this approach also recycles impurities such as arsenic, antimony, bismuth, and lead. As the levels of these impurities build up, their concentration in the matte increases as well, and this eventually means high impurity levels in the blister copper. This in turn makes electrolytic refining more difficult. The recycle of dust also decreases furnace capacity, which hurts productivity (Parra & Parada, 2008). As impurity levels in

TABLE 21.4 Composition of Dusts from the Rönnskär Smelter of Boliden, Mass % (Samuelsson & Carlsson, 2001)

Dust source	Collection device	Cu	Fe	Zn	Pb	Sb	As	S	Sn
Roaster	ESP	20.0	30.0	5.7	2.5	0.26	1.40	11.5	<0.01
Smelter	ESP	20.2	25.6	12.3	3.0	0.05	1.88	7.5	0.02
Converter	Baghouse	8.3	1.56	15.1	20.2	0.21	4.51	8.6	23.6
Slag cleaning	Baghouse	1.96	1.22	33.3	16.6	0.92	11.5	1.8	20.0

concentrates increase, the need to find a way of treating smelter dust to separate the copper from the impurities has increased.

Table 21.4, taken from data published by Samuelsson and Carlsson (2001), provides typical compositions of dust from various locations in the coppermaking process at the Rönnskär smelter of Boliden. Dusts from concentrate roasting and smelting feature high copper and sulfur levels, the result of concentrate particles reporting directly to the dust stream. Dusts from converting and slag cleaning have lower copper levels, and higher levels of lead, antimony, and arsenic. As a result, direct recycle of the roaster and smelter dusts is more feasible, and the incentive to treat converter and slag cleaning dusts to remove impurities is higher.

Numerous hydrometallurgical and pyrometallurgical process have been developed to treat coppermaking dusts. The biggest concern with these processes is recovering the arsenic in a stable form for disposal. An example is the process for treating Teniente furnace dust described by Ichimura, Tateiwa, Almendares, & Sanchez (2007). The flowsheet (Fig. 21.3) is designed to generate a copper electrolyte solution that can be sent to the electrorefining circuit, a copper sulfide concentrate that is returned to the smelter, purified zinc sulfate for sale, and crystalline ferric arsenate suitable for disposal.

21.7. USE OR DISPOSAL OF SLAG

Chapter 11 described the processing of smelting slags to recover the copper content. These processes leave behind a substantial amount of material to be disposed of or otherwise used. About 25 million tonnes of slag are generated per year worldwide (Gorai, Jana, & Premchand, 2003). Over the years, several applications for smelting slag have been found. These include:

- *Cement and concrete* (Pavez, Rojas, Palacios, & Sanchez, 2004; Shi, Meyer, & Behnood, 2008). The iron content of copper slag makes it a useful replacement for iron powder in cement clinker production. Copper slag has also been used as a partial substitute for Portland cement in concrete, with mixed effects on properties. The most beneficial aspect of using copper slag as a replacement is the reduction in energy use required to produce the concrete.
- *Abrasive grit* (Flynn & Susi, 2004; Pavez et al., 2004).Copper slag is a common abrasive for blasting operations. Its advantages include higher hardness and density than

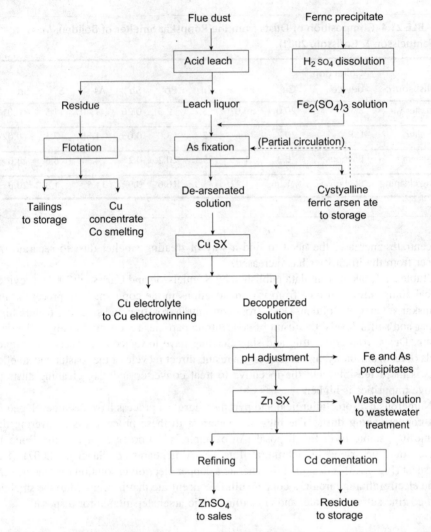

FIGURE 21.3 Process flowsheet for treatment for Teniente furnace dust (after Ichimura et al., 2007).

other abrasives, and a lack of free silica (a workplace hygiene hazard). The biggest disadvantage is the presence of hazardous metal content, which can potentially contaminate job sites.
- *Land reclamation* (Lim & Chu, 2006). Copper slag is denser than the sand normally used for fill, and has a better particle-size distribution. The friction angle is higher as well, enhancing slope and foundation stability.
- *Iron ore* (Cao et al., 2009). The iron content of smelter slag is 40–60% (Shi et al., 2008), which is higher than that of many iron ores. However, the iron is largely present as fayalite (Fe_2SiO_4), which is difficult to concentrate. Experiments with reverse magnetic flotation have upgraded the iron content to a level, which might be suitable for some ironmaking processes. González, Parra, Klenovcanova, Imris, and Sánchez (2004) smelted copper slag with carbon to generate a molten Fe-1.9 Cu-5 C alloy, but this has too much copper to be of value to steelmakers.

- *Ceramic tile* (Marghussian & Maghsoodipoor, 1999). The use of copper slag in tile production improves the product strength and acid resistance. However, the sulfur content results in SO_2 emission and bloating during firing, limiting the amount that can be used.

Other potential uses of copper slag include roofing granules, hot-mix asphalt, and cutting tools (Gorai et al., 2003). However, all of the applications described here are cost-sensitive, and the cost of transporting slag from a remote location to a manufacturing facility often makes its use uncompetitive. As a result, most slag is still landfilled after processing.

21.8. SUMMARY

MoS_2 occurs in economic quantities in many porphyry Cu deposits. It is recovered to MoS_2 flotation concentrates by (a) floating MoS_2 and Cu—Fe—S minerals together in a bulk Cu—Mo concentrate, then (b) depressing Cu—Fe—S minerals and floating MoS_2 in a Mo—Cu flotation separator plant. Typical Mo recovery to final MoS_2 concentrate is ~70% (80% to the bulk concentrate and 90% of that to the final MoS_2 concentrate).

Other byproducts from copper production include the slime from electrorefining, dust recovered from baghouses and electrostatic precipitators, and smelting slag. All of these can be treated to recover their copper content, and the remaining material is also valuable. The slime in particular contains significant levels of gold, silver, selenium, and tellurium. The recovery of these metals has a significant impact on the profitability of copper production facilities.

REFERENCES

Amelunxen, P., & Amelunxen, R. (2009). Moly plant design considerations. *Proceedings of the SME annual meeting*. Littleton, CO: SME.

Becker, E. (2006). Modernisation of precious metals refining at Norddeutsche Affinerie AG. *World of Metallurgy — Erzmetall, 59*, 87—94.

Cao, H., Zhang, L., Wang, C., Fu, N., Sui, Z., & Feng, N. (2009). Study of selectively separating iron constituents in copper smelting slags. In *EPD congress 2009* (pp. 939—941). Warrendale, PA: TMS.

Chen, T. T., & Dutrizac, J. E. (2005). Mineralogical characterization of a copper anode and the anode slimes from the La Caridad Copper Refinery of Mexicana de Cobre. *Metallurgical and Materials Transactions B, 36B*, 229—240.

Chevron Phillips. (2009). *Molyflo® flotation oil*. www.cpchem.com/bl/specchem/en-us/Pages/MOLYFLO FlotationOil.aspx.

Cytec. (2011). *Cytec develops new reagent to increase copper and moly recovery at Latin American mine*. www.cytec.com/specialty-chemicals/PDFs/Mineralprocessing/MCT-1126-A.pdf.

Flynn, M. R., & Susi, P. (2004). A review of engineering control technology for exposure generated during abrasive blasting operation. *Journal of Occupational and Environmental Hygiene, 1*, 680—687.

González, C., Parra, R., Klenovcanova, A., Imris, I., & Sánchez, M. (2004). Reduction of Chilean copper slags: a case of waste management project. *Scandinavian Journal of Metallurgy, 34*, 143—149.

Gorai, B., Jana, R. K., & Premchand, M. (2003). Characteristics and utilization of copper slag — a review. *Resources Conservation and Recycling, 39*, 299—313.

Hait, J., Jana, R. K., & Sanyal, S. K. (2009). Processing of copper electrorefining anode slime: a review. *Transactions of the Institution of Mining and Metallurgy, Section C, 118*, 240—252.

Hoffmann, J. E. (2008). The world's most complex metallurgy revisited. *World of Metallurgy — Erzmetall, 61*, 6—13.

Ichimura, R., Tateiwa, H., Almendares, C., & Sanchez, G. (2007). Arsenic immobilization of Teniente furnace dust. In D. Rodier & W. Adams (Eds.), *Copper 2007 — Vol. VI: Sustainable development, HS&E, and recycling* (pp. 275—288). Montreal: CIM.

Kim, D., Wang, S., & Brees, D. (2010). Sustainable developments in copper anode slimes — wet chlorination-processing. In *Copper 2010, Vol. 4: Electrowinning and refining* (pp. 1393—1402). Clausthal-Zellerfeld, Germany: GDMB.

Kurokawa, H., Asano, S., Sakamoto, K., Heguri, S., & Hashikawa, T. (2007). Chlorine leaching technology of precious metals in copper anode slimes. In B. Davis & M. Free (Eds.), *General abstracts: Extraction and processing* (pp. 95—100). Warrendale, PA: TMS.

Lim, T.-T., & Chu, J. (2006). Assessment of the use of spent copper slag for land reclamation. *Waste Management & Research, 24*, 67—73.

Marghussian, V. K., & Maghsoodipoor, A. (1999). Fabrication of unglazed floor tiles containing Iranian copper slag. *Ceramics International, 25*, 717—722.

Moats, M., Robinson, T., Davenport, W., Karcas, G., & Demetrio, S. (2007). Electrolytic copper refining — 2007 world tankhouse operating data. In G. E. Houlachi, J. D. Edwards & T. G. Robinson (Eds.), *Cu2007, Vol. V: Copper electrorefining and electrowinning* (pp. 195—241). Montreal: CIM.

Montenegro, V., Sano, H., & Fujisawa, T. (2011). Recirculation of high arsenic content copper smelting dust to smelting and converting processes. *Minerals Engineering*, in press.

Parra, R., & Parada, R. (2008). Minor element control by flue dust treatment in copper smelting. In B. Mishra, C. Ludwig & S. Das (Eds.), *REWAS 2008* (pp. 1027—1038). Warrendale, PA: TMS.

Pavez, O., Rojas, F., Palacios, J., & Sanchez, M. (2004). A review of copper pyrometallurgical slags utilization. In S. R. Rao (Ed.), *Waste processing and recycling in mineral and metallurgical industries V* (pp. 291—298). Montreal: CIM.

Samuelsson, C., & Carlsson, G. (2001). Characterization of copper smelter dusts. *CIM Bulletin, 1051*, 111—115.

Shi, C., Meyer, C., & Behnood, A. (2008). Utilization of copper slag in cement and concrete. *Resources Conservation and Recycling, 52*, 1115—1120.

Thuy, M., Antrekowitsch, H., Antrekowitsch, J., & Pesl, J. (2004). Recovery of valuable metals from anode slimes of the secondary copper industry. *BHM: Berg- und hüttenmännische Monatshefte, 149*, 9—15.

Torres, F., Soto, F., & Amelunxen, P. (2009). A kinetics approach to the design of the moly plant for the Codelco Andina Phase 2 Expansion Project. In P. Amelunxen, W. Kracht & R. Kuyvenhoven (Eds.), *VI international mineral processing seminar — Procemin 2009* (pp. 335—345). Santiago: Universidad de Chile.

SUGGESTED READING

Gorai, B., Jana, R. K., & Premchand, M. (2003). Characteristics and utilization of copper slag — a review. *Resources Conservation and Recycling, 39*, 299—313.

Hait, J., Jana, R. K., & Sanyal, S. K. (2009). Processing of copper electrorefining anode slime: a review. *Transactions of the Institution of Mining and Metallurgy, Section C, 118*, 240—252.

Hernandez-Aguilar, J. R., & Basi, J. (2009). Improving column flotation cell operation in a copper/molybdenum separation circuit. *CIM Journal, 1*(3), 165—175.

Hoffmann, J. E. (2008). The world's most complex metallurgy revisited. *World of Metallurgy — Erzmetall, 61*, 6—13.

Morales, R., Gimenez, P., Arru, M., & Authievre, J. (2003). Molybdenum recovery evolution at Codelco Chile, El Teniente Division. In A. Casali, C. Gómez & R. Kuyvenhoven (Eds.), *V international mineral processing seminar — Procemin 2008* (pp. 189—196). Santiago: Universidad de Chile. downloads.gecamin.cl/cierre_eventos/procemin2008/prsntcns/00136_00662_pr.pdf.

Woodcock, J. T., Sparrow, G. J., Bruckard, W. J., Johnson, N. W., & Dunne, R. (2007). Plant practice: sulfide minerals and precious metals. In M. C. Fuerstenau, G. Jameson & R.-H. Yoon (Eds.), *Froth flotation, a century of innovation* (pp. 781—817). Littleton, CO: SME.

Chapter 22

Costs of Copper Production

This chapter:

(a) Describes the investment and production costs of producing copper metal from ore
(b) Discusses how these costs are affected by such factors as ore grade, process route, and inflation
(c) Indicates where cost savings might be made in the future.

The discussion centers on mine, concentrator, smelter, and refinery costs. Costs of producing copper by leach/solvent extraction/electrowinning and from scrap are also discussed.

The cost data have been obtained from published information and personal contacts in the copper industry. They were obtained during early 2011 and are expressed in 2011 USA dollars. The data are directly applicable to plants in South and North America. They are thought to be similar to costs in other parts of the world.

Investment and operating costs are significantly affected by inflation. Fortunately, inflation has been small from 2009 to 2011. It will probably be low in developed countries during the next few years, but it is already increasing in China.

Inflation in the mining and extraction industry is represented by the Marshall and Swift Mining and Milling Equipment Cost Index (Fig. 22.1). The basic equation for using this index for identical equipment is:

$$\frac{\text{Cost}_{(\text{year A})}}{\text{Cost}_{(\text{year B})}} = \frac{\text{Index}_{(\text{year A})}}{\text{Index}_{(\text{year B})}} \qquad (22.1)$$

The investment and operating costs in this chapter are at the *study estimate* level, which is equivalent to an accuracy of ±30%. Data with this accuracy can be used to examine the economic feasibility of a project before spending significant funds for piloting, market studies, land surveys, and property acquisitions (Christensen & Dysert, 2005).

22.1. OVERALL INVESTMENT COSTS: MINE THROUGH REFINERY

Table 22.1 lists study estimate investment costs for a mine/concentrator/smelter/refinery complex designed to produce electrorefined cathodes from 0.5% Cu ore. These costs are for a *greenfield* complex starting on a virgin site with construction beginning April 1, 2011. (It should be noted that construction of completely integrated complexes is now rare. Typically, new mines and concentrators are built in South America and new smelters/refineries are built in coastal Asia.) The investment costs are expressed in

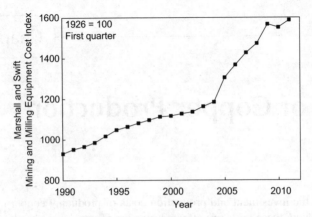

FIGURE 22.1 Marshall and Swift mining and milling equipment cost index from 1990 to 2011 (Chemical Engineering, 2011). Steep inflation from 2003 to 2008 and flatter inflation from 2009 to 2011 are notable.

terms of investment cost per annual tonne of product copper. This is defined by the equation:

$$\text{plant cost} = \text{investment cost per annual tonne of copper} \\ \times \text{plant capacity, tonnes of copper per year} \quad (22.2)$$

This equation shows, for example, that the investment in a concentrator which (a) costs $10,000 per annual tonne of copper, and (b) produces 200 000 tonnes of Cu-in-concentrate per year will be $2,000 × 10^6 ($2 billion).

Table 22.1 indicates that the fixed capital investment for a complex that produces electrorefined copper from 0.5% Cu ore is ~$30,000 per annual tonne of copper. To this must be added working capital to cover the initial operating expenses of the complex (about 10% of fixed capital investment; Peters & Timmerhaus, 1968). It means that a new mine/mill/smelter/refinery complex which is to produce 200 000 tonnes of copper per year will cost ~$6.5 billion ($33,000 × 200 000).

TABLE 22.1 Fixed Investment Costs for a Copper Extraction Complex, Starting with 0.5% Cu Ore. The Costs are at the Study Estimate Level of Accuracy

Facility	Fixed investment cost ($U.S. per tonne of Cu per year)
Mine (open pit)	10,000
Concentrator (including water acquisition and recycle)	10,000
Smelter (Outotec flash furnace smelting/converting), including sulfuric acid plant	9,000
Electrolytic refinery (excluding precious metals refinery)	1,000
Total	30,000

22.1.1. Variation in Investment Costs

Mine investment costs vary considerably between mining operations. This is due to (a) differences in ore grades, mine sizes, mining method, topography, and ground condition, and (b) infrastructure requirements for electric power and water. Underground mine development costs considerably more than open-pit mine development, per annual tonne of mined ore. This, and the high cost of operating underground explain why underground ore bodies must contain (a) higher ore grade than open-pit ore bodies and/or (b) valuable byproducts such as Ag, Au, Co, Mo, and Ni.

Ore grade has a direct effect on mine investment costs. Consider (for example) two identical ore bodies, one containing 0.5% Cu ore and the other 1% Cu ore. Achievement of an identical annual production of Cu requires that the 0.5% Cu ore be mined at twice the rate of the 1% Cu ore. This, in turn, requires about (a) twice as much plant and equipment (e.g. trucks), and (b) twice as much investment. The same is true for concentrators and heap leach pads.

A significant investment cost for new concentrators is electric energy and water production/distribution. This is particularly true in South America, where virtually all new Cu mines are in high, dry desert locations.

Smelter investment costs, per annual tonne of copper production, are influenced by concentrate grade rather than by ore grade. The higher the % Cu in the concentrate, the smaller the smelter (and smelter investment) for a given annual production of copper. High-grade concentrates also minimize smelter operating costs (such as materials handling costs, fuel consumption costs, gas handling costs) per tonne of copper.

Refinery investment costs are not affected by mine/concentrator/smelter characteristics. This is because copper refineries treat 99.5% Cu anodes, irrespective of the preceding processes.

22.1.2. Economic Sizes of Plants

Mines can be economic at any size, depending upon the Cu grade of their ore. Thus, copper mines operate at production rates between 10 000 tonnes of ore per day (a high-grade operation) to 250 000 tonnes per day (a large open-pit operation; ICSG, 2010).

Concentrators vary similarly. A new large concentrator unit typically consists of a semi-autogenous grinding mill, high pressure grinding rolls, two ball mills, and a flotation circuit. It can treat up to 170 000 tonnes of ore per day (Sartain, 2010). Larger concentrators consist of multiples of this basic concentrating unit.

Smelters are almost always large because their minimum economic output is that of a single, fully used high intensity smelting furnace. These furnaces typically smelt up to 4 000 tonnes of concentrate per day.

Copper refineries are usually sized to match the anode output of an adjacent smelter. The advantage of having a smelter and refinery at the same site is the presence of shared site facilities, particularly for anode casting and anode scrap re-melting. A few refineries treat the anodes from several smelters. The largest electrorefineries produce up to 1 100 tonnes of copper cathodes per day (ICSG, 2010).

22.2. OVERALL DIRECT OPERATING COSTS: MINE THROUGH REFINERY

Direct operating (cash) costs (excluding depreciation, capital repayment, and income taxes) for mining/concentrating/smelting/electrorefining are given in Table 22.2. The

TABLE 22.2 Direct Operating Costs for Producing Electrorefined Copper Cathodes from a 0.5% Cu Ore (Assuming 90% Cu Recovery). Maintenance is Included. The Costs are at the Study estimate Level

Activity	Direct operating cost ($USA per kg of Cu)
Open-pit mining, 0.5% Cu ore @ $2.5/tonne of ore	0.5
Concentration from 0.5% Cu ore to 30% Cu concentrate at shipping point, including tailings disposal @ $5/tonne of mined ore	1.0
Smelting @ $80/tonne of 30% Cu concentrate including sulfuric acid production	0.3
Electrolytic refining, excluding precious metals recovery	0.1
Local services and overhead	0.1
Total direct operating cost	2.0

table shows that direct operating costs for the major steps are, in descending order, concentration, open-pit mining, smelting, and electrorefining. Overall direct operating costs for extraction are ~$2/kg of copper.

22.2.1. Variations in Direct Operating Costs

The operating costs that vary most are those for mining and concentrating. The amounts of ore, which must be handled by these operations, per tonne of Cu, vary directly with ore grade — and this significantly affects operating costs. Underground mining can also cost up to five times as much as open-pit mining, which must be offset by (a) higher ore grade and (b) the presence of byproduct metals in underground ore.

22.3. TOTAL PRODUCTION COSTS, SELLING PRICES, PROFITABILITY

The total cost of producing copper from ore is made up of (a) direct operating costs and (b) finance (indirect) costs, including interest and capital recovery. A reasonable estimate for (b) is 12% of the total capital investment per year. Based on a fixed capital investment of $33,000 (including 10% working capital) per annual tonne of copper, this is equivalent to $4,000 per tonne of copper ($4/kg). Thus the direct ($2) plus indirect ($4) operating costs of producing electrorefined copper in a new operation total approximately $6/kg. For a new operation to be profitable, the selling price of copper must exceed these costs.

Mines and plants that have been in operation for many years may have repaid much of their original capital investment. In this case, direct operating costs (plus refurbishing) are the main cost component. This type of operation will be profitable at selling prices of ~$3/kg of copper.

In summary, the price-profit situation is:

(a) At copper selling prices above $6/kg, copper extraction is profitable and expansion of the industry is encouraged. Underground ore bodies containing about 1.5 % Cu are viable, as are open-pit ore bodies containing about 0.5% Cu.
(b) At selling prices below about ~$3/kg, some mines and plants are unprofitable. Some operations begin to shut down.

These costs and prices all refer to April 1, 2011. They will increase at about the same rate as the cost index in Fig. 22.1.

The 2011 selling price of copper was about $9/kg (LME, 2011) so that total operating and finance costs were met in almost all cases. Virtually all operations were profitable and the industry expanded rapidly.

22.3.1. Byproduct Credits

Many Cu ore bodies contain Au and Ag (Chadwick, 2010; Kendrick, Baum, Thompson, Wilkie, & Gottlieb, 2003; Supomo et al., 2008). These metals follow Cu during concentration, smelting, and refining. They are recovered during electrorefining (with some additional treatment; see Chapter 21) and sold. Other ore bodies contain MoS_2 (Section 21.1), which is recovered in the concentrator and usually sold. The credits (sales minus extra costs for recovery) for these byproducts should be included in all project evaluations (Freeport-McMoRan Copper & Gold, 2011). Sulfuric acid is also often a valuable byproduct, especially when the smelter is near leach/solvent extraction/electrowinning operations.

22.4. CONCENTRATING COSTS

The investment costs of constructing a Cu concentrator are of the order of $10,000 per annual tonne of copper (Table 22.1). This means that a concentrator producing 200 000 tonnes of copper-in-concentrate will cost ~$2 billion (Section 22.1).

Table 22.3 breaks concentrator investment costs into major cost components, expressed as a percentage of total investment cost. The largest cost for this

TABLE 22.3 Concentrator Investment Costs. Investment Costs for a Copper Concentrator by Section, Expressed as a Percentage of the Total Investment Cost. Control Equipment Costs are Encluded in Each Section

Section	Percent of total investment cost
Ore handling, storage, conveying equipment	10
Semi-autogenous grinding mill, high pressure grinding rolls, ball mills, and size classifiers	25
Flotation cells and associated equipment	15
Water acquisition/distribution system, dewatering/filtering equipment, tailings dam, concentrate loading facilities	50
Total	100

TABLE 22.4 Direct Operating Costs of Producing 30% Cu Concentrate from 0.5% Cu Ore. Ore Cost is not Included

Activity	Cost per tonne of ore ($USA)
Crushing, conveying, storage	0.50
Semi-autogenous grinding, ball mill grinding, size classification	3.10
Flotation	0.60
Dewatering, filtering, drying, storage, and loading of concentrate	0.20
Tailings disposal, effluent control, water recycle	0.30
Local services and overhead (accounting, clerical, environmental, human resources, laboratory, management, property taxes, safety)	0.10
Total	4.80

concentrator is water acquisition, which indicates that it is in a desert region. Grinding is second.

Concentrator direct operating costs (Tables 22.4 and 22.5) are of the order of $4.8/tonne of ore, which is equivalent to about $1.1/kg of Cu (assuming 0.5% Cu ore and 90% Cu recovery). Grinding is by far the largest operating cost, followed by flotation.

TABLE 22.5 Concentrator Operating Costs by Cost Component. Ore Cost is not Included

Component	Percent of concentrating cost
Electric energy	20
Water (new + recycled)	10
Operating and maintenance labor	10
Maintenance and operating supplies	25
Reagents and lime	10
Grinding media	15
Local services and overhead (accounting, clerical, environmental, human resources, laboratory, management, safety	10
Total	100

Chapter | 22 Costs of Copper Production

Grinding and flotation costs vary markedly for different ores. Grinding costs are high for hard primary ores and low for secondary (altered) ores. Flotation costs are low for simple Cu sulfide ores. They increase with increasing ore complexity.

22.5. SMELTING COSTS

The investment cost of a new Outotec flash furnace/flash converter smelter is ~$9,000 per annual tonne of copper (Table 22.1). A smelter designed to produce 200 000 tonnes of new anode copper per year will cost, therefore, about $1,800 \times 10^6$ — about $2 billion.

Table 22.6 breaks this investment cost into its major components. About 70% of the investment goes into concentrate handling/smelting/converting/anode casting and about 30% into gas handling/sulfuric acid manufacture.

In 2011, there are three major intensive smelting processes available for installing in new smelters or for modernizing old smelters — Outotec flash, Isasmelt, and Ausmelt. Each has been installed extensively in since 2005. Each appears to be competitive for new and replacement smelting units.

Table 22.7 shows the direct costs of operating an oxygen-enriched Outotec flash smelting/flash converting smelter. The total is about $80 per tonne of concentrate after sulfuric acid and energy credits. For a 30% Cu concentrate, this is equivalent to about $0.3/kg of new copper anodes. Table 22.8 breaks down these direct operating costs into labor, fuel, oxygen, and supplies. Energy is shown to be the largest cost.

TABLE 22.6 Smelter Investment Costs. Investment Costs of a Flash Furnace/Flash Converter Smelter by Section, Expressed as a Percentage of Total Fixed Investment Cost. The Costs include Installation and Housing of the Units. Control Systems are also Included

Item	Percent of smelter cost
Concentrate handling and drying, including delivery of dry concentrate to smelting furnace	10
Oxygen plant	10
Flash furnace	15
Flash converter, including matte granulation, and crushing	15
Cu-from-slag recovery equipment (electric furnace or flotation) including barren slag disposal	10
Anode furnaces and anode casting equipment	10
Gas handling system including waste heat boilers, electrostatic precipitators, and sulfuric acid plant	30
Total	100

TABLE 22.7 Smelter Operating Costs by Activity. Direct Operating Costs for Producing Anodes from 30% Cu Concentrate in a Flash Smelting/Flash Converting Smelter, Including Maintenance. Concentrate Cost is not Included

Activity	Cost, $U.S. per tonne of concentrate
Concentrate reception, storage, and delivery to dryer	5
Flash furnace smelting including concentrate drying, gas handling, and delivery of 70% Cu crushed matte granules to flash converting	20
Flash converting including delivery of molten copper to anode furnaces	20
Cu recovery from smelting slag	15
Anode-making including desulfurization and deoxidation of molten copper, anode casting and loading for transport to electrorefinery	5
Sulfuric acid plant including acid storage and loading of rail cars and trucks. Costs of treating acid plant blowdown and credit for sulfuric acid are included	10
Local overhead (accounting, clerical, environmental, human resources, laboratory, management, property taxes, safety)	5
Total	80

TABLE 22.8 Smelter Operating Costs by Cost Component. Expenditures on Manpower, Utilities, and Supplies in a Flash Smelting/Flash Converting Smelter, by Percentage. Concentrate Cost is not Included

Component	Percent of smelting cost
Operating & maintenance manpower, including supervision	20
Electricity, including electricity used for making oxygen and sulfuric acid)	30
Hydrocarbon fuel	20
Flux and refractories	10
Maintenance supplies	15
Local services and overhead (accounting, clerical, environmental, human resources, laboratory, management, property taxes, safety)	5
Total	100

22.6. ELECTROREFINING COSTS

The investment cost of a new electrorefinery using stainless-steel cathode technology is ~$1,000 per annual tonne of electrorefined cathodes. This means that a refinery producing 200 000 tonnes per year of cathodes will cost of the order of 200×10^6.

The relative investment costs of various sections of a refinery are shown in Table 22.9. The production electrorefining section (including stainless-steel blanks) is by far the largest investment cost component of the refinery.

The direct cost of producing electrorefined cathodes in an electrolytic refinery is ~$0.12/kg of cathode copper, Table 22.10. The largest component of that cost is energy, Table 22.11.

22.7. PRODUCTION OF COPPER FROM SCRAP

Chapter 18 shows that copper scrap varies in grade from 99.5+% Cu (manufacturing wastes) to 5% Cu (recycled mixed-metal scrap). The high-grade manufacturing wastes require only reclamation, melting, casting, and marketing, which costs about $0.2/kg of copper. Low-grade scrap, on the other hand, requires reclamation, sorting, smelting, refining, and marketing, which costs about $0.6/kg of copper (Table 22.2). Intermediate-grade scrap treatment lies between these two extremes.

For scrap recovery to be profitable, the difference between refined copper sales price and scrap purchase price must exceed these treatment charges. When this is not so, scrap is held off the market. In 2011, many smelters are actively seeking scrap and scrap collection is very profitable.

TABLE 22.9 Investment Costs of Components in an Electrolytic Copper Refinery Expressed as a Percentage of the Total Fixed Investment Cost

Component	Percent of total fixed investment cost
Anode reception, weighing, straightening, lug milling, sampling equipment	10
Production electrorefining equipment, including stainless-steel blanks, polymer concrete cells, transformers, rectifiers, electrical distribution system	55
Electrolyte circulation and purification equipment, including filters, heaters, pumps, storage tanks, reagent addition equipment, electrorefining cells	15
Cathode handling equipment including stripping, washing, weighing, sampling, and bundling equipment	5
Anode (and purchased) scrap melting and anode casting equipment: including Asarco shaft furnace, holding furnace, pouring equipment (including mass control system) & mold-on-wheel caster	15
Total	100

TABLE 22.10 Direct Operating Costs Including Maintenance, for Producing Electrorefined Cathode Plates from 99.5% Cu Anodes in a Stainless Steel Mother Blank Electrorefinery. Anode Cost is not Included

Activity	Cost, $U.S. per kg of cathode Cu
Anode reception, weighing, straightening, lug milling, delivery to tankhouse	0.01
Production electrorefining, including cell cleaning, electrolyte purification, reagent addition, delivery of cathodes to washing and delivery of slimes to Cu/precious metal recovery plant	0.06
Cathode handling, including stripping, washing, weighing, quality control, and delivery to loading docks	0.01
Anode scrap washing and melting, purchased scrap melting, anode casting, and anode delivery to tankhouse	0.03
Local services and overhead (accounting, clerical, environmental, human resources, laboratory, management, property taxes, safety)	0.01
Total	0.12

22.8. LEACH/SOLVENT EXTRACTION/ELECTROWINNING COSTS

The investment and operating costs of heap leach/solvent extraction/electrowinning plants are listed in Tables 22.12 and 22.13. The costs are shown to be considerably lower than those for conventional concentration/smelting/refining complexes. This accounts for the rapid adoption of leaching around the world, especially in Chile.

TABLE 22.11 Expenditures on Electrorefinery Manpower, Electricity, and Supplies (Excluding Anodes and Scrap), by Percentage. Anode Cost is not Included

Component	Percentage of electrorefining cost
Operating & maintenance manpower, including supervision	20
Electricity	40
Maintenance, excluding labor	15
Natural gas	15
Local services and overhead (accounting, clerical, environmental, human resources, laboratory, management, property taxes, safety)	10
Total	100

TABLE 22.12 Fixed Investment Costs for a Heap Leach/Solvent Extraction/Electrowinning Plant. The Plant Produces Copper Cathode Plates Ready for Shipment from 0.5% Cu Oxide Ore. Ir-,Ta-Oxide—Coated Ti Anodes (Sandoval, Clayton, Dominguez, Unger, & Robinson, 2010), 316L Stainless Steel Cathodes, and Polymer Concrete Cells are Used. Mine Investment Cost is not Included. Sulfuric Acid Plant Investment is not Included

Component	$U.S. per annual tonne of copper
Heap leach system, including leach pad, crusher, agglomerating drum, on-off heap building and removal equipment, piping, pumps, solution collection ponds, etc.	1000
Solvent extraction plant, including mixer—settlers, pumps, piping, storage tanks, and initial first fill extractant and diluent	500
Electrowinning plant, including electrical equipment, polymer concrete cells, Ir-,Ta-oxide—coated Ti anodes, 316L stainless-steel cathodes, cranes, cathode stripping, washing, and handling equipment	1000
Utilities and infrastructure, including water supply	500
Total	3000

TABLE 22.13 Direct Operating Costs of a Heap Leach/Solvent Extraction/Electrowinning System. The Plant Produces Copper Cathode Plates Ready for Shipment from 0.5% Cu Oxide Ore. Ir-,Ta-oxide—Coated Ti Anodes, 316L Stainless-Steel Cathodes, and Polymer Concrete Cells are Used. Ore Cost is not Included

Item	$/kg of copper
Heap leach operation, including crushing, acid curing, agglomeration, on-off heap construction/removal, solution delivery and collection, excluding sulfuric acid	0.25
Sulfuric acid	0.25
Solvent extraction plant operation, including maintenance	0.05
Reagent make-up: extractant, diluent, guar, and $CoSO_4 \cdot 7H_2O$	0.05
Electrowinning tankhouse operation, delivering cathode plates to loadout platform	0.15
Local services and overhead (accounting, clerical, environmental, human resources, laboratory, management, property taxes, safety)	0.05
Total	0.80

Unfortunately, chalcopyrite ore (the world's largest source of copper) cannot yet be economically processed by heap leach/solvent extraction/electrowinning (Chapter 15). Chalcopyrite ores must usually be treated by conventional concentration/smelting/refining, irrespective of cost.

The small investment requirement of leach/solvent extraction/electrowinning plants is due to the small equipment and infrastructure requirements of these processes. Specifically, leaching and solvent extraction require much less equipment than concentrating, smelting, converting, and anode casting.

An interesting aspect of pyrometallurgical and hydrometallurgical copper extraction is sulfuric acid production and use. Hydrometallurgical copper extraction requires sulfuric acid (Chapter 15), and pyrometallurgical copper extraction produces it (Chapter 12). Companies with both processes benefit significantly from this synergistic effect, especially if the operations are located close to each other.

22.9. PROFITABILITY

The key to a profitable mine-to-market copper operation is, of course, a large, high Cu-grade ore body. Such an ore body maximizes copper production per tonne of ore mined, moved, and processed. Optimal use of an ore body requires that each part of the ore body be processed by its most efficient method — leaching or concentrating/smelting. Separation of the ore body into milling ore, leaching ore, leaching waste, and unleachable waste is crucial for optimal utilization of the resource.

22.10. SUMMARY

The total direct plus indirect cost of producing electrorefined copper from ore by conventional mining/concentration/smelting/refining is in the range of $3–$6 per kg of copper. The total direct plus indirect cost of producing electrowon copper cathodes from oxide and chalcocite ores (including mining) is $1–$2 per kg of copper.

Copper extraction is distinctly profitable when the selling price of copper is above $6/kg. It is unprofitable for some operations when the selling price falls below $3/kg. At the former price, the industry tends to expand. At the latter, it begins to contract.

The selling price of copper in mid-2011 is ~$9/kg. Virtually all existing operations are profitable and the industry is expanding.

REFERENCES

Chadwick, J. (2010). Grasberg concentrator. *International Mining, 27*(5), 8–20. http://www.infomine.com/publications/docs/InternationalMining/Chadwick2010p.pdf.

Chemical Engineering. (2011). (McGraw-Hill Publishing Company, New York, NY), data obtained from July issues, 1990–2011.

Christensen, P., & Dysert, L. R. (2005). Cost estimate classification system — as applied in engineering, procurement and construction for the process industries. *AACE International Recommended Practice*. http://www.costengineering.eu/Downloads/articles/AACE_CLASSIFICATION_SYSTEM.pdf 18R–97.

Freeport-McMoRan Copper & Gold. (2011). *Freeport-McMoRan copper and gold 4th quarter 2010 earnings conference call, Jan 20, 2011.* http://www.fcx.com/ir/2011present/FCX_4Q10CC.pdf.

ICSG. (2010). *Directory of copper mines and plants 2008 to 2013.* Available through http://www.icsg.org/index.php?option=com_content&task=view&id=4&Itemid=64.

Kendrick, M., Baum, W., Thompson, P., Wilkie, G., & Gottlieb, P. (2003). The use of the QemSCAN automated mineral analyzer at the Candelaria concentrator. In C. O. Gomez & C. A. Barahona (Eds.), *Copper/Cobre 2003, Vol. III: Mineral processing* (pp. 415–430). Montreal: CIM.

London Metal Exchange. (2011). *Copper price graphs*. http://www.lme.com/copper_graphs.asp.

Peters, M. S., & Timmerhaus, K. D. (1968). *Plant design and economics for chemical engineers* (2nd ed.). New York, NY: McGraw-Hill Book Company.

Sandoval, S., Clayton, C., Dominguez, S., Unger, C., & Robinson, T. (2010). Development and commercialization of an alternative anode for copper electrowinning. In *Copper 2010, Vol. 4: Electrowinning and −refining* (pp. 1635–1647). Clausthal-Zellerfeld, Germany: GDMB.

Sartain, C. (2010). *Xstrata copper: Driving the organic growth strategy*. http://www.xstrata.com/assets/pdf/xcu_speech_201004071_world_copper_conference.pdf.

Supomo, A., Yap, E., Zheng, X., Banini, G., Mosher, J., & Partanen, A. (2008). PT Freeport Indonesia's mass-pull control strategy for rougher flotation. *Minerals Engineering, 21*, 808–816.

SUGGESTED READING

AME Group. (2011). *Escondida, Chile (Copper Mine)*. http://www.ame.com.au/Mines/Cu/Escondida.htm.

Index

A

Abrasive grit, slag for, 423–424
Absorption towers, sulfuric acid manufacture, 226, 226t
Acid coolers, 227
Acid cure, heap leaching, 296
Acid mist suppression, electrowinning of copper, 356, 357t
Acid plant blowdown, 212, 213f
Agglomeration, heap leaching, 296–297, 297f
  optimum, 297
Agitation leaching, 286
  copper losses in, 304
  electrowinning of copper from, 362–368
  industrial data for, 305t–307t
  leaching methods compared to, 316t
  oxide minerals in, 303–304
  sulfide minerals in, 304
Air Liquide Shrouded Injector (ALSI), 142–143
Aldoxime-ketoxime mixtures, 327
Aldoximes, 327
Alloys, copper, 1
  scrap, 12
Anode slimes, 418–422
  composition of, 418–421, 419t–420t
  electrorefining and Cu recovery from, 253, 256–257, 260–261, 267
  Gold, recovery from anode slimes, 253
  Selenium, recovery from anode slimes, 253, 253f
  Silver, recovery from anode slimes, 253, 253f
  Tellurium, recovery from anode slimes, 253f, 255
  treatment process, 421–422
    flowsheet for, 422f
  wet chlorination process for, 421–422
Anodes, electrorefining
  antimony in, 245–247
  arsenic in, 245–247
  casting, 241–243, 242f, 258
    continuous, 244–245, 243f, 244f, 246t, 247f
    industrial data for, 246t
    molds for, 243
    preparation in, 244, 258
    uniformity of thickness in, 243–244
  composition of, 258
  copper, 241–243, 242f
    industrial range of, 252t
  flowsheet, refining, 253f

Hazelett Contilanod, 243f, 244–245, 244f, 246t, 247f
  passivation and, 268
  purification of, 266
  scrap and reject, 245
  slimes from, Cu recovery from, 253, 256–257, 260–261, 267
Anodes, electrowinning, 351–353
  insulator strips on, 353
  life expectancy of, 352–353
  new, 368–369
  reaction of, 10
Anterite, 19t
Antimony
  in anodes, 245–247
  in electrorefining, 254
Aqueous-continuous phase, 339
Aqueous-in-organic entrainment, 328–329
Arsenic
  in anodes, 245–247
  in electrorefining, 254–255
  TSL and, 163
Asarco shaft furnaces, 402
  industrial data on, 406t
ASTM specifications, copper, 397, 398t
Ausmelt batch converting, 143
Ausmelt Continuous Copper Converting (C3), 150
Ausmelt/Isasmelt smelting
  basic operations of, 153–154
  coal and, 156
  feed materials for, 156, 158t–161t
  industrial data on, 158t–161t
  Mitsubishi process compared to, 165–174
  offgas in, 155–156
  oxygen enrichment in, 156
  secondary smelting, 395
Auto shredder residue (ASR), 382f, 383–384
Autogenous grinding mills (AG mills), 36
Automobile scrap, 382–384
  flowsheet for, 382f
Avitone, 264, 265
Azurite, 19t

B

Ball mills, 36
  industrial data for, 37t–42t
Bar and rod casting, 404–409, 404f, 405f, 407t–408t

441

Barren aqueous raffinate, 281
Bath smelting, 83–84. *See also* Ausmelt/Isasmelt smelting; Top Submerged Lance smelting
BAYQIK reactor, 230
Bedding, 97
Billet casting, 403f, 404
Bin and feed system, in flash furnace matte smelting, Outotec, 97–98
Bin-onto-belt blending, 96
Bismuth behavior
  anodes and brittleness in, 134
  in electrorefining, 254–255
Black copper
  converting, 395–396
  flowsheet for, 394f
  secondary copper smelting to, 393–395, 394f
Blast heater, in flash furnace matte smelting, Outotec, 98
Blasting, 31
  fragmentation in, 33, 33f
  ore-size determination and, 32–34
  ore-toughness measurements for, 34
  process of, 32, 33f
Blister copper, 6f, 6–7
  electrorefining of, 7–8
  in Mitsubishi process converting furnace, 169
  production of (converting), 127
Boliden Norzink process, 212
Bornite, 19t
Branched-chain alcohols, 55
Brochantite, 19t
Bustle pipe, 136

C

Cable scrap, 380–385
  flowsheet for, 381f
Calcium ferrite slag ($CaO-Cu_2O-FeO_3$), 79–81, 79f
  advantages and disadvantages of, 80–81
  flash converting and, 146
CaO-base slag
  controlling, 175
  Mitsubishi process converting furnace and, 168–169
Casting copper, 10
  anodes, 241–243, 242f
    continuous (Hazelett Contilanod), 243f, 244–245, 244f, 247f, 246t
    industrial data for, 246t
    molds for, 243
    preparation in, 244, 245
    uniformity of thickness in, 243–244
  Hazelett, 243f, 244–245, 244f, 246t, 247f, 405
    industrial data for, 246t, 407t–408t
  machines for, 403–410
    bar and rod, 404–409, 404f, 405f, 407t–408t
    billet casting, 403f, 404

Conticast casting, 409, 410f
  industrial data for, 407t–408t
  lubrication of, 409
  metal level in, 409
  Properzi casting, 405
  Southwire casting, 404f, 405f, 407t–408t
  vertical, 409
oxygen-free, 409, 410f
product grades and quality of, 397–399, 398t
strip, 409–410
Catalyst for SO_2 to SO_3 conversion, 214–215
Catalytic converter, 224, 224f
Cathodes. *See also* Melting of cathode copper
  electrorefining
    composition of, 258–259
    copper industrial range of, 252t
    copper starter sheet for, 258f, 259
    current densities in, 268
    current efficiencies in, 268–269
    edge strips for, 259
    purity maximization for, 267
    stainless steel blanks as, 258, 258f
  electrowinning, 351–360
    copper starter sheets for, 351
    current density of, 355–356, 355f, 356f
    purity, maximization of, 360–361
    reaction of, 9
    SELE cathode holder, 355, 356f
    size of, 351
    stainless steel, 352f
  purity of
    chemical factors affecting, 267–268
    electrical factors affecting, 268–269
    physical factors affecting, 267
Cells
  electrorefining, 259–260
    electrical connection in, 260
    polymer concrete, 260
  electrowinning
    design of, 353–355
    polymer concrete, 353
    schematic of, 354f
Cement, slag as, 423
Ceramic tile, slag for, 425
Chalcocite (Cu_2S), 1, 19t
  dump and heap leaching and, 290
Chalcopyrite ($CuFeS_2$), 1, 19t
  pressure oxidation leaching of, 304–308
    elevated temperature in, 308–315
    processes for, 310t–311t
Chelation, 325, 325f
Chemically-refined tough pitch copper (CRTP), 399
Chevrons, 330–331
Chile Continuous Converting (CCC), 150
China, Cu demand in, 13, 15t, 29, 65t, 145, 147, 230, 254t, 310t–311t, 391, 398t
Chingola direct-to-copper smelting, 180f, 181t

Index

Chloride in electrolytes
  electrorefining and, 261, 264, 265
  electrowinning and, 360
Chrysocolla, 19t
Cleaner flotation cells, 56
  industrial data on, 58t–64t
Cleaner-scavenger flotation cells, 56
  industrial data on, 58t–64t
Coal, 112
  Ausmelt/Isasmelt smelting and, 156
Cobalt
  in electrorefining, 254–255
  electrowinning electrolyte addition of, 360
Coke, 112
Collectors, flotation, 52–55, 54f
Column flotation cells, 56–65, 66f
  industrial data on, 65t
Comminution. See also Crushing; Grinding
  feed to froth flotation and, 47
  high-pressure roll crushing for, 46, 46f
  recent developments in, 46–47
  stages of, 31
Computational fluid dynamics (CFD), 83–84
Concentrate burner, in flash furnace matte smelting
  Inco, 104, 105f
    feed system, 106
  Outotec, 95, 95f
Concentrates
  blending system, 96–97
  Cu content worldwide in, 57t
  in direct-to-copper smelting, 181t
  drying
    rotary dryer, 97
    steam dryer, 97
  maximizing grades of, 193
  Noranda smelting process and dried, 113t
  particle size of, 35–36, 35f
  Peirce–Smith converting and smelting, 142
  production of, 31
  SEM for optimizing, 47, 47f
  throughput rate, 100–101
Concentration of Cu ores
  costs for
    investment, 427–433, 428t, 431t
    operating, 431–433, 432t
    variation in, 429
  flowsheet of, 31, 32f
  by froth flotation, 4, 4f
  industrial data for, 37t–42t, 57t, 58t–64t
  plant locations for, 17–29
  zero water discharge plants and, 68
Concrete, slag as, 423
Construction and demolition waste (C&D), 375
Conticast casting machine, 409, 410f
Continuous casting of copper, Hazelett process
  anodes, 244–245, 243f, 244f, 246t, 247f
  industrial data for, 246t

Continuous converting. See Flash converting; Mitsubishi process; Noranda continuous converting
Continuous phase, 339
Control
  Direct-to-copper smelting, 186
  electrorefining electrolyte addition agent, 265
  flash furnace matte smelting, Outotec, 100f, 100–101
    concentrate throughput rate, 100–101
    matte grade, 96
    reaction shaft and hearth, 101
    slag, SiO_2 composition, 101
    temperature, 101
  flotation, 64–65
    in-stream chemical analysis for, 65–67
  grinding, 36–43
    hydrocyclone controls for, 42–43, 43f
    instrumentation and strategies for, 44f, 43–46, 45t
    particle-size control loop in, 44, 44f
  melting of cathode copper, hydrogen and oxygen, 403
  Mitsubishi process, 169–175, 173t–175t
    temperature, 169, 173t
  Noranda continuous converting, 149
  Noranda smelting process, 116–117, 117f
    temperature, 116
  Peirce–Smith converting
    slags and flux, 139
    temperature, 137–138
  Teniente matte smelting, 120
    matte and slag composition, 120
    matte and slag depth, 120
    temperature, 120
Converting furnace, Mitsubishi process, 168–169
  blister copper in, 169
  CaO-base slag in, 168–169
Converting of copper matte, 5–7, 127. See also Peirce–Smith converting
  chemistry in, 127
  continuous, 165
  converters
    Ausmelt, 143
    Ausmelt C3, 150
    Chile Continuous Converting, 150
    flash, 144, 144f, 145t, 146
    Hoboken, 143, 144
    ISACONVERT, 150
    ISASMELT, 143
    Mitsubishi, 143
    Noranda, 143, 147–149, 147f, 148t
  coppermaking stage in, 127–128
  $Cu-Cu_2-S$ system, 131–134, 137f
  offgases from, 206–208, 206t–207t
  products of, 127

Converting of copper matte (*Continued*)
   reactions in, 129−134
   scrap additions in, 393
   slag-forming stage in, 127
   stages of, 6
   with TSL smelting, 164−165
Cooling jackets, in flash furnace matte smelting, Outotec, 94
Copper (Cu). *See also* Casting copper; Melting of cathode copper; Scrap, copper recovery from
   anode slimes and recovery of, 253, 256−257, 260−261, 267
   anodes, 241−244, 242f
   anodes and cathodes industrial range for, 252t
   black
      converting, 395−396
      flowsheet for, 394f
      secondary copper smelting to, 393−395, 394f
   Cu−Fe−S extracting, 2−8
   chemically-refined tough pitch, 399
   China demand for, 29
   in concentrates around world, 57t
   costs of production of, 430−431
   electrochemical processing of, 12
   electrolytic tough pitch, 10, 397−398, 400t
   extraction costs for
      investment, 427−430, 428t
      operating, 430, 430t
      variation in, 429
   fayalite slags solubility of, 85f
   fire-refined high conductivity, 399
   impurities of electrical resistivity of, 399f
   melting point of, 127
   minerals of, 14−17, 19t
      air bubbles floating, 51, 52t
      grind size and liberation of, 35f, 35−36
      in heap leaching, 287t
   oxygen-free, 10, 399
   Peirce−Smith converting, blowing, 139−140
   price of, 29, 29t
   production of
      by country, 15t−16t
      growth in, 13, 13f, 29
   profitability of, 430−431, 438
   rough, 396
   secondary
      characteristics of, 391
      processing in primary smelters, 391−393, 391f
      Top Submerged Lance smelting furnace for, 395
   in slag concentration, direct-to-copper smelting, 185, 186, 187
   slags, losses of, 191
   sources of, 1−2
   unalloyed classes of, 397−398
   uses of, 13, 13f
      United States, 14t
Copper alloys, 1
Copper in slags, 191−193
   decreasing
      Cu concentration in slag minimized for, 193−194
      pyrometallurgical slag settling/reduction for, 194−197, 196t
      slag generation minimized for, 193
      slag minerals processing for, 197−201, 198t−200t
Copper losses
   agitation leaching and, 304
   flotation tailings, 58t−59t
   slag, 191
      minimizing, 193, 194−197, 197−201
Copper recovery from scrap. *See* Scrap, copper recovery from
Cost index (mining equipment), 427, 428f
Costs of copper extraction, 427−430, 428t
Counter-current decantation (CCD), 303−304
Covellite, 19t
Crud formation, in solvent extraction, 337−339
Crushing, 31
   eccentric crusher for, 36
   flowsheet for, 32f
   gyratory crusher for, 34, 34f
   high-pressure roll, 46, 46f
   industrial data for, 37t−42t
   process of, 35
Cuprite, 19t
Current efficiency
   electrorefining, 268−269
   electrowinning, 361
Cu-sulfide minerals, 51, 69
Cut-off grade, 14, 17

D

Decoppering, 421
Deep dispersion gate (DDG), 331, 331f
Dewatering, 67−68
Digenite, 19t
Diluents for solvent extraction, 327
   aliphatic compared to partially aromatic, 328t
   loss of, 339
Direct-to-copper smelting, 7, 188
   advantages and disadvantages of, 179
   chemical reactions in, 184−185
   Chingola, 180f, 181t
   concentrates in, 181t
   control, 186
   Cu recovery from slags in, 186−187
      Głogów, 187
      Olympic Dam, 187
   Cu-in-slag concentration in, 185, 186
      limitation of, 187

Index

flash furnace matte smelting, Outotec and, 182–184, 182t–184t
   differences between, 185
flowsheet for, 180f
foaming, avoidance of, 186
Głogów, 181t, 185
ideal process for, 179–182, 182f
impurity behavior in, 187–188, 188t
Olympic Dam, 180f, 181t, 185
slags from, 182t
temperature in, 182
Dispersed phase, 339
Doré bullion, 421, 422f
Double absorption acid plant, 218, 218t–222t, 224, 226, 229t, 232
Drip emitters, 300
Drying concentrate
   dewatering and, 67–68
   fluid bed dryer, 106
   rotary dryer, 97
   steam dryer, 97
Dump leaching, 286, 287–301
   bacterial action in, optimum activity for, 290
   chalcocite and, 290
   chemistry of, 288–290
   industrial, 301
   leaching methods compared to, 316t
   reactions in, 289
   sulfide mineral reactions in, 289
   uses of, 288, 288f
Dust
   composition of, 423t
   electrostatic precipitation of, 211
   matte smelting and, 83–84
   sulfuric acid manufacture catalysts accumulation of, 216
   treatment of, 422–423
      flowsheet for, 424f
Dust recovery and recycle system, in flash furnace matte smelting
   Inco, 106
   Outotec, 98–99
Dynamic heaps, 295, 296

E

Eccentric crusher, 36
Eddy-current separation (ECS), 382–383
Electric slag-cleaning furnace, 195f, 194
   improvements in design of, 197
   industrial details of, 196t
   Mitsubishi process, 167–168, 167f
   offgases from, 207
   Teniente slag cleaning furnace, 197, 198t
Electrical conductivity
   matte, 81, 82
   slag, 79

Electrolytes
   electrorefining
      addition agent control for, 265
      addition agents to, 262–265
      chloride added to, 261, 264, 265
      cobalt addition to, 360
      composition of, 261–265, 262t
      conductivity of, 261, 263t
      filtering, 266
      purification of, 266–267
   electrowinning, 356–359
      additives for, 360, 360f
      conductivity of, 357, 358t
      guar addition to, 360, 360f
      impurities in, 359
      iron in, 358
      manganese in, 358–359
      organic burn in, 357
      reduction potentials and, 359t
      solvent extraction and flowrate determination for, 335–336
      maximizing purity of, 328–329
   spent, 281–282
Electrolytic tough pitch copper (ETP), 10, 397–398, 399, 400t
Electronic scrap treatment, 384–385
Electrorefining of copper. *See also* Anodes, electrorefining
   anode passivation and, 268
   antimony in, 254–255
   arsenic in, 254–255
   bismuth in, 254–255
   of blister copper, 7–8
   cathodes
      composition of, 258–259
      copper industrial range of, 252t
      copper starter sheet for, 258f, 259
      edge strips for, 259
      purity maximization for, 267–269
      stainless steel blanks as, 258, 258f
   cells, 259–260
      electrical connection in, 260
      polymer concrete, 260
   chemistry of, 252–257, 253f
   cobalt in, 254
   costs for
      investment, 435, 435t
      operating, 435, 436t
   current densities in, 268
   current efficiencies in, 268–269
   electrochemical potentials of elements in, 255t
   electrolytes in
      addition agent control for, 265
      addition agents to, 262–265
      chloride added to, 261–267, 264, 265
      composition of, 261–267, 262t

Electrorefining of copper (*Continued*)
  conductivity of, 261, 263t
  filtering, 266
  purification of, 266–267
 energy consumption minimized in, 269
 equipment in, 257, 257f
 flowsheet for, 253f
 gold in, 253
 grain-refining agents in, 264–265, 265f
 impurity behavior in, 256–257, 256t
 industrial data for, 269–273, 270t–273t
 iron in, 254
 lead and tin in, 254
 leveling agents in, 263–264, 264f
 nickel in, 254
 oxygen in, 255–256
 passivation of anodes in, 268
 plant locations for, 19, 23t–24t, 25f
 platinum-group metals in, 253
 process of, 251–252
 production cycle in, 260–261, 261f
 recent developments and emerging trends in, 274–275
 secondary copper smelting and, 396
 selenium in, 253
 silver in, 255
 slimes from anodes, Cu recovery from, 253, 256–257, 260, 267
 sulfur in, 254
 tellurium in, 253
 voltage in, 260
 world production in, 15t–16t
Electrostatic precipitation of dust, 211
Electrowinning of copper, 1, 3f
 acid mist suppression in, 356, 357f
 from agitation leaching, 362–368
 anodes in, 351–353
  insulator strips on, 353f
  life expectancy of, 352–353
  new, 368–369
  reaction of, 10
 cathodes in, 351–360
  copper starter sheets for, 351
  current density of, 355, 355f, 356f
  purity, maximization of, 360–361
  reaction of, 9
  SELE cathode holder, 355, 356f
  size of, 351
  stainless steel, 352f
 cells in
  design of, 353–355
  polymer concrete, 353
  schematic of, 354f
 chemistry of, 349–350
 costs of
  investment, 436–438, 437t
  operating, 436–438, 437t

 crane automation facilities for, 368–369
 current density in, 355, 355f, 356f
 developments in, 368–369
 electrolyte, 356–359
  additives for, 360, 360f
  chloride in, 360
  cobalt addition to, 360
  conductivity of, 357, 358t
  guar addition to, 360, 360f
  impurities in, 359
  iron in, 358
  manganese in, 358–359
  organic burn in, 357
  reduction potentials and, 359t
 energy and
  maximizing efficiency of, 361–362
  requirements of, 350f, 350–351
 equipment and, 351–360
 flowsheet for, 281–282, 282f
 industrial data for, 362, 363t–367t
 plant locations for, 26, 26t–27t, 28f
 process of, 349
 reactions in, 349–350
 world production in, 15t–16t
End-of-life vehicles (ELV), 375, 382–384
Explosive load, 34
External arising scrap, 374f, 375–379
Extractants for solvent extraction, 325–327, 325f
 classes of, 327
 concentration determined for, 333–334
 efficiency of, 334, 335f
 loss of, 339
 in organic phase, 335
 properties of, 326t

F

Faraday's Law, 361
Fayalite slags, 75–78
 Cu solubility in, 85f
Ferric cure, heap leaching, 297
Ferrosilicon, 195
Ferrous calcium silicate (FCS), 80–81
Fire refining of molten copper, 7
 casting anodes, 241, 242f
  continuous (Hazelett), 243f, 244, 244f, 246t, 247
  industrial data for, 246t
  molds for, 243
  preparation in, 244
  uniformity of thickness in, 243
 chemistry of, 240
 in hearth-refining furnaces, 240
 hydrocarbons for, 241
 impurity behavior and removal in, 245
 oxygen contents in
  removal of, 241
  at various stages, 237t

Index

in rotary refining furnace, 237, 238f
   industrial data for, 239t
   sequence in, 238–240
secondary copper smelting and, 396
sulfur contents in
   removal of, 240
   at various stages, 237t
Fire-refined high conductivity copper (FRHC), 399
Flash converting, 89, 144, 144f, 145t
  calcium ferrite slag in, 146
  chemistry reactions in, 146
  no matte layer in, 146
  productivity of, 146
Flash furnace matte smelting, 5, 5f, 83–107
  disadvantages of, 155
  goals of, 89
  impurities distribution in, 85t
  products of, 89
Flash furnace matte smelting, Inco, 106–107
  components of, 104, 104f
  concentrate burner in, 104, 105f
    feed system of, 106
  dust recovery systems in, 106
  equipment in, 105
  fluid bed dryer in, 106
  matte tapholes in, 105
  offgas uptake in, 105
    cooling of, 106
  Outotec compared to, 107
  process of, 103
  slag tapholes in, 105
  water cooling in, 104
Flash furnace matte smelting, Outotec, 89–104
  bin and feed system in, 97
  blast heater in, 98
  concentrates and
    blending system in, 96
    burner in, 95, 95f
  construction details of, 90
  control, 100, 100f
    concentrate throughput rate, 100–101
    matte grade, 100–101
    reaction shaft and hearth, 101
    slag, SiO_2 composition, 101
    temperature, 101
  cooling jackets in, 94
  design of, 89, 90f
  developments and trends in, 103
  dimensions and production details of, 91t–92t, 93t–94t
  direct-to-copper smelting with, 182, 182t–184t
    differences between, 185
  dust recovery and recycle system in, 98–99
  element distribution in, 102t
  equipment used in, 96
  hydrocarbon fuel burners in, 95–96

impurity behavior during, 101, 102t
   non-recycle of, 102
Inco compared to, 107
main features of, 90
matte tapholes in, 96
operation of, 99
   startup and shutdown in, 99
   steady-state, 99
oxygen plant in, 98
rotary dryer in, 97
slag tapholes in, 96
steam dryer in, 97
waste heat boiler in, 98
Flotation
  cells, 51, 52f
    cleaner, 56
    cleaner-scavenger, 56
    column, 56, 65t, 66f
    industrial data on, 58t–65t
    mechanical, 56
    re-cleaner, 56
    rougher-scavenger, 56
  chemistry, 52
  collectors, 52, 54f
  control, 64
    in-stream chemical analysis for, 65
  dewatering and, 67
  flowsheets for, 53f
  froth
    comminution for feed to, 47
    concentration of Cu ores by, 4, 4f
    frothers for, 51, 52f, 55
    principles of, 51, 52f
  gold, 68–69
  grinding particle-size control for, 36–43
    hydrocyclone controls for, 42–43, 43f
    instrumentation and strategies for, 43, 44f, 45t
    particle-size control loop in, 44, 44f
  machine vision systems for, 67
  modifiers, 54–55, 55f
  molybdenite recovery and, 415
    flowsheet for, 418f
  operation of, 64
  ores of copper special procedures of, 55–56
  products, 67–68
  reagents, 415, 417t
  selectivity in, 53–54
  sensors, 64–65, 67t
  separations and, 68
  tailings
    copper losses and, 58t–59t
    disposal of, 68
Flotation recovery of Cu from slag, 197–201
  industrial data for, 199t–200t
Flowsheets
  anode slime treatment process, 422f

Flowsheets (*Continued*)
  automobile scrap, 382f
  black copper, 394f
  cable scrap, 381f
  concentration of Cu ores, 31, 32f
  crushing, 32f
  direct-to-copper smelting, 180f
  dust treatment, 424f
  electrorefining, 253f
  electrowinning of copper, 281–282, 282f
  flotation, 53f
  grinding, 32f
  Haldor Topsøe WSA, 228f
  hydrometallurgical extraction of Cu, 281, 282f–283f
  leaching, 3f, 281, 282f
  materials cycle, 373–375, 374f
  Mitsubishi process, 166f
  molybdenite flotation, 418f
  scrap copper general, 11f
  solvent extraction, 281, 282f, 324f
  sulfuric acid manufacture, 209f
    SO_2 to SO_3 oxidation, 223f
    SO_3 absorption, 223f
Fluid bed dryer, 106
Fluxes
  adding less, 193
  $CaCO_3$ for Mitsubishi process, 165–166, 168
  cost of, 83
  Peirce–Smith converting control of, 139
  silica, for smelting, 73, 78f, 79
Foaming, direct-to-copper smelting avoiding, 186
Fragmentation
  in blasting, 33, 33f
  explosive load and optimal, 34
  improving, 34
Froth flotation
  comminution for feed to, 47
  concentration of Cu ores by, 4, 4f
  frothers for, 51, 52f, 55
  principles of, 51, 52f
Frothers, 51, 52f, 55
Fuel-fired slag cleaning furnaces, 197, 198t

G

Gangue, 281
Gaspé mobile carriage punchers, 137f
Głogów direct-to-copper smelting, 185, 181f
  Cu recovery from slags, 187
Glues, protein colloid bone, 263, 264f
Gold, in electrorefining, 253
Gold flotation, 68
Grain-refining agents, 264–265, 265f
Granulator, 380
Grinding, 32, 35–36
  autogenous, 36
  ball mills for, 36
  industrial data for, 37t–42t
  flowsheet for, 32f
  grind size and copper mineral liberation in, 35, 35f
  industrial data for, 37t–42t
  ore-throughput control for, 44, 44f
  particle-size control of flotation for, 36
    hydrocyclone controls for, 42–43, 43f
    instrumentation and strategies for, 43, 44f, 45t
    particle-size control loop in, 44, 44f
  semi-autogenous mills, 36, 42f
    industrial data for, 37t–42t
Guar, 360, 360f
Gyratory crusher, 34, 34f

H

Haldor Topsøe WSA, 227
  flowsheet for, 228f
  Sulfacid compared to, 229t
Hazelett casting, 405
  anodes, 243f, 244, 244f, 247f
    industrial data for, 246t
  industrial data for, 407t–408t
Heap leaching, 8, 286, 286f, 287
  acid cure in, 296
  agglomeration in, 296–297, 297f
    optimum, 297
  bacterial action in, 289, 289f
    optimum activity for, 290
  chalcocite and, 290
  chemistry of, 288
  copper minerals in, 287t
  costs of
    investment, 436, 437t
    operating, 436, 437t
  ferric cure in, 297
  future developments in, 315
  heaps for
    aeration of sulfide, 298
    construction of, 295, 295f
    dynamic, 295, 296
    impermeable base of, 296
    irrigation of, 298–300
    multi-lift permanent, 295–296, 295f
    ore stacking on, 297, 298f, 299f
  industrial data for, 290, 291t–294t
  leaching methods compared to, 316t
  lixiviant in, 295f, 299, 300
  optimum conditions for, 300
  ore preparation in, 296
  reactions in, 289
  solvent extraction from, 8, 9f
  sulfide mineral reactions in, 289
  uses of, 288
Hearth furnace refining, 240
Heat exchangers, 223f, 226, 227

Index

High-density polyethylene (HDPE), 296
High-pressure roll crushing, 46, 46f
Hoboken converter, 143, 144
Home scrap (run around scrap), 373–374, 374f
Hydrocarbon fuel burners, in flash furnace matte smelting, Outotec, 95
Hydrocyclone particle size control, 42–43, 43f
Hydrogen, 403
Hydrometallurgical extraction of Cu, 8. *See also* Electrowinning of copper; Leaching; Solvent extraction
    comparison of methods for, 316t
    flowsheets for, 281, 282f, 283f
    future developments in, 315
Hydroxyl ion, 54

I

Impurity behavior
    direct-to-copper smelting and, 187, 188t
    in electrorefining, 256, 256t
    electrowinning electrolyte and, 359
    fire refining of molten copper and, 245
    flash furnace matte smelting, Outotec and, 101, 102t
        non-recycle of, 102–103
    Noranda smelting process and, 115, 115t
    Peirce–Smith convertor and, 134, 135t
    Teniente matte smelting and, 120, 121t
    Top Submerged Lance smelting and, 163
Industrial electrical equipment waste (IEW), 379
Industrial non-electrical equipment waste (INEW), 379
In-situ leaching, 286–287
Institute of Scrap Recycling Industries (ISRI), 379
Intermediate leaching solution (ILS), 286
Internal arising scrap, 374–375, 374f
Iron
    in electrorefining, 254
    in electrowinning electrolyte, 358
    ore, slags for, 424
ISACONVERT converting, 150
ISASMELT batch converting, 143, 150
Isasmelt furnaces, 395–396
Isasmelt smelting. *See* Ausmelt/Isasmelt smelting

K

Kaldo furnace, 395
Ketoximes, 327

L

Lance smelting. *See* Ausmelt/Isasmelt smelting; Top Submerged Lance smelting
Land reclamation, slag for, 424
Law of Mass Action, 324
Le Chatelier's Principle, 324

Leaching, 1. *See also* Agitation leaching; Dump leaching; Heap leaching; Pressure oxidation leaching; Vat leaching
    chemistry of, 282
        sulfide minerals, 283, 285f
    comparison of methods for, 316t
    costs of
        investment, 436, 437t
        operating, 436, 437t
    flowsheet, 3f, 281, 282f
    future developments in, 315
    in-situ, 286–287
    methods for, 285
    percolation, 286
    plant locations for, 26, 26t–27t, 28f
    residue, 281
Lead, in electrorefining, 254
Leveling agents, electrorefining, 263–264, 264f
Lixiviant, 281
    in heap leaching, 295f, 299, 300
Low-density polyethylene (LDPE), 296
Lug blocks, 245

M

Machine vision systems for flotation, 67
Magnetite, slags and
    CaO based, 79–80, 79f
    SiO_2 based, 75f
Malachite, 19t
Manganese, in electrowinning, 358–359
Marshall and Swift Mining and Milling Equipment Cost Index, 427, 428f
Materials cycle, 373–379, 374f
Matte (Cu–Fe–S)
    electrical conductivity of, 81, 82
    extracting Cu from, 2–8
    flash furnace matte smelting and
        Inco, 105
        Outotec, 96
    grade control, 100–101
    immiscible liquid compositions in, 75t
    from matte smelting, 84
    melting points of, 80f, 81
    Mitsubishi process and grade of, 169
    Noranda smelting process grade of, 115
    Noranda smelting process separating slag and, 114–115
    settling rates for, 192, 192t
    slag and, 74–82
    specific gravity of, 81
    surface tension of, 82
    Teniente matte smelting and high grade, 121
    Teniente matte smelting control
        composition, 120
        depth, 120
    viscosity of, 81

Matte smelting. *See also* Flash furnace matte smelting
    dust and, 83—84
    matte from, 84
    objective of, 4
    offgas from, 86
    processes of, 83—84
    products of, 82, 86
    purpose of, 73—74, 82
    reactions in, 73—74, 82—83
    settling region in, 84
    slags from, 75—78, 76t—77t, 84—86
    tap holes in, 84
McCabe—Thiele diagram, 334, 335f
"Meatballs", 383, 383f
Mechanical flotation cell, 56—64
Melting of cathode copper, 10—11
    furnace types for, 399—402, 401f
        Asarco, 402, 406t
        burners of, 400—401
        induction, 402
        Properzi/Luxi, 402
        Southwire shaft, 402
    hydrogen measurement and control in, 403
    oxygen measurement and control in, 403
    product grades and quality of, 397—399, 398t
    slag production from, 402
    technology for, 399—403
Minerals of copper, 14—17, 19t
    air bubbles floating, 51, 52
    grind size and liberation of, 35—36, 35f
    in heap leaching, 287t
Mining copper ores. *See also* Blasting; Crushing
    costs of, 427, 428f
        investment, 427—430, 428t
        operating, 430, 430t
        variation in, 429
    cut-off grades, 14, 17
    economic sizes of plants in, 429—430
    profitability of, 438
    world production in, 15t—16t
        by highest capacity, 17t
Mitsubishi process
    Ausmelt/Isasmelt smelting compared to, 165—174
    $CaCo_3$ flux in, 165—166, 168
    control, 169—175, 173t—175t
    converting furnace in, 168—169
        blister copper in, 169
        CaO-base slag in, 168—169
    electric slag-cleaning furnace, 167—168, 167f
    flowsheet for, 166f
    industrial data on, 170t—173t
    lances in, 166
    matte grade in, 169
    offgas in, 167
    smelting furnace in, 166—167, 167f
    temperature
        control, 169, 173t
        rise in, 167
    in 2000s, 174—175
Mitsubishi top-blown converter, 143
Mixer-settler, solvent extraction, 329, 329f
    designs for
        mixer, 329—330, 330f
        settler, 330—331, 331f
    picket fences in, 330—331
    reverse-flow settlers, 331, 331f
Modifiers, in flotation, 54—55, 55f
Molecular sieve oxygen plants, 98
Molybdenite recovery
    flotation and, 415—417
        flowsheet for, 418f
    operation of, 415
    optimization of, 417—418
    product content in, 416t
Monel tube, 94
Multi-lift permanent heaps, 295, 295f, 295—296
Municipal solid waste (MSW), 379

N

New scrap, 374—375, 374f
Nickel, in electrorefining, 254—255
Noranda continuous converting, 143, 147—149, 147f
    chemical reactions in, 147
    control, 149
    industrial data for, 148t
    reaction mechanisms in, 148—149
    SiO_2 based slag in, 149
Noranda smelting process, 111—114, 112f
    campaign life and repair of, 121
    concentrates, dry through tuyeres, 113t
    control, 116—117, 117f
        temperature, 116
    impurity behavior in, 115, 115t
    industrial data on, 112t
    matte and slag separation in, 114—115
    matte grade choice in, 115
    operation of, 116—117
    production rate enhancement in, 117
    scrap smelting and, 115—116
    tuyere pyrometer in, 116, 117f

O

Obsolete scrap, 374f, 375—379
Offgases
    in Ausmelt/Isasmelt smelting, 155—156
    drying, 212—214, 214f
    from electric slag-cleaning furnaces, 207
    flash furnace matte smelting, Inco uptake of, 105
        cooling of, 106
    from matte smelting, 86

in Mitsubishi process, 167
Peirce−Smith converters collecting, 136−137, 138f
from smelting and converting processes, 206−207, 206t−207t
sulfuric acid manufacture from, 208
Teniente matte smelting and heat recovery of, 122
treatment of smelter, 208−212
    acid plant blowdown in, 212, 213f
    electrostatic precipitation of dust in, 211
    gas cooling and heat recovery in, 210−211, 210f
    mercury removal in, 211−212
    spray cooling for, 211
    water quenching, scrubbing, cooling in, 211
Old scrap (obsolete scrap), 375−379, 374f
Olivine slags, 80−81
Olympic Dam direct-to-copper smelting, 180f, 181t, 185
    Cu recovery from slags, 187
Ores of copper. *See also* Concentration of Cu ores; Run-of-mine ore
    %Cu, 1, 1f
    cut-off grade, 14, 17
    extraction processes for, 2f
    flotation special procedures for, 55−56
Ore-size determination, 34
Ore-throughput control, for grinding, 44−46, 44f
Ore-toughness measurements, 34
Organic burn, 357
Organic-continuous phase, 339
Outotec. *See* Flash furnace matte smelting; Outotec
Oxygen
    in electrorefining, 255−256
    fire refining stages and contents of, 239t
        removal of, 241
    melting of cathode copper, measurement and control of, 403
Oxygen enrichment
    in Ausmelt/Isasmelt smelting, 156
    Peirce−Smith converting and, 140
    shrouded blast injection and, 142−143
Oxygen plant, in flash furnace matte smelting, Outotec, 98
Oxygen-free copper, 10, 399
    casting, 409, 410f

P

Passivation of electrorefining anodes, 268
Pebbles, oversize, 36
Peirce−Smith converting, 5, 6f, 128−143
    alternatives to, 143−150
        recent developments in, 150
    campaign life maximization for, 142
    concentrate smelting in, 142
    converter in, 128f, 129f, 134f
    copper blow, 139−140
    end point determinations for, 139−140
    environmental concerns with, 232
    improvements in, 142−143
    impurity behavior and, 134, 135t
    industrial data for, 130t−133t, 135t
    offgas collection in, 136−137, 138f
    oxygen enrichment of blast, 140
    oxygen-enriched air in, 135
    productivity maximization for, 140−142
        solids melting, 141−142
    reactions in, 129−134
    scrap additions, 141−142
        pneumatic injection of, 143
    shell design for, 143
    slag and, 128
        blow, 139
        control, 139
        formation rate, 139
    temperature
        choice of, 138
        control, 137
        measurement of, 138, 138f
    TSL compared to, 164
    tuyere, 129f
        accretion buildup and, 136−137, 136f
        flowrate in, 135
        shrouded blast injection for, 142−143
    tuyere pyrometer and, 138f
Percolation leaching, 286. *See also* Dump leaching; Heap leaching; Vat leaching
Phase diagrams
    $CaO-FeO_x-SiO_2$, 80f
    $Cu-S$, 133f
    $Fe-O-S-SiO_2$, 74f
Picket fences, 330−331
Platinum-group metals, in electrorefining, 253
Polymer concrete cells
    electrorefining, 260
    electrowinning, 353
Polyvinyl chloride (PVC), 296, 353
Post-consumer scrap, 374f, 375−379
Pourbaix diagrams
    $Cu-Fe-S-O-H_2O$ system, 285f
    $Cu-S-O-H_2O$ system, 284f
Pregnant leach solution (PLS), 281
    collection of, 300
    minimizing solids in, 338
    trains of, 332
Pressure oxidation leaching (POx), 286
    chalcopyrite and
        concentrates, 304−308
        processes for, 310t−311t
        temperature elevation in, 308−315
    high-temperature, high-, 309
    industrial data on, 312t−314t
    leaching methods compared to, 316t

Pressure oxidation leaching (POx) (*Continued*)
    medium-temperature, medium-, 309—315
Pressure swing absorption (PSA), 98
Price, of copper, 29, 29f
Process wastes, 379
Production of copper
    by country, 15t—16t
    growth in, 13, 13f, 29
Prompt industrial scrap, 374—375, 374f
Properzi casting machine, 405
Properzi/Luxi furnaces, 402
Protein colloid bone glues, 263—264, 264f

R

Re-cleaner flotation cells, 56
    industrial data on, 62t—64t
Recycling of copper scrap, 11—12, 11f. *See also*
    Scrap; copper recovery from
Refining, electrolytic, 251. *See also* Electrorefining
Refining locations
    map of largest, 25f
    production statistics for, 15t—16t
Refractories. *See specific smelting processes*
Reverse-flow settlers, 331, 331f
Rod production, bar casting and, 404—409, 404f,
    405f, 407t—408t
Rotary dryers, 97
Rotary refining furnace, fire refining of molten
    copper in, 237—240, 238f
    industrial data for, 239t
    sequence in, 238—240
Rough copper, 396
Rougher-scavenger flotation cells, 56
    industrial data on, 58t—64t
Run around scrap, 373—374, 374f
Run-of-mine ore (ROM), 288
    pretreatment of, 296

S

Scanning electron microscopy (SEM), 47, 47f
Scrap, copper recovery from, 1
    anode furnaces and, 393
    anodes and, 245
    auto shredder residue for, 382f, 383—384
    automobile, 382f, 382—384
    cable, 380—385, 381f
    converting of copper matte and, 393
    electronic, 384—385
    external arising, 375—379, 374f
    flowsheet for general, 11f
    government regulations of, 391
    grades of, 376t, 377t—378t
    home (run around), 373—374, 374f
    internal arising, 374—375, 374f
    "meatballs", 383, 383f
    new, 374—375, 374f
    Noranda smelting process and, 115—116

old (obsolete), 374f, 375—379
Peirce—Smith converting and, 141—142
    pneumatic injection of, 143
post-consumer, 375—379, 374f
process wastes and, 379
profitability of, 435
prompt industrial, 374f, 374—375
secondary materials industries and, 375
in smelting furnaces, 392, 392f
wire, 380—385
Secondary copper
    characteristics of, 391
    processing in primary smelters, 392, 392f
    Top Submerged Lance smelting furnace for,
        395—396
Secondary copper smelting
    to black copper, 393—395, 394f
    fire refining and electrorefining after, 396
    high-grade, 393
    products of, 395
    types of, 392
Secondary materials industries and scrap, 375
SELE cathode holder, 355, 356f
Selectivity, in flotation, 53—54
Selenium, in electrorefining, 253—254
Semi-autogenous grinding mills (SAG mills),
    36, 42f
    industrial data for, 37t—42t
Sensors, for flotation, 64—67, 67t
Short-circuiting, 269
Shrouded blast injection, 142—143
Silver, in electrorefining, 255
Single absorption acid plant, 222t—223t, 224, 229t
Slag-cleaning furnaces. *See* Electric slag-cleaning
    furnace
Slags. *See also* specific slags
    for abrasive grit, 423—424
    CaO based
        magnetite and, 79—80, 79f
        viscosity of, 80
    for cement and concrete, 423
    for ceramic tile, 425
    Cu losses in, 191
        minimizing, 193—201
    decreasing copper in slag
        minimizing Cu concentration in,
            193—194
        minimizing slag generation for, 193
        pyrometallurgical slag settling/reduction for,
            194—197, 196t
        slag minerals processing for, 197—201,
            198t—200t
    from direct-to-copper smelting, 182t
        Cu-in-slag concentration of, 185, 186, 187
    flash furnace matte smelting and
        Inco, 105
        Outotec, 96

Index

for iron ore, 424
for land reclamation, 424
melting of cathode copper production of, 402
Noranda smelting process separating matte and, 114–115
Peirce–Smith converting and, 128
  blow, 139
  control, 139
  formation rate, 139
SiO_2 based
  electrical conductivity of, 79
  Fe–O–S–SiO_2 phase diagram for, 748
  flash furnace matte smelting, Outotec control and, 101
  immiscible liquid compositions in, 75t
  magnetite and, 75f
  matte and, 74–82
  from matte smelting, 75–78, 84–86, 76t–77t
  melting point of, 79f
  molecular structure of, 74, 75f
  Noranda continuous converting and, 149
  specific gravity of, 79
  surface tension of, 79
  viscosity of, 78, 79
Teniente matte smelting control
  composition, 120
  depth, 120
uses or disposal for, 423–425
Slags, Cu recovery from
  direct-to-copper smelting and, 186–187
    Glogów, 187
    Olympic Dam, 187
  flotation and, 197–201
    industrial data for, 199t–200t
Slimes. See Anode slimes
Slurry level sensor, 44
Smelting. See also Secondary copper smelting, specific processes
  bath, 83–84
  costs for
    investment, 427–430, 428t, 433–435, 433t
    operating, 434t
    variation in, 429
  decreasing copper in slag and, 193
  flash, 83–84
  Mitsubishi process, 165–167
  Noranda process, 111–114, 112f
  offgas from, 206–208, 206t–207t
  Peirce–Smith converting and concentrate, 142
  plant locations for, 19, 19t–21t, 22f
  purpose of, 73
  scrap use in, 392, 392f
  secondary copper processing in primary, 391–393, 392f
  silica and, 73, 78f, 79
  Teniente matte smelting, 117–118, 118f
  world production in, 15t–16t

Solvent extraction (SX), 1, 3f, 323–343
  aqueous-in-organic entrainment in, 328–329
  chemistry of, 324–325
  circuit configurations in, 331–333, 332f
    alternative approach to, 336
    wash stage included in, 333
  costs of
    investment, 436–438, 437t
    operating, 436–438, 437t
  crud formation in, 337–339
  diluents for, 327–328
    aliphatic compared to partially aromatic, 328t
    loss of, 339
  electrolytes and
    flowrate determination for, 335–336
    purity maximization in, 328–329
  equilibrium stripped organic Cu concentration in, 334, 335f
  equipment for, 329–331, 329f
  extractants for, 325–327, 325f
    classes of, 327
    concentration determined for, 333–334
    efficiency of, 334, 335f
    loss of, 339
    in organic phase, 335
    properties of, 326t
  flowsheet for, 281, 282f, 324f
  from heap leaching, 8–9, 9f
  industrial data on, 339–343, 340t–343t
  mixer-settler for, 329–330, 329f
    mixer designs, 329–330, 330f
    picket fences in, 330–331
    reverse-flow settlers, 331, 331f
    settler designs, 330–331, 331f
  operational stability of, 336–337
  organic phase composition in, 325–328
  parallel circuit in, configuration of, 332f, 333
  phase continuity in, 339
  plant locations for, 28f, 26, 26t–27t
  process of, 323–324, 324f
    extraction, 323
    stripping, 323
  series circuit in
    configuration of, 331–333, 332f
    quantitative design of, 332f, 333–336, 334t
    series-parallel circuits compared to, 336, 337t
  series-parallel circuit in
    configuration of, 332f, 333
    series circuits compared to, 336, 337t
  stripping isotherms and, 334, 335f
Southwire casting machine, 404f–405f
  industrial data for, 407t–408t
Southwire shaft furnace, 402

Specific gravity
  matte, 81
  separator, 380–381, 381f
  slag, 79
Spray cooling, 211
Sprinklers, 300
Stainless steel cathode blanks
  copper starter sheet for, 258f, 259
  edge strips for, 259
  electrorefining, 258, 258f
  electrowinning, 352f
Steam dryers, 97
Strip casting, 409–410
Stripping isotherms, solvent extraction and, 334, 335f
Submerged tuyere smelting
  Noranda, 111–114, 112f
  Teniente, 117–118, 118f
  Vanyukov, 122–123, 122f
Sulfacid, 228–229
  Haldor Topsøe WSA compared to, 229t
Sulfide heaps, aeration of, 298
Sulfide minerals, 51
  in agitation leaching, 304
  dump and heap leaching reactions with, 289
  leaching and, 283–285, 285f
Sulfur
  alternative products with, 231
  capture and fixation of, 205
    future improvements in, 231
  in copper, Cu-S phase diagram, 133f
  in electrorefining, 254–255
  fire refining stages and contents of, 239t
    removal of, 240
  melting point of, 309
Sulfur dioxide (SO_2), 205t
Sulfuric acid concentration
  lixiviant (heap leaching), 295f, 299, 300
  pregnant leach solution collection, 281, 300, 332, 338
  raffinate, 331–332, 332f, 334, 335f
Sulfuric acid manufacture, 208–231
  acid plant blowdown in, 212, 213f
  alternative methods for, 227–229
  catalysts for, 214–215
    dust accumulation in, 216
    industrial arrangement of, 215, 215f
    temperature of, 225–226
  catalytic converter for, 224, 224f
  double absorption, 224
  industrial data for, 218t–221t
  electrostatic precipitation of dust in, 211
  flowsheet for, 209f
  SO_2 to SO_3 oxidation and SO_3 absorption, 223f
  gas cooling and heat recovery for, 210–211, 210f
    spray cooling, 211
  Haldor Topsøe WSA and Sulfacid compared to conventional, 229t
  mercury removal in, 211
  offgases and, 208
    drying in, 212–214, 214f
  product grades of, 227
  reaction paths in, 224–225, 225f
    characteristics of, 225–226
  recent and future developments in, 229–231
  single absorption, 224
    industrial data for, 222t–223t
  SO_2 to SO_3 oxidation, 214–217
    conversion equilibrium curve for, 216, 217f, 225f
    Cs-promoted catalyst in, 216
    flowsheet for, 223f
    heat exchangers for, 226, 227, 223f
    heat recovery during, 230–231
    ignition and degradation temperatures in, 215–216
    industrial data for, 218t–223t
    maximizing, 229–230
    reaction paths in, 224–226, 225f
  SO_3 absorption, 216–217
    absorption towers for, 226, 226t
    acid cooling and, 227
    flowsheet for, 223f
    optimum absorbing acid composition for, 217, 224f, 225f
  water quenching, scrubbing, cooling in, 211
Surface tension
  of matte, 82
  of slag, SiO_2 based, 79

T

Tailings from flotation
  Cu content (ore flotation), 58t–59t
  disposal of, 68
Tellurium, in electrorefining, 253–254
Temperature
  in direct-to-copper smelting, 182
  flash furnace matte smelting, Outotec control of, 101
  high-pressure oxidation leaching, high-, 309
  medium-pressure oxidation leaching, medium-, 309
  Mitsubishi process and
    control, 169, 173t
    rise in, 167
  Noranda smelting process control of, 116
  Peirce—Smith converting and
    choice of, 138
    control, 137–138
    measurement of, 138, 138f
  pressure oxidation leaching of chalcopyrite and elevated, 308–315
  sulfuric acid manufacture catalysts and, 225–226

Index

sulfuric acid manufacture SO_2 to SO_3 oxidation
    ignition and degradation, 215−216
  Teniente matte smelting control of, 120
Teniente matte smelting, 117−118, 118f
  campaign life and repair of, 121
  control, 120
    matte and slag composition, 120
    matte and slag depth, 120
    temperature, 120
  cooling of, 121
  impurity behavior distribution in, 120−121, 121t
  industrial data on, 112t, 119t
  matte grade, high in, 121
  offgas heat recovery in, 122
  operation of, 118−119
  seed matte, 117−118
  SO_2 capture efficiency of, 121
Teniente slag cleaning furnace, 197, 198t
Tenorite, 19t
Thickeners, 67−68
Thiourea, 264
Tin, in electrorefining, 254
Top Submerged Lance smelting (TSL), 150
  arsenic and, 163
  capacity in, 155, 156f
  copper converting using, 164−165
  furnace, 156−163, 157f
  impurity elimination in, 163
  international locations of, 164, 164f
  lance tip in, 156−163, 162f
  mechanisms of, 163
  Peirce−Smith converting compared to, 164
  secondary copper and, 395
  startup and shutdown of, 163
  versatility of, 156
Tough pitch copper. See Electrolytic tough pitch copper
Tuyere
  Noranda smelting furnace, 112f
  Noranda smelting process and dried concentrates through, 113t
  Peirce−Smith converter, 129f
    accretion buildup and, 136−137, 136f
    flowrate in, 135
    shrouded blast injection for, 142−143
  punching, 136
  Teniente smelting furnace, 118f
  Vanyukov smelting furnace, 122f
Tuyere pyrometer
  Noranda smelting process and, 116, 117f
  Peirce−Smith converting and, 138f

U

Unalloyed copper classes, 397−398
Uses, copper and copper alloys, 13, 13f
  United States, 14t

V

V_2O_5 catalysts for sulfuric acid manufacture, 214−215, 215f
  ignition and degradation temperatures in, 215−216
Vacuum pressure swing absorption (VPSA), 98
Vanyukov smelting, 122−123, 122f
  industrial data for, 123t
  process of, 122
  stationary furnace in, 122−123
Vat leaching, 286, 301−303, 303f
  industrial data on, 301t−302t
  leaching methods compared to, 316t
Vertical casting machines, 409
Viscosity
  of matte, 81
  of slag
    CaO based, 80
    SiO_2 based, 78, 79

W

Wash stage, solvent extraction, 333
Waste from electrical and electronic equipment (WEEE), 375
Waste heat boiler
  in flash furnace matte smelting, Outotec, 98
  gas cooling in, 210f
Waste minerals, 281
Wet chlorination process for anode slimes, 421
Wet gas Sulfuric Acid (WSA), 208
  Haldor Topsøe, 227−228
  flowsheet for, 228f
  Sulfacid compared to, 229t
Wet gas Sulfuric Acid-Double Condensation (WSA-DC), 230
White metal (high grade matte), 127, 133f
Wire and cable scrap, 380−385
Wobblers, 300

X

Xanthate, 53
X-ray fluorescence analysis, 65, 67t

Z

Zero water discharge plants, 68

图书在版编目（CIP）数据

铜提取冶金：第5版：英文／（美）马克·施莱辛格(Mark E. Schlesinger)等著. --长沙：中南大学出版社，2017.10
ISBN 978-7-5487-3006-4

Ⅰ.①铜… Ⅱ.①马… Ⅲ.①炼铜－英文 Ⅳ.①TF811

中国版本图书馆CIP数据核字(2017)第242391号

铜提取冶金(第5版)
TONG TIQU YEJIN (DI 5 BAN)

Mark E. Schlesinger　等著

□责任编辑	史海燕		
□责任印制	易红卫		
□出版发行	中南大学出版社		
	社址：长沙市麓山南路	邮编：410083	
	发行科电话：0731-88876770	传真：0731-88710482	
□印　装	湖南众鑫印务有限公司		
□开　本	720×1000　1/16	□印张 30.5	□字数 801 千字
□版　次	2017年10月第1版	□2017年10月第1次印刷	
□书　号	ISBN 978-7-5487-3006-4		
□定　价	142.00 元		

图书出现印装问题，请与经销商调换